Lecture Notes in Artificial Intelligence 9362

Subseries of Lecture Notes in Computer Science

More information about this series at http://www.springer.com/series/1244

Juanzi Li · Heng Ji · Dongyan Zhao
Yansong Feng (Eds.)

Natural Language Processing and Chinese Computing

4th CCF Conference, NLPCC 2015
Nanchang, China, October 9–13, 2015
Proceedings

Springer

Editors
Juanzi Li
Tsinghua University
Beijing
China

Dongyan Zhao
Peking University
Beijing
China

Heng Ji
Rensselaer Polytechnic Institute
Troy, NY
USA

Yansong Feng
Peking University
Beijing
China

ISSN 0302-9743   ISSN 1611-3349   (electronic)
Lecture Notes in Artificial Intelligence
ISBN 978-3-319-25206-3   ISBN 978-3-319-25207-0   (eBook)
DOI 10.1007/978-3-319-25207-0

Library of Congress Control Number: 2015951416

LNCS Sublibrary: SL7 – Artificial Intelligence

Printed on acid-free paper

Springer International Publishing AG Switzerland is part of Springer Science+Business Media
(www.springer.com)

Organization

Organizing Committee

General Chairs

Zhi Tang   Peking University, China
Kam-Fai Wong   The City University of Hong Kong, SAR China

Program Committee Co-chairs

Juanzi Li   Tsinghua University, China
Heng Ji   Rensselaer Polytechnic Institute, USA

Area Co-chairs

Fundamentals on CIT

Ru Li   Shanxi University, China
Olivia Kwong   The City University of Hong Kong, SAR China

Applications on CIT

Yanjun Ma   Baidu, China
Wenjie Li   Hong Kong Polytechnic University, SAR China

Machine Translation

Deyi Xiong   Soochow University, China
Boxing Chen   National Research Council, Canada

Web Ming and Big Data

Minlie Huang   Tsinghua University, China
Aixin Sun   Nanyang Technological University, Singapore

Machine Learning for NLP

Wei Gao   Qatar Research Insitute, Qatar
Wei Liu   IBM Research, USA

Knowledge Acquisition

Haofen Wang   East China Univeristy of Science and Technology,
                              China
Yulan He   Aston University, UK

NLP for Social Networks

Xing Xie   Microsoft Research, China
Yizhou Sun   Northeastern University, USA

NLP for Search and Ads

Bin Wang   Institute of Information Engineering, CAS, China
Honging Wang   University of Virginia, USA

QA and Information Extraction

Bing Qin   Harbin Institute of Technology, China
Yu Hong   Rensselaer Polytechnic Institute, USA

Panel Chair

Min Zhang   Soochow University, China

Demo Co-chairs

Kun Zhang   Sogou Inc., China
Liangcai Gao   Peking University, China

Organization Co-chairs

Dongyan Zhao   Peking University, China
Mingwen Wang   Jiangxi Normal University, China

Sponsorship Co-chairs

Ming Zhou   Microsoft Research, China
Shuanhu Bai   Sina Weibo, China

Evaluation Chair

Xiaojun Wan   Peking University, China

Publicity Chair

Ruifeng Xu   HIT Shenzhen Graduate School, China

ADL/Tutorial Co-chairs

Xuanjing Huang   Fudan Univeristy, China
Yiqun Liu   Tsinghua University, China

Publication Chair

Yansong Feng   Peking University, China
Zhichun Wang   Beijing Normal University, China

Website Chair

Aixia Jia   Peking University, China

Program Committee

Bing, Lidong   Carnegie Mellon University, USA
Cao, Hailong   Harbin Institute of Technology, China
Chang, Baobao   Peking University, China
Chang, Kai-Wei   Microsoft Research, USA
Che, Wanxiang   Harbin Institute of Technology, China
Chen, Boxing   National Research Council, Canada
Chen, Wenliang   Soochow University, China
Chen, Yubo   Institute of Automation, Chinese Academy of Sciences,
                                    China
Chen, Lin   University of Illinois at Chicago, USA
Chen, Jiajun   Nanjing University, China
Chen, Yidong   Xiameng University, China
Cheng, Gong   Nanjing University, China
Cheng, Li   Xinjiang Institute of Ecology and Geography, CAS,
                                    China
Cui, Lei   Microsoft Research, China
Dai, Xin-Yu   Nanjing University, China
Deng, Zhi-Hong   Peking University, China
Dou, Zhicheng   Renmin University, China
Du, Yajun   Xihua University, China
Du, Jinhua   Xi'an University of Technology, China
Duan, Nan   Microsoft Research, China
Duan, Xiangyu   Soochow University, China
Duan, Huizhong   Walmart Lab, USA
Eck, Matthias   Facebook, USA
Feng, Yansong   Peking University, China
Feng, Shi   Northeastern University, USA
Fu, Guohong   Heilongjiang University, China
Gao, Wei   Qatar Research Institute, Qatar
Ge, Tao   Peking University, China
Gu, Quanquan   University of Virginia, USA
Guo, Weiwei   Columbia University, USA
He, Yulan   Aston University, UK
He, Zhongjun   Baidu, China
He, Ben   University of Chinese Academy of Sciences, China
He, Yifan   New York University, USA
He, Xiaodong   Microsoft Research, USA
Hong, Yu   Rensselaer Polytechnic Institute, USA
Zan, Hongying   Zhengzhou University, China

Hu, Yunhua	Taobao, China
Hu, Qinmin	East China Normal University, China
Huang, Minlie	Tsinghua University, China
Huang, Hongzhao	Rensselaer Polytechnic Institute, USA
Huang, Xuanjing	Fudan University, China
Huang, Zhongqiang	Raytheon BBN Technologies, USA
Huang, Shujian	Nanjing University, China
Ji, Heng	Rensselaer Polytechnic Institute, USA
Jiang, Jing	Singapore Management University, Singapore
Jiang, Peng	University of Illinois at Urbana-Champaign, USA
Jiang, Shengyi	Guangdong University of Foreign Studies, China
Jin, Peng	Leshan Normal University, China
Kang, Shiyong	Ludong University, China
Kit, Chunyu	City University of Hong Kong, SAR China
Kong, Fang	Soochow University, China
Ku, Lun-Wei	IIS, Academia Sinica, China
Kwong, Olivia	City University of Hong Kong, SAR China
Lam, Wai	Chinese University of Hong Kong, SAR China
Lan, Man	East China Normal University, China
Li, Juanzi	Tsinghua University, China
Li, Wenjie	Hong Kong Polytechnic University, SAR China
Li, Ru	Shanxi University, China
Li, Hao	Rensselaer Polytechnic Institute, USA
Li, Fangtao	Google, USA
Li, Qi	Rensselaer Polytechnic Institute, USA
Li, Haibo	City University of New York, USA
Li, Wu-Jun	Nanjing University, China
Li, Chenliang	Wuhan University, China
Li, Shoushan	Soochow University, China
Li, Maoxi	Jiangxi Normal University, China
Li, Yanen	LinkedIn Inc., USA
Li, Peng	Institute of Information Engineering, CAS, China
Li, Qingsheng	Anyang Normal University, China
Li, Sujian	Peking University, China
Lian, Defu	University of Electronic Science and Technology of China, China
Liao, Xiangwen	Fuzhou Univeristy, China
Lin, Chenghua	University of Aberdeen, UK
Liu, Wei	IBM Research, USA
Liu, Kang	Institute of Automation, Chinese Academy of Sciences, China
Liu, Zhiyuan	Tsinghua University, China
Liu, Qi	University of Science and Technology of China, China
Liu, Maofu	Wuhan University of Science and Technology, China
Liu, Yang	Shandong University, China
Liu, Haiming	University of Bedfordshire, UK

Liu, Bingquan	Harbin Institute of Technology, China
Liu, Qun	Dublin City University, Ireland
Liu, Yang	Tsinghua University, China
Liu, Shujie	Microsoft Research, China
Liu, Dexi	JiangXi University of Finance and Economics, China
Lv, Yajuan	Baidu, China
Ma, Yanjun	Baidu, China
Ma, Jun	Shandong University, China
Min, Bonan	Raytheon BBN Technologies, USA
Ni, Yuan	IBM Research, China
Pan, Jeff	University of Aberdeen, UK
Peng, Min	Wuhan University, China
Qi, Haoliang	Heilongjiang Institute of Technology, China
Qian, Mingjie	University of Illinois at Urbana-Champaign, USA
Qin, Bing	Harbin Institute of Technology, China
Qin, Tao	Microsoft Research, China
Qiu, Xipeng	Fudan University, China
Qu, Weiguang	Nanjing Normal University, China
Ruan, Tong	East China Univeristy of Science and Technology, China
Shen, Huawei	Institute of Computing Technology, CAS, China
Song, Dongjin	University of California at San Diego, USA
Song, Xuan	Tokyo University, Japan
Song, Rou	Beijing Language and Culture University, China
Su, Jinsong	Xiameng University, China
Sugiyama, Kazunari	National University of Singapore, Singapore
Sun, Aixin	Nanyang Technological University, Singapore
Sun, Yizhou	Northeastern University, USA
Sun, Chengjie	Harbin Institute of Technology, China
Sun, Guangzhong	University of Science and Technology of China, China
Sun, Weiwei	Peking University, China
Tang, Duyu	Harbin Institute of Technology, China
Tang, Jiliang	Yahoo!, USA
Wang, Haofen	East China Univeristy of Science and Technology, China
Wang, Hongning	University of Virginia, USA
Wang, Bin	Institute of Information Engineering, CAS, China
Wang, Yan	Columbia University, USA
Wang, Lu	Cornell University, USA
Wang, Xun	NTT Communication Science Laboratories, Japan
Wang, Longyue	Dublin City University, Ireland
Wang, Xiaoling	East China Normal University, China
Wang, Zhichun	Beijing Normal University, China
Wang, Zhiguo	IBM Research, USA
Wang, Bo	Tianjin University, China
Wang, Chi	Microsoft Research, USA

Wang, Mingwen   Jiangxi Normal University, China
Wang, Quan   Institute of Information Engineering, CAS, China
Wang, Xiaojie   Beijing University of Posts and Telecommunications,
                             China
Wei, Zhongyu   University of Texas at Dallas, USA
Wong, Derek F.   University of Macau, SAR China
Wu, Yunfang   Peking University, China
Wu, Hua   Baidu, China
Wu, Yuanbin   Fudan University, China
Wu, Jian   Pennsylvania State University, USA
Wu, Xiaofeng   Dublin City University, Ireland
Xia, Rui   Nanjing University of Science and Technology,
                             China
Xia, Yunqing   Tsinghua University, China
Xiang, Bing   IBM Research, USA
Xiang, Ren   University of Illinois at Urbana-Champaign, USA
Xiao, Xinyan   ICT, CAS, China
Xiao, Tong   Northeastern University, China
Xie, Xing   Microsoft Research, China
Xiong, Deyi   Soochow University, China
Xiong, Zhongyang   Chongqing University, China
Xu, Jun   ICT, Chinese Academy of Sciences, China
Xu, Jun-Ming   University of Wisconsin-Madison, USA
Xu, Ruifeng   Harbin Institute of Technology Shenzhen Graduate
                             School, China
Xu, Weiran   Beijing University of Posts and Telecommunications,
                             China
Xu, Jinan   Beijing Jiaotong University, China
Xu, Peng   Google, USA
Xun, Endong   Beijing Language and Culture University, China
Ye, Zheng   York University, Canada
Yin, Hongzhi   University of Queensland, Australia
Yu, Hongqing   University of Bedfordshire, UK
Yuan, Nicholas   Microsoft Research, China
Zhang, Min   Soochow University, China
Zhang, Dongdong   Microsoft Research, China
Zhang, Qi   Fudan University, China
Zhang, Yue   Singapore University of Technology and Design,
                             Singapore
Zhang, Longkai   Peking University, China
Zhang, Lei   University of Illinois at Chicago, USA
Zhang, Min   Tsinghua University, China
Zhang, Yangsen   Beijing Information Science and Technology
                             University, China
Zhang, Chengzhi   Nanjing University of Science and Technology, China

Zhang, Jiajun	Institute of Automation, Chinese Academy of Sciences, China
Zhang, Joy Ying	Facebook, USA
Zhang, Kevin	Beijing Institute of Technology, China
Zhao, Shiqi	Baidu, China
Zhao, Hai	Shanghai Jiaotong University, China
Zhao, Wayne Xin	Renmin University, China
Zhao, Zhou	Zhejiang University, China
Zhao, Dongyan	Peking University, China
Zhao, Yanyan	Harbin Institute of Technology, China
Zhao, Jun	Institute of Automation, Chinese Academy of Sciences, China
Zhou, Deyu	Southeast University, China
Zhou, Mianwei	Yahoo!, USA
Zhou, Ming	Microsoft Research, China
Zhou, Dong	Hunan Univeristy of Science and Technology, China
Zhou, Yu	Chinese Academy of Sciences, China
Zhou, Guangyou	Central China Normal University, China
Zhu, Feida	Singapore Management University, Singapore
Zhu, Hengshu	Baidu, China
Zhu, Jiaqi	Institute of Software, CAS, China

Organizers

Organized by

China Computer Federation, China

Hosted by

Jiangxi Normal University   State Key Laboratory of Digital Publishing

Publishers:

ACTA Scientiarum Naturalium   Lecture Notes in Artificial Intelligence
Universitatis Pekinensis   Springer

In Cooperation with:

Asian Federation of Natural Language Processing

Sponsoring Institutions

Microsoft Research Asia

Sogou Inc.

Mingbo Education

Wanfang Data

speechocean

SpeechOcean

Contents

Fundamentals on Language Computing

A Maximum Entropy Approach to Discourse Coherence Modeling 3
  Rui Lin, Muyun Yang, Shujie Liu, Sheng Li, and Tiejun Zhao

Transition-Based Dependency Parsing with Long Distance Collocations. 12
  Chenxi Zhu, Xipeng Qiu, and Xuanjing Huang

Recurrent Neural Networks with External Memory for Spoken
Language Understanding 25
  Baolin Peng, Kaisheng Yao, Li Jing, and Kam-Fai Wong

Improving Chinese Dependency Parsing with Lexical Semantic Features 36
  Lvexing Zheng, Houfeng Wang, and Xueqiang Lv

Machine Translation and Multi-Lingual Information Access

Entity Translation with Collective Inference in Knowledge Graph 49
  Qinglin Li, Shujie Liu, Rui Lin, Mu Li, and Ming Zhou

Stochastic Language Generation Using Situated PCFGs 64
  Caixia Yuan, Xiaojie Wang, and Ziming Zhong

Machine Learning for NLP

Clustering Sentiment Phrases in Product Reviews by Constrained
Co-clustering 79
  Yujie Cao, Minlie Huang, and Xiaoyan Zhu

A Cross-Domain Sentiment Classification Method Based on Extraction
of Key Sentiment Sentence. 90
  Shaowu Zhang, Huali Liu, Liang Yang, and Hongfei Lin

Convolutional Neural Networks for Correcting English Article Errors 102
  Chengjie Sun, Xiaoqiang Jin, Lei Lin, Yuming Zhao, and Xiaolong Wang

NLP for Social Media

Automatic Detection of Rumor on Social Network 113
  *Qiao Zhang, Shuiyuan Zhang, Jian Dong, Jinhua Xiong,
  and Xueqi Cheng*

Multimodal Learning Based Approaches for Link Prediction
in Social Networks   123
   Feng Liu, Bingquan Liu, Chengjie Sun, Ming Liu, and Xiaolong Wang

Sentiment Analysis Based on User Tags for Traditional Chinese Medicine
in Weibo   134
   Junhui Shen, Peiyan Zhu, Rui Fan, Wei Tan, and Xueyan Zhan

Predicting User Mention Behavior in Social Networks   146
   Bo Jiang, Ying Sha, and Lihong Wang

Convolutional Neural Networks for Multimedia Sentiment Analysis   159
   Guoyong Cai and Binbin Xia

Applications on Language Computing

An Adaptive Approach to Extract Characters from Digital Ink Text
in Chinese Based on Extracted Errors   171
   Hao Bai

Context-Dependent Metaphor Interpretation Based on Semantic
Relatedness   182
   Chang Su, Shuman Huang, and Yijiang Chen

Context Vector Model for Document Representation:
A Computational Study   194
   Yang Wei, Jinmao Wei, and Hengpeng Xu

NLP for Search Technology and Ads

Refine Search Results Based on Desktop Context   209
   Xiaoyun Li, Ying Yu, and Chunping Ouyang

Incorporating Semantic Knowledge with MRF Term Dependency Model
in Medical Document Retrieval.   219
   Zhongda Xie, Yunqing Xia, and Qiang Zhou

A Full-Text Retrieval Algorithm for Encrypted Data in Cloud
Storage Applications   229
   Wei Song, Yihui Cui, and Zhiyong Peng

How Different Features Contribute to the Session Search?   242
   Jingfei Li, Dawei Song, Peng Zhang, and Yuexian Hou

Web Mining

Beyond Your Interests: Exploring the Information Behind User Tags   257
Weizhi Ma, Min Zhang, Yiqun Liu, Shaoping Ma, and Lingfeng Chen

Nonparametric Symmetric Correspondence Topic Models for Multilingual
Text Analysis   270
Rui Cai, Miaohong Chen, and Houfeng Wang

Knowledge Acquisition and Information Extraction

Mining RDF from Tables in Chinese Encyclopedias   285
*Weiming Lu, Zhenyu Zhang, Renjie Lou, Hao Dai, Shansong Yang,
and Baogang Wei*

Taxonomy Induction from Chinese Encyclopedias by Combinatorial
Optimization.   299
*Weiming Lu, Renjie Lou, Hao Dai, Zhenyu Zhang, Shansong Yang,
and Baogang Wei*

Recognition of Person Relation Indicated by Predicates   313
Zhongping Liang, Caixia Yuan, Bing Leng, and Xiaojie Wang

Target Detection and Knowledge Learning for Domain Restricted
Question Answering   325
Mengdi Zhang, Tao Huang, Yixin Cao, and Lei Hou

Short Papers

An Improved Algorithm of Logical Structure Reconstruction
for Re-flowable Document Understanding   339
Lin Zhao, Ning Li, Xin Peng, and Qi Liang

Mongolian Inflection Suffix Processing in NLP: A Case Study   347
Xiangdong Su, Guanglai Gao, Yupeng Jiang, Jing Wu, and Feilong Bao

Resolving Coordinate Structures for Chinese Constituent Parsing   353
Yichu Zhou, Shujian Huang, Xinyu Dai, and Jiajun Chen

P-Trie Tree: A Novel Tree Structure for Storing Polysemantic Data   362
Xin Zhou

Research on the Extraction of Wikipedia-Based Chinese-Khmer Named
Entity Equivalents.   372
Qing Xia, Xin Yan, Zhengtao Yu, and Shengxiang Gao

Bilingual Lexicon Extraction with Temporal Distributed Word
Representation from Comparable Corpora   380
   Chunyue Zhang and Tiejun Zhao

Bilingually-Constrained Recursive Neural Networks with Syntactic
Constraints for Hierarchical Translation Model   388
   Wei Chen and Bo Xu

Document-Level Machine Translation Evaluation Metrics Enhanced
with Simplified Lexical Chain   396
   Zhengxian Gong and Guodong Zhou

Cross-Lingual Tense Tagging Based on Markov Tree Tagging Model   404
   Yijiang Chen, Tingting Zhu, Chang Su, and Xiaodong Shi

Building a Large-Scale Cross-Lingual Knowledge Base from
Heterogeneous Online Wikis   413
   Mingyang Li, Yao Shi, Zhigang Wang, and Yongbin Liu

Refining Kazakh Word Alignment Using Simulation Modeling Methods
for Statistical Machine Translation   421
   Amandyk Kartbayev

A Local Method for Canonical Correlation Analysis   428
   Tengju Ye, Zhipeng Xie, and Ang Li

Learning to Rank Microblog Posts for Real-Time Ad-Hoc Search   436
   Jing Li, Zhongyu Wei, Hao Wei, Kangfei Zhao, Junwen Chen,
   and Kam-Fai Wong

Fuzzy-Rough Set Based Multi-labeled Emotion Intensity Analysis
for Sentence, Paragraph and Document   444
   Chu Wang, Shi Feng, Daling Wang, and Yifei Zhang

What Causes Different Emotion Distributions of a Hot Event?
A Deep Event-Emotion Analysis System on Microblogs   453
   Yanyan Zhao, Bing Qin, Zhenjiang Dong, Hong Chen, and Ting Liu

Deceptive Opinion Spam Detection Using Deep Level Linguistic Features . . .   465
   Changge Chen, Hai Zhao, and Yang Yang

Multi-sentence Question Segmentation and Compression for Question
Answering   475
   Yixiu Wang, Yunfang Wu, and Xueqiang Lv

A User-Oriented Special Topic Generation System for Digital Newspaper . . .   484
   Xi Xu, Mao Ye, Zhi Tang, Jian-Bo Xu, and Liang-Cai Gao

Shared Task (Long Papers)

Exploiting Heterogeneous Annotations for Weibo Word Segmentation
and POS Tagging   495
  Jiayuan Chao, Zhenghua Li, Wenliang Chen, and Min Zhang

Entity Recognition and Linking in Chinese Search Queries   507
  *Jinwei Yuan, Yan Yang, Zheng Jia, Hongfeng Yin, Junfu Huang,
and Jie Zhu*

BosonNLP: An Ensemble Approach for Word Segmentation
and POS Tagging   520
  Kerui Min, Chenggang Ma, Tianmei Zhao, and Haiyan Li

Research on Open Domain Question Answering System   527
  Zhonglin Ye, Zheng Jia, Yan Yang, Junfu Huang, and Hongfeng Yin

Overview of the NLPCC 2015 Shared Task: Chinese Word Segmentation
and POS Tagging for Micro-blog Texts   541
  Xipeng Qiu, Peng Qian, Liusong Yin, Shiyu Wu, and Xuanjing Huang

Overview of the NLPCC 2015 Shared Task: Entity Recognition
and Linking in Search Queries   550
  Yansong Feng, Zhe Han, and Kun Zhang

Overview of the NLPCC 2015 Shared Task: Weibo-Oriented Chinese
News Summarization.   557
  Xiaojun Wan, Jianmin Zhang, Shiyang Wen, and Jiwei Tan

Overview of the NLPCC 2015 Shared Task: Open Domain QA   562
  Nan Duan

Short Task (Short Papers)

Word Segmentation of Micro Blogs with Bagging   573
  Zhenting Yu, Xin-Yu Dai, Si Shen, Shujian Huang, and Jiajun Chen

Weibo-Oriented Chinese News Summarization via Multi-feature
Combination.   581
  Maofu Liu, Limin Wang, and Liqiang Nie

Linking Entities in Chinese Queries to Knowledge Graph   590
  *Jun Li, Jinxian Pan, Chen Ye, Yong Huang, Danlu Wen,
and Zhichun Wang*

A Hybrid Re-ranking Method for Entity Recognition and Linking
in Search Queries 598
     Gongbo Tang, Yuting Guo, Dong Yu, and Endong Xun

Author Index 607

Fundamentals on Language Computing

A Maximum Entropy Approach to Discourse Coherence Modeling

Rui Lin[1(✉)], Muyun Yang[1], Shujie Liu[2], Sheng Li[1], and Tiejun Zhao[1]

[1] Machine Intelligence & Translation Lab,
Harbin Institute of Technology, Harbin 150001, China
{linrui,ymy}@mtlab.hit.edu.cn, {lisheng,tjzhao}@hit.edu.cn
[2] Microsoft Research, Harbin, China
shujliu@microsoft.com

Abstract. This paper introduces a maximum entropy method to Discourse Coherence Modeling (DCM). Different from the state-of-art supervised entity-grid model and unsupervised cohesion-driven model, the model we proposed only takes as input lexicon features, which increases the training speed and decoding speed significantly. We conduct an evaluation on two publicly available benchmark data sets via sentence ordering tasks, and the results confirm the effectiveness of our maximum entropy based approach in DCM.

Keywords: Discourse coherence · Maximum entropy · Sentence ordering

1 Introduction

A high performance discourse coherence model (DCM) is important for natural language processing and generating tasks of multi-sentence document. Coherence, both logically and syntactically, makes a text meaningful [17]. For a well-written text, if we keep the word order in each sentence and swap the sentences randomly, the original text could be totally unreadable. The following two examples will show what coherence is:

- Example 1: You want my name? My name is John Smith.
- Example 2: You want my name? I am 24 years old now.

Each of the two simple texts contains only two sentences. Considering each of the four sentences above, we can see that all of them are correct both logically and syntactically. But regarding the two adjacency sentence as a whole text, example 1 is coherent and easy to understand while the example 2 makes people confused and unable to get the point. Therefore, coherence is essential in generating readable text with reasonable sentence order. The discourse coherence is a key requirement for text generating system so that it is widely used in natural language processing and generating applications such as: statistical machine translation [1], discourse generation [2] and summary [3].

© Springer International Publishing Switzerland 2015
J. Li et al. (Eds.): NLPCC 2015, LNAI 9362, pp. 3–11, 2015.
DOI: 10.1007/978-3-319-25207-0_1

Owing to the importance of discourse coherence modeling, a variety of coherence theories have been developed since 1980s. Halliday and Hasan aruge that text is not consists of irrelevant sentences in 1980: each sentence plays with important role regards to the whole[4]. Beaugrande and Dressle[5] point out 7 basic features of a coherent discourse in 1981, which are :cohesion, coherence, intentionality, acceptability, informativity, situationality, and intertextuality, in which the cohesion and coherence are meaningful in natural language processing. A quite influential theory, Rhetorical Structure Theory (RST)[6], is proposed by Mann and Thompson. RST defines 25 relations that govern clause interdependencies and ordering. These relations can be represent as a tree structure. Cristea and Romary propose Veins Theory which is also based on RST[7]. Another influential theory is Centering Theory (CT)[8] proposed by Grosz et al. CT use entity to capture the coherence between sentences in a document. Besides these method, there are also many others such as Dscourse Representation Theory (DRT) proposed by Kamp[9].

Based on these theories, some computable approaches have been developed. Barilay and Lapata[10] propose an entity-grid method which is a recent popular method based on CT to do discourse coherence modeling. Their method capture the information on the role the entity plays to judge whether a document is coherence or not. Many following efforts adopt the same framework and extend the entity-grid method by adding useful features such as discourse relations[11], multiple rank[12], named entities[13] and graph model[14]. Louis and Nenkova do coherence modeling using Hidden Markov Model (HMM) with syntactic features[15] . Xu et al.[16] proposed an unsupervised cohesion-driven method. It can get a satisfying performance when coreference resolution. Li and Hovy propose a recurrent and recursive neural network coherence model which improves the accuracy significantly[17].

All these method mentioned above need a high cost preprocess such as syntax parsing and coreference resolution which makes the training inefficient. In the other hand, most of these methods regard the whole document as inseparable element which can't capture the inner connection with high performance. And it is not suitable for statistical machine translation or text generating for there is no completed sentence or discourse for the decoding period in these tasks.

In this paper, we propose a maximum entropy based discourse coherence modeling method. Our maximum entropy based method modeling the discourse coherence with lexicon features instead of extract features from the whole document. In contrast to the previous methods, our approach can train the discourse coherence model without preprocess such as syntax parsing and coreference resolution. In this sense, out method is language independent since it uses only lexicon feature. Compared with the state-of-the-art methods in the same benchmarks, our approach also performs a better result as well as a faster training speed.

The rest of this paper is organized as follows: Section 2 introduces work related to doing discourse coherence modeling. Section 3 introduces how we do discourse coherent modeling with maximum entropy model. Section 5 presents our experiments and the results. Finally, we conclude in Section 7.

2   Related Work

Early works about discourse coherence model are described by linguists, such as RST, CT and DRT. These works only illustrate some concepts of discourse modeling and few of them are computable methods.

A recent popular approach is entity-grid method proposed by Barzilay and Lapata [10]. Their method represent a document with an entity-grid. An entity-grid is a table that each row of it denotes a sentence and each column of it denotes an entity. The element in the entity-grid has four states: O stands for object; S stands for Subject; X stands for neither object nor subject and – stands for absent. They get a discourse feature vector by counting the state transition frequency. A support vector machine (SVM) is used to judge whether a document is coherent or not. This approach can add many other features such as discourse relations[11], multiple rank[12], named entities[13] and graph model[14]. But it need high cost preprocess and is not suitable for all tasks.

Louis and Nenkova[15] propose a HMM based coherence model which is different from entity-grid methods. They use syntax features which convert the sentences in the document into production type. By clustering these productions they get some classed and regard them as hidden state. The productions are regarded as observation and the document is regard as a sequence data. They train a HMM to compute the coherent. Their approach only uses syntax features but ignore the semantics features.

Besides the supervised approaches mentioned above, an unsupervised method, cohesion driven approach, is proposed by Xu et al.[16]. Their method divides a sentence into two parts, theme and rheme. The coherent score of adjacent sentences is computed via thematic progression. The coherent of the whole document is computed with each score of adjacent sentences. This method can easily applied but the accuracy is lower than the supervised approaches and it also need high cost preprocess.

To further improve the performance of the discourse coherence model, Li and Hovy[17] propose a neural network coherence model. Their method examines a recurrent and a recursive neural network to train sentence embedding to represent a sentence. And a neural network classifier takes a slide window of these embeddings as input to compute the coherent probability of the window of sentences. After sliding all the sentences in the document, the coherent can be computed from the score of each window. This approach has the state-of-the-art performance but it is a deep learning method so the training and decoding speed is quite slow. It is a method with high computational complexity.

3   Maximum Entropy Based Discourse Coherence Model

Beaugrande and Dressle[5] point out 7 basic features of a coherent discourse, in which, cohesion and coherence explains the relationship between sentences in the same document. Therefore, in a coherence document, the words in the current sentence are chosen depending on the previous sentences.

3.1   Discourse Model

To model the coherence of sentences in the document D, which contains sentences $S_1, S_2, \ldots, S_n$, we need to maximize the objective function as fellow:

$$p(D) = p(S_1, S_2, \ldots, S_n) = p(S_1) \cdot p(S_2|S_1) \cdot \ldots \cdot p(S_n|S_1, S_2, \ldots, S_{n-1}) \tag{1}$$

Where $p(D)$ denotes the probability of the coherence of document D. From the Eq.(1) we can find that this objective function is almost the same as the objective function of language modeling. The difference is that we compute the probability of the document here while language model computes the probability of sentence. It is too complex to compute the probability directly with Eq.(1). So we limit the history length as language model does. After introduced this feature, the equation can be simplified as follow:

$$p(D) \approx p(S_1) \cdot p(S_2|S_1) \cdot p(S_3|S_2) \cdot \ldots \cdot p(S_n|S_{n-1}) = \prod_{k=1}^{n} p(S_k|h) \tag{2}$$

Where h denotes the history. Like language modeling, the longer the history length is, the closer to original objective function this function will be. But considering the sparsity of the sentence, the longer history will make the model over-fitting. So our model uses bigram history. To improve the performance of the model, the sentence also should be simplified as a vector. Considering the computational complexity, we use bag of words to represent a sentence in our model. So the Eq.(2) can be simplified as follow:

$$p(D) \approx \prod_{k=1}^{n} p(BoWS_k|BoWh) \tag{3}$$

where $BoWS_k$ denotes the bag of words of k-th sentence. To maximize the $p(D)$, we can maximize each multiplier separately. And bag of words can be easily converted into a feature vector. Here we introduce the maximum entropy model to model the discourse.

3.2   Maximum Entropy Based Discourse Coherence Model

Ep.(2) shows the similarity between discourse coherence model and language model. In this paper, we introduce the language model approach to capture the discourse coherence. Ep.(3) shows the bag of words representation of sentences, conventional n-gram language model is not suitable for this. Maximum entropy language model can capture more information and we can add any features to maximum entropy model[18]. So we decide to introduce maximum entropy model to doing discourse coherence modeling.

For a maximum entropy language model, the probability of current word w given history h is computed as follow:

$$p(w|h) = \frac{1}{Z(h)} \cdot \exp\left(\sum_i \lambda_i f_i(h, w)\right) \tag{4}$$

where $Z(h)$ denotes the normalization factor of history, f_i denotes the i-th feature functions. Many different features can be added into maximum entropy language model, both n-gram features and long distance trigger. The more features added, the more information will be captured. We can get a good performance with an appropriate feature set. Combining Ep.(2), Ep.(3) and Ep.(4), we can compute the conditional probability of sentence as follow:

$$p(S_k|S_{k-1}) = {}^1\!/\!{}_{Z(S_{k-1})} \cdot \exp(\Sigma_i \lambda_i f_i(S_{k-1}, S_k))$$

$$= {}^1\!/\!{}_{Z(BoWS_{k-1})} \cdot \exp(\Sigma_i \lambda_i f_i(BoWS_{k-1}, BoWS_k)) \quad (5)$$

where $p(S_k|S_{k-1})$ denotes the probability of coherence while current sentence is S_k given history S_{k-1}, different from maximum entropy language model. Because of the difficultness representation of sentence and the data sparsity, we use bag of words to represent a sentence.

To better capture the discourse information, we introduce two feature functions $f_{w_k}(BoWS_{k-1}, BoWS_k)$ and $f_{w_{k-1}}(BoWS_{k-1}, BoWS_k)$ as follow:

$$f_{w_k}(BoWS_{k-1}, BoWS_k) = \begin{cases} 1 & w_k \in BoWS_k \\ 0 & w_k \notin BoWS_k \end{cases} \quad (6)$$

$$f_{w_{k-1}}(BoWS_{k-1}, BoWS_k) = \begin{cases} 1 & w_{k-1} \in BoWS_{k-1} \\ 0 & w_{k-1} \notin BoWS_{k-1} \end{cases} \quad (7)$$

Ep.(6) captures the information of current sentence while Ep.(7) captures the information of history. We can model the discourse coherence with this two feature functions well. We can get the coherent score of adjacent sentences after training the model. And the coherent score of the whole document can be computed with each adjacent score multiplied.

3.3   Model Training

We train our model with an open source maximum entropy tool, Maxent [文献或者 链接]. The positive examples are sampled from the sentence pairs of the original documents while the negative examples are sampled from the sentence pairs of the permutation randomly. The ratio of positive example to negative examples is about 1. The training algorithm is default as L-BFGS, maximum number of iteration is 300.

4   Experiment

We conduct a sentence ordering task with two different corpora to evaluate our model. Sentence ordering is to find the original ordered document from a pair of articles. An article pair consists of one original document order and a random permutation of the sentences from the same document. Our approach is predicated on the assumption that the original article is always more coherent than a random permutation, which has been verified in Lin et al.'s work[11]. We use the accuracy to evaluate the performance of sentence ordering task. Accuracy defines as the ratio of the correct number of pair to the total number of pair.

4.1   Dataset

Following the former experimental settings[10-17], the two different corpora we use for evaluation is from the Barzilay and Lapata[10] which contains original documents and generated permutation of the documents. One corpora contain reports about earthquake from the Associated Press and the other contains reports on airplane accidents from the National Transportation Safety Board. Each document contains about 10 sentences with clear structure. The information of the dataset are shown in the below table.

Table 1. Dataset Information

	Original	Permutation
Earthquake Train	100	2035
Earthquake Test	99	1956
Accident Train	100	2100
Accident Test	100	1986

From the Table 1 we can see that there is a total of 2135 training documents and 1956 test pairs of articles for earthquake corpare. For accdent corpora, there is a total of 2200 training documents and 1986 test pairs of articles.

4.2   Model Comparison

Table 2 shows the performance of our approach and other related work, include:

Recurrent and Recursive Neural Network Coherence Model: Li and Hovy[17] propose a neural network coherence model which obtains the best performance. They use a recurrent or recursive neural network to convert a sentence to a tree structure with word embedding. And the root node is a vector of a sentence. Another neural network classifier is applied to generate the probability of the coherence of the slide window. Comparing to former approaches, they model can be learned without feature engineering. The results are taken directly from Li and Hovy's paper[17].

Entity-Grid Model: This model is proposed by Barzilay and Lapata[10] in 2005. Only the neural network model gets a better performance than this method considering the average accuracy. This approach obtains good performance when coreference resolution, expressive syntactic information and salience-based features are incorporated. They use the n-gram transition as the feature vector and apply a SVM ranker to judge which document is better. The results are taken directly from Barzilay and Lapata's paper[10].

HMM: Hidden Markov Coherence Model proposed by Louis and Nenkova[15] capture the hidden state transition probability in the coherent context using syntactic features. They use a production to represent a sentence. And the productions are clustered as the hidden state. The results are taken directly from Louis and Nenkova's paper[15].

Table 2. Sentence ordering task experimental result

	Accident	Earthquake	Average
Entiy-Grid[10]	**0.904**	0.872	0.888
HMM[15]	0.822	0.938	0.880
Cohesion-driven[16]	0.886	0.848	0.867
Recursive[17]	0.864	**0.976**	0.920
Recurrent[17]	0.840	0.951	0.895
ME model	0.877	0.973	**0.925**
CIME model	0.870	0.970	0.920
ME + Entity-Grid	0.877	0.973	**0.925**

Cohesion-Driven Model: Xu et al.[16] propose a Cohesion-driven discourse coherence model. They divide a sentence into two parts, theme and rheme. The coherent score of adjacent sentences is the cosine similarity computed by thematic progression. The coherent of the whole document is computed with each score of adjacent sentences. The performance will be increased with coreference resolution applied. The results are taken directly from Xu et al.'s paper[16].

As can be seen in Table 1, our maximum entropy based approach outperforms all existing baselines and obtains a state-of-art performance. Our method's accuracy is a little low than the neural network model in earthquake corpora and gets a better performance in average.

Comparing to other baselines, our method can process the data without any cost preprocess such as syntax parsing and coreference resolution. So our model can be added into any other natural language processing application easily. And the training and decoding speed is much quicker than other methods.

Comparing to the recurrent and recursive neural network method, the maximum entropy based approach we proposed has little training and decoding cost. The training and decoding speed is 30x~60x faster than the recurrent and recursive neural network method. While the maximum entropy model is similar to two layer neural network. So our model can be easily extended to a deep learning model.

We also train a case insensitive maximum entropy (CIME) model with lowercased corpora. Comparing to the original case sensitive maximum entropy model we can find that the case sensitive model outperforms the case insensitive model a little bit. This is because the uppercase name entity plays an important role in discourse coherence modeling so it should be treated as a feature.

To better evaluate our model, we also extend the entity-grid method with our model. For the entity-grid method, it generates a feature vector for each document. We compute the coherence score for each document with our model and integrate this score into the feature vector of entity-grid method. We can find that after adding our features, the accuracy of entity-grid method stays the same as our maximum entropy model.

It means that our maximum entropy model has the strong features that makes the model has an extremely high accuracy on training data. The maximum entropy model is more suitable for the sentence ordering tasks than entity-grid model.

5   Conclusion

In this paper, we compare the existing discourse coherence modeling methods and the discourse coherence application. We conduct a maximum entropy based discourse coherence model without cost preprocessing such as coreference resolution or syntax parsing. Experiment with sentence ordering task, our model can get a good performance with only lexicon features applied in two different corpora. The training and decoding for our model are also efficient.

In the future, we try to apply our maximum entropy based discourse coherence model to statistical machine translation or discourse generation, hope to get a better performance. Also we decide to convert the maximum entropy model to two-layer neural network model and adding other neural network techniques to improve our model's performance.

Acknowledgments. Thanks for the three anonymous reviewers for their efforts. This paper is supported by the project of National Natural Science Foundation of China (Grant No. 61272384, 61370170 &61402134).

References

1. Tu, M., Zhou, Y., Zong, C.: Enhancing grammatical cohesion: generating transitional expressions for SMT. In: 52nd Annual Meeting of the ACL, Baltimore, USA (2014)
2. Prasad, R., Bunt, H.: Semantic relations in discourse: the current state of ISO 24617-8. In: Proceedings 11th Joint ACL-ISO Workshop on Interoperable Semantic Annotation (ISA-11), pp. 80–92 (2015)
3. Lin, Z.H., Liu, C., Ng, H.W., Kan, M.Y.: Combining coherence models and machine translation evaluation metrics for summarization evaluation. In: Proceedings of the ACL. Association for Computational Linguistics, Jeju, pp. 1006–1014 (2012)
4. Halliday, M.A.K., Hasan, R.: Text and context: aspects of language in a social-semiotic perspective. Sophia Linguistica 6, 4–91 (1980). Working Papers in Linguistics Tokyo
5. De Beaugrande, R.A., Dressler, W.U.: Introduction to text linguistics. Longman, London (1981)
6. Mann, W.C., Thompson, S.A.: Rhetorical structure theory: Toward a functional theory of text organization. Text 8(3), 243–281 (1988)
7. Cristea, D., Ide, N., Romary, L.: Veins theory: A model of global discourse cohesion and coherence. In: Proceedings of the 17th international conference on Computational linguistics. Association for Computational Linguistics, vol. 1, pp. 281–285 (1998)
8. Grosz, B.J., Joshi, A.K., Weinstein, S., et al.: Centering: A Framework for Modelling the Local Coherence of Discourse. Computational Linguistics 21(2), 203–225 (1995)
9. Kamp, H., Kamp, H.: Discourse Representation Theory: What it is and Where it Ought to Go. Natural Language at the Computer 320(1), 84–111 (1988)

10. Barzilay, R., Lapata, M.: Modeling local coherence: an entity-based approach. Computational Linguistics **34**(1), 1–34 (2008)
11. Lin, Z.H., Ng, H.T., Kan, M.Y.: Automatically evaluating text coherence using discourse relations. In: Proceedings of the ACL. Association for Computational Linguistics, Portland, pp. 997–1006 (2011)
12. Feng, V.W., Hirst, G.: Extending the entity-based coherence model with multiple ranks. In: Proceedings of the EACL. Association for Computational Linguistics, Avignon, pp. 315–324 (2012)
13. Eisner, M., Charniak, E.: Extending the entity grid with entity-specific features. In: Proceedings of the 49th Annual Meeting of the Association for Computational Linguistics: Human Language Technologies, short papers, vol. 2. Association for Computational Linguistics, pp. 125–129 (2011)
14. Guinaudeau, C., Strube, M.: Graph-based Local Coherence Modeling. In: ACL, vol. 1, pp. 93–103 (2013)
15. Louis, A., Nenkova, A.: A coherence model based on syntactic patterns. In: Proceedings of the EMNLP-CNLL. Association for Computational Linguistics, Jeju, pp. 1157–1168 (2012)
16. Xu, F., Zhu, Q., Zhou, G., et al.: Cohesion-driven Discourse Coherence Modeling. Journal of Chinese Information **28**(3), (2014)
17. Li, J., Hovy, E.: A model of coherence based on distributed sentence representation. In: Proceedings of the EMNLP (2014)
18. Rosenfeld, R.: A maximum entropy approach to adaptive statistical language modelling. Computer Speech & Language **10**(3), 187–228 (1996)

Transition-Based Dependency Parsing with Long Distance Collocations

Chenxi Zhu, Xipeng Qiu$^{(\boxtimes)}$, and Xuanjing Huang

Shanghai Key Laboratory of Intelligent Information Processing,
School of Computer Science, Fudan University, Shanghai, China
{13210240078,xpqiu,xjhuang}@fudan.edu.cn

Abstract. Long distance dependency relation is one of the main challenges for the state-of-the-art transition-based dependency parsing algorithms. In this paper, we propose a method to improve the performance of transition-based parsing with long distance collocations. With these long distance collocations, our method provides an approximate global view of the entire sentence, which is a little bit similar to top-down parsing. To further improve the accuracy of decision, we extend the set of parsing actions with two more fine-grained actions based on the types of arcs. Experimental results show that our method improve the performance of parsing effectively, especially for long sentence.

1 Introduction

Dependency parsing uses dependency representation of syntactic structure, which directly reflects relationships among the words in a sentence. In recent years, greedy transition-based dependency parsing systems are widely used in a variety of practical tasks, especially for web-scale data, because it runs fast and performs accurately [10,14].

However, it is well known that transition-based parsing systems tend to have lower accuracy for long sentence. The main reason is that they greedily execute shift-reduce actions, and all of their decisions are very local, only two or three input tokens are available to the parser. This leads to error propagation and worse performance when predicting long arcs [9]. For example, when an arc is constructed between two words, the modifier word should have no other child anymore. But it is difficult to check this condition especially in long distance dependency.

An improved approach is to use beam search instead of greedy decoding, in combination with a globally trained model that tries to minimize the loss over the entire sentence instead of a locally trained classifier that tries to maximize the accuracy of single decisions [16]. However, the drawback with this approach is that parsing speed is proportional to the size of the beam, which means that the most accurate transition-based parsers are not nearly as fast as the original greedy transition-based parsers.

Another line of research tries to retain the efficiency of greedy classifier-based parsing by utilizing richer non-local information [17]. However, these non-local

© Springer International Publishing Switzerland 2015
J. Li et al. (Eds.): NLPCC 2015, LNAI 9362, pp. 12–24, 2015.
DOI: 10.1007/978-3-319-25207-0_2

features is still in a context window and just use extra complex structures on the left context. Yamada and Matsumoto [14] also reported that the dependency accuracy depends on the context length, and the longer the right context contributes the performance. However, they also found that the accuracy becomes worse when the length of right context is larger than 5. One reason behind this lies that not all features included in the longer context are effective for parsing.

Therefore, it is important to utilize the non-local information without introducing noises to improve the parsing accuracy of long distance arcs.

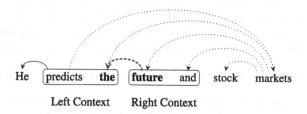

Fig. 1. An example of local decision, with the red dashed arc identifying wrong decision. "the" and "future" are the target nodes. The solid arcs are dependency relations which are already constructed. The black dotted arcs represent the gold (desired) dependency relations.

In this paper, we deal with the predicting problem of the long distance dependency in two ways. One is to utilize the non-local information by incorporating the long distance collocations, which avoids introducing extra noise information. Another is to use more reasonable actions to improve the classification accuracy. The proposed method costs up to $O(n \log(n))$ extra time complexity to find the collocation information. However, our method still runs in almost linear time in practice. The experiments show that our method leads to significant improvements in parsing accuracy, especially for long sentence.

Specifically, we make the two following contributions:

- We improve the transition-based bottom-up parsing with an approximate global view, which greatly reduces the locality of decisions. By utilizing the long distance collocations, our method provides a rough outline of the entire sentence, which is a little bit similar to top-down parsing.
- In addition to the standard parsing actions in transition-based parsing, we provide two extra fine-grained actions to avoid confusion between parsing actions and actual arcs, which can also improve the accuracy of classifier-based parsing decision.

2   Transition-Based Dependency Parsing

In a typical transition-based parsing process, an input sentence is usually processed from left to right based on shift-reduce algorithm. A set of actions are defined to construct a parse tree [10,14].

We develop our method based on Yamada's algorithm [14], which performs multi-pass scans of a partially built dependency structure. At each point, it focuses on a pair of adjacent nodes in the partial dependency structure and uses a classifier to decide whether to create a dependency link between them or to shift the focus to the next pair of heads.

There are three following deterministic actions:

Shift. No construction of dependency between the target nodes, and the position of focus simply moves to the right.

Left. A dependency relation is constructed between two neighboring nodes where the right node of target nodes becomes a child of the left one.

Right. A dependency relation is constructed between two neighboring nodes where the left node of target nodes becomes a child of the right one.

With these three actions, Yamada' parsing algorithm consists of three steps:

1. Estimate an appropriate parsing action by a classifier with features extracted from contextual information surrounding the target nodes.
2. Execute the estimated action.
3. While there is no construction in one pass, we use the action (**Left** or **Right**) with maximum score from the substitute actions.

The pseudo-code of Yamada's parsing algorithm is shown in Algorithm 1.

We use T to represent the sequence of nodes consisting of m elements $t_m (m \leq n)$, each element of which represents the root node of a sub-tree constructed in the parsing process (initially each t_i is a word w_i).

During the execution, a parsing action $y_i \in \{\textbf{Right}; \textbf{Left}; \textbf{Shift}\}$ is estimated for the focus nodes pair $\langle t_i, t_{i+1} \rangle$. The appropriate parsing action is usually decided by one or more classifiers. The contextual features x are extracted from the context surrounding the focus nodes pair.

Meanwhile, we use y_i' to record the substitute action if $y_i = \textbf{Shift}$. y_i' is the action with second largest score s_i' given by a classifier.

If all actions are **Shift** in one pass(i.e., $no\_construction = true$) we select the action y_j' at position j ($j = \arg\max_i s_i'$) and make the construction.

The complexity of the algorithm is $O(n^2)$ in the worst case because it takes up to $n-1$ passes to construct a complete dependency tree. However, it runs in linear time in practice.

3   Incorporating Long Distance Collocations

In transition-based parsing systems, all of the decisions are very local, and the strict left-to-right order implies that, while the feature set can use rich structural information from the left of the current attachment point, it is also very restricted in information to the right context. Traditionally, only the next two or three input tokens are available to the parser. This limited look-ahead window leads to error propagation and worse performance for long distant dependencies.

```
input  : Input Sentence: (w₁, w₂ ··· wₙ)
output: dependency tree T and |T| = 1
Initialize:
i = 1 ;
T = w₁, w₂ ··· wₙ ;
no_construction = true ;
while |T| ≥ 1 do
   if i == |T| then
      if no_construction == true then
         find the substitute action y'ⱼ with largest score j = arg maxᵢ s'ᵢ
         construct(T, j, y'ⱼ);
      end
      no_construction = true ;
      i = 1 ;
   end
   get the contextual features x;
   estimate the substitute action y;
   construction(T, i, y) ;
   if y == Left or Right then
      no_construction = false ;
   else
      estimate the substitute action y'ᵢ and its score s'ᵢ ;
   end
end
return T ;
```

Algorithm 1. Parsing Algorithm

As an example, Figure 1 shows a wrong decision due to the restriction of locality. The words "the" and "future" are assumed as the target nodes, and the length of context window is 1 (just for illustration). The decision prefers to make **Right** action if the word "markets" is unseen. But if we know that "future" and "markets" are collocation and often occur together, a more reasonable decision should be made.

To utilize the non-local information without introducing noises, we incorporate the long distance collocation information to help the greedy parsing decision.

In this paper, we use the term *collocation* (see [6], chapter 5) in a rather loose sense, which refers to any pair of words w_1 and w_2 with dependency relation in a sentence. The occurrence of w_1 in a text makes the appearance of w_2 in the same text more likely. w_1 and w_2 do not need to be consecutive, but they should have a dependency relation.

3.1 Collocations Extraction

Firstly, we search the entire training corpus to build the collocations between words. Here, we just use the first-order dependency relation. Relations that include a preposition are "collapsed" by directly connecting the head and the child of the preposition.

Given a sentence $s = w_1, \cdots, w_n$, if two words w_i, w_j have a dependency relation, they are regarded as a candidate collocation. Thus, we can build a collocation dictionary to store all the candidate collocations by scanning the entire training corpus.

Besides the word forms, we could also build a dictionary to store the collocation (pos_i, pos_j) of part-of-speech (POS) tags for each collocation (w_i, w_j).

In order to score the textual association strength between words and obtain the reliable collocations, we calculate the (pointwise) mutual information for each pair of words which have dependency relation in our training corpus.

Mutual information is one of many measures that seems to be roughly correlated to the degree of semantic relation between words. The mutual information between two words w_1 and w_2 is given by: $I(w_1, w_2) = \frac{Pr(w_1, w_2)}{Pr(w_1)Pr(w_2)}$.

Notice that, directionality is not taken into account (i.e., both (w_l, w_2) and (w_2, w_1) are counted as occurrences of the same bigram).

3.2   Features with Collocation Information

Since the traditional features are extracted from local context, the decision lacks foresight for global information. We wish to improve the decision in each parsing step with collocation information.

Considering an intermediate step in parsing process, the sequence of nodes T consists of m elements $t_m (1 < m \leq n)$, and the focus nodes pair is $\langle t_i, t_{i+1} \rangle$.

The left context is defined as the nodes on the left side of the target nodes: $t_l (l < i)$, and the right context is defined as those on the right: $t_r (i + 1 < r)$. Context length (l, r) represents the numbers of nodes within the left and right contexts.

We find the collocation information outside the context window. For a node t_j $(1 \leq j < i - l$ or $i + r + 1 < j \leq m)$ outside the context, if (t_j, t_i) or (t_j, t_{i+1}) is in collocation dictionary, we add a collocation feature.

The collocation feature can be represented as quadruple (v_1, p_1, v_2, p_2), in which v_1 is the one of the target nodes, $p_1 \in \{L, R\}$ denotes that v_1 is left or right node, v_2 is the collocation node outside the context length, and $p_2 \in \{L, R\}$ denotes that v_2 occurs in the left or right side of v_1.

For the example in Figure 1, the collocation features include (future,R,markets,R), (the,L,markets,R) and so on.

With a discriminative classifier, these collocation features are assigned the corresponding weights according their usefulness.

3.3   Speedup

In each parsing decision, our method need find the collocation information of the target nodes. The parser will search every token in the sentence and then search for its potential relevant token, which is relative slow since the entire sentence is scanned. Fortunately, we can speedup this search process with preprocessing.

When parsing a sentence, we build an index for all collocations in the beginning. The collocation index records position of the token and the positions of its related tokens in the sentence. These two tokens must be included in the word/POS collocation dictionary.

Thus, when extracting the collocation features, our parser will easily use the collocation index to find the relevant token of target nodes as global feature, and it will not cost extra complexity.

After executing a **Left** or **Right** action, we find the position of relevant token of the modifier node, and then delete the modifier node from the collocation index.

By this preprocess, our method just need $O(nlog(n))$ extra time complexity.

4   Enhanced Actions with Fine-Grained "Shift"

Due to the bottom-up regulation, a dependency relation $h \leftarrow d$ can be constructed only after all modifiers of the d have been constructed.

For the example in Figure 2, the decision prefers to make **Left** action between two target words "able" and "handle" if the word "information" is unseen. But if we know that "handle" and "information" are collocation and often occur together, a more reasonable decision (**Shift**) should be made.

Although both the examples in Figure 1 and 2 should execute **Shift** action, they are different essentially. In Figure 1, there is no dependencies relation between the target nodes. However in Figure 1, there is dependency relation between the target nodes, but the modifier node has not been a complete subtree yet.

Therefore, **Shift** action in Yamada's algorithm should be distinguished in two cases. These two cases are more intuitively shown in Figure 3.

ABC is a word sequence in Figure 3. **Shift** action is executed for pair of (A, B) for both cases at the first pass. However, for case (b), it need predict to take **Left** action at node pair (A, B) when the arc between (B, C) is constructed. The difference between two predictions is whether C has been attached as a dependent of node B. This might cause a degradation of performance since that the prediction is often made with linear classifier. The features extracted from

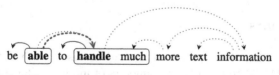

be [able] to [handle much] more text information

Left Context   Right Context

Fig. 2. Another example of wrong local decision, with the red dashed arc identifying wrong decision. "able" and "handle" are the target nodes. The solid arcs are dependency relations which are already constructed. The black dotted arcs represent the gold (desired) dependency relations.

(a)   (b)

Fig. 3. Two Cases of Shift Action

this two situations have a lot of overlap. So the prediction of (A, B) is very difficult and may cause error propagation. If it predicts **Left** action on (A, B) at the first pass, the predict head of C will be never right due to the bottom-up regulation.

The main reason of the problem is that the parser ignores the potential relation between the target nodes when it predicts a **Shift** action. In order to cope with this problem, Yamada and Matsumoto [14] suggested to divide the action **Shift** into two kinds of actions, **Shift'** and **Wait** for two cases respectively. These new actions are the same behavior in parsing process. However, they do not report any experimental result for their suggestion.

In give more unambiguous definition of parsing actions, we expand the actions by adding two extra actions: **Shift-left** and **Shift-right**. The new set of parsing actions is as follows.

Left let the right node become a child of the left one.
Right let the left node become a child of the right one.
Shift let the point of focus move to the right. And there is no relation between target nodes.
Shift-right let the point of focus move to the right. And there is a right relation between target nodes but we can just shift because dependent child still have further dependent children in the sentence.
Shift-left let the point of focus move to the right. And there is a left relation between target nodes.

Compared with standard set of actions of Yamada's algorithm, we split the original **Shift** action into **Shift**, **Shift-left** and **Shift-right**, which is used to distinguish the relation between the target node pair. The **Shift-left** and **Shift-right** actions can be also called **pseudo Shift** actions.

5   Experiments

In the implementation of our proposed method, we use Margin Infused Relaxed Algorithm (MIRA) [3] to train the parameters of model. Following [2], an averaged strategy is used to avoid the overfitting problem. The training iterations of all parsers are up to 50. The baseline method in our experiments is Yamada's algorithm with MIRA as learning method instead of support vector machine.

5.1 Datasets

To empirically demonstrate the effectiveness of our approach, we use two datasets from different languages (English and Chinese) in our experimental evaluation.

English. For English, we convert the Penn WSJ Treebank (WSJ) [7] constituency trees to dependencies using Yamadas head rules. We then train on the standard PTB split with sections 2-21 as training, section 22 as validation, and section 23 as test. This test data was used in several other previous works, enabling mutual comparison with the methods reported in those works.

Chinese. For Chinese, we use the Chinese dataset in the CoNLL 2009 shared task [5].

5.2 Parsing Accuracy

Three measures (Unlabeled Attachment Score (UAS), Labeled Attachment Score (LAS) and Complete Match (CM)) are used to evaluate the parsing performances. In the evaluation, we consistently excluded punctuation marks.

Table 1. Accuracy on English Dataset

	UAS	LAS	CM	Learning Method
Yamada & Matsumoto [14]	90.4	-	38.4	support vector machine
Nivre [11]	87.1	84.4	30.4	memory based learning
Goldberg and Elhadad [4]	89.7	-	37.5	structured perceptron
Baseline(Yamada's algorithm)	90.42	89.26	37.58	MIRA
Baseline + collocation (this work)	90.61	89.43	38.53	MIRA
Baseline + collocation + actions (this work)	**90.91**	**89.77**	**40.03**	MIRA

Table 2. Accuracy on Chinese Dataset

	UAS	LAS	CM	Learning Method
Maltparser	82.31	80.64	28.17	support vector machine
Top CoNLL 2009[12] (Transition-based)	81.22	79.19	-	support vector machine
Baseline(Yamada's algorithm)	83.18	81.51	29.54	MIRA
Baseline + collocation (this work)	83.50	81.73	30.13	MIRA
Baseline + collocation + actions (this work)	**83.95**	**82.25**	**31.18**	MIRA

The experiments results are shown in Table 1 and 2.

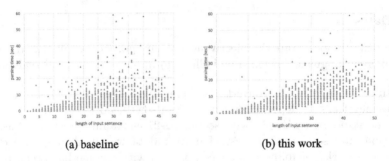

(a) baseline   (b) this work

Fig. 4. Parsing time for sentences.

For English dataset, our method achieves 5.1% and 4.7% error reductions in UAS and LAS respectively over baseline, and also outperforms the state-of-the-art greedy transition-based parsing methods.

For Chinese dataset, our method achieves 4.5% and 4% error reductions in UAS and LAS respectively over baseline, and also outperforms the state-of-the-art greedy transition-based parsing methods.

5.3   Parsing Time

Our utilization of the collocations also concerns the speed of our method. Here, we compare its computational time with baseline method on English WSJ dataset.

Figure 4 shows plots of the parsing times for all sentences in the test data. The average parsing time for our method was 7.1 sec, whereas that for baseline method was 5.9 sec. Although the worst-case time complexity the baseline method is $O(n^2)$, worst-case situations (e.g., all words having heads on their left) did not appear frequently. Our method costs extra $O(n \log(n))$ time to find the collocations, but it still runs in (almost) linear-time for most cases.

5.4   Result Analysis with Length Factors

It is well known that transition-based parsing systems tend to have lower accuracy for long sentence. Therefore, we evaluate the effects of our method according to different lengths of sentences and arcs. These detailed comparisons are made between our method and baseline on English WSJ dataset.

Figure 5a shows the accuracies of two methods relative to sentence length (in bins of size 10: 110, 1120, etc.). We can find that our method performs better than baseline on sentences of all lengths. Although our method is to utilize the long distance information, it is helpful for shorter sentences.

Figure 5b shows the dependency recalls of two methods relative to dependency lengths, which measures the percentage of gold standard arcs of length d that are correctly predicted. The length of a dependency from word w_i to word w_j is simply equal to $|i - j|$. We can find that our method is better for long dependency arc.

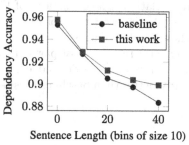

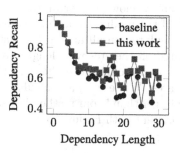

(a) Accuracy relative to sentence length

(a) Accuracy relative to sentence length

(b) Dependency recall relative to gold dependency length.

(b) Dependency recall relative to gold dependency length.

Fig. 5. Performances with different length.

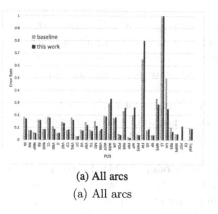

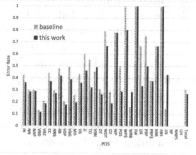

(a) All arcs

(a) All arcs

(b) Long arcs (length ¿ 5)

(b) Long arcs (length ¿ 5)

Fig. 6. Error rates for different parts of speech.

These two behaviors can be explained using the same reason: the long distance collocation is useful for local decision in the greedy parsing procedure and are less tend to error propagation.

5.5   Result Analysis with Different POS Tags

We also evaluate the effects of our method according to different parts of speech. Longer dependencies typically represent modifiers of the root or the main verb in a sentence. Shorter dependencies are often modifiers of nouns such as determiners or adjectives or pronouns modifying their direct neighbours.

These detailed comparisons are made between our method and baseline on the same English WSJ dataset.

Figure 6 shows the error rates of the two models relative to the part of speech of the modifier word in a dependency relation. The error rates of Figure 6a is calculated on all arcs in test dataset, while the error rates of Figure 6b is calculated on the arcs with length larger than 5 in test dataset.

We see that our method has slightly lower error rate for most parts of speech on all arcs. While we just calculate the error rates on the dependencies with length larger than 5, our method is significantly better than baseline. This results are consistent with Figure 5b.

6   Related Works

There are several methods are proposed to avoid the drawback of local decision for the transition-based parsing.

Cheng et al. [1] try to deal with this problem by selecting the words which has not been processed as the global features. However, their method uses all words out of local context and suffers from the noise features which are brought by the unrelated words.

Goldberg and Elhadad [4] build a dependency tree with easy-first non-directional parsing strategy, which iteratively selects the best pair of neighbors to connect at each parsing step. Their methods can capture some, though not all, long-distance relations and therefore incorporate more global information than left-to-right deterministic algorithm.

Zhang and Nivre [17] utilize the richer non-local information to transition-based parsing. However, these non-local features is still in a context window and just use extra complex structures on the left context. The information outside context window is still unseen.

Zhang et al. [15] use a dependency language model (DLM) [13] to enrich the feature representations for the graph-based models [8]. However, the features based on the DLM are difficult to incorporate into the transition-based parser because DLM needs a constructed parsing tree.

7   Conclusion

In this paper, we have proposed a method to improve the accuracy of transition-based parser by utilizing long distance collocations, which achieves 5.1% and 4.5% error reductions in unlabeled attachment score over baseline on English and Chinese datasets respectively. The idea of utilizing collocation information is useful for both long and short sentences.

In future, we will extend the first-order collocation to higher-order and extract multi-word collocations, which should involve more useful information for parsing decision.

Acknowledgments. We would like to thank the anonymous reviewers for their valuable comments. This work was partially funded by the National Natural Science Foundation of China (61472088), National High Technology Research and Development Program of China (2015AA015408), Shanghai Science and Technology Development Funds (14ZR1403200).

References

1. Cheng, Y., Asahara, M., Matsumoto, Y.: Chinese deterministic dependency analyzer: examining effects of global features and root node finder. In: Proceedings of the Fourth SIGHAN Workshop on Chinese Language Processing, pp. 17–24 (2005)
2. Collins, M.: Discriminative training methods for hidden markov models: theory and experiments with perceptron algorithms. In: Proceedings of the 2002 Conference on Empirical Methods in Natural Language Processing (2002)
3. Crammer, K., Dekel, O., Keshet, J., Shalev-Shwartz, S., Singer, Y.: Online passive-aggressive algorithms. Journal of Machine Learning Research **7**, 551–585 (2006)
4. Goldberg, Y., Elhadad, M.: An efficient algorithm for easy-first non-directional dependency parsing. In: Human Language Technologies: The 2010 Annual Conference of the North American Chapter of the Association for Computational Linguistics, pp. 742–750 (2010)
5. Hajič, J., Ciaramita, M., Johansson, R., Kawahara, D., Martí, M., Màrquez, L., Meyers, A., Nivre, J., Padó, S., Štěpánek, J., et al.: The CoNLL-2009 shared task: syntactic and semantic dependencies in multiple languages. In: Proceedings of the Thirteenth Conference on Computational Natural Language Learning: Shared Task, pp. 1–18 (2009)
6. Manning, C., Schütze, H.: Foundations of statistical natural language processing. MIT Press (1999)
7. Marcus, M.P., Marcinkiewicz, M.A., Santorini, B.: Building a large annotated corpus of english: The penn treebank. Computational linguistics **19**(2), 313–330 (1993)
8. McDonald, R., Crammer, K., Pereira, F.: Online large-margin training of dependency parsers. In: Proceedings of the 43rd Annual Meeting on Association for Computational Linguistics, pp. 91–98 (2005)
9. McDonald, R.T., Nivre, J.: Characterizing the errors of data-driven dependency parsing models. In: EMNLP-CoNLL, pp. 122–131 (2007)
10. Nivre, J.: An efficient algorithm for projective dependency parsing. In: Proceedings of the 8th International Workshop on Parsing Technologies (IWPT) (2003)
11. Nivre, J., Scholz, M.: Deterministic dependency parsing of english text. In: Proceedings of the 20th International Conference on Computational Linguistics, p. 64 (2004)
12. Ren, H., Ji, D., Wan, J., Zhang, M.: Parsing syntactic and semantic dependencies for multiple languages with a pipeline approach. In: Proceedings of the Thirteenth Conference on Computational Natural Language Learning: Shared Task, pp. 97–102 (2009)
13. Shen, L., Xu, J., Weischedel, R.M.: A new string-to-dependency machine translation algorithm with a target dependency language model. In: ACL, pp. 577–585 (2008)
14. Yamada, H., Matsumoto, Y.: Statistical dependency analysis with support vector machines. In: Proceedings of the International Workshop on Parsing Technologies (IWPT), vol. 3 (2003)

15. Zhang, M., Chen, W., Duan, X., Zhang, R.: Improving graph-based dependency parsing models with dependency language models. IEEE Transactions on Audio, Speech, and Language Processing **21**(11), 2313–2323 (2013)
16. Zhang, Y., Clark, S.: A tale of two parsers: investigating and combining graph-based and transition-based dependency parsing using beam-search. In: Proceedings of the Conference on Empirical Methods in Natural Language Processing, pp. 562–571 (2008)
17. Zhang, Y., Nivre, J.: Transition-based dependency parsing with rich non-local features. In: Proceedings of the 49th Annual Meeting of the Association for Computational Linguistics: Human Language Technologies, short papers, vol. 2, pp. 188–193 (2011)

Recurrent Neural Networks with External Memory for Spoken Language Understanding

Baolin Peng[1]([⊠]), Kaisheng Yao[2], Li Jing[1], and Kam-Fai Wong[1]

[1] Department of Systems Engineering and Engineering Management,
The Chinese University of Hong Kong, Hong Kong, Hong Kong
{blpeng,jingli,kfwong}@se.cuhk.edu.hk
[2] Microsoft Research, Hong Kong, Hong Kong
kaisheny@microsoft.com

Abstract. Recurrent Neural Networks (RNNs) have become increasingly popular for the task of language understanding. In this task, a semantic tagger is deployed to associate a semantic label to each word in an input sequence. The success of RNN may be attributed to its ability to memorise long-term dependence that relates the current-time semantic label prediction to the observations many time instances away. However, the memory capacity of simple RNNs is limited because of the gradient vanishing and exploding problem. We propose to use an external memory to improve memorisation capability of RNNs. Experiments on the ATIS dataset demonstrated that the proposed model was able to achieve the state-of-the-art results. Detailed analysis may provide insights for future research.

1 Introduction

Neural network have recently demonstrated promising results on many natural language processing tasks [2,6]. Specifically, recurrent neural networks (RNNs) based methods have shown strong performances, in language modeling [16], language understanding [25], and machine translation [7,5] tasks.

The goal of a language understanding (LU) system is to associate words with semantic meanings [17]. For example, in the sentence "Please book me a ticket from HK to Seattle", a LU system would tag "HK" as the departure-city of a trip and "Seattle" as its arrival city. The widely used approaches include conditional random fields (CRFs) [19,13], support vector machine [12], and, more recently, RNNs [25,14].

A RNN consists of an input, a recurrent hidden layer, and an output layer. The input layer reads each word and the output layer produces probabilities of semantic labels. The success of RNNs can be attributed to the fact that RNNs, if successfully trained, can relate the current prediction with input words that are several time steps away. However, RNNs are difficult to train, because of the gradient vanishing and exploding problem [3]. The problem also limits RNNs' memory capacity because error signals may not be able to back-propagated far enough.

© Springer International Publishing Switzerland 2015
J. Li et al. (Eds.): NLPCC 2015, LNAI 9362, pp. 25–35, 2015.
DOI: 10.1007/978-3-319-25207-0_3

There have been two lines of researches to address this problem. One is to design learning algorithms that can avoid gradient exploding, e.g., using gradient clipping [18], and/or gradient vanishing, e.g., using second-order optimization methods. Alternatively, researchers have proposed more advanced model architectures, in contrast to the simple RNN that uses, e.g., Elman architecture [8]. Specifically, the long short-term memory (LSTM) [11,9] neural networks have three gates that control flows of error signals. The recently proposed gated recurrent neural networks (GRNN) [5] may be considered as a simplified LSTM with fewer gates.

Along this line of research on developing more advanced architectures, this paper focuses on a novel neural network architecture. Inspired by the recent works in Graves et al. [10] and Sukhbaatar et al. [20], we extend the simple RNN to that with an external memory. The external memory stores the past hidden layer activities, not only from the current sentence but also from past sentences. To predict outputs, the model uses input observation together with a content retrieved from the external memory. The proposed model performs strongly on a common language understanding dataset and achieves new state-of-the-art results.

2   Background

2.1   Language Understanding

A language understanding system predicts an output sequence with tags such as named-entity given an input sequence words. Often, the output and input sequences have been aligned. In these alignments, an input may correspond to a null tag or a single tag. An example is given in Table 1.

Table 1. An example of language understanding. Label names have been shortened to fit. Many words are labeled null or '-'.

book	a	flight	from	Hong Kong	to	Seattle
-	-	-	-	Dpt-city	-	Arv-city

Given a T-length input word sequence x_1^T, a corresponding output tag sequence y_1^T, and an alignment A, the posterior probability $p(y_1^T|A, x_1^T)$ is approximated by

$$p(y_1^T|x_1^T) \approx \prod_{t=1}^{T} p(y_t|x_{t-k}^{t+k}), \tag{1}$$

where k is the size of a context window and t indexes the positions in the alignment.

2.2 Simple Recurrent Neural Networks

The above posterior probability can be computed using a RNN. A RNN consists of an input layer x_t, a hidden layer h_t, and an output layer y_t. In Elman architecture [8], hidden layer activity h_t is dependent on both the input x_t and also recurrently on the past hidden layer activity h_{t-1}.

Because of the recurrence, the hidden layer activity h_t is dependent on the observation sequence from its beginning. The posterior probability is therefore computed as follows

$$p(y_1^T | x_1^T) \approx \prod_{t=1}^{T} p(y_t | x_1^t)$$

$$= \prod_{t=1}^{T} p(y_t | h_t, x_t) \tag{2}$$

where the output y_t and hidden layer activity h_t are computed as

$$y_t = g(h_t), \tag{3}$$
$$h_t = \sigma(x_t, h_{t-1}). \tag{4}$$

In the above equation, $g(\cdot)$ is softmax function and $\sigma(\cdot)$ is sigmoid or tanh function. The above model is denoted as simple RNN, to contrast it with more advanced recurrent neural networks described below.

2.3 Recurrent Neural Networks Using Gating Functions

The current hidden layer activity h_t of a simple RNN is related to its past hidden layer activity h_{t-1} via the nonlinear function in Eq. (4). The non-linearity can cause errors back-propagated from h_t to explode or to vanish. This phenomenon prevents simple RNN from learning patterns that are spanned with long time dependence [18].

To tackle this problem, long short-term memory (LSTM) neural network was proposed in [11] with an introduction of memory cells, linearly dependent on their past values. LSTM also introduces three gating functions, namely input gate, forget gate and output gate. We follow a variant of LSTM in [9].

More recently, a gated recurrent neural network (GRNN) [5] was proposed. Instead of the three gating functions in LSTM, it uses two gates.

One is a reset gate r_t that relates a candidate activation with the past hidden layer activity h_{t-1}; i.e.,

$$\hat{h}_t = tanh(W_{xh}x_t + W_{hh}(r_t \odot h_{t-1})) \tag{5}$$

where $\hat{h}_t$ is the candidate activation. W_{xh} and W_{hh} are the matrices relate the current observation x_t and the past hidden layer activity. $\odot$ is element-wise product.

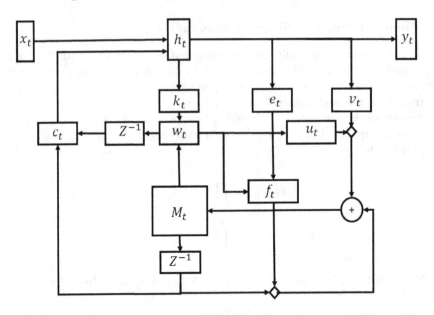

Fig. 1. The RNN-EM model. The model reads input x_t and outputs y_t. Its hidden layer activity h_t depends on the input and the model's memory content retrieved in c_t. f_t and u_t are the forget and update gates. k_t, e_t and v_t each denote key, erase and new content vector. M_t is the external memory. w_t is the weight and it is a function of k_t and M_t. Z^{-1} denotes a time-delay operator. The diamond symbol $\diamond$ denotes diagonal matrix multiplication.

The second gate is an update gate z_t that interpolates the candidate activation and the past hidden layer activity to update the current hidden layer activity; i.e.,

$$h_t = (1 - z_t) \odot h_{t-1} + z_t \odot \hat{h}_t. \tag{6}$$

These gates are usually computed as functions of the current observation x_t and the past hidden layer activity; i.e.,

$$r_t = \sigma(W_{xr}x_t + W_{hr}h_{t-1}) \tag{7}$$
$$z_t = \sigma(W_{xz}x_t + W_{hz}h_{t-1}) \tag{8}$$

where W_{xr} and W_{hr} are the weights to observation and to the past hidden layer activity for the reset gate. W_{xz} and W_{hz} are similarly defined for the update gate.

3   The RNN-EM Architecture

In this section, We introduce simple RNN with an external memory. Figure 1 illustrates the proposed model, which we denote it as RNN-EM. Same as with the simple RNN, it consists of an input layer, a hidden layer and an output layer. However, instead of feeding the past hidden layer activity directly to the hidden layer as with the simple RNN, one input to the hidden layer is from a content of an external memory. RNN-EM uses a weight vector to retrieve the content from the external memory to use in the next time instance. The element in the weight vector is proportional to the similarity of the current hidden layer activity with the content in the external memory. Therefore, content that is irrelevant to the current hidden layer activity has small weights. All of the equations to be described are with their bias terms, which we omit for simplicity of descriptions. We implemented RNN-EM using Theano [1,4].

3.1   Model Input and Output

The input to the model is a dense vector $x_t \in R^{d \times 1}$. In the context of language understanding, x_t is a projection of input words, also known as word embedding.

The hidden layer reads both the input x_t and a content c_t vector from the memory. The hidden layer and output layer activities are computed as follows

$$h_t = tanh(W_{ih}x_t + W_c c_t) \tag{9}$$

$$y_t = g(W_{ho}h_t) \tag{10}$$

$W_{ih} \in R^{p \times d}$ is the weight to the input vector. $c_t \in R^{m \times 1}$ is the content from a read operation to be described in Eq. (15). $W_c \in R^{p \times m}$ is the weight to the content vector. where W_{ho} is the weight to the hidden layer activity and $g(\cdot)$ is softmax function. Notice that in case of $c_t = h_{t-1}$, the above model is simple RNN.

3.2   External Memory Read

RNN-EM has an external memory $M_t \in R^{m \times n}$. It can be considered as a memory with n slots and each slot is a vector with m elements. Similar to the external memory in computers, the memory capacity of RNN-EM may be increased if using a large n.

The model generates a key vector k_t to search for content in the external memory. Though there are many possible ways to generate the key vector, we choose a simple linear function that relates hidden layer activity h_t as follows

$$k_t = W_k h_t \tag{11}$$

where $W_k \in R^{m \times p}$ is a linear transformation matrix. Our intuition is that the memory should be in the same space of or affine to the hidden layer activity.

We use cosine distance $K(u, v) = \frac{u \cdot v}{\|u\|\|v\|}$ to compare this key vector with contents in the external memory. The weight for the c-th slot $M_t(:, c)$ in memory M_t is computed as follows

$$\hat{w}_t(c) = \frac{\exp \beta_t K(k_t, M_t(:, c))}{\sum_q \exp \beta_t K(k_t, M_t(:, q))} \tag{12}$$

where the above weight is normalized and sums to 1.0. β_t is a scalar larger than 0. It sharpens the weight vector when β_t is larger than 1.0. Conversely, it smooths or dampens the weight vector when β_t is between 0.0 and 1.0.

We use the following function to obtain β_t; i.e.,

$$\beta_t = \log(1 + \exp(W_\beta h_t)) \tag{13}$$

where $W_\beta \in R^{1 \times p}$ maps the hidden layer activity h_t to a scalar.

Importantly, we also use a scalar coefficient g_t to interpolate the above weight estimate with the past weight as follows:

$$w_t = (1 - g_t)w_{t-1} + g_t \hat{w}_t \tag{14}$$

This function is similar to that in the gated RNN, except that we use a scalar g_t to interpolate the weight updates and the gated RNN uses a vector to update its hidden layer activity. The memory content is retrieved from the external memory at time $t - 1$ using

$$c_t = M_{t-1} w_{t-1}. \tag{15}$$

3.3 External Memory Update

RNN-EM generates a new content vector v_t to be added to its memory; i.e,

$$v_t = W_v h_t \tag{16}$$

where $W_v \in R^{m \times p}$. We use the above linear function based on the same intuition in Sec. 3.2 that the new content and the hidden layer activity are in the same space of or affine to each other.

RNN-EM has a forget gate and update gate as follows:

$$f_t = 1 - w_t \odot e_t \tag{17}$$
$$u_t = w_t. \tag{18}$$

where $e_t \in R^{n \times 1}$ is an erase vector, generated as $e_t = \sigma(W_{he} h_t)$. Notice that the c-th element in the forget gate is zero only if both read weight w_t and erase vector e_t have their c-th element set to one. Therefore, memory cannot be forgotten if it is not to be read and can only be updated if it is to be read.

With the above described two gates, the memory is updated as follows

$$M_t = diag(f_t)M_{t-1} + diag(u_t)v_t \tag{19}$$

Notice that when the number of memory slots is small, it may have similar performances as a gated RNN. Specifically, RNN-EM subsumes GRNN as a special case.

Table 2. F1 scores (in %) on ATIS.

Method	F1 score
CRF [15]	92.94
simple RNN [25]	94.11
CNN [23]	94.35
LSTM [24]	94.85
GRNN	94.82
RNN-EM	95.25

3.4   Dataset

In order to compare the proposed model with alternative modelling techniques, we conducted experiments on a well studied language understanding dataset, Air Travel Information System (ATIS) [22,21]. The training part consists of 4978 sentences and 56590 words. There are 893 sentences and 9198 words for test. The number of semantic label is 127, including the common null label. We use lexicon-only features in experiments.

3.5   Comparison with the Past Results

The input x_t in RNN-EM has a window size of 3, consisting of the current input word and its neighbouring two words. We use the AdaDelta method to update gradients [26]. The maximum number of training iterations was 50. Hyper parameters for tuning included the hidden layer size p, the number of memory slots n, and the dimension for each memory slot m. The best performing RNN-EM had 100 dimensional hidden layer and 8 memory slots with 40 dimensional memory slots.

Table 2 lists performance in F1 score of RNN-EM, together with the previous best results of alternative models in the literature. These results are optimal in their respective systems. Since there are no previous results from GRNN, we use our own implementation of it for this study. The previous best result was achieved using LSTM. A change of 0.38% of F1 score from LSTM result is significant at the 90% confidence level. Results in Table 2 show that RNN-EM is significantly better than the previous best result using LSTM.

3.6   Analysis on Convergence and Averaged Performances

Results in the previous sections were obtained with models using different sizes. This section further compares neural network models given that they have approximately the same number of parameters, listed in Table 3. We use AdaDelta [26] gradient update method for all these models.

Figure 2 plots their training set entropy with respect to iteration numbers. To better illustrate their convergences, entropy values have been converted to their logarithms. The results show that RNN-EM converges to lower training entropy

Table 3. The size of each neural network models.

Model	#Hidden	# of Parameters
simple RNN	115	$\approx 7.4 * 10^3$
LSTM	50	$\approx 7.5 * 10^3$
GRNN	60	$\approx 7.4 * 10^3$
RNN-EM[†]	100,40 × 8	$\approx 7.3 * 10^3$

[†] 100 dimensional hidden layer, 40 dimensional slot with 8 slots.

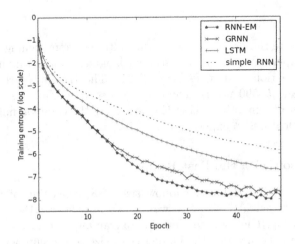

Fig. 2. Convergence of training entropy.

than other models. RNN-EM also converges faster than the simple RNN and LSTM.

Furthermore, we repeated ATIS experiments for 10 times with different random seeds for each model. We evaluated their performances after their convergences. Table 4 lists their averaged F1 scores, together with their maximum and minimum F1 scores. A change of 0.12% is significant at the 90% confidence level, when comparing against LSTM result. Results in Table 4 show that RNN-EM, on average, significantly outperforms LSTM. The best performance by RNN-EM is also significantly better than the best performing LSTM.

3.7   Analysis on Memory Size

We fixed the dimension of memory slots to 40 and varied the number of slots. Table 5 lists their test set F1 scores. The best performing RNN-EM was with $n = 8$. Notice that RNN-EM with $n = 1$ performed better than the simple RNN with 94.09% F1 score. This can be explained as using gate functions in RNN-EM, which are absent in simple RNNs. RNN-EM with $n = 1$ also performed

Table 4. The maximum, minimum and averaged F1 scores (in %) by neural network models.

Method	Max	Min	Averaged
simple RNN	94.09	93.64	93.80
LSTM	94.81	94.62	94.73
GRNN	94.70	94.32	94.61
RNN-EM	95.22	94.71	94.96

Table 5. Test set F1 scores (in %) and training set entropy by RNN-EM with different slot numbers.

slot number n	1	2	4	8	16
F1 score	94.67	94.87	94.91	95.22	94.75
entropy$\times 10^3$	2.23	1.96	1.91	1.90	2.05

slot number n	32	64	128	256	512
F1 score	94.87	94.77	94.57	94.84	94.53
entropy$\times 10^3$	2.16	2.30	2.36	3.43	6.10

similarly as the gated RNN with 94.70% F1 score in Table 4, because of these gate functions.

Memory capacity may be measured using training set entropy. Table 5 shows that training set entropy is decreased initially with n increased from 1 to 8, showing that the memory capacity of the RNN-EM is improved. However, the entropy is increased with n further increased. This suggests that memory capacity of RNN-EM cannot be increased simply by increasing the number of slots. A large n may introduce noise to RNN-EM. We plan to conduct future research on mechanisms to increase memory capacity.

4 Related Work

The RNN-EM is along the same line of research in [10,20] that uses external memory to improve memory capacity of neural networks. Perhaps the closest work is the Neural Turing Machine (NTM) work in [10], which focuses on those tasks that require simple inference and has proved its effectiveness in copy, repeat and sorting tasks. NTM requires complex models because of these tasks. The proposed model is considerably simpler than NTM and can be considered as an extension of simple RNN. Importantly, we have shown through experiments on a common language understanding dataset the promising results from using the external memory architecture.

5 Conclusion and Discussion

In this paper, we have proposed a novel neural network architecture, RNN-EM, that uses external memory to improve memory capacity of simple recurrent neural networks. On a common language understanding task, RNN-EM achieves new state-of-the-art results and performs significantly better than the previous best result using long short-term memory neural networks. We have conducted experiments to analyze its convergence and memory capacity. These experiments provide insights for future research directions such as mechanisms of accessing memory contents and methods to increase memory capacity.

Acknowledgement. This work is partially supported by General Research Fund of Hong Kong (417112), RGC Direct Grant (417613). We would like to thank anonymous reviewers for the useful comments.

References

1. Bastien, F., Lamblin, P., Pascanu, R., Bergstra, J., Goodfellow, I.J., Bergeron, A., Bouchard, N., Bengio, Y.: Theano: new features and speed improvements. In: Deep Learning and Unsupervised Feature Learning NIPS 2012 Workshop (2012)
2. Bengio, Y., Ducharme, R., Vincent, P., Janvin, C.: A neural probabilistic language model. Journal of Machine Learning Research **3**, 1137–1155 (2003)
3. Bengio, Y., Simard, P.Y., Frasconi, P.: Learning long-term dependencies with gradient descent is difficult. IEEE Transactions on Neural Networks **5**(2), 157–166 (1994)
4. Bergstra, J., Breuleux, O., Bastien, F., Lamblin, P., Pascanu, R., Desjardins, G., Turian, J., Warde-Farley, D., Bengio, Y.: Theano: a CPU and GPU math expression compiler. In: Proceedings of the Python for Scientific Computing Conference (SciPy), June 2010, Oral Presentation
5. Cho, K., van Merrienboer, B., Gülçehre, Ç., Bahdanau, D., Bougares, F., Schwenk, H., Bengio, Y.: Learning phrase representations using RNN encoder-decoder for statistical machine translation. In: EMNLP, pp. 1724–1734 (2014)
6. Collobert, R., Weston, J.: A unified architecture for natural language processing: deep neural networks with multitask learning. In: ICML, pp. 160–167 (2008)
7. Devlin, J., Zbib, R., Huang, Z., Lamar, T., Schwartz, R.M., Makhoul, J.: Fast and robust neural network joint models for statistical machine translation. In: ACL, pp. 1370–1380 (2014)
8. Elman, J.: Finding structure in time. Cognitive Science **14**(2), 179–211 (1990)
9. Graves, A., Mohamed, A., Hinton, G.E.: Speech recognition with deep recurrent neural networks. In: ICASSP, pp. 6645–6649 (2013)
10. Graves, A., Wayne, G., Danihelka, I.: Neural turing machines. CoRR abs/1410.5401 (2014)
11. Hochreiter, S., Schmidhuber, J.: Long short-term memory. Neural Computation **9**(8), 1735–1780 (1997)
12. Kudo, T., Matsumoto, Y.: Chunking with support vector machines. In: NAACL (2001)
13. Lafferty, J.D., McCallum, A., Pereira, F.C.N.: Conditional random fields: probabilistic models for segmenting and labeling sequence data. In: ICML, pp. 282–289 (2001)
14. Mesnil, G., Dauphin, Y., Yao, K., Bengio, Y., Deng, L., Hakkani-Tur, D., He, X., Heck, L., Tur, G., Yu, D., Zweig, G.: Using recurrent neural networks for slot filling in spoken language understanding. IEEE/ACM Trans. Audio, Speech, and Language Processing **23**(3), 530–539 (2015)

15. Mesnil, G., He, X., Deng, L., Bengio, Y.: Investigation of recurrent-neural-network architectures and learning methods for language understanding. In: INTERSPEECH (2013)
16. Mikolov, T., Karafiát, M., Burget, L., Cernocký, J., Khudanpur, S.: Recurrent neural network based language model. In: INTERSPEECH, pp. 1045–1048 (2010)
17. de Mori, R.: Spoken language understanding: a survey. In: ASRU, pp. 365–376 (2007)
18. Pascanu, R., Mikolov, T., Bengio, Y.: On the difficulty of training recurrent neural networks. In: ICML, pp. 1310–1318 (2013)
19. Raymond, C., Riccardi, G.: Generative and discriminative algorithms for spoken language understanding. In: INTERSPEECH, pp. 1605–1608 (2007)
20. Sukhbaatar, S., Szlam, A., Weston, J., Fergus, R.: Weakly supervised memory networks. CoRR abs/1503.08895 (2015). http://arxiv.org/abs/1503.08895
21. Tur, G., Hakkani-Tr, D., Heck, L.: What's left to be understood in ATIS? In: IEEE Workshop on Spoken Language Technologies (2010)
22. Wang, Y.Y., Acero, A., Mahajan, M., Lee, J.: Combining statistical and knowledge-based spoken language understanding in conditional models. In: COLING/ACL, pp. 882–889 (2006)
23. Xu, P., Sarikaya, R.: Convolutional neural network based triangular CRF for joint intent detection and slot filling. In: ASRU, pp. 78–83 (2013)
24. Yao, K., Peng, B., Zhang, Y., Yu, D., Zweig, G., Shi, Y.: Spoken language understanding using long short-term memory neural networks. In: IEEE SLT (2014)
25. Yao, K., Zweig, G., Hwang, M., Shi, Y., Yu, D.: Recurrent neural networks for language understanding. In: INTERSPEECH, pp. 2524–2528 (2013)
26. Zeiler, M.D.: ADADELTA: an adaptive learning rate method (2012). arXiv:1212.5701

Improving Chinese Dependency Parsing with Lexical Semantic Features

Lvexing Zheng[1(✉)], Houfeng Wang[1], and Xueqiang Lv[2]

[1] Key Laboratory of Computational Linguistics, Peking University,
Ministry of Education, Beijing, China
zhengulxin@163.com
[2] Beijing Key Laboratory of Internet Culture and Digital Dissemination Research,
Beijing, China

Abstract. Lexical semantic information plays an important role in supervised dependency parsing. In this paper, we add lexical semantic features to the feature set of a parser, obtaining improvements on the Penn Chinese Treebank. We extract semantic categories of words from HowNet, and use them as semantic information of words. Moreover, we investigate the method to compute semantic similarity between Chinese compound words, and obtain semantic information of words which did not record in HowNet. Our experiments show that unlabeled attachment scores can increase by 1.29%.

Keywords: Lexical semantic features · HowNet · Semantic similarity · Chinese compound words

1 Introduction

Due to the data sparseness problem, the lexical information from a Treebank for a lexicalized parser could be insufficient, and the parser is mainly based on part-of speech (POS) information. However, the low granularity of POS limits the performance of the parser. Using lexical semantic information to solve data sparseness problem has become an interesting research means with the emergence of lexical semantic resources such as WordNet (Fellbaum,1998), HowNet (Dong et al., 2003), CiLin (Che et al., 2010). Several research works have tried to test the intuition that lexical semantic information should help parsing, as a word can be generalized to semantic classes in a lexicalized parser (i.e. Bengoetxea et al., 2014; Agirre et al., 2011; Xiong et al., 2005).

In general, a word would be more syntactically similar to the other if they were more semantically similar to each other. For example, we may observe in our training data that "大学生" (undergraduate student) often occurs as the subject of word "阅读" (read). We assume that the word pair "小学生" (elementary school student) and "阅读" (read) do not appear in the training data, but "小学生" is semantically similar to word

J. Li et al. (Eds.): NLPCC 2015, LNAI 9362, pp. 36–46, 2015.
DOI: 10.1007/978-3-319-25207-0_4

"大学生". If we incorporate their semantic information into the parser, it can help the parser to predict that they have the same attachment preferences in dependency tree.

In this paper, we present a simple and effective method for incorporating lexical semantic information into the parser. Instead of substituting words with their semantic classes (Agirre et al., 2011), we add lexical semantic information to the feature set of the parser. When lexical information cannot help recognize relation of two words, our approach makes it back off to semantic information. For example, word bi-gram information $<word_i, word_j>$ can back off to $<sense_i, word_j>$, $<word_i, sense_j>$ and $< sense_i, sense_j>$.

Moreover, many Chinese words (called **UNword** for short) hardly recorded in a semantic dictionary, so we cannot directly obtain their semantic information. We investigate the construction of Chinese compound words, and propose an approach to computing semantic similarity between the **UNword** and the word recorded in the dictionary. Then, we obtain the most similar semantic information of the **UNword**, and incorporate it into the parser.

As our baseline parser, we use MSTParser (McDonald et al., 2005; McDonald and Pereira, 2006). We extract semantic categories at various granularity levels in HowNet. We present a set of experiments in dependency parsing of the Penn Chinese Treebank 5.1 (Xue et al., 2000). The results show that a significant improvement in performance is achieved when lexical semantic information is incorporated into the parser.

2   Framework

In this section, we extract semantic categories in HowNet (subsection 2.1). In subsection 2.2, we describe the semantic feature templates used by the parser. In subsection 2.3, we present an approach to computing semantic similarity between the **UNword** and the word recorded in HowNet.

2.1   Extracting Semantic Categories

We use the HowNet3.0 dictionary to extract semantic categories, which covers 66,181 words defined by sememes.

HowNet (HN): Each sememe defined in the HowNet is regarded as a semantic category. The typical relation between different categories is hypernym-hyponym. Through the hypernym ladders, we can extract semantic categories at various granularity levels HN1, HN2, and HN3. HN1 stands for semantic categories at the first level, so HN2 and HN3 is the second and third level. Since a word may have many senses in HowNet, we obtain semantic category of the word by two ways. The first one is to choose the first sense of the word in HowHet by following the work (Xiong et al., 2005, Agirre et al, 2011), and call them HN1-1st, HN2-1st, and HN3-1st. The second method is to automatically rank senses (Agirre et al, 2011), but the experiments show that improvement is worse than the first method. Instead of that, we extract all senses of a word to form a new category, and call them HN1-all, HN2-all, and HN3-all. Table 1 shows information about words and semantic categories.

Table 1. Size and coverage of words and semantic categories.

	Data	HN1 -1st	HN2 -1st	HN3 -1st	HN1- all	HN2- all	HN3 -all
Words in train	43799 0	372974					
Words in test	50319	43240					
Words in both	47172	42218					
Categories in train		910	3985	5838	3766	7383	9043
Categories in test		714	1921	2669	2077	3398	3983
Categories in both		714	1817	2498	2018	3201	3714

In Table 1, words of train data have a great coverage of test data, so atomic features (McDonald et al., 2005) such as word unigrams are less likely to be sparse, but the higher-order features are poor for parsing. For example, the dependency coverage of word bi-grams feature ($<word_i, word_j>$) is only 50.09%. As can be seen, HowNet has a great coverage of train and test data, and lexical semantic information can help recognize relation of two words. In our experiments, we choose HN1/2/3-st and HN1/2/3-all as our external lexical semantic sources.

2.2 Semantic Feature Templates

We extend the baseline 1-order and second-order features in (McDonald et al., 2005; McDonald and Pereira, 2006) by introducing lexical semantic information into the parser. The feature templates are shown in table 2.

In Table 2, we incorporate lexical semantic information into the parser by using bi-gram and surrounding features which almost follow the bastline feature set. For example, we change the baseline features (<p-word, c-word>) into features <p-word, c-sense>, <p-sense, c-word> and <p-sense, c- sense>. Then <p-word, c-word> can back off to <p-word, c-sense>, <p-sense, c-word> and <p-sense, c-sense>, when it do not exist in train data.

In Table 3, Following the work (McDonald and Pereira, 2006), we also include conjunctions between these lexical semantic features and the direction and distance from sibling j to sibling k.

It is notable that if one word is not recorded in HowNet, we substitute the sense of this word with its POS tag. We try to substitute all of this word with "no-sense", but the dependency accuracy of the parser cannot increase. We also try to incorporate single-sense features into the parser, but it works worse than baseline. The main reason may be that the features induce dependency ambiguities of single-word.

Table 2. Lexical semantic features used by MSTParser-1-order. p-word: word of parent node in dependency tree. c-word: word of child node. p-pos: POS of parent node. c-pos: POS of child node. p-sense: the semantic class of parent node. c-sense: the semantic class of child node. p-sense+1: sense to the right of parent in sentence. p-sense-1: sense to the left of parent. c-sense+1: sense to the right of child. c-sense-1: sense to the left of child.

Bi-gram Semantic Features	*\<p-sense/p-pos, c-sense/c-pos\>;* *\<p-word/p-sense, c-word/c-sense\>;* *\<p-sense, c-word\>; \<p-sense, c-sense\>;* *\<p-sense/p-pos, c-pos\>; \<p-pos, c-sense/c-pos\>;* *\<p-word, c-word/c-sense\>; \<p-word/p-sense, c-word\>* *\<p-sense, c-pos\>; \<p-pos, c-sense\>; \<p-word, c-sense\>;*
Surrounding Semantic Features	*\<p-sense, p-sense+1, c-sense-1, c-sense\>;* *\<p-sense-1, p-sense, c-sense-1, c-sense\>;* *\<p-sense, p-sense+1, c-sense, c-sense+1\>;* *\<p-sense-1, p-sense, c-sense, c-sense+1\>;*

Table 3. Lexical semantic features used by MSTParser-2-order. x_i-sense: the sense of the ith word in sentence. x_k: the sibling node k of x_i. x_j: the sibling node j of x_i.

second-order Semantic Features	*\<x_k-sense, x_j-sense\>; \<x_k-sense, x_j-pos\>;* *\<x_k-word, x_j-sense\>; \<x_k-sense, x_j-word\>;* *\<x_k-pos, x_j-sense\>; \<x_i-sense, x_k-sense, x_j-sense\>;*

2.3   Semantic Similarity

The next problem is to incorporate **UNword** (not recorded in HowNet) semantic information into dependency parsing. Many researchers measure semantic similarity between two words with a large number of contexts in which the two words occurring, what is called "distributional similarity" (McCarthy et al., 2004). However, due to the data sparseness problem, there is obvious limitation for this measure (Agirre et al, 2011).

Instead, we investigate the construction of Chinese words, and try to compute semantic similarity between **UNword** and word recorded in HowNet. Then, we can obtain the most similar semantic class of **UNword**.

Compounding is the most important method to form new words in Chinese word-formation system (Yan, 2007). The compound words follow the endocentric principles, and a basic framework for the semantic construction of compound words is studied (Yan, 2007), that is *one supplementary semantic component (affix) + one head semantic component (root) = one word meaning*. For the sake of convenience, we call them "affix" and "root". The root could be consistent with word meaning, and two words containing the same root could have similar meanings.

The problem we face is to automatically split a compound word into the affix and the root. There are many types of compound words in Chinese, but modification-center is the most common structure and its root is at end of the word (Yan, 2007).

For convenience, we regard the shared part of two words as a root, and look for it at the end of two words. Thus, after the root of a word is found, the remaining part becomes its affix. So, we denote a word w_i as (a_{w_i}, r_{w_i}), where a_{w_i} stands for the affix of the word w_i, and r_{w_i} stands for the root. For a pair of words, we present them as follows:

(小学生(elementary school student)/w_1); (大学生(undergraduate school student)/w_2) ->

(小(small)/a_{w_1}, 学生(student)/r_{w_1}); (大(big)/a_{w_2}, 学生(student)/r_{w_2});

(学生(student)/w_3); (大学生(undergraduate school student)/w_2)->

(Ø/a_{w_3}, 学生(student)/r_{w_3}); (大(big)/a_{w_2}, 学生(student)/r_{w_2});

Taking the principle of compound word segmentation, we need to consider two situations:

1. Both the affixes of two word are not Ø, such as "小学生 *(elementary school student)*, 大学生 *(undergraduate school student)*". Intuitively, the word pair "小学生, 大学生" are likely to have the same attachment preferences. There are two reasons for it. One reason is that the meaning of the root "学生" (student) with any prefix   is very similar to the word "学生" (student), and belonging to the semantic class *"human_人|*study_学"* in HowNet. So, we reckon that the more words in a semantic class share a root, the word with this root is the more likely to belong to this semantic class.

   The other reason is that the affixes "小 *(small)*" and "大 *(big)*" both show the age of the root "学生" (student), and they have very semantic similar that the majority of pairs of compound words are synonymous, which have the same root and the affixes "小" and "大". So, we can use synonyms relation in a semantic dictionary to measure the similarity between them. We predict that the more pairs of compound words with same root and affixes (a_{w_i}, a_{w_j}) are synonymous in a semantic dictionary, thus affixes a_{w_i} and a_{w_j} are the more semantic similar.

2. One word is the root of another, such as "学生 *(student)*, 大学生 *(undergraduate school student)*". Unlike (1), the affix of one word must to be Ø. But, if we regard Ø as a special affix, the situation 2 is the similar to 1. A pair of compound words with affixes a_{w_i} and Ø is the more semantic similar, if the more pairs of compound words with same root and affixes $(a_{w_i}, Ø)$ are synonymous.

So, to compute semantic similarity between two compound words, there are two factors need to be considered: the effect of the root on the whole similarity in a semantic class and the similarity between two affixes of a pair of words. In this paper, we integrate two factors to similarity evaluation which is shown in equation (1).

$$Sim(uw_k, w_{ij}) = Sim((a_{uw_k}, r_{uw_k}), (a_{w_{ij}}, r_{w_{ij}}))$$
$$= \lambda(r_{w_{ij}}) * AffixSim(a_{uw_k}, a_{w_{ij}})$$

$$where:\ r_{uw_k} = r_{w_{ij}}\ and\ r_{w_{ij}} \neq Ø \tag{1}$$

Here $Sim(uw_k, w_{ij})$ represents semantic similarity between **UNword** k and word j with the semantic class i. a_{uw_k} stands for the affix of the word uw_k, and r_{uw_k} stands for the root. $\lambda(r_{w_{ij}})$ represents the degree of effect of the root $r_{w_{ij}}$ in the semantic class i. $AffixSim(a_{uw_k}, a_{w_{ij}})$ represents the semantic similarity between the affixes a_{uw_k} and $a_{w_{ij}}$.

The $\lambda(r_{w_{ij}})$ is estimated in equation (2).

$$\lambda\left(r_{w_{ij}}\right) = \frac{C^2_{Sum_{class_i}\left(r_{w_{ij}}\right)}}{\sum_x C^2_{Sum_{class_x}\left(r_{w_{ij}}\right)}} \tag{2}$$

Here $Sum_{class_i}(r_{w_{ij}})$ indicates the number of words containing the *root* r_{w_i} in the semantic class i. $C^2_{Sum_{class_i}(r_{w_{ij}})}$ means the number of word pairs which root are $r_{w_{ij}}$ in the semantic class i. $\sum_x C^2_{Sum_{class_x}(r_{w_{ij}})}$ indicates the number of all pairs of synonyms which roots are $r_{w_{ij}}$. So, if $\lambda(r_{w_{ij}})$ is closer to 1, the fewer effects have on the whole similarity, and the word uw_k is more likely to belong to the semantic class i.

The $AffixSim(a_{uw_k}, a_{w_{ij}})$ is estimated in equation (3).

$$AffixSim(a_{uw_k}, a_{w_{ij}})$$
$$= \frac{\sum_{r_{w_x} \neq \emptyset} count((a_{uw_k}, r_{w_x}) = (a_{w_{ij}}, r_{w_x}))}{\sum_{r_{w_x} \neq \emptyset} count((a_{uw_k}, r_{w_x}) = (a_{w_{ij}}, r_{w_x})) + \sum_{r_{w_x} \neq \emptyset} count((a_{uw_k}, r_{w_x}) \neq (a_{w_{ij}}, r_{w_x}))}$$
$$where\ words\ (a_{uw_k}, r_{w_x})\ and\ (a_{w_{ij}}, r_{w_x}) \in HowNet \tag{3}$$

Here (a_{uw_k}, r_{w_x}) means the word is split into the affix a_{uw_k} and the root r_{w_x}. $count((a_{uw_k}, r_{w_x}) = (a_{w_{ij}}, r_{w_x}))$ indicates the number of the pairs in which the words (a_{uw_k}, r_{w_x}) and $(a_{w_{ij}}, r_{w_x})$ are synonymous. $count((a_{uw_k}, r_{w_x}) \neq (a_{w_{ij}}, r_{w_x}))$ means the number of the pairs in which the words are not synonymous.

Therefore, we obtain the most similar semantic information of **UNword** k by equation (4), and respectively extend our external lexical semantic sources (HN1/2/3-st, HN1/2/3-all) with the semantic class of **UNword** k. Here $SemClass(w_{ij})$ means the semantic class i of the word w_{ij}, $POS(uw_k)$ means part-of speech of word uw_k recorded in CTB, and $POS(w_{ij})$ is recorded in HowNet. NR means proper nouns, and CD means cardinal numbers. In many cases, NR which includes personal names and transliterated words cannot fit the semantic construction of compound words. Recognizing the dependency relation of CD is already used by simple rules in the baseline parser.

$$SemClass(uw_k) = SemClass(argmax_{(w_{ij})} Sim(uw_k, w_{ij}))$$

$$where \ Sim(uw_k, w_{ij}) \geq \theta, POS(uw_k) = POS(w_{ij}),$$
$$and \ POS(uw_k) \neq "NR" and "CD"; \ otherwise \ SemClass(uw_k) = \emptyset \qquad (4)$$

3   Experiment

3.1   Data set

We use Penn Chinese Treebank 5.1 (Xue et al., 2000) as data set in the experiments. The Penn Chinese Treebank 5.1 (CTB) is phrase structure Treebank, and we use the toolkit Penn2Malt (Johansson and Nugues, 2007) to transfer them to dependency treebank. To balance each resource in train set, development set and test set, we follow Duan's work (Duan, 2007), and split the data set as in table 4.

Table 4. The division of CTB data set

	CTB files	Number of sentences
Train set	001-815, 1001-1136	16,091
Development set	886-931, 1148-1151	803
Test set	816-885, 1137-1147	1,910

3.2   Experimental Results

We use MSTParser (McDonald et al., 2005; McDonald et al., 2006) as our basic parser. It represents global, exhaustive graph-based parsing that finds the highest scoring directed spanning tree in a graph. The parser can be trained using first or second order models, and we use default options.

We use both labeled attachment score (LAS) and unlabeled attachment score (UAS) to evaluate the all experiments, and punctuation is included in all evaluation metrics. We consider three options:

1. We substitute words with their semantic classes in the process of training and testing the parser using our external lexical semantic sources (HN1/2/3-st, HN1/2/3-all). The results for parsing are given in Table 5.

Table 5. Parsing results by substituting words with their semantic classes.

	LAS	UAS
MSTParser-1order(baseline)	78.86	80.95
HN1-st	77.71	79.96
HN2-st	78.23	80.42
HN3-st	78.44	80.65
HN1-all	78.68	80.89
HN2-all	78.69	80.82
HN3-all	78.69	80.87

In Table 5,　by substituting words with their semantic classes, the results are not superior to　the baseline. Though the parser knows that words are synonymous, it can help the parser to predict that they have the same attachment preferences in dependency tree. However, there are many words belonging to the same semantic class in a semantic dictionary, but they have not the same attachment preferences or have some different attachment preferences in dependency parsing. For example, the words "执法" (enforce the law) and "执法必严" (strictly enforce the law) are in the same semantic class "*conduct_实施*" in HowNet. But the word "执法必严" (strictly enforce the law) has different attachment preference compared to "执法" (enforce the law), and do not have an adverb. So, this method causes lexical information loss and cannot improve parsing performance.

2. We extend the baseline features with the semantic feature template (Table 2), then train and test MSTParser-1order by using our sources (HN1/2/3-st, HN1/2/3-all). Experimental results show that the best performance of MSTParser-1order model is obtained with HN2-all. Due to the fact that the MSTParser-2order model extends the MSTParser-1order with the second-order features, we only incorporate HN2-all into 2order model with the semantic feature templates (Table 2 and 3). Moreover, based on section 3.3, we can obtain the most similar semantic information of **UN-words** by equation (4), and extend our external lexical semantic source HN2-all, and call it HN2-all-E for short. Then, the MSTParser-1order and 2order model need to retrain by using HN2-all-E. In equation 4, we set threshold $\theta = 0.9$. The results of semantic class of **UNwords** are shown in Table 6 and the experimental results are given in Table 7.

Table 6. Examples of semantic class of UNwords using HN2-all

UNword	The most similar Word	Semantic class in HN2-all	Similarity
西南郊 (southwestern suburbs)	西郊 (western suburbs)	part_部件 /%place_地方	1.0
系列赛 (series of competitions)	公开赛 (open championship)	fact_事情 /compete_比赛	0.9848
乍看 (glance)	观看 (watch)	look_看	0.6000
乐于 (be happy to)	位于 (locate)	situated_处于	0.1429

Table 7 shows the performance of the baseline that is extended with the semantic feature templates. We can see that in all cases our external lexical semantic sources improve over the baseline. Using HN2-all, UAS of MSTParser-1order and 2order respectively increase by 1.21%, 1.16%. The main purpose of adding the semantic feature into the parser is that if lexical dependency information of two words is sparse, and it can back off to lexical semantic dependency information, and the experimental results prove its validity. Compare with the method of substituting words with their semantic classes, this method do not loss high-frequency word and word-pair dependency information in general, and it can help recognize relation of lower-frequency word and word-pair. For example, if a high-frequency verb always does not have an object (OBJ), the parser can also consider this situation.

Table 7. Parsing results by extending the baseline features with the semantic feature templates.

	θ	LAS	UAS
MSTParser-1order(baseline)	–	78.86	80.95
1order + HN1-st	–	79.79	81.85
1order + HN2-st	–	79.95	82.05
1order + HN3-st	–	79.98	82.00
1order + HN1-all	–	80.06	82.09
1order + HN2-all	–	80.09	**82.16(+1.21)**
1order + HN3-all	–	79.92	82.00
1order + HN2-all-E	0.9	80.18	82.28
MSTParser-2order(baseline)	–	80.85	83.04
2order + HN2-all	–	82.08	**84.20(+1.16)**
2order + HN2-all-E	0.9	**82.19(+1.34)**	**84.33(+1.29)**

In Table 6 and 7, the number of **UNwords** is 11,071 in train data and 1,467 in test data, and our approach (section 3.3) adds 1,199 (10.83%) words with semantic class in train data, and 192 (13.09%) words in test data. Compared with HN2-all, table 7 shows UAS of MSTParser-2order increases by 0.13% using HN2-all-E. So, it proves our approach can effectively obtain the semantic information of **UNwords**. Due to the fact that the lexical semantic information is auxiliary information for parsing, the improvement is not significant.

3. We include the results of ZPar-dep (Zhang and Nivre,2011) and neural network model of Chen and D.Manning (2014) for comparison, and the experimental results are given in Table 8. As we can see, compared with ZPar-dep (Zhang and Nivre, 2011) and neural network model (Chen and D.Manning, 2014), our approach gets the best UAS.

Table 8. Parsing results with HN2-all-E and comparion with high performance models

	LAS	UAS	LAS(excluding punctuations)	UAS(excluding punctuations)
2order + HN2-all-E	82.19	**84.33**	84.27	**86.36**
ZPar-dep (Zhang and Nivre, 2011)	–	–	84.40	86.00
Neural Network Model (Chen and D.Manning, 2014)	82.40	83.90	–	–

4   Related Work

Agirre et al. (2008) used semantic classes to help parsing. Later, they extended the test and successfully introduced WordNet classes in a dependency parser (Agirre et al., 2011). MacKinlay et al. (2012) investigated the addition of semantic annotations in the form of word sense hypernyms, in HPSG parse ranking.

Ciaramita and Attardi (2007) showed that adding semantic features extracted by a named entity tagger (such as PERSON or MONEY) improved the accuracy of a dependency parser. Candito and Seddah (2010) studied statistical parsing of French, where terminal forms were replaced by more general symbols, particularly clusters of words obtained through unsupervised clustering. The results showed that word clusters had a positive effect.

Apart from these, there have been other attempts to solve the data sparseness problem, Koo et al. (2008) and Suzuki et al. (2009) presented a semi-supervised method for training dependency parsers, using word clusters derived from a large unannotated corpus as features.

5   Conclusion

In this paper, we present a simple and effective method for incorporating lexical semantic information into the parser. We can relieve sparse data problem by extending with the semantic feature sets, and obtain the most similar semantic information of words which are not recorded in the lexical semantic resource. Experiments on CTB dataset show our approach achieves significant improvement. Our approach is only a preliminary work and has much future work to do. The considered future work includes incorporating word sense disambiguation method and deep research on basic framework for the semantic structure of Chinese words.

Acknowledgement. Our work is supported by National High Technology Research and Development Program of China (863 Program) (No. 2015AA015402), National Natural Science Foundation of China (No.61370117 & No.61333018), Major National Social Science Fund of China (No.12&ZD227) and the Opening Project of Beijing Key Laboratory of Internet Culture and Digital Dissemination Research (ICDD201501).

References

1. Fellbaum, C. (ed.): WordNet: An Electronic Lexical Database. MIT press, Cambridge (1998)
2. Dong, Z., Dong, Q.: Hownet [EB/OL]. http://www.keenage.com, 2003201211
3. Che, W., Li, Z., Liu, T.: LTP: a Chinese language technology platform. In: Proceedings of the Coling 2010: Demonstrations, Beijing, China, 2010.08, pp. 13–16 (2010)
4. Bengoetxea, K., Agirre, E., Nivre, J., Zhang, Y., Gojenol, K.: On WordNet semantic classes and dependency parsing. In: Proceedings of ACL 2014, Baltimore, Maryland, pp. 649–655 (2014)

5. Agirre, E., Bengoetxea, K., Gojenola, K., Nivre, J.: Improving dependency parsing with semantic classes. In: Proceedings of ACL 2011, Poorland, Oregon, pp. 699–703 (2011)
6. Xiong, D., Li, S., Liu, Q., Lin, S.-X., Qian, Y.: Parsing the Penn Chinese treebank with semantic knowledge. In: Dale, R., Wong, K.-F., Su, J., Kwong, O.Y. (eds.) IJCNLP 2005. LNCS (LNAI), vol. 3651, pp. 70–81. Springer, Heidelberg (2005)
7. McDonald, R., Crammer, K., Pereira, F.: Online large-margin training of dependency parsers. In: Proceedings of the 43rd Annual Meeting on Association for Computational Linguistics, pp. 91–98. Association for Computational Linguistics (2005)
8. McDonald, R., Pereira, F.: Online learning of approximate dependency parsing algorithms. In: EACL. Citeseer (2006)
9. Xue, N., Xia, F., Huang, S., Kroch, A.: The bracketing guidelines for the perm Chinese treebank (3.0) (2000)
10. McCarthy, D., Koeling, R., Weeds, J., Carroll, J.: Finding predominant senses in untagged text. In: Proc. of the 42nd Annual Meeting of the ACL, Barcelona, Spain, pp. 280–287 (2004)
11. Yan, H.: The Study on the Semantic structure of Compound Words in Contemporary Chinese. Capital Normal University, Beijin (2007)
12. Johansson, R., Nugues, P.: Extended constituent-to-dependency conversion for English. In: Proc.of 16th Nordic Conference on Computational Linguistics (NODALIDA), pp. 105–112 (2007)
13. Duan, X., Zhao, J., Xu, B.: Probabilistic models for action-based Chinese dependency parsing. In: Kok, J.N., Koronacki, J., Lopez de Mantaras, R., Matwin, S., Mladenič, D., Skowron, A. (eds.) ECML 2007. LNCS (LNAI), vol. 4701, pp. 559–566. Springer, Heidelberg (2007)
14. Zhang, Y., Nivre, J.: Transition-based dependency parsing with rich non-local features. In: Proceedings of the 49th Annual Meeting of the Association for Computational Linguistics: Human Langauge Technologies, pp. 188–193. Association for Computational Linguistics, Portland (2011)
15. Chen, D., Manning, C.: A fast and accurate dependency parser using neural networks. In: Proceedings of the 2014 Conference on Empirical Methods in Natural Language Processing (EMNLP), pp. 740–750. Association for Computational Linguistics, Doha, Qatar, October 2014
16. Agirre, E., Baldwin, T., Martinez, D.: Improving parsing and PP attachment performance with sense information. In: Proceedings of ACL08: HIT, Columbus, Ohio, pp. 317–325 (2008)
17. MacKinlay, A., Dridan, R., McCarthy, D., Baldwin, T.: The effects of semantic annotations on precision parse ranking. In: First Joint Conference on Lexical and Computational Semantics (*SEM), Montreal, Canada, pp. 228–236 (2012)
18. Ciaramita, M., Attardi, G.: Dependency parsing with second-order feature maps and annotated semantic information. In: Proceedings of the 10th International Conference on Parsing Technology (2007)
19. Candito, M., Seddah, D.: Parsing word clusters. In: Proceedings of the NAACLHLT 2010 First Workshop on Statistical Parsing of Morphologically-Rich Languages, LosAngeles, CA, USA, pp. 76–84 (2010)
20. Koo, T., Cameras, X., Collins, M.: Simple semi-supervised dependency parsing. In: Proceedings of ACL-08: HLT, Columbus, Ohio, pp. 595–603 (2008)
21. Suzuki, J., Isozaki, H., Cameras, X., Collins, M.: An empirical study of semi-supervised structured conditional models for dependency parsing. In: Proceedings of the 2009 Conference on Empirical Methods in Natural Language Processing, Singapore, pp. 551–560 (2009)

Machine Translation and Multi-Lingual Information Access

Entity Translation with Collective Inference in Knowledge Graph

Qinglin Li[1], Shujie Liu[2]([⊠]), Rui Lin[3], Mu Li[2], and Ming Zhou[2]

[1] Shanghai Jiaotong University, Shanghai, China
v-lqing@microsoft.com
[2] Microsoft Research Asia, Beijing, China
{shujliu,muli,mingzhou}@microsoft.com
[3] Harbin Institute of Technology, Harbin, China
linrui@mtlab.hit.educ.n

Abstract. Nowadays knowledge base (KB) has been viewed as one of the important infrastructures for many web search applications and NLP tasks. However, in practice the availability of KB data varies from language to language, which greatly limits potential usage of knowledge base. In this paper, we propose a novel method to construct or enrich a knowledge base by entity translation with help of another KB but compiled in a different language. In our work, we concentrate on two key tasks: 1) collecting translation candidates with as good coverage as possible from various sources such as web or lexicon; 2) building an effective disambiguation algorithm based on collective inference approach over knowledge graph to find correct translation for entities in the source knowledge base. We conduct experiments on movie domain of our in-house knowledge base from English to Chinese, and the results show the proposed method can achieve very high translation precision compared with classical translation methods, and significantly increase the volume of Chinese knowledge base in this domain.

Keywords: Knowledge base · Machine translation · Collective learning

1 Introduction

Knowledge bases are structured databases that store facts concerning entities and relationships about the world, and they are widely used in NLP applications, such as question answering systems, search engines, and expert systems [12,3,2,14]. Knowledge base construction is attracting more and more attention from researchers in both academia and industry. Most knowledge bases are mainly compiled in one language, making it hard to adapt them in applications of other languages. For example, in freebase[1], there are 23M entities, but only 1% of them are Chinese entities, which limits the application of freebase for Chinese language processing. In order to extend the application of a knowledge

---

[1] http://www.freebase.com/

© Springer International Publishing Switzerland 2015
J. Li et al. (Eds.): NLPCC 2015, LNAI 9362, pp. 49–63, 2015.
DOI: 10.1007/978-3-319-25207-0_5

base, in this paper, we propose a method to translate a knowledge base from one language to another.

In contrast to conventional machine translation of sentences, knowledge base translation poses two particular challenges. For sentence translation, translation equivalence for phrases can be extracted from a bilingual corpus that is mined from the web. But for knowledge base translation, many entities are rare words/phrases (for example, *Hobbit* and *Philomena*, which are in movie names), and as such cannot be covered by this method. We collect translation candidates from multiple sources, including the output of machine translation engines, traditional dictionaries, translation candidates mined from web. All these results are collected as translation candidates and then ranked by our model to determine the best one.

Fig. 1. Translation disambiguation for *The Matrix*. The mathematical entity *The Matrix* should be translated as 矩阵 (ju'zhen, meaning *matrix*), while the movie name *The Matrix* should be translated as 黑客帝国 (hei'ke'di'guo, meaning *Empire of the hackers*).

Given the translation candidates, to get the correct translation, we should do translation disambiguation. For example in Figure 1, the entity *The Matrix* can be translated into 矩阵 (ju'zhen, matrix) if it is a mathematical term, and 黑客帝国 (hei'ke'di'guo, Empire of the hackers) if it is a movie name. When the phrase *The Matrix* is surrounded by mathematical words in a sentence, conventional SMT models can select the correct translation 矩阵 (ju'zhen, matrix) by taking the surrounding words into consideration, or 黑客帝国 (hei'ke'di'guo, Empire of the hackers) if the sentence is about movies. Unfortunately, the entities are only words/phrases, which do not have such contextual information. In this paper, we propose taking the surrounding entities into consideration. The phrase *The Matrix* should be translated into 矩阵 (ju'zhen, matrix) if its neighbor nodes are mathematical entities, and 黑客帝国 (hei'ke'di'guo, Empire of the hackers) if its neighbors are movie names, movie actors, or directors, as shown in Figure 2.

Based on the above discussion, we propose a graph-based collective inference method for knowledge base translation. We first collect all the translation candidates with as good converage as possible from various sources, and calculate all the features for each translation candidate to run collective inference method to get the best configuration of the graph. We conducted experiments on a knowledge base translation task in the movie domain from English to Chinese, and the results indicate that our proposed method can significantly improve performance compared over baseline systems.

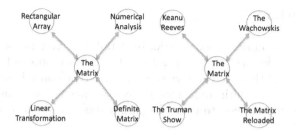

Fig. 2. Illustration of two *The Matrix* entities in the knowledge graph.

We will first introduce related work in section 2, and then we will illustrate our model in section 3, including the multi-source candidate generation, collective inference, model features, and model training. The experiments are explained in detail in section 4, followed by the conclusion and future work.

2 Related Work

[7] proposed an integrated method approach to extract an entity translation dictionary from a bilingual corpus. They first extract entities from the bilingual corpus independently for each language, and then a statistical model is used to align the entities. An iterative process is applied to extract entity pairs with high alignment probability and the entity pairs are used to improve the entity annotation quality for both languages. [8] proposed a method to improve entity translation by combining a transliteration approach with web mining, using web information as a source to complement transliteration and using transliteration to guide and enhance web mining. [1] proposed a two-step method to translate entities: first they generate a ranked list of translation candidates using bilingual and monolingual resources, and then all the candidates are rescored using monolingual clues. All the above methods treat the entities independently and try to build an entity translation table, instead of translate a knowledge base. The entity translation table can be used as a source of translation candidates collection of our method. [5] proposed a graph-based method to create a multilingual knowledge base by linking wordnet to wikipedia. Different from our method which translate a knowledge base from one language to another, they try to build a knowledge base based on a semi-structured data base, and merge entities expressed in different languages.

There are also many approaches that apply a graph-based collective method to different tasks. [6] apply a graph-based collective method to entity linking tasks, in which global interdependence between different entity linking decisions can be modeled and entities that are name mentions can be inferred jointly. [11] apply a graph-based collective method to POS tasks. They use a similarity graph to encourage similar n-grams to have similar POS tags. [10] apply a graph-based collective method to learn sentence translation consensus, in which each node is a phrase and similar phrases are connected with each other.

3   Our Method

In this section, we illustrate our model in detail. We first use a multi-source candidate generation method to generate all the translation candidates (shown in section 3.1), and then we conduct collective inference on a knowledge graph (in section 3.2 and 3.3) with confidence and consistency features (in section 3.4). We use a three-step method to train the parameters (shown in section 3.5).

3.1   Multi-Source Candidates Generation

Since the entity names in the knowledge base often contain new and rare words that are hard to find in a translation dictionary, we generate our translation candidates from multiple sources.

- Translation dictionary from **semi-structured web sites** containing movie information. Certain web sites collect movie information including movie names (some of them are in both languages), actors, and directors. We mined translation pairs from *BaiduBaike*[2], *Wikipedia*[3], *VeryCD*[4], *Douban*[5], and *IMDBCN*[6].
- **Traditional Dictionaries**. We collect translation candidates from the two dictionaries: Oxford dictionaries[7] and Longman contemporary English dicitonaries[8].
- We mined **parenthetical translation pairs** as another source, following [9].
- Translation results of **MT engine**. We fed the entity name into a state-of-the-art machine translation engine to get the translation result. It may not contain certain new words, but it's still useful for the coverage of our system.

3.2   Collective Inference with Neighbors

We find the correct translation results from the translation candidates of each entity by collective inference from any neighbors.

The basic principle of our model is that the translation of entities should be consistent with each other, which means, if there is a relationship between entity e_1 and e_2 in the knowledge base, there should also be the same kind of relationship between the translations of e_1 and e_2. For example, the director of *The Matrix* in the knowledge base should be translated into the director of 黑客帝国 (hei'ke'di'guo, Empire of the hackers). And the entity of the mathematical term *The Matrix* should be translated into 矩阵 (ju'zhen, matrix), instead of 黑客帝国 (hei'ke'di'guo, Empire of the hackers), since all its neighbors are mathematics terms.

---

[2] http://baike.baidu.com/
[3] http://baike.baidu.com/
[4] http://www.verycd.com/
[5] http://www.douban.com/
[6] http://www.imdb.cn/
[7] http://oxforddictionaries.com/
[8] http://www.ldoceonline.com/

Let E be the set of entities in a knowledge base. For $\forall e \in E$, let $T(e)$ be the translation candidates of e. Based on the translation consistency principle, we propose a collective inference model, as shown in Equation 1.

$$\max_{t_i \in T(e_i)} \left(\sum_{e_i \in E} g(t_i, e_i) + \sum_{e_i, e_j \in E} f(t_i, t_j, e_i, e_j) \right) \qquad (1)$$

There are two parts in our model. The first part is $g(t_i, e_i)$, which is called translation confidence only based on the entity itself, without considering its neighbors. The translation confidence $g(t_i, e_i)$ is a linear combination of several confidence features:

$$g(t_i, e_i) = \sum_k \theta_k g_k(t_i, e_i) \qquad (2)$$

where $g_i(t_i, e_i)$ is the confidence feature and θ_i is the corresponding feature weight.

The second part is $f(t_i, t_j, e_i, e_j)$, called translation consistency, which is the measure of translation consistency between two entities connected with a link in the knowledge graph. Translation consistency $f(t_i, t_j, e_i, e_j)$ is defined as:

$$f(t_i, t_j, e_i, e_j) = \sum_{k'} \theta_{k'} f_{k'}(t_i, t_j, e_i, e_j) \qquad (3)$$

where $f_{k'}(t_i, t_j, e_i, e_j)$ is the consistency feature and $\theta_{k'}$ is the corresponding feature weight.

The first part plays a role in the translation model while the second part plays the role the same as translation model for conventional SMT. In the the second part, translation consistency features can be treated as a special bi-gram language model in a graph, playing a similar role to n-gram language models used for sentence translation.

3.3   Translation Graph

To solve the collective inference problem, we have built a translation graph based on the English knowledge base. Each node in this graph corresponds to an entity in the knowledge base. And we also create an edge between a pair of nodes in the graph if there is a relationship between the corresponding entities in the knowledge base.

For each node in the graph, we use our multi-source candidate generation method to mine the translation candidates. We add a factor node g to each node to model the translation confidence, and a factor node f to each edge to model the translation consistency. An example of a translation graph is illustrated by Figure 3.

In Figure 3, the movie entity *The Matrix* is linked with four entities, including the actor *Keanu Reeves*, the directors *The Wachowskis*, and another two films, *The Truman Show* and *The Matrix Reloaded*. For each entity node, we collect

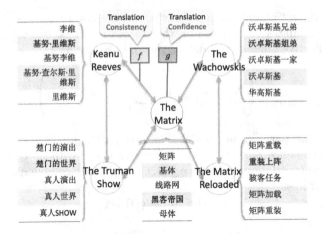

Fig. 3. An example of a translation graph.

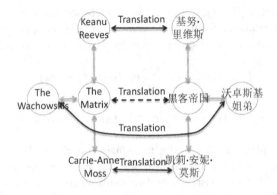

Fig. 4. An example of the confidence propagation in the translation graph.

the translation candidates, which are listed in the frame attached to the node, and the candidate in bold is the correct translation of the entity. To rank the translation candidates of the node *The Matrix* , we calculate the translation confidence features g, which are not related with the neighbors, and also the translation consistency features f, which are based on the best translations of the neighbor nodes. Then we use the trained feature weights to rank the list and get the best one as the result.

We use label propagation for inference. When feature weights are trained, starting from anchor nodes (whose translations are fixed and correct), label propagation can propagate the translation confidence from the anchor nodes to its neighbors and to the neighbors of the neighbors. For example, in Figure 4, the translation confidence of the directors *The Wachowskis* can be propagated to the movie *The Matrix*. And also, the translation confidence of the directors can be propagated to the actor nodes *Carrie-Anne Moss* and *Keanu Reeves*, and then propagated to the node *The Matrix*. Our model will consider all the confidences received to rank the candidate list and get the final translation result. The propagation is an iteration process, and will be terminated when a pre-set limit is reached.

3.4   Features

Translation confidence features, which are defined in the g node, are listed as following:

- **Source** refers to where the candidate is from, e.g. a dictionary or SMT.
- **Bilingual Concurrence** is the concurrence ratios ($p_{bicol}(e|t)$ and $p_{bicol}(t|e)$) of the entity name and the translation candidates, which is calculated from a large corpus. $p_{bicol}(e|t)$ is defined as:

$$p_{bicol}(e|t) = \frac{\#(e,t)}{\sum_{e'} \#(e',t)} \tag{4}$$

  where e is the English entity name and t is the translation candidate, and $\#(e,t)$ is the count of the pair (e,t) shown in the corpus. $p_{bicol}(t|e)$ is calculated in a similar way. The corpus we used to calculate the pair numbers is a corpus mined from the web using the indicator of brackets [9], which contains sentences like:
  The Matrix 母体为什么不能用核能发 电？
  有关电影《黑客帝国》*(The Matrix)* 资深人士进
  From these two sentences, we can find the move entity name and the translation.
- **Translation Score** is the word translation probabilities: $p_{smt}(e|t)$ and $p_{smt}(t|e)$. We use IBM-Model1[4] to calculate these scores. $p_{smt}(e|t)$ is defined as:

$$p_{smt}(e|t) = \frac{1}{(l_t+1)^{l_e}} \prod_{j=1}^{l_e} \sum_{i=0}^{l_t} t(e_j|t_i) \tag{5}$$

  l_e and l_t are the length of e and t. $t(e_j|t_i)$ is the word translation proability of word e_j given t_i, and this proability can be calculated with a big bilingual corpus. $p_{smt}(t|e)$ can be computed in a similiar way.

Relation consistency is the first translation consistency feature we used for f. This means $\forall e_i, e_j$ with relationship $\mathcal{R}$, whether the translation t_i, t_j has the same relationship $\mathcal{R}$. Figure 5 shows the consistency of node *The Matrix* and its neighbors. If we know that the three neighbors *The Wachowskis*, *Keanu Reeves* and *Carrie-Anne Moss* have corresponding translations 沃卓斯基姐弟, 基努?里维斯 and 凯莉?安妮?莫斯 as shown in Figure 5, and also, from semi-structured data, we can have the actor-movie and director-movie relationships similiar as in the English knowledge base, we will have more confidence that the movie name in the same position, which is 黑客帝国 (hei'ke'di'guo,

Empire of the hackers) , could be the translation of the movie node *The Matrix*. The relation consistency feature is defined as:

$$f_{kb}(t_i, t_j, e_i, e_j) = \begin{cases} 1, & \text{if} \quad t_i \in \mathcal{R}(t_j) \\ & \text{and} \quad e_i \in \mathcal{R}(e_j) \\ 0, & \text{otherwise} \end{cases} \quad (6)$$

where $\mathcal{R}(t_j)$ is the set of neighbors for node t_j with relation $\mathcal{R}$. If t_i is the translation of e_i, and e_j is the neighbor of e_i with relation $\mathcal{R}$, while t_j is the neibhor of t_i with relation $\mathcal{R}$, then the feature for the translation t_j of the node e_j is set to 1, otherwise, the feature value is 0.

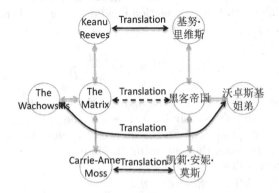

Fig. 5. Relation consistency of *The Matrix*.

We can also determine consistency features such as phrase concurrence using web search. For example, if we feed the query 凯莉 基努 黑客帝国 (黑客帝国 is the correct translation for the movie entity *The Matrix*) into a search engine, we will get about 108,000 results, but if the query is 凯莉 基努 矩阵 (矩阵 is not the correct translation), we will get only about 4,630 results. Which means, the correct translation of an entity is likely to be shown with the correct translation of its neighbors. One problem with using the search engine to calculate this feature is that the time cost is very high, so instead of using the search engine, we calculate this feature with a large monolingua corpus and the feature is called **cooccurence consistency**. Cooccurence consistency features are calculated in a smilar way to the bilingual cooccurence features.

3.5 Training

To train our model, we adopt a three-step training method, which is comprised of the pre-training for confidence feature weights, pre-training for consistency feature weights, and joint training of total feature weights.

In the pre-training phase for confidence feature weights, our model only uses the confidence features with the consistency features removed. In this case, each

---

Algorithm 1. Modified SampleRank

Input: $q: y \rightarrow y$: MCMC transition kernel;
$w{:}y \rightarrow R$: performance metric;
D_a: the anchor set;
D_t: the training set;

Output: $\frac{1}{T}\sum_{t=1}^{T}\theta_t$

1 **Initialization:** $\theta_0 \leftarrow \theta_{init}$; $t = 0$;

2 **while** *pre-set limitation not reached* **do**

3     $t{+}{+}$; $D_q \leftarrow D_a$; $\theta_t = \theta_{t-1}$;

4   y_t: initial configuration in $y(x)$ with θ_t;

5   **while** D_q *is not empty* **do**

6        $e \leftarrow \text{pop}(D_q)$;

7   **for** $e' \in \mathcal{N}(e)$ *and* $e' \notin D_q$ **do**

8   Attemp an update for e' : $y_{e'} \leftarrow q(\cdot|y_{e'})$

9   **if** $e' \in D_t$ **then**

             $y^+ = argmax_{y \in y_t, y_{t+1}} w(y)$

10              $y^- = argmin_{y \in y_t, y_{t+1}} w(y)$

             $\nabla = \phi(y^+) - \phi(y^-)$

             if $\theta_t \nabla < w(y^+) - w(y^-)$ *and* $w(y_t) \neq w(y_{t+1})$ **then**

11                 $\theta_t = \theta_t + \eta_t \nabla$

12   **end**

13   **end**

14           $\text{push}(D_q, e')$;

15   **end**

16   **end**

17 **end**

---

entity is independent from others and the feature weights can be trained using linear model methods, such as perception, logistic regression, or SVM. We take the pairs of entities with the correct translations as positive samples and the pairs of entities with other translation candidates as negative samples. The trained confidence feature weights will be used to pre-train the consistency feature weights.

During the pre-training of consistency feature weights, we fix the confidence feature weights and use a graph-based training method to tune the consistency feature weights. Since the confidence feature weights are fixed, it is faster and easier to tune the model. Based on the pre-training of translation confidence and consistency feature weights, we use the same graph-based training method to tune the confidence and consistency feature weights jointly.

To train our model, we adopt sample rank [13] for the pre-training of consistency features and the joint training of all the features. Sample rank is a rank-based learning method that uses MCMC inference to train factor graphs with an atomic gradient. The following is the adopted SampleRank alogrithm shown in Algorithm1. As shown, we first initialize the feature weight vector θ as θ_{init}. For the pre-training of consistency feature weights, the confidence feature weights of θ_{init} are set to be the results of the pre-training of the confidence feature weights, and they are fixed during the SampleRank training. For all feature weight training, confidence and consistency feature weights are set as the results of the two pre-training phases.

In addition to the initialization, another difference with traditional SampleRank is the order of the samples for training. Since we have a large number of *Anchor Nodes* (whose translation are given, will be introduced in Section 4), we modify the traditional SampleRank algorithm to leverage them. From the *Anchor Nodes* (in line 6), the translation confidence is propagated to the neighbors, and the translation of neighbor nodes are updated. If the neighbor node is a training node, it is used to update the model. The translation confidence of *Anchor Nodes* will be propagated to other nodes in the graph via the neighbors, until we update the whole graph. The training iteration will continue until a pre-set limitation is reached.

4   Experiments

4.1   Dataset

We construct a movie domain dataset by first gathering $3.2M$ English entities from the movie domain of an inhouse knowledge base. Each entity consists of a list of names in English, and their relationships with other entities. Some of the entities have IMDB IDs, making it possible to match the English entities with Chinese translations. This creat $124,908$ *golden-match* pairs of English and Chinese translation pairs, which will be used as **Anchor Nodes** We build our translation graph using these 3.2M entities and the relationships between them. We sample 2,000 english movie entities and manually search their Chinese translations using web search engines. These 2,000 pairs are split into two groups: **Training**, and **Test**. The **Training** nodes are used to run SampleRank for feature weights tuning. Model performance is evaluated with the **Test** nodes. The data statistics is shown in Table 1.

Table 1. Data statistics.

#Nodes	#Anchors	#Training	#Test
3.2M	125K	1K	1K

Our monolingual cooccurence data is collected from movie web sites, not only the pages for movies, but also the forums. Such web pages contain information about movies and actors. We extract sentences from these pages and indexed them to speed up the feature calculations. Our monolingual cooccurence data contains 7.6M sentences and 114M words.

Our bilingual cooccurence data is actually the byproduct of our implementation of [9]. We mine the sentences containing brackets from movie web sites.

Table 2. Performance comparison.

Features		SMT	Perceptron	LR	LK_SVM	Our Method
All		48.1%	87.1%	87.8%	87.9%	92.8%
g	-Source	-	83.0%	84.8%	84.7%	88.1%
	-TransProb	-	83.8%	85.9%	85.9%	90.3%
	-Category	-	84.4%	86.6%	85.7%	91.5%
	-BilColoc	-	83.2%	84.7%	84.1%	89.2%
f	-Consist$_{Col}$	-	-	-	-	88.7%
	-Consist$_{Rel}$	-	-	-	-	88.6%

The subsequent process of [9] may introduce errors and discard useful translation pairs. We use the unprocessed data to calculate the bilingual cooccurence features. The sentences are also indexed to speed up the feature calculation. Our bilingual cooccurence data contains 456K sentences and 5M words.

4.2 Baselines

We compare our method with several baselines. The first is the output of machine translation. The machine translation engine we used is an inhouse implemented state-of-the-art SMT system. We also use three classification methods as baselines: Perceptron, Linear Regression (LR) and Support Vector Machine (SVM). To train these three baselines, we generate the candidates and calculate the translation confidence features. From anchor and training nodes, the pairs of correct translations and the entity names are used as positive samples and the pairs of incorrect translation candidates and the entity names as negative samples to train the classifiers.

4.3 Translation Result

The performance of our method and the baseline systems are shown in Table 2. It is not surprising that the performance of machine translation was only 48.1%, since many movie names are not translated literally. For example, the Chinese name for the movie *The Matrix* is 黑客帝国 (hei'ke'di'guo, Empire of the hackers). And also, without the context information, SMT cannot do translation disambiguation. The performance of the linear classifiers are similar, with a variance of less than 1.0%. Among them, Perceptron is the worst and the Linear_Kernel SVM (LK_SVM) was the best. We also try the radial basis function (RBF) kernel and the performance is even worse than LK_SVM.

Our method can outperform the LK_SVM (the best of the baselines) by over 4.9% on accuracy. For our SampleRank training, we use LK_SVM as the training method for the pre-training of confidence feature weights, which is the initial feature weights of pre-training of consistency feature weights. Our model can then leverage the results of LK_SVM and can also leverage the links of entities and learn the translation consensus.

All the baseline systems, SMT, Perceptron, LR and LK_SVM cannot use the information of the neighbors, since the translation of the neighbors are not given, so that, the baseline systems have no information to learn the tanslation consistency of the entities in the graph. Our proposed method can model the translation consistency using two consistency features, relation consistency and cooccurence consistency, and try to find the most consistent translation for a entity by taking its neighbors' translation into consideration.

Table 3. Performance for the popular movies.

	Top500	Top5000
Precision	97.0%	95.2%
Recall	98.5%	96.3%

Our method can generate about 483K movie translation for 3.2M movie names, in another word, the translation rate of our method is about 15.1%. We check the entities which cannot find the translation, most of them are very old movies, which are not well known for Chinese people, and donnot have the Chinese translation, for example, a movie *The Legacy* taken in 1978. Some of them are not very famous, for example, a movie named *Quantized Love* taken in 2014. It is not possible to chech whether there are translations for all of the 3.2M movies, so we randomly sample 400 movie names and manually search the web to find the Chinese translation. Out of these 400 movies, at most 85 Chinese translations can be found by human labeler, which means by using existing resources, the upper bound for translation rate is about 21%, so that the translation recall of our method is about $\frac{15.1\%}{21\%} = 72\%$. We also check the coverage for the most popular 500 and 5,000 movies listed in the web *http://www.imdb.com/*, and the result is shown in Table 3. As shown in Table 3, our method can achieve high precision and recall for the popular movies.

4.4   Training Phase Gain

To test the three step training method, we evaluate the performance after each of them and the results are shown in Table 6. inited by the results of LK_SVM (pre_confidence), the pre-training of consistency feature weights (pre_consistency) improves the performance about 3.7%, and the joint training of all the features (joint training) can further improve the performance about 1.2% to achieve a total improvement of over 4.9%.

4.5   Feature Gain

We remove the features one by one from the feature list to test the feature contribution to the performance, and the results are shown in Figure 2. For the confi-

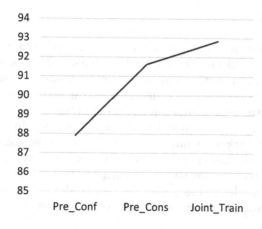

Fig. 6. Performance gain for training phases.

dence features, **Source** is the most important. If we remove the **Source** feature, all of the methods drop more than 3 points. Bilingual Cooccurence(BilColoc) is the second most important confidence feature. Compared with the confidence features, for our method, consistency features are also important. Both the cooccurence consistency feature (**Consist**$_{Col}$) and relation consistency feature (**Consist**$_{Rel}$) are important. Relation consistency feature is even stronger than the **Source** feature, which is the most important one of the confidence features.

4.6   Source Gain

We also remove the sources one by one from the candidate generation sources to test how the source contribution affects the performance. The results are shown in Table 4. From Table 4, we find that the semi-structured web sites (**-Website**) are the most important source. When they are removed from the source list, the performance drops significantly. The second most important one is parenthetical translation pairs (**-Parenth**) mined from web pages. Even the the translation results from SMT engines (**-MT**) contribute the least compared with the three other sources, when they are removed the performance drops more than 1.5 points.

Table 4. Translation candidates source contribution. **-Website** is the performance after removing the translation pairs of semi-structured web sites. **-Dict** is the results after removing the traditional dictionary. **-Parenth** was the results with the parenthetical translation pairs removed, and **-MT** shows the results after the SMT engine has been removed.

-Website	-Dict	-Parenth	-MT
81.6%	90.1%	89.7%	91.2%

5   Conclusion

In contrast to conventional sentence translation, knowledge base translation suffers from two main problems. The first is data sparsity. To handle this problem, we mine translation candidates from several different sources. The second problem is that for traditional sentence translation, we can use the surrounding words/phrases to select the best translation and do word/phrase disambiguation. For entities in a knowledge base, we donot have this information. In order to perform translation disambiguation, we have proposed a graph-based collective inference method to take the translation of the surrounding entities into consideration, and learn the translation consensus for all the entities. We conducte experiments on a movie domain knowledge base, and the results show that our method can improve the translation performance significantly, compared with several strong baselines.

References

1. Al-onaizan, Y., Knight, K.: Translating named entities using monolingual and bilingual resources. In: Proceedings of the 40th Meeting of the Association for Computational Linguistics (ACL02), pp. 400–408 (2002)
2. Bao, J., Duan, N., Zhou, M., Zhao, T.: Knowledge-based question answering as machine translation. In: Proceedings of the 52nd Annual Meeting of the Association for Computational Linguistics. Long Papers, vol. 1, pp. 967–976. Association for Computational Linguistics, Baltimore, June 2014
3. Berant, J., Chou, A., Frostig, R., Liang, P.: Semantic parsing on Freebase from questionanswer pairs. In: Proceedings of the 2013 Conference on Empirical Methods in Natural Language Processing, pp. 1533–1544. Association for Computational Linguistics, Seattle, Washington, October 2013
4. Brown, P.F., Pietra, S.D., Pietra, V.J.D., Mercer, R.L.: The mathematics of statistical machine translation: Parameter estimation. Computational Linguistics 19(2), 263–311 (1993)
5. de Melo, G., Weikum, G.: UWN: a large multilingual lexical knowledge base. In: Proceedings of the 50th Annual Meeting of the Association for Computational Linguistics, pp. 151–156. Association for Computational Linguistics, Stroudsburg (2012)
6. Han, X., Sun, L., Zhao, J.: Collective entity linking in web text: a graph-based method. In: Proceedings of the 34th International ACM SIGIR Conference on Research and Development in Information Retrieval, SIGIR '11, pp. 765–774. ACM, New York (2011)
7. Huang, F., Vogel, S.: Improved named entity translation and bilingual named entity extraction. In: Proceedings of Fourth IEEE International Conference on Multimodal Interfaces. Institute of Electrical & Electronics Engineers (IEEE) (2002)
8. Jiang, L., Zhou, M., Chien, L.F., Niu, C.: Named entity translation with Web mining and transliteration. In: Proceedings of the 20th International Joint Conference on Artificial Intelligence, pp. 1629–1634 (2007)
9. Lin, D., Zhao, S., Van Durme, B., Paşca, M.: Mining parenthetical translations from the Web by word alignment. In: Proceedings of ACL-08: HLT, pp. 994–1002. Association for Computational Linguistics, Columbus, June 2008

10. Liu, S., Li, C.-H., Li, M., Zhou, M.: Learning translation consensus with structured label propagation. In: Proceedings of the 50th Annual Meeting of the Association for Computational Linguistics. Long Papers, vol. 1, pp. 302–310. Association for Computational Linguistics, Jeju Island, July 2012
11. Subramanya, A., Petrov, S., Pereira, F.: Efficient graph-based semi-supervised learning of structured tagging models. In: Proceedings of the 2010 Conference on Empirical Methods in Natural Language Processing, pp. 167–176. Association for Computational Linguistics, Cambridge, October 2010
12. Talukdar, P.P., Pereira, F.: Experiments in graph-based semi-supervised learning methods for class-instance acquisition. In: Proceedings of the 48th Annual Meeting of the Association for Computational Linguistics, pp. 1473–1481. Association for Computational Linguistics, Uppsala, July 2010
13. Wick, M., Rohanimanesh, K., Culotta, A., Mccallum, A.: Samplerank: learning preference from atomic gradients. In: NIPS WS on Advances in Ranking (2009)
14. Yao, X., Van Durme, B.: Information extraction over structured data: question answering with Freebase. In: Proceedings of the 52nd Annual Meeting of the Association for Computational Linguistics. Long Papers, vol. 1, pp. 956–966. Association for Computational Linguistics, Baltimore, June 2014

Stochastic Language Generation
Using Situated PCFGs

Caixia Yuan[✉], Xiaojie Wang, and Ziming Zhong

Beijing University of Posts and Telecommunications, Beijing 100876, China
{yuancx,xjwang,zimingzhong}@bupt.edu.cn

Abstract. This paper presents a purely data-driven approach for generating natural language (NL) expressions from its corresponding semantic representations. Our aim is to exploit a parsing paradigm for natural
language generation (NLG) task, which first encodes semantic representations with a situated probabilistic context-free grammar (PCFG), then
decodes and yields natural sentences at the leaves of the optimal parsing
tree. We deployed our system in two different domains, one is response
generation for a Chinese spoken dialogue system, and the other is instruction generation for a virtual environment in English language, obtaining
results comparable to state-of-the-art systems both in terms of BLEU
scores and human evaluation.

Keywords: Natural language generation · Meaning representation ·
Situated PCFG

1 Introduction

Natural language generation (NLG) is the task of constructing natural-language
sentence from formal, abstract meaning representation (MR) (Reiter and Dale,
2000). Depending on the application at hand, the meaning representation can
have various forms such as database records, domain knowledge bases, geo-
information. It is generally assumed that the core tasks of language generation
process can be split up into two stages: (1) content selection, which decides what
meanings to express, and (2) surface realization, which expresses those meanings
using natural language (Belz and Kow, 2009). Over the past decade, statistical
methods for NLG have received considerable attention (e.g., Wong and Mooney,
2007; Belz, 2008; Konstas and Lapata, 2012; McKinley and Ray, 2014). However, this prior work is mostly based on hand-crafted generation rules, which
are extensive, but also expensive. Furthermore, although it is a consensus that
at a rather abstract level natural language generation can benefit a lot from its
counterpart natural language understanding (NLU), the problem of leveraging
NLU resources for NLG remains pretty much open.

In this paper, we propose a data-driven natural language generation model
which exploits a PCFG parser to assist natural language generation. The basic
idea underlying our method is that the generated sentence is licensed by a

© Springer International Publishing Switzerland 2015
J. Li et al. (Eds.): NLPCC 2015, LNAI 9362, pp. 64–75, 2015.
DOI: 10.1007/978-3-319-25207-0_6

context-free-grammar, and thus can be deduced from a parsing tree which encodes hidden structural linkage between meaning representation and its sentence expression. We operate in a setting in which we are given a set of records, where each record is a data pair consisting of a structured meaning representation and its natural language sentence. A situated PCFG - i.e., a PCFG with the context of application-specific concepts, is learned from data pairs and then used to guide generation processes for other previously unseen meaning representations. Table 1 exemplifies two records from the two applications at hand.

The strength of our approach is that it allows generation process to be represented as an optimization problem within a tree structure, without concerns about how the surfacial words are ordered and selected, and without the need to manually define PCFG derivations, which is one of the most important prerequisites in work of (Belz and Kow, 2009) and (Konstas and Lapata, 2012). We demonstrate the versatility and effectiveness of our method on (1) response generation for a situated Chinese spoken dialogue system (SDS)[1] for booking meeting rooms, and (2) GIVE (Generating Instructions in Virtual Environments)[2] challenge, within which a NLG module generates a sequence of English instructions that will help in a "treasure hunt" task in a virtual 3D environment.

Table 1. Examples of meaning representation input as a structured database and its corresponding natural language expression. Each meaning representation has several fields, each field has a value.

(a) SDS

Meaning representation	action1 confirm	object1 budget	value11 2,000	value12 2,500	action2 request	object2 date	value21 null	value22 null
Text	您的预算在2,000元到2,500元之间，请问您在哪天开会？ (Your budget is between 2,000 yuan and 2,500 yuan. When is the meeting scheduled?)							

(b) GIVE

Meaning representation	action move	direction slightly right	visible null	adj blue	object button	reference true	adj2 green	ref-object button
Text	Go to the blue button near to a green button. It should be in front of you slightly to the right.							

2   Related Work

Over the past decade, there has been a surge of interest in statistical techniques for natural language generation, a methodology that was largely inspired by the blossom of statistical natural language processing. Statistical NLG mainly follows two streams of research. The one is to introduce statistics at the sentence

---

[1] A demo can be found at http://www.aidc.org.cn:8008/WebContent/

[2] More about GIVE challenge can be seen at http://www.give-challenge.org/research/

generation level by training a model which refines or reranks candidate outputs of a handcrafted generator. A pioneering work is Langkilde and Knight's Nitrogen systems (Langkilde and Knight, 1998), which first generates a candidate set of sentences and then reranks them using an n-gram language model trained on news articles. Langkilde and Knight proved that the statistical post-processor yielded more fluent outputs and reduced the need for deep, hand-crafted grammars. In order to produce more customerized outputs, Walker et al. investigate a trainable sentence planner on the basis of feedback from users (Walker et al., 2002). The major drawbacks of such "overgenerate and rank" approach are their inherent computational cost and not grammatically informed.

The second stream of research has focused on introducing statistics at the generation decision level by training models that find the set of generation parameters maximizing an objective function, e.g., generating the most likely context-free derivations (Belz, 2008; Konstas and Lapata, 2012), or maximizing the expected reward using reinforcement learning (Rieser and Lemon, 2009; Dethlefs and Cuayahuitl, 2014). While such methods do not suffer from overgeneration problem, they still require a set of handcrafted generation rules or reward functions to derive a generation decision space within which an optimal sentence can be deduced statistically. Our model is closest to (Konstas and Lapata, 2013) who reformulates the Markov structure between a world state and a string of text depicted in (Liang, et al., 2009) into a set of CFG rewrite rules, and then deduces the best derivation tree for a set of database records. Although this Markov structure can capture a few elements of rudimentary syntax, it is essentially not linguistic grammars. Thus the sentences produced by this model are usually ungrammatically informed (for instance, its 1-best model outputs grammatically illegal sentences like "Milwaukee Phoenix on Saturday on Saturday on Saturday on Saturday"). (Konstas and Lapata, 2013) claims that dependency structure is an efficient complementary to CFG grammar, and incorporates dependency information between words into the reranking procedure to boost the performance.

Although conceptually related to (Konstas and Lapata, 2013), our model directly learns more grammatical rewrite rules from hybrid syntactic trees whose non-terminal nodes are comprised of phrasal nodes inheriting from a syntactic parser and conceptual nodes designed for encoding target meaning representation. Therefore, the learning aspect of two models is fundamentally different. We have a single CFG grammar that applies throughout, whereas they train different CFG grammar and dependency grammar respectively.

3   Problem Formulation

3.1   The Grammar

Following most previous works in this area (Liang, et al., 2009; Konstas and Lapata, 2013), we use the term record r to refer to a (m, w) pair. Each meaning representation m is described as several fields f, each field has value $f.v$. As exemplified in Table 1, each m in GIVE system has eight fields: action, direction, visible, adj (adjunct), object, reference, adj2 (adjunct 2) and ref-object (reference object). Each field has a specific value. The value can be a string (e.g., blue, button), a numeric quantity (e.g., 2000, 2500), or null. The text is simply a sequence of words $w = (w_1, ..., w_{|w|})$.

Our goal is to learn a PCFG for paraphrasing a MR with NL expression. As mentioned in Section 2, prior research on CFG based natural language generation has mainly focused on relatively simple grammar, either hand-crafted grammars for deterministic parsing (Belz and Kow, 2009), or probabilistic regular grammar describing Markov dependency (Konstas and Lapata, 2013) among fields and word strings. In order to generate more grammatical sentence, the established grammar should capture recursive structure of phrases. Meanwhile, in order to generate sentence expressing target meanings, the grammar should also capture concept embeddings corresponding to desired meaning fields. Under this framework, the situated PCFG grammar we used for generation can be described as a 6-tuple:

$$G = \langle N_p, N_c, T, S, L, \lambda \rangle \tag{1}$$

where N_p is a finite set of non-terminal symbols inheriting from a phrase structure parser, N_c is a finite set of concept symbols corresponding with record fields, T is a finite set of NL terminal symbols (words), $S \in Np$ is a distinguished start symbol, L is a lexicon which consists of a finite set of production rules, and λ is a set of parameters that define a probability distribution over derivations under G.

3.2 Grammar Induction

In this section, we present a learning procedure for the proposed grammar described above. The input to the learning algorithm is a set of training sentences paired with their correct meaning representations (as illustrated in Table 1). The output from the learning algorithm is a PCFG describing both phrase and concept embeddings. The learning algorithm assumes that a phrase structure parser is available, but it does not require any prior knowledge of the MR syntax.

To describe the grammar learning procedure, we start with an example. Consider the NL sentence in Table 1(a). We first analyze its phrase structure using a syntactic parser whose non-terminals are syntactic categories (e.g., NP, VP and QP) and part-of-speech tags (e.g., PN, DEG and NN). The parser we used for GIVE and SDS are both the Stanford Parser[3]. Figure 1(a) outlines the partial parser tree of sentence in Table 1(a).

The meaning of the sentence is then integrated by adding conceptual symbols of its subparts into the parser tree. Figure 1 (b) shows a hybrid parse tree of Figure 1 (a). Here the non-terminal symbols in bold, BUDGET, VAL1 and VAL2, represent domain-specific concepts corresponding to fields *budget*, *value*1 and *value*2. To get the hybrid parse tree, we first align phrases in the NL with the actual MR fields mentioned using the model of (Liang, et al., 2009) which is learned in an unsupervised manner using EM to produce which words in the

---

[3] http://nlp.stanford.edu/software/lex-parser.shtml

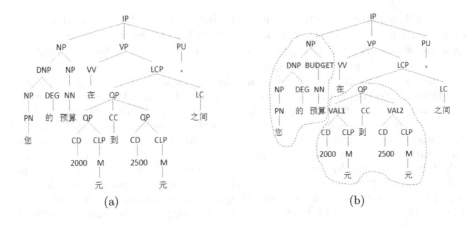

Fig. 1. Example of (a) a syntactic tree and (b) its corresponding hybrid tree from which the situated PCFG defined in Formula (1) is constructed. The subtree circled by dotted line contains conceptual node and its terminal derivations.

text were spanned by the fields. The aligned pairs are recorded in a temporary dictionary. Then for each phrase in the dictionary, we find the minimal subtree spanning it, and modify its ancestor node attached directly below the subtree's root node to the conceptual symbol of its aligned field. All ancestor nodes keep unchanged for phrases not in the alignment dictionary. The central characteristic of a tree structured representation is that component concept appears as a node in a tree, with its word realizations as terminal nodes derived by it. For example, the concept BUDGET has a terminal node "预算 (budget)", and VALUE1 "2,000元 (2,000 yuan)", these could then form part of the representation for the sentence "您的预算在2,000元到2,500元之间。(Your budget is between 2,000 yuan and 2,500 yuan.)" The use of a recursive hybrid syntactic and conceptual structure is one characteristic that distinguishes the proposed grammar from earlier work in which meaning is represented by logical forms or regular grammars (Lu and Ng, 2011; Konstas and Lapata, 2013).

Given hybrid trees, N_p, N_c, T, S and the set of derivations that are possible are fixed, we only need to learn a probabilistic model parameterized by λ. Since the correct correspondence between NL words and MR fields is fully accessible, i.e., there is a single deterministic derivation associated with each training instance, model parameter λ can be directly estimated from the training corpus by counting. Because the derived trees output by parser can be noisy, we need to process them to obtain cleaner PCFG rules. We compare the 3-best trees produced by the Stanford Parser, and prune off inconsistent components voted by majorities when extracting and counting rules.

3.3   Decoding

Our goal in decoding is to find the most probable sentence $\hat{s}$ for a given meaning expression m:

$$\hat{s} = g(\operatorname*{argmax}_{D \ s.t. \ m(D)=m} P(D|G) \cdot \ln(|D| + 1)) \tag{2}$$

where g is a function that takes as input a derivation tree D and returns $\hat{s}$, $m(D)$ refers to the meaning representation of a derivation D, and $P(D|G)$ is product of weights of the PCFG rules used in a derivation D, the factor $\ln(|D|+1)$, offers a way to compensate the output sentence length $|D|$.

A conventional CKY-style decoder (Kasami, 1965; Younger, 1967) is not applicable to this work since the fields of MR do not exhibit a linear structure. We use a basic decoding introduced in (Konstas and Lapata, 2013) which is essentially a bottom-up chart-parsing algorithm. It first fills the diagonal cell of the chart with the top scoring words emitted by the unary productions of the type $A \rightarrow \alpha$, where A is a non-terminal symbol, and α is a terminal word. The extracted grammar is binarized such that decoding takes cubic time with respect to the sentence length.

In order to search among exponentially many possible generations for a given input, it would be preferable if we added to the chart a list of the top k words and production rules, and thus produced a k-best list of derivations at the root node, yielding k-best sentences at the leaf nodes. We do this k-best decoding using the lazy algorithm introduced in (Huang and Chiang, 2005) which delays the whole k-best calculation until after parsing. Then an external language model can be applied to rerank the k-best derivations. We examine two ways of intersecting language model, one is to rerank directly the k-best sentences after those sentences are generated (Langkilde and Knight, 1998), the other is to rerank derived partial trees in a timely manner with cube pruning (Huang and Chiang, 2005; Konstas and Lapata, 2013).

4   Empirical Evaluation

This section presents our experimental setup for assessing the performance of our model. We give details on our dataset, model parameters, the metric used for comparison and experimental results.

4.1   Data Set

We conducted experiments on a Chinese spoken dialogue system (SDS) for meeting room booking. Our NLG module receives structured input from dialogue management (DM) module and generates natural language response to user. The structured input includes dialogue actions (e.g., greet, request, confirm), objects (e.g., date, budget, location) and object values which can be a null. DM delivers to NLG at most two actions at a time. The SDS corpus consists of 1,406 formal meaning representations, along with their Chinese NL expressions written by 3 Chinese native speakers. The average sentence length for the 1,406-example data set is 15.7 Chinese words. We randomly select 1,000 record pairs as training data, and the remaining 406 as testing data.

In order to assess the generation performance across different domains and different languages, we conducted experiments on the platform provided by the Challenge on Generating Instructions in Virtual Environments (GIVE), a theory-neutral, end-to-end evaluation effort for NLG systems. The NLG model generates a sequence of English NL instructions guiding users performing a "treasure hunt" task in a virtual 3D environment. We obtain 63 American English written discourses in which one subject guided another in a treasure hunting task in the spirit of the GIVE-2 virtual worlds. In order to ensure quality of the gold data, we preprocessed the corpus to delete non-sense instructions (i.e., sentences not related with any environmental parameters or operation instructions, e.g., "lol.", "what?"), correct spelling mistakes and tokenize the abbreviations (e.g., "srry", "u"). Finally, 1,159 NL and MR pairs from 50 discourses are used for training, and 294 pairs from the remaining 13 discourses for testing. The average sentence length for the 1,453 sentences is 7.8 English words.

4.2   Evaluation Metric

To evaluate the quality of the generated sentences, the BLEU score (Papineni et al., 2002) is computed by comparing system-generated sentences with human-written sentences. Specifically, the BLEU score is the geometric mean of the precision of n-grams of various lengths, multiplied by a brevity penalty factor that penalizes candidate sentences shorter than the reference sentences. BLEU has a fairly good agreement with human judgment and has been used to evaluate a variety of language generation systems (Angeli et al., 2010; Konstas and Lapata, 2012).

In addition, we evaluated the generated text via a human judgment as designed in (Angeli et al., 2010). Human evaluators were presented with a meaning representation and were asked to rate its corresponding NL expression along two dimensions: fluency (is the text grammatical and overall understandable?) and semantic correctness (does the meaning conveyed by the NL sentences correspond to meaning representations?). Human evaluators used a five point rating scale where a high number indicates better performance. The averaged score of three difference human evaluators was computed.

4.3   Results

In order to compare our work with previous related work, we implement the method of (Konstas and Lapata, 2013) on our datasets. The BLEU scores of different systems are summarized in Table 2.

Table 2 compares BLEU scores achieved using the situated grammar described in Section 3.1 and 3.2 with that using the grammar described in (Konstas and Lapata, 2013). 1-BEST signifies results obtained from the basic decoder.

Table 2. BLEU scores on SDS and GIVE.

system	SDS	GIVE
1-BEST-Konstas	9.32	10.49
k-BEST-Konstas	19.14	21.70
k-BEST-LM-Konstas	21.85	24.26
1-BEST-Our	30.88	31.07
k-BEST-Our	31.82	30.26
k-BEST-LM-Our	31.96	31.21

k-BEST and k-BEST-LM are results obtained respectively from reranking after generation and reranking during generation described in Section 3.3. Here we set k=20 without more fine-tuning work. Since the training data and the sentence length of our applications are relatively small, we used a bigram language model with add-one smoothing. Regarding these results, one point should be noted that the sentence length N is not restricted as a fixed number, while varying from 1 to a length of the longest sentence in the training data. The sentences with different length are overall sorted to obtain the 1-BEST and the k-BEST.

From Table 2, we find that differences in BLEU scores between 1-BEST-Konstas and 1-BEST-Our are statistically significant (9.32 vs. 30.88 in SDS domain, and 10.49 vs. 31.07 in GIVE domain). Since the only difference between these two results is the grammars used, we have reason to justify that the situated grammar learnt from the phrase-concept-hybrid trees is superior for modeling NL and MR correspondence to that used in (Konstas and Lapata, 2013).

It is interesting to notice that k-BEST-Konstas observes substantial increase in performance compared to 1-BEST-Konstas in both two domains, while k-BEST-Our only achieves a slight increase compared to 1-BEST-Our. The same observation also happens for the timely reranking k-BEST-LM-Konstas and k-BEST-LM-Our. As reported in (Konstas and Lapata, 2013), statistical language model offers potentially significant advantages for the sequential Markov grammar. Meanwhile, these results verify the robustness of the proposed method. Another major advantage of our method over method of (Konstas and Lapata, 2013) is that it does not require any prior knowledge of the MR syntax for training. Therefore, transplanting our method to other NLG application is relatively easy.

Overall, k-BEST-LM performs better than k-BEST, but the improvement is moderate. In practice, k-BEST-LM is more commonly used due to its computational efficiency and integrity.

Table 3 shows the human ratings for each system and the gold-standard human-authored sentences. On both Chinese and English domains our system is significantly better than the 1-BEST-Konstas baseline in terms of grammatical coherence and semantic soundness.

In order to evaluate the quality of our generation system in a practical view, we implement our English NLG system on the GIVE platform.

10 undergraduate volunteers are enrolled to play the "treasure hunt" game following the NL instructions our system generated. We collected 103 turns of game in total, 58 turns are guided by instructions produced by 1-BEST-Our, 45 turns by k-BEST-LM-Our.

Table 3. Human ratings for syntactic fluency (SF) and semantic correctness (SC).

system	SDS		GIVE	
	SF	SC	SF	SC
1-BEST-Konstas	2.29	1.94	2.04	2.47
k-BEST-LM-Konstas	3.91	3.12	3.82	3.49
1-BEST-Our	4.36	3.95	4.02	3.88
k-BEST-LM-Our	4.34	4.33	4.22	4.05
HUMAN	4.76	4.89	4.58	4.13

GIVE challenge adopts several objective metrics so as to measure the success of instructions in a situated interaction scenario. Results of the objective metrics can be induced automatically from log files. Table 4 outlines objective metrics used in GIVE-2.5 challenge (Striegnitz, et al., 2011).

Table 4. Objective measures used in GIVE-2.5 challenge.

binary task success: Percent of the player get the trophy.
duration: Time in seconds from the end of the tutorial until retrieval of the trophy.
instructions: Number of instructions produced by the NLG system.
words: Number of words used by the NLG system.

Table 5 shows the comparison of our systems with the work of (Dethlefs and Cuayahuitl, 2014) which is one of state-of-the-art systems evaluated on GIVE challenge. In terms of task success, k-BEST-LM-Our outperforms 1-BEST-Our by 7%, while both of them are less than that of Dethlefs. But it is a consensus that, besides the quality of generated instructions, there are many other subjective factors influencing the task success rate, for example, whether the player is a native English speaker, to what content the player is familiar with the game.

It is worthy to notice that both 1-BEST-Our and k-BEST-LM-Our generate significantly less interactions which guarantees much shorter interaction time to finish a task. Averagely, 1-BEST-Our generates 10.4 words per instruction and k-BEST-LM-Our 14.0 words, while Dethlefs's system generates 9.8 words for each instruction. Our systems produce many instructions such as "click the green button in front of you to the left of the lamp", "turn right into the room ahead". In contrast, as reported (Striegnitz et al., 2011; Dethlefs and Cuayahuitl, 2014), most other GIVE systems output much shorter instructions such as "click green", "turn right". Although brief enough, such instructions will lead more false actions due to lack of necessary reference information.

Table 5. Objective metrics for our systems compared with systems of (Dethlefs and Cuayahuitl, 2014).

metric	Dethlefs	1-BEST-Our	k-BEST-LM-Our
binary task success	0.80	0.67	0.74
duration	700	261	491
instructions	312.3	105.3	160.7
words	3075.6	1093	2249

5 Conclusions

We have presented a PCFG-based natural language generation method. In particular, the method learns situated PCFG rules from hybrid phrase-concept trees automatically augmented from the output of an existing syntactic parser. A compelling advantage of the proposed method is that it does not rely on prior knowledge of the MR syntax for training. We have shown the competitive results across different application domains in both Chinese and English language. Future extensions include deploying more efficient decoding algorithms, and richer structural features to rerank the derivations.

Acknowledgments. This work was partially supported by Natural Science Foundation of China (No. 61202248, No. 61273365), Discipline Building Planing 111 Base Fund (No. B08004) and Engineering Research Center of Information Networks, Ministry of Education.

References

1. Langkilde, I., Knight, K.: Generation that exploits corpus based statistical knowledge. In: Proceedings of the 36th Annual Meeting of the Association for Computational Linguistics (ACL), pp. 704–710 (1998)
2. Reiter, E., Dale, R.: Building natural language generation systems. Cambridge University Press, New York (2000)
3. Walker, M.A., Rambow, O., Rogati, M.: Training a sentence planner for spoken dialogue using boosting. Computer Speech and Language 16(3–4), 409–433 (2002)
4. Angeli, G., Liang, P., Klein, D.: A simple domain-independent probabilistic approach to generation. In: Proceedings of the 2010 Conference on Empirical Methods in Natural Language Processing, Cambridge, MA, pp. 502–512 (2010)
5. Kim, J., Mooney, R.: Generative alignment and semantic parsing for learning from ambiguous supervision. In: Proceedings of the 23rd Conference on Computational Linguistics, Beijing, China, pp. 543–551 (2010)

6. Konstas, I., Lapata, M.: Concept-to-text generation via discriminative reranking. In: Proceedings of the 50th Annual Meeting of the Association for Computational Linguistics: Human Language Technologies, Jeju, South Korea, pp. 369–378 (2012)

7. Konstas, I., Lapata, M.: A Global Model for Concept-to-Text Generation. Journal of Artificial Intelligence Research **48**(2013), 305–346 (2013)

8. Ratnaparkhi, A.: Trainable Approaches to Surface Natural Language Generation and Their Application to Conversational Dialog Systems. Computer Speech and Language **16**(3–4), 435–455 (2002)

9. Rieser, E., Lemon, O.: Natural language generation as planning under uncertainty for spoken dialogue systems. In: Proceedings of the 12th Conference of the European Chapter of the ACL, Athens, Greece, pp. 683–691 (2009)

10. Huang, L., Chiang, D.: Better k-best parsing. In: Proceedings of the 9th International Workshop on Parsing Technology, Vancouver, British Columbia, pp. 53–64 (2005)

11. Liang, P., Jordan, M., Klein, D.: Learning semantic correspondences with less supervision. In: Proceedings of the Joint Conference of the 47th Annual Meeting of the ACL and the 4th International Joint Conference on Natural Language Processing of the AFNLP, Suntec, Singapore, pp. 91–99 (2009)

12. Lu, W., Ng, H.T.: A probabilistic forest-to-string model for language generation from typed lambda calculus expressions. In: Proceedings of the 2011 Conference on Empirical Methods in Natural Language Processing, Edinburgh, Scotland, UK, pp. 1611–1622 (2011)

13. Wong, Y.W., Mooney, R.: Generation by inverting a semantic parser that uses statistical machine translation. In: Proceedings of the Human Language Technology and the Conference of the North American Chapter of the Association for Computational Linguistics, Rochester, NY, pp. 172–179 (2007)

14. McKinley, N., Ray, S.: A decision-theoretic approach to natural language generation. In: Proceedings of the 52nd Annual Meeting of the Association for Computational Linguistics, Baltimore, Maryland, USA, pp. 552–561 (2014)

15. Dethlefs, N., Cuayahuitl, H.: Hierarchical reinforcement learning for situated natural language generation. Natural Language Engineering **21**(03), 391–435 (2014)

16. Belz, A.: Automatic Generation of Weather Forecast Texts Using Comprehensive Probabilistic Generation-Space Models. Natural Language Engineering **14**(4), 431–455 (2008)

17. Belz, A., Kow, E.: System building cost vs. output quality in data-to-text generation. In: Proceedings of the 12th European Workshop on Natural Language Generation, Athens, Greece, pp. 16–24 (2009)

18. Gargett, A., Garoufi, K., Koller, A., Striegnitz K.: The GIVE-2 corpus of giving instructions in virtual environments. In: Proceedings of the 7th Conference on International Language Resources and Evaluation (LREC), Valletta, Malta (2010)

19. Striegnitz, K., Denis, A., Gargett, A., Garoufi, K., Koller, A., Theune, M.: Report on the second challenge on generating instructions in virtual environments (GIVE-2.5). In: Proceedings of the 13th European Workshop on Natural Language Generation (ENLG), Nancy, France, pp. 270–279 (2011)

20. Chen, Q., Manning, C.D.: A fast and accurate dependency parser using neural networks. In: Proceedings of the 2014 Conference on Empirical Methods in Natural Lan-guage Processing (EMNLP), Doha, Qatar, pp. 740–750 (2014)

21. Levy, R., Manning, C.D.: Is it harder to parse Chinese, or the Chinese Tree-bank? In: Proceedings of the ACL 2003, Sapporo, Japan, pp. 439–44 (2003)

22. Kasami, T.: An efficient recognition and syntax analysis algorithm for context-free languages. Tech. rep. AFCRL-65-758, Air Force Cambridge Research Lab, Bedford, Mas-sachusetts (1965)
23. Younger, D.H.: Recognition and parsing for context-free languages in time n3. Information and Control **10**(2), 189–208 (1967)
24. Papineni K., Roukos S., Ward, T., Zhu, W.: BLEU: a method for automatic evaluation of machine translation. In: Proceedings of the 40th Annual Meeting of the Association for Computational Linguistics, Philadelphia, PA, pp. 311–318 (2002)
25. Benotti, L., Denis, A.: Giving instructions in virtual environments by corpus-based selection. In: Proceedings of the 12th Annual Meeting of the Special Interest Group on Discourse and Dialogue, Portland, Oregon, pp. 68–77 (2011)

Machine Learning for NLP

Clustering Sentiment Phrases in Product Reviews by Constrained Co-clustering

Yujie Cao[✉], Minlie Huang, and Xiaoyan Zhu

State Key Laboratory of Intelligent Technology and Systems,
National Laboratory for Information Science and Technology,
Department of Computer Science and Technology, Tsinghua University,
Beijing 100084, People's Republic of China
caoyujieboy@163.com, {aihuang,zxy-dcs}@tsinghua.edu.cn

Abstract. Clustering sentiment phrases in product reviews is convenient for us to get the most important information about one product directly through thousands of reviews. There are mainly two components in a sentiment phrase, the aspect word and the opinion word. We need to cluster these two parts simultaneously. Although several methods have been proposed to cluster words or phrases, limited work has been done on clustering two-dimensional sentiment phrases. In this paper, we apply a two-sided hidden Markov random field (HMRF) model on this task. We use the approach of constrained co-clustering with some priori knowledge, in a semi-supervised setting. Experimental results on sentiment phrases extracted from about 0.7 million mobile phone reviews show that this method is promising for this task and our method outperforms baselines remarkably.

Keywords: Sentiment analysis · Sentiment phrase clustering · Constrained co-clustering · Sentiment extraction

1 Introduction

The product reviews on the Internet can give both the sellers and the buyers very useful information. However, it's not convenient for us to go through a vast number of reviews to get the information we want. One of the solutions is to cluster sentiment phrases in product reviews. A sentiment phrase is a short phrase that consists of an aspect (feature) word and an opinion word. Nowadays many famous shopping websites or product review websites all provide the clustered sentiment phrases, as exemplified by Fig 1.

商品详情	包装和参数	累计评价 425	月成交记录662件	电器城服务详情

| 与描述相符 4.8 ★★★★★ | 大家都写过 | 手机不错(86) 性能强大(20) | 机子是正品(55) 耗电量快(18) | 服务好(46) 手机一般(7) | 外型不错(40) | 物流快(32) |

Fig. 1. Clustered Sentiment Phrase (from www.taobao.com)

© Springer International Publishing Switzerland 2015
J. Li et al. (Eds.): NLPCC 2015, LNAI 9362, pp. 79–89, 2015.
DOI: 10.1007/978-3-319-25207-0_7

Clustering sentiment phrases in product reviews is different from ordinary text clustering in two ways. Firstly, clustering objects are different. Sentiment phrases in product reviews are not only too short to contain as much information as documents, but also in a fixed format with an aspect (feature) word followed by an opinion word. Secondly, it needs to be clustered in two dimensions simultaneously. We cluster the phrases that are similar both in "aspect" dimension and "opinion" dimension.

To the best of our knowledge, there is limited prior work on clustering two-dimensional sentiment phrases. Typical words or phrases clustering is one-way clustering [1, 2]. And there are some works about clustering aspect-related phrases [3, 4]. Inspired by Li et al. [5] and Song et al. [6, 7], we adopt the idea of using word co-occurrence frequency and applying constrained co-clustering to our task. Word co-occurrence frequency is the frequency of the co-occurrence of an aspect word together with an opinion word. By using this we can find that usually similar aspect words have similar frequency distributions along the opinion words, and vice versa.

Constrained co-clustering is based on Information-theoretic co-clustering (ITCC), and incorporates constraints by a two-sided HMRF regularization [6, 7]. Dhillon et al. pointed out that the ideal co-clustering is one that can minimize the mutual information loss between the original random variables and the clustered ones [8]. In ITCC, the clustering of one dimension enhances that of the other dimension. This method is proper to cluster the two parts of a sentiment phrase simultaneously. However, ITCC cannot solve the problems such as that many dissimilar words of one dimension have similar distribution along the other dimension. And another problem it cannot cope with is that the opinion words opposite in sentiment share similar distribution along the aspect words. So it's preferable to take prior knowledge about clusters into consideration.

Constrained co-clustering leverages the constraints from human-labeled data or the constraints derived from the unlabeled data automatically. We use it to incorporate pairwise constraints into the ITCC, which produces better results as presented in Section 3.

We target Chinese sentiment phrases in this paper. Our experiment was conducted based on a vast number of sentiment phrases extracted from 0.7 million product reviews on mobile phones.

To sum up, the main contributions of this paper are:

- We study the problem of clustering sentiment phrases in product reviews by constrained co-clustering.
- We adopt additional constraints to help the clustering results, and most of the constraints are derived from the unlabeled data automatically.
- Experiment results shows that our method outperforms the ITCC. Our constraints produce better results.

2   Methodology

2.1   Information-Theoretic Co-clustering

Information-theoretic co-clustering (ITCC) is proposed by Dhillon et al. [8]. Different from ordinary one way clustering, information-theoretic co-clustering can solve the simultaneously clustering of two-dimensional data. In the process of ITCC, to determine the row cluster (column) prototype, we have to make use of the information of column (row) clustering. In other words, the clustering of two dimensions can enhance each other.

The ideal co-clustering is one with minimum mutual information loss [8]. The loss in mutual information can be written in the form of the KL divergence of the joint distribution of X and Y, which is p(X, Y), and its approximation q(X, Y). $\hat{X}$ and $\hat{Y}$ are the resulting cluster sets.

$$I(X;Y) - I(\hat{X};\hat{Y}) = KL(p(X,Y) \| q(X,Y)) \tag{1}$$

where q is an approximation of p:

$$q(x,y) = p(\hat{x},\hat{y})p(x \mid \hat{x})p(y \mid \hat{y}), \text{ where } x \in \hat{x}, y \in \hat{y} \tag{2}$$

The objective function for loss in mutual information can be written as follows:

$$KL(p(X,Y) \| q(X,Y)) = \sum_{\hat{x}} \sum_{x \in \hat{x}} p(x) KL(p(Y \mid x) \| q(Y \mid \hat{x})). \tag{3}$$

The algorithm of the information-theoretic co-clustering [8] is to minimize the above function.

Here, p(X, Y) is the words co-occurrence frequency table, where each row represents an aspect word and each column represents an opinion word. Input the joint distribution of X and Y, p(X, Y), and the target row-cluster and column-cluster numbers r and c, we can get a clustering result that simultaneously clusters rows and columns.

	清晰	清楚	高	低	耐用	持久
屏幕	0.2	0.2	0	0	0	0
价钱	0	0	0.2	0.2	0	0
电池	0	0	0	0	0.1	0.1

⇩

	清晰	清楚	高	低	耐用	持久
屏幕	0.2	0.2	0	0	0	0
价钱	0	0	0.2	0.2	0	0
电池	0	0	0	0	0.1	0.1

Fig. 2. An example of co-clustering

From the above example we can see that ITCC is proper to deal with our clustering sentiment phrases, satisfying the demand of clustering two-dimensional simultaneously according to the word co-occurrence frequency. However, it cannot cope with the following two problems. Firstly, some different aspect words have similar distributions over opinion words such that they cannot be separated by the algorithm. Secondly, some opinion words which are opposite in sentiment have similar distributions over aspect words. For instance, "quick" and "slow" always share almost the same distribution over their common aspect words. So the two words cannot be separated apart.

To address these problems, we propose further to incorporate constraints into this framework.

2.2   Constrained Co-clustering

Constrained co-clustering can add the benefits of constrained clustering to information-theoretic co-clustering, which is proposed by Song et al. [6, 7]. By taking into consideration some prior knowledge about the clusters, the problem can be addressed in a semi-supervised manner. This prior knowledge can guide the clustering process with better results.

The constrained co-clustering is a two-sided HMRF regularized ITCC model [6, 7]. The constraints are formulated by using HMRF for both dimensions. In the HMRF, some pairwise constraints are added to both dimensions.

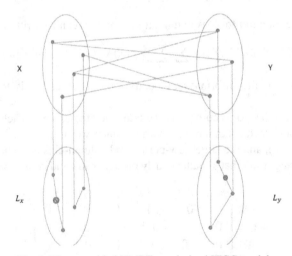

Fig. 3. The two-sided HMRF regularized ITCC model

In Fig 3 the two-sided HMRF regularized ITCC model is exemplified. L_x is the latent label set for X and L_y is the latent label set for Y in the HMRF. The latent labels actually denote the clusters index, and the lines between latent labels are the pairwise constraints.

Compared to the objective function of ITCC, the objective function is added the penalty for violating pairwise constraints between points.

For a latent label l_{x_1} for x_1, the must-link set is denoted as M_{x_1}, and the cannot-link set is denoted as C_{x_1}. The penalty for violating a must-link is:

$$V(x_1, x_2 \in M_{x_1}) = a_{1,2} D(p(Y \mid x_1) \| p(Y \mid x_2)) \cdot I(l_{x_1} \neq l_{x_2}) \tag{4}$$

And the penalty for violating a cannot-link is:

$$V(x_1, x_2 \in C_{x_1}) = \overline{a}_{1,2} (D_{max} - D(p(Y \mid x_1) \| p(Y \mid x_2))) \cdot I(l_{x_1} = l_{x_2}) \tag{5}$$

where I($*$) here is an indicator function, and I(true) = 1, I(false) = 0 . $a_{1,2}$ and $\overline{a}_{1,2}$ are tradeoff parameters. D here measures the distance between points. D_{max} is the maximum value for all Ds.

With constraints, we can guide the clustering results with the prior knowledge. We can solve the two problems in ITCC by adding pairwise links to a certain degree.

2.3   Constraints

Manually labeling data is expensive. So we try to generate the constraints automatically without asking the user to label a large scale of data. However, the number of aspect words in sentiment phrases are usually limited. So manually adding pairwise constraints among aspect words will result in a cheap promotion in the clustering result. We put a must-link among the aspect words which describe the same feature (e.g.: "price", "cost" and "worth"), and put a cannot-link among those describing different features (e.g.: "price" and "screen").

The number of opinion words are usually much more than the number of aspect words. We use two ways to generate constraints among these huge number of opinion words. One way is to put cannot-links between opinion words which are opposite in sentiment polarity by making full use of the sentiment information of the phrases and the words. We can get the sentiment polarity ("+1" for "positive" and "-1" for negative) of a sentiment phrase by putting it into a SVM classifier, which has been trained by a vast number of semi-structured product reviews with sentiment polarity tags from the Internet. We suppose that the opinion word contained by the sentiment phrase share the same sentiment polarity. By adding up the polarity of all the sentiment phrases that contains one particular opinion word, we can get the sentiment polarity of this opinion word. If the summed-up polarity is above 0, then this opinion word will be regarded as a positive one, and otherwise a negative one. With the information of the polarity of all the opinion words, cannot-links can be added between every two opinion words with opposite sentiment polarity.

The second way of generating constraints automatically is to take advantage of the words in common. Opinion words having common adjective words are likely to be put into one group, such as "clear" and "very clear". Here we should also take negative words into consideration. If two opinion words share common adjective but one of them contains a negative word, such as "no" or "not" in it, then they may not be put in a common group. Otherwise, if they share common adjective and both of them contain or neither of them contains negative words, then they may be put into a common group.

Note that these constraints may not be all correct. But together with the constrained co-clustering algorithm and mostly correct constraints, the clustering will have a remarkable promotion as soon shown later.

2.4 Sentiment Phrase Extraction

Sentiment Phrases are extracted from product reviews. With a product review, first of all, we split it with some punctuations such as "......, 。; ? ! ,.;?!" into short sub-sentences. Next, word segmentation will be conducted in each sub-sentence.

With the sentences after word segmentation, we will check that whether they contain any one of the aspect words obtained in advance. If so, we look up for the first adjective word after that aspect word. Once we find the first adjective word, we get a phrase between the aspect word and this adjective word, and we will continue looking up in the sentence until the end of this short sentence. If the next word is an adjective word, then it will be added up to the phrase we get and the looking up process continues. If the next word is a noun and it is also at the end of the sentence, then it will be added to the phrase. If the next word is a verb, then it will be added in and the looking up process stops. When we get the phrase, we only preserve those which contain less than 5 words.

3 Experiment Results

3.1 Data Preparation

The details of our review corpus are given in Table 1.

Table 1. Statistics of the review corpus

#Products	8
#Reviews	708,450
#Aspects	17

These reviews were crawled from the following websites: www.jd.com, www.pcpop.com, www.it168.com, www.zol.com, weibo.com.

The number of the sentiment phrases extracted from these reviews is 206,793. After duplicate removal, there are 7,263 phrases left, containing 64 unique aspect words and 2,941 opinion words.

The ground-truth of the clustering was labeled manually, containing 336 clusters.

3.2 Evaluation Metrics

We adapt three measures, Purity, Normalized Mutual Information (NMI), and Adjusted Rand Index (ARI) for performance evaluation.

Given a data set D with N items in D, we suppose its gold-standard partition with a total number of J clusters is $G = \{g_1, g_2, ..., g_J\}$. A clustering algorithm partitions D into K clusters and $R = \{r_1, r_2, ..., r_K\}$ is the clustering result.

Purity: The purity of the entire clustering result is calculated by:

$$Purity(G, R) = \frac{1}{N} \sum_k \max_j |r_k \cap g_j|,$$

where each individual item in this addition equation is the intersection of one cluster r_k and the gold-standard cluster by which the majority of r_k are contained.

We can see that if the clustering results perfectly match the gold-standard clusters, the purity is 1. However, it doesn't mean that if the purity is 1 the result is perfect. If K is relatively large, say K = N (the size of D) with only one element on every clusters, the purity is also 1. So by using purity as a measure, there exists a trade-off between the quality of the clustering and the cluster number.

NMI: NMI is the mutual information of the clustering result and the gold-standard clusters divided by half of the sum of their entropies. The NMI of the entire clustering result is calculated by:

$$NMI(G,R) = \frac{I(G,R)}{[H(G)+H(R)]/2},$$

where $I(G,R)$ is the mutual information of G and R, $H(G)$ and $H(R)$ is the entropies of G and R.

It can overcome the disadvantage of purity when K is very large, for $H(R)$ will increase as the K increases. Generally speaking, a larger NMI usually means a better result of clustering.

ARI: ARI is the adjusted form of Rand Index. Rand Index (RI) will give punishment to false positive decisions and false negative decisions, where false positive decision means that a pair of elements in different clusters in G was put into one cluster in R and false negative is similar. ARI is the difference of the RI and its expected value under the null hypothesis [9]. The ARI is calculated by:

$$ARI(G,R) = \frac{2(N_{00}N_{11} - N_{01}N_{10})}{(N_{00}+N_{01})(N_{01}+N_{11})+(N_{00}+N_{10})(N_{10}+N_{11})}.$$

N_{00} is the number of element pairs that are in different clusters both in G and R. N_{11} is that in same cluster both in G and R. N_{10} is that in same cluster in G but in different clusters in R. And N_{01} is similar but opposite to N_{10} in G and R.

ARI measures the degree of agreement between the gold-standard result and the clustering result by checking every pair of elements.

3.3   Evaluation Results

Comparison to Different Constraints. We apply our approach with different constraints and make comparisons among them. The given cluster number of aspect words is empirically set to 36, and that of opinion words is 100. The clusters with less than 5 sentiment phrases in the clustering results are all merged into two big clusters according to their sentiment polarity (+1 and -1) in the end.

We compare methods with different constraints. These methods are listed as follows.

- **Information-Theoretic   Co-clustering   (ITCC):** Information-theoretic co-clustering without any constraints. Here we use ITCC as a baseline.

- **Information-Theoretic Co-clustering with Aspect Constraints (ITCCAC):** Constrained Information-theoretic co-clustering with constraints only on aspect words. These constraints are human-labeled.
- **Information-Theoretic Co-clustering with Opinion Constraints (ITCCOC):** Constrained Information-theoretic co-clustering with constraints only on opinion words, including cannot-links from sentiment polarity and constraints from words in common.
- **Information-Theoretic Co-clustering with Aspect and Opinion Constraints (ITCCAOC):** Constrained Information-theoretic co-clustering with constraints both on aspect and opinion words.

Table 2. Comparisons among co-clustering with different constraints with #aspect cluster = 36 and #opinion cluster = 100

	NMI	Purity	ARI
ITCC	0.707	0.469	0.204
ITCCAC	0.742	0.547	0.240
ITCCOC	0.806	0.651	0.426
ITCCAOC	**0.859**	**0.795**	**0.566**

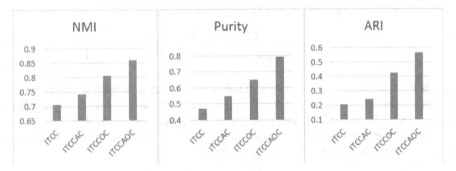

Fig. 4. NMI, Purity and ARI of the comparison

From the above table we can see that co-clustering without any constraints has the worst performance, with the lowest scores in three measures. By adding constraints to aspect words, the three measures all have a promotion. By adding constraints to opinion words, the three measures have a bigger promotion, for the number of constraints in opinion words are much more than that of aspect words so that they have stronger impact on the clustering result. By adding constraints to both the aspect words and the opinion words, the result is the best.

We can draw a conclusion that adding proper constraints to both the aspect and opinion words can help do a better job in clustering sentiment phrases.

Comparison to Other Clustering Methods. We further compare our approach with other clustering methods. We choose these clustering methods in the experiment setting of [6, 7]. The given cluster number of aspect words is empirically set to 36, and that of opinion words is 100. These methods are listed as follows. Kmeans, constrained Kmeans (CKmeans) [10], Tri-factorization of Semi-NMF (STriNMF) [11], and constrained Tri-factorization of Semi-NMF (CSTriNMF) [11]. Kmeans and CKmeans are one-way clustering methods. We apply them to both two dimensions (the aspect words and the opinion words) separately. Then we assemble sentiment phrases clusters with the aspect clusters and the opinion clusters. STriNMF, CSTriNMF, Information-Theoretic Co-Cluestering (ITCC) and constrained Information-Theoretic Co-Clustering (CITCC) are co-clustering methods. CKmeans, CSTriNMF and CITCC are all clustering methods with constraints.

Table 3. Comparison to different clustering methods

	NMI	Purity	ARI
STriNMF	0.685	0.421	0.187
CSTriNMF	0.694	0.446	0.192
Kmeans	0.708	0.459	0.215
CKmeans	0.773	0.586	0.362
ITCC	0.707	0.469	0.204
CITCC	**0.859**	**0.795**	**0.566**

From the above table we can see that we apply CITCC to our task can get the best results. Our constraints can improve the performance in KMeans and ITCC, but can do little favor to STriNMF and it can make a greater promotion in CITCC than other constrained clustering methods.

4   Related Works

The related works are in three parts: co-clustering, constrained co-clustering, words and phrases clustering.

Co-clustering algorithms deal with two-dimensional clustering. The two-dimensional data can be modeled in a co-occurrence matrix and the clustering problem can be solved by matrix factorization [12]. And it can also be modeled in a bipartite graph form and the clustering problem can be solved by graph partition [13]. It can be modeled in a joint distribution of two discrete random variables and information theory are used to partition the two sets of variables [8].

Constrained co-clustering incorporates prior knowledge constraints to co-clustering to have a promotion. Wang et al. [11] proposed a constrained co-clustering by matrix factorization, and so is Shi et al. [14]. The objective functions of all these constrained co-clustering methods are all sum squared residue-based in Euclidean distance. However, Song et al. [6, 7] use ITCC framework which use KL divergence and proposed constrained co-clustering method. It's more proper for sparse and high dimensional data.

Words and phrases clustering is also very much related to our work. Matsuo et al. [15] proposed a method of using web search engines as a corpus to perform a graph-based word clustering. SanJuan et al. [16] proposed a method for clustering phrases based on general lexico-syntactic relations without prior knowledge. Zhai et al. [3] proposed their EM based unsupervised methods for aspect expressions clustering. Zhao et al. [4] used a soft constraint with the PR framework to cluster aspect-related phrases.

5   Conclusion

Clustering sentiment phrases in product reviews is an important and useful task for sentiment analysis. In order to cluster sentiment phrases, this paper applies constrained co-clustering and incorporates rich constrained knowledge. We obtain most of the constraints automatically. Experiments show that our constraints are proper and useful and our method is superior to the baselines.

References

1. Yutaka, M., Takeshi, S., Koki, U., and Mitsuru, I.: Graph-based word clustering using a web search engine. In: Proceedings of the 2006 Conference on Empirical Methods in Natural Language Processing (EMNLP), pp. 542–550 (2006)
2. Lin, D., Wu, X.: Phrase clustering for discriminative learning. In: Proceedings of the Joint Conference of the 47th Annual Meeting of the ACL and the 4th International Joint Conference on Natural Language Processing of the AFNLP (ACL-IJCNLP), pp. 1030–1038 (2009)
3. Zhai, Z., Liu, B., Xu, H., Jia, P.: Clustering product features for opinion mining. In: Proceedings of the 4th ACM International Conference on Web Search and Data Mining, pp. 347–354 (2011)
4. Zhao, L., Huang, M., Chen, H., Cheng, J., Zhu, X.: Clustering aspect-related phrases by leveraging sentiment distribution consistency. In: Proceedings of the 2014 Conference on Empirical Methods in Natural Language Processing (EMNLP), pp. 1614–1623 (2014)
5. Li, H., Abe, N.: Word clustering and disambiguation based on co-occurrence data. In: Proceedings of the 17th International Conference on Computational Linguistics, pp. 749–755 (1998)
6. Song, Y., Pan, S., Liu, S.: Constrained co-clustering for textual documents. In: Proceedings of the 24th AAAI conference on Artificial Intelligence, pp. 581–586 (2010)
7. Song, Y., Pan, S., Liu, S., Wei, F., Zhou, M.X., Qian, W.: Constrained text coclustering with supervised and unsupervised constraints. IEEE Trans. Knowl. Data Eng. 25(6), 1227–1239 (2013)
8. Dhillon, I.S., Mallela, S., Modha, D. S.: Information-Theoretical Coclustering. In: Proceedings of the Ninth ACM SIGKDD Int'l Conf. Knowledge Discovery and Data Mining (KDD 2003), pp. 89–98 (2003)
9. Wagner, S., Wagner, D.: Comparing clusterings - an overview. Technical report 2006-04, Faculty of Informatics, Universitat Karlsruhe (TH) (2006)
10. Basu, S., Bilenko, M., Mooney, R. J.: A probabilistic framework for semi-supervised clustering. In: Proceedings of the 10th ACM SIGKDD Int. Conf. on Knowledge Discovery and Data Mining, pp. 59–68 (2004)

11. Wang, F., Li, T., Zhang, C.: Semi-supervised clustering via matrix factorization. In: Proceedings of SIAM Int'l Conf. Data. Mining (SDM), pp. 1–12 (2008)
12. Ding, C., Li, T., Peng, W., Park, H.: Orthogonal nonnegative matrix trifactorizations for clustering. In: Proceedings of the 12th ACM SIGKDD International Conference on Knowledge Discovery and Data Mining, pp. 126–135 (2006)
13. Dhillio, I.S.: Co-clustering documents and words using bipartite spectral graph partitioning. In: Proceedings of the 7th ACM SIGKDD International Conference on Knowledge Discovery and Data Mining (KDD-2001), pp. 269–274 (2001)
14. Shi, X., Fan, W., Yu, P.S.: Efficient semi supervised spectral co-clustering with constraints. In: Proceedings of IEEE 10th International Conf. Data Mining (ICMD), pp. 1043–1048 (2010)
15. Matsuo, Y., Sakaki, T., Uchiyama, k., Ishizuka, M.: Graph based word clustering using web search engine. In: Proceedings of the 2006 Conference on Empirical Methods in Natural Language Processing (EMNLP), pp. 542–550 (2006)
16. SanJuan, E., Fidelia I.: Phrase clustering without document context. In: Proceedings of the 28th European Conference on Information Retrieval, pp. 494–497 (2006)

A Cross-Domain Sentiment Classification Method Based on Extraction of Key Sentiment Sentence

Shaowu Zhang[1,2], Huali Liu[1], Liang Yang[1], and Hongfei Lin[1(✉)]

[1] School of Computer Science and Technology, Dalian University of Technology,
Dalian 116024, China
{zhangsw,hflin}@dlut.edu.cn, {liuhuali,yangliang}@mail.dlut.edu.cn
[2] School of Computer Science and Engineering,
Xinjinag University of Finance and Economics, Urumqi 830012, China

Abstract. Cross-domain sentiment analysis focuses on these problems where the source domain and the target domain are from different domains. However, traditional sentiment classification approaches usually perform poorly to address cross-domain problems. So, this paper proposed a cross-domain sentiment classification method based on extraction of key sentiment sentence. Firstly, based on the observation that not every part of the document is equally informative for inferring the sentiment orientation of the whole document, the concept of key sentiment sentence was defined. Secondly, taking advantage of three properties: sentiment purity, keyword property and position property, we construct heuristic rules, and combine with machine learning to extract key sentiment sentence. Then, data is divided into key and detail views. Integrating two views effectively can improve performance. Finally, experimental results show the superiority of our proposed method.

Keywords: Cross-domain · Key sentiment sentence · Multi-view ensemble

1 Introduction

With the rapid development of the Internet and the rise of social networking platforms, more and more people choose to express their views and opinions on the network. Subsequently, a large amount of text data with potential value emerges. Mining sentiment orientation of these data is of importance for scientific significance and commercial value. Under this situation, sentiment analysis has become one of the most active research areas in natural language processing and attracted a lot of research interest in recent years.

Sentiment analysis is the process of automatically detecting whether a text expresses a positive or negative semantic orientation. On this issue of vital importance, traditional supervised machine learning approaches have been shown promising and effectiveness [1–3]. However, these successful studies are based on

© Springer International Publishing Switzerland 2015
J. Li et al. (Eds.): NLPCC 2015, LNAI 9362, pp. 90–101, 2015.
DOI: 10.1007/978-3-319-25207-0_8

an assumption, namely training data and test data are independent and identically distributed. But, in practice, data drawn from different domains is difficult to meet the independent and identically distributed condition. For example, in book reviews, *"obscure"*, *"relevant"* show strong sentiment intensity, while rarely appear in hotel reviews. And equally, *"tatty"*, *"comfortable"* are usually used to comment on hotel and indicate strong sentiment orientation, but hardly occur in book reviews. Thus, how to solve the problem that the performance of classifier reduced caused by the difference of the distributions of training data and testing data, is a core task of cross-domain sentiment analysis.

In cross-domain sentiment analysis field, for supervised learning techniques, most classifiers are domain-specific. So, it becomes challenging to adapt a classifier trained on one domain to another. To address this problem, Blitzer et al. [4] proposed structural correspondence learning (SCL). *SCL*, selected a set of pivot features to model the relationship between pivot features and non-pivot features and to link the source and target domains. Also, Bollegala et al. [5] proposed a feature-level Sentiment Sensitive Thesaurus(SST) method. Firstly, they created a sentiment sensitive thesaurus. And then used the thesaurus to expand feature vectors for classification. Finally experiment proved the created sentiment sensitive thesaurus can accurately capture words that express similar sentiments. In addition, Li et al. [6] performed active learning for cross-domain sentiment analysis. They selected informative samples based on *QBC* strategy by the source and target classifiers. And combination-based to make decision. At last results showed comparable performances to in-domain method.

Follow the ideas of mentioned paper above, we proposed a cross-domain sentiment classification method based on extraction of *key sentiment sentence*. We divide a document into *key sentiment sentences* and *detailed sentences*. *Key sentiment sentence* is brief, discriminative and usually the most representative sentence in a document for the overall sentiment orientation and opinion. While *detailed sentence* is often more details, trivial, expressed complexly and ambiguously. Then, *key view* and *detail view* are constructed by *key sentiment sentences* and *detailed sentences* respectively. Through effectively integrating two views, we can not only acquire knowledge from domain-independent part to a large extent, but also bridge the distribution gap from *detail view* by domain-dependent part to a certain extent. Experimental results show that our proposed method achieve a very impressive improvement across three domains on 6 tasks.

The remainder of this paper is organized as follows. Section 2 introduces the problem setting and related concepts. Section 3 describes our proposed method in detail. Section 4 presents experimental results and discusses related parameters. At last, we conclude the paper and outline the future work in Section 5.

2   Problem Setting and Related Concepts

In this section, we introduce the definition of the problem that will be discussed below and several concepts related to our proposed method.

2.1 Problem Setting

Source Domain: $S = \{(x_i, y_i)\}_{i=1}^{n_s}$ refers to a set of labeled instances from a certain domain. Here, x_i is the i_{th} labeled instances. Particularly, it denotes a piece of product review in this paper. y_i denotes the sentiment label which is assigned for the i_{th} instances, and $y_i \in \{+1, -1\}$, here the sentiment labels $+1$ and -1, respectively, denote positive and negative. In addition, n_S indicates the number of labeled instances in source domain.

Target Domain: $T = \{(x_i)\}_{i=1}^{n_T}$ refers to a set of unlabeled instances from a certain domain, which is different from source domain but is related to source domain. Here, x_i is the i_{th} unlabeled instances. n_T denotes the number of unlabeled instances in target domain.

Cross-Domain Sentiment Classification: cross-domain sentiment classification seeks to generalize a model trained on source domain and uses it to infer the sentiment label of unlabeled instances in target domain.

2.2 Related Concepts

Domain-Independent Feature: If some features appear frequently in both source and target domains, and their sentiment orientations are coherent across different domains, they are *domain-independent*. Such as *"like"*, *"hate"*.

Domain-Dependent Feature: Contrary to *domain-independent feature*, if some features usually occur in a certain domain, whereas hardly occur in other domains, or their sentiment orientations are not coherent even opposite between source and target domains, they are *domain-dependent*. Such as *"obscure"* in book reviews, *"short battery life"* in computer reviews.

Sentiment Contribution: *Sentiment contribution* is the importance of sub-sentence in a document for predicting the overall sentiment label.

Key Sentiment Sentence: *Key sentence* for short, refers to the sentence that can substantially express the sentiment orientation of a document. Generally, the *sentiment contribution* of *key sentence* is larger than the other sentences.

Detailed Sentence: *Detailed sentence*, mainly refers to the sentence that is about what happened or attribute of products and so on. Usually, the *sentiment contribution* of *detailed sentence* is relative smaller compared with *key sentence*.

3 Cross-Domian Sentiment Classification Method Based on Extraction of Key Sentiment Sentence

In this section, we describe how to extract *key sentence*, and then present the process of integrating each view to promote the performance of base classifier for cross-domain sentiment classification.

3.1 Key Sentiment Sentence Extraction Algorithm

As a special issue of text classification, sentiment classification has received the considerable attention recently. However, due to the personalization and arbitrariness of human expression, sentiment classification, aiming at discriminating sentiment orientation of text, is more complicated than the ordinary text classification. Take a *NB* review for example, as shown in Tab. 1.

Table 1. *NB* reviews with mixture sentiments

Overall sentiment label	Sub-sentence content	Sub-sentence label
+1	<s1>Inadequate: USB interface is too few,and is left!	-1
	<s2>Installing system is a little trouble, trouble...	-1
	<s3>But,overall it is pretty good price.	+1

In this review, although there are some shortcomings(such as *"few"*, *"trouble"*), the overall sentiment orientation towards *NB* is positive, which can be seen from <s3>. In addition, compared with the other sentences, <s3> is shorter and unambiguous, contains *"overall"* and at the end of this review, intuitively it is much more important for inferring the whole sentiment label, namely the *sentiment contribution* of <s3> is larger. Thus, on the basis of the different *sentiment contribution* for each sub-sentence, the whole sentiment label of this review above is positive. However, those negative words may cause the classifier to discriminate the whole sentiment orientation negative, which leads to misclassification unfortunately.

To solve this specific problem, some scholars have found that when predicting sentiment orientation of a document, different sentences in the same document have different sentiment contribution [7]. Although the sentiment orientations of different sentences may be different, seeking out those having larger sentiment contribution as *key sentences*, then making a distinction between *key sentences* and *detailed sentences* will be helpful for inferring the overall sentiment orientation. Inspired by above, this paper proposed a two-stage extraction algorithm combining heuristic rules with machine learning. In the first stage, we make use of heuristic rules to construct the initial training sets for *key view* and *detail view*. In the second stage, we utilize the initial training sets to extract *key sentences* for remaining documents based on machine learning.

The heuristic rules in the first stage mainly extract three factors that would impact on the sentiment contribution of sentences, namely, *sentiment purity*, *keyword property* and *position property*.

Sentiment Purity. As a representation of the overall sentiment orientation of a document, *key sentence* should contain quite pure sentiment orientation. Therefore, we define *sentiment purity* as a factor to measure the *sentiment contribution* of a sentence. Higher *sentiment purity* means greater *sentiment contribution* and more likely to be a *key sentence*. The formula of *sentiment purity* score function is shown as below:

$$sentiPurity(s_i) = e^{\frac{|\sum_{w \in s_i} polarity(w)|}{|s_i|}}.$$ (1)

Here, $|s_i|$ is the number of words in a sentence. $polarity(w)$ denotes the sentiment label of the word w, if the label of w is positive, then $polarity(w) = 1$, label is negative, $polarity(w) = -1$, otherwise, $polarity(w) = 0$.

In this paper, we utilize the sentiment vocabulary ontology published by Dalian University of Technology [8] to judge whether a word is sentiment vocabulary or not. From formula (1), we can see that *sentiment purity* is inversely proportional to the length of a sentence, indicating that the more brief, the more likely a sentence is a *key sentence*. Besides, the more simple the sentiments are, $|\sum polarity(w)|$ is relatively larger, the higher the *sentiment purity* score is. It illustrates that a quite simple sentence is more likely to be *key sentence*. Based on the habits of human expression, when individuals tend to describe the details, sentiments usually are complex and mixed, whereas sentiments are unambiguity and concise when presenting overall sentiment orientation.

Keyword Property. Inspired by human language habits, *key sentence*, laying the sentiment tone of the whole document, should contain summary keywords generally, such as *"overall"*, *"in my opinion"* and so on. Therefore, summary keywords are also a critical factor for extraction of *key sentence*. We selected the top 15 terms as the keywords by term frequency in the first and last sentences. The formula of *keyword property* score function is shown as below:

$$keyWordNum(s_i) = \sum_{w \in s_i} I_K(w).$$ (2)

Here, K is the set of keywords used in this paper. $I_K(w)$ is the indicator function, when $w \in K$, $I_K(w) = 1$, otherwise $I_K(w) = 0$.

Position Property. It can be found that people often present their opinions at the beginning of a document, and make a summary in the end generally. So, *position property* is a key factor that cannot be ignored. The formula of *position property* score function is shown as below:

$$position(s_i) = i^2 - n \times i + 100.$$ (3)

Here, n presents the number of sub-sentence in a document. i denotes i_{th} sub-sentence, and $i \in [1, n]$. Constant term "100" guarantees the *position property* score is a positive number. Formula (3) is a parabola with opening upwards, the symmetry axis at the very center of the document($n/2$), which ensures that the beginning and end sentences have larger advantage, namely relatively high *position property* score. On the contrary, the *position property* score of these sentences in the middle of a document is quite low.

In summary, we normalize scores of three factors above, and work out the weighted sum of these normalized scores. Then we get the *sentiment contribution* score of each sentence based on the formula (4).

$$sentiContri(s_i) = w_1 \times sentiPurity(s_i) + w_2 \times keyWordNum(s_i) \quad (4)$$
$$+w_3 \times position(s_i).$$

Here, w_1, w_2 and w_3 deonte the weights of the corresponding factors respectively. We get the value of each weight by *Analytic Hierarchy Process* [9]. First, construct judgement matrix based on ratio scale(Tab. 2). Then, obtain the weight value by finding the eigenvector of the judgement matrix. Finally, $[w_1, w_2, w_3] = [0.17, 0.53, 0.30]$.

Table 2. The judgement matrix

three factors	sentiment Purity	keyword Property	position Property
sentiment Purity	1	1/3	1/2
keyword Property	3	1	2
position Property	2	1/2	1

For thresholds δ_1 and δ_2 , if $sentiContri(s_i) > \delta_1$, s_i is divided into *key view*, whereas if $sentiContri(s_i) < \delta_2$, s_i is divided into *detail view*. Thus, the initial training sets of two views can be drawn.

Then, we take advantage of the initial training sets of two views, and employ machine learning algorithm to extract *key sentences*. Each iteration selects these sentences with high-confidence to add to corresponding view. Repeat iteration until all reviews have been extracted *key sentences*. Then, the remaining sentences are considered as *detail sentences*. The procedure of our algorithm is shown in Tab. 3.

3.2 Multi-view Ensemble

Ensemble learning has demonstrated impressive capacities to improve the performance of base learning algorithm [11,12]. Ensemble classifier refers that combining the classifiers that have independent decision-making ability together. It has been proved that the ensemble classifier is more predictive than the single classifier normally.

After dividing data into two independent and complementary views, *key view* and *detail view*, we train two base classifiers f_{key} and f_{detail} on these two views separately. Effectively integrating the two base classifiers, can learn the knowledge that a single base classifier cannot learn. In this paper, two well-known simple fixed rules [13] for integrating the set of base classifiers into a combining classifier will now be described as follows.

Table 3. The procedure of *key sentence* extraction

Input:The training data(in the source domain)
Output:The *key view* and the *detail view*.

Procedure:
1: Segment reviews into sentences. And segment Sentence into words via NLPIR [10].
2: Apply heuristic rules to construct the initial training sets for *key/detail view*.
3: Train a classifier f with the initial training sets for the two views.
4: Employ f to classify the remaining sentences. And put top-confident sentences into corresponding view.
5: Repeat 4, 5 until all the reviews have been extracted *key sentence*.
6: Return *key view* and *detail view*.

Product Rule. *Product rule* is to combine base classifiers by multiplying the posterior possibilities and using the multiplied possibility for decision.

$$p_i = \prod_{f=1}^{n} p_f(c_i|x) \qquad y = arg \max_i p_i \qquad (5)$$

Sum Rule. *Sum rule* is to combine base classifiers by summing the weighted posterior possibilities and using the accumulative possibility for decision.

$$p_i = \sum_{f=1}^{n} w_f p_f(c_i|x) \qquad y = arg \max_i p_i \qquad (6)$$

4   Experiments

In this section, we will present experiments on a real-world dataset to prove the effectiveness of the method we proposed. Besides, we also discuss the sensitivity of related parameter.

4.1   Dataset Description

The dataset used in our experiments is Chinese sentiment corpus published by TanSongbo [14]. It contains a collection reviews of three products: *book, hotel,* and *NB*. So we can construct 6 cross-domain sentiment classification tasks: *book→hotel, book→NB, hotel→book, hotel→NB, NB→book, NB→hotel,* where the domain before an arrow is source domain and the domain after an arrow is target domain. For the sake of convenience, we use *ChnSentiCorp* to denote this dataset. The detail of the dataset is shown in Tab. 4.

Fig. 1 demonstrates the sentence distribution of *key view* and *detail view*. As can be seen from Fig. 1, the number of sentences in *detail view* is more than that in *key view* in three domains. Moreover, in *book* and *hotel,* the number of sentences in *detail view* is twice as many as that in *key view,* while the number

Table 4. Scale of the datasets

Domain	Positive	Neative	AveLength	AveSentence	Features
book	2000	2000	73	3.7	17517
hotel	2000	2000	72	4.4	13835
NB	2000	2000	30	2.3	6565

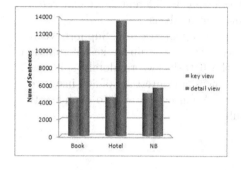

Fig. 1. Sentence number in each view

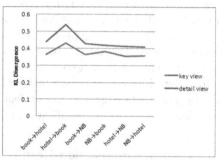

Fig. 2. KL divergence across domains

of sentences in *key view* is not much different from that in *detail view* in *NB*. Through analysis and comparison, there are a lot of sentences about storylines in *book* and details of what happened in *hotel*. However, in *NB*, there is mainly simple description about attributes of *NB*. Therefore, in *NB*, the number of sentences in two views is relatively balance than that in *book* and *hotel*. Besides, we can find that the average sentence number of *NB* is 2.3 from Tab. 4, which proved that in *NB*, the sentences in two views is more balance again.

In order to verify the distribution gap in *key view* is much smaller, we calculate the Kullback-Leibler (KL) [15] divergence of the same view in different domains. The closer two distributions are, the smaller KL divergence is. Fig. 2 shows the KL divergence curves between different domains. It can be seen that the KL divergence of *detail view* is much larger than that of *key view* overall, which reveals that the distribution among *key view*s is much closer. Besides, it coincides with our intuition that features in *key view* is mainly *domain-independent* while *domain-dependent* features usually appear in *detail view*.

4.2 Baselines Setting

In order to evaluate the effectiveness of extracting *key sentence* for cross-domain sentiment classification, we will compare our proposed method with several other methods in this sub-section.

No Transf. No Transf denotes that we train a classifier using source domain and directly apply the classifier on target domain. Just because there is not any transfer learning, this method can be considered as a lower bound.

SCL. SCL, short for structural correspondence learning, is proposed by Blitzer et al. [4]. We follow the details described in Blitzers thesis to implement *SCL*.

GraphOA. GraphOA applied the graph-ranking algorithm using the accurate labels of source domain as well as the pseudo labels of target domain for cross-domain sentiment classification.

Product Rule. In order to verify the robustness and effectiveness of our proposed view mining strategy, we employ *product rule* that is different from *sum rule* to integrate multi-view for cross-domain sentiment analysis.

For *No Transf, SCL, product rule* and *sum rule*, we use Support Vector Machine (SVM) [16] as the basic classifier. We choose linear kernel function for experiments. For all experiments on the 6 tasks, we randomly split each domain data into a training set of 3200 instances and a test set of 800 instances. We use information gain to select features. Accuracy is used to evaluate the cross-domain sentiment classification result.

4.3   Overall Comparision Results

The performances of different methods are shown in Tab. 5.

Table 5. Accuracy comparison of different methods

methods	book→hotel	hotel→book	book→NB	NB→book	hotel→NB	NB→hotel
No Transf	54.88%	60.00%	56.25%	54.88%	72.25%	72.50%
SCL	69.30%	80.42%	69.85%	65.38%	80.55%	73.80%
GraphOA	71.45%	76.08%	74.43%	79.30%	78.68%	79.80%
product rule	84.57%	79.97%	79.90%	72.90%	84.20%	86.08%
sum rule	85.98%	80.95%	80.98%	74.18%	84.88%	86.43%

Fig. 3 and Tab. 5 show the results of different methods together. Compared with *No Transf, SCL, GraphOA* and *product rule*, our method *sum rule* get significant improvements in all tasks. It is not surprising to get this result. The most intuitional reason is that capturing common information between domains can find a reasonable representation for cross-domain sentiment classification, and it will reduce the gap of distributions. It is coincident with our analysis that *No Transf* can be seen as a lower bound. Besides, *sum rule* performs better than other methods including *SCL*, which is well-known and classic method. On the one hand, the performance of *SCL* highly relies on pivot features, but heuristically selecting method might not guarantee the best performance always. On the other hand, it reveals that extraction of *key sentence* is significative for cross-domain sentiment classification. For *GraphOA*, except for the task *NB→book*, the accuracy of *sum rule* is higher than that of *GraphOA*. Moreover, average accuracy increased 5.6% as a whole. The reason might be that *GraphOA* is based on the instance similarity. However, in instance-level sentiment classification, the dataset is relatively sparse. In this case, it is hard to guarantee each test instance and all instances which have higher similarity have the same sentiment label. So, the performance of *GraphOA* is not so superior compared with the other methods. But, *GraphOA* presents its advantage on *NB→book* and *NB→hotel* tasks. Especially on the *NB→hotel* task, the performance is the best among all methods. The analysis indicates that *NB* review is quite short, and thus it becomes easy to detect the similarity with those instances in other domains accurately.

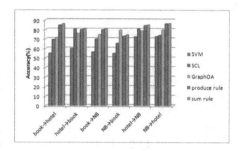
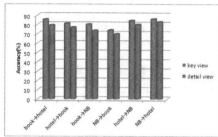

Fig. 3. Sum rule vs. other algorithms   **Fig. 4.** Performance of *key/detail view*

As for *product rule*, *sum rule* has about 1% increase. From the formula (5) and (6), we can observe that *sum rule* allocates different weights for each base classifier respectively, whereas *product rule* does not. To reduce the gap between different domains is the core aim for cross-domain sentiment classification. As proved before, distribution gap of *key view* is much smaller than that of *detail view*. Therefore, in theory *key view* should be allocated much larger weight. Above all, it is reasonable that *sum rule* is more suitable for our proposed method.

Fig. 4 shows the performance of base classifiers trained on *key view* and *detail view*. Clearly, the classifier trained on *key view* is better than that on *detail view*. Compared with *detail view*, *key view* improves about 4.8%. This coincides with the analysis that the distribution gap of *key view* is much smaller than that of *detail view*.

4.4 Parameter Sensitivity Analysis

In this part, we discuss the effect of parameter w_{key} which is the weight of *key view* on the performance of ensemble classifier.

In Fig. 5, we present the weight-dependent accuracy curves on 4 tasks. From these figures, we can see that although each set of results obtained the best integrated performance at different w_{key}, w_{key} is more than 0.5 at the optimal performance. For example, for the task *NB→hotel*, when the value of w_{key} is nearly 0.65, accuracy is the best, the value of w_{key} is about 0.75 on the task *hotel→NB* as well. Thereby this verified the conclusion that allocating much larger weight for *key view* than *detail view* will be of benefit to cross-domain sentiment classification. Clearly, *sum rule* is much more suitable for our proposed

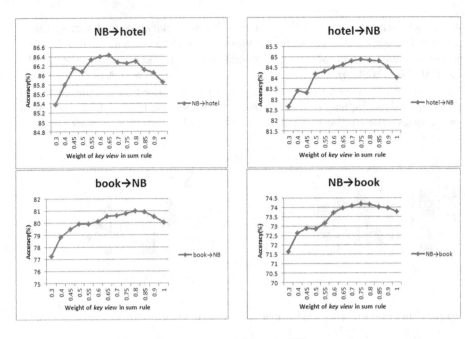

Fig. 5. Parameter sensitivity on different tasks

method. Consequently, our proposed view mining strategy based on extraction of *key sentence* is valid for cross-domain sentiment classification.

5   Conclusion

In this paper, we proposed a cross-domain sentiment classification based on extraction of *key sentiment sentence*. Firstly, we construct heuristic rules, and use them to initialize the training sets for two views, *key view* and *detail view* respectively. Secondly, machine learning is adopted to extract *key sentiment sentences* for all reviews. Thus data is divided into *key view* and *detail view*. Finally, two views are integrated effectively for cross-domain sentiment classification.

In the future, we are planning to detect deep properties of sentence for extraction of *key sentiment sentence*. In addition, multi-view ensemble strategy *product rule* does not distinguish the importance of integrated views, how to improve this rule to make it have a distinguish ability for different views is also future research content.

Acknowledgments. This work is partially supported by grant from the Natural Science Foundation of China (No. 61277370, 61402075), Natural Science Foundation of Liaoning Province, China (No.201202031, 2014020003), State Education Ministry and The Research Fund for the Doctoral Program of Higher Education (No.20090041110002), the Fundamental Research Funds for the Central Universities.

References

1. Pang, B., Lee, L., Vaithyanathan, S.: Thumbs up? sentiment classification using machine learning techniques. In: Proceedings of the ACL-02 Conference on Empirical Methods in Natural Language Processing, pp. 79–86 (2002)
2. Singh, V.K., Piryani, R., Uddin, A., Waila, P., et al.: Sentiment analysis of textual reviews; Evaluating machine learning, unsupervised and SentiWordNet approaches. In: 5th International Conference on Knowledge and Smart Technology (KST), pp. 122–127. IEEE Press (2013)
3. Mohammad, S.M., Kiritchenko, S., Zhu, X.: NRC-Canada: building the state-of-the-art in sentiment analysis of tweets. In: Proceedings of the Second Joint Conference on Lexical and Computational Semantics (SEMSTAR 2013) (2013)
4. Blitzer, J., McDonald, R., Pereira, F.: Domain adaptation with structural correspondence learning. In: Proceedings of the 2006 Conference on Empirical Methods in Natural Language Processing, pp. 120–128 (2006)
5. Bollegala, D., Weir, D., Carroll, J.: Cross-domain sentiment classification using a sentiment sensitive thesaurus. IEEE Transactions on Knowledge and Data Engineering **25**(8), 1719–1731 (2013)
6. Li, S., Xue, Y., Wang, Z., et al.: Active learning for cross-domain sentiment classification. In: Proceedings of the Twenty-Third International Joint Conference on Artificial Intelligence. AAAI Press, pp. 2127–2133 (2013)
7. Lin, Z., Tan, S., Cheng, X.: Sentiment classification analysis based on extraction of sentiment key sentence. Journal of Computer Research and Development **49**, 2376–2382 (2012)
8. Xu, L., Lin, H., Zhao, J.: Construction and analysis of emotional corpus. Journal of Chinese Information Processing **22**, 116–122 (2008)
9. Saaty, T.L.: Decision making with the analytic hierarchy process. International Journal of Services Sciences **1**(1), 83–98 (2008)
10. NLPIR. http://ictclas.nlpir.org
11. Chalothom, T., Ellman, J.: Simple Approaches of Sentiment Analysis via Ensemble Learning, pp. 631–639. Springer, Heidelberg (2015)
12. Xia, R., Zong, C.: A Pos-based Ensemble Model for Cross-domain Sentiment, pp. 195–197. Citeseer (1981)
13. Alexandre, L.A., Campilho, A.C., Kamel, M.: On combining classifiers using sum and product rules. Pattern Recognition Letters **22**, 1283–1289 (2001)
14. Chinese Sentiment Corpus. http://www.searchforum.org.cn/tansongbo/corpusenti.htm
15. Kullback, S., Leibler, R.A.: On information and sufficiency. The Annals of Mathematical Statistics, 79–86 (1951)
16. Chang, C., Lin, C.: LIBSVM: a library for support vector machines. ACM Transactions on Intelligent Systems and Technology (TIST) **2**, 27 (2011)

Convolutional Neural Networks for Correcting English Article Errors

Chengjie Sun[1]([✉]), Xiaoqiang Jin[1], Lei Lin[1],
Yuming Zhao[2], and Xiaolong Wang[1]

[1] Harbin Institute of Technology, Harbin 150001, Heilongjiang, China
cjsun@insun.hit.edu.cn
[2] Northeast Forestry University, Harbin 150040, Heilongjiang, China

Abstract. In this paper, convolutional neural networks are employed for English article error correction. Instead of employing features relying on human ingenuity and prior natural language processing knowledge, the words surrounding the context of the article are taken as features. Our approach could be trained both on an error annotated corpus and an error non-annotated corpus. Experiments are conducted on CoNLL-2013 data set. Our approach achieves 38.10 % in F1, and outperforms the best system (33.40 %) that participates in the task. Experimental results demonstrate the effectiveness of our proposed approach.

Keywords: Convolutional neural networks · Article error correction · Deep learning

1 Introduction

Grammatical Error Correction (GEC) for non-native English language learners has gained more and more attention with the developing of the Natural Language Processing (NLP), machine learning and big-data techniques [1]. Given an English essay written by a learner of English as a Second Language (L2) or English as a Foreign Language (EFL), the task of GEC is to detect and correct the grammatical errors present in the essay, and return the corrected essay.

The most representative approaches for GEC use machine learning-based classifier. However, features employed in these approaches either rely on human ingenuity or prior NLP knowledge. It takes a lot of manpower and lead to the propagation of errors in the existing tools of NLP. And different errors usually employ different features, that is to say, to correct a new type of error, features need to be extracted again. Besides, lacking of training data often prohibits a robust statistical model to be trained.

In this paper, we focus on the article error since it is one of the most difficult challenges faced by non-native speakers. A simple convolutional neural networks with one layer of convolution and pooling is employed to solve article error. The model takes words surrounding the context of article as input and outputs the

© Springer International Publishing Switzerland 2015
J. Li et al. (Eds.): NLPCC 2015, LNAI 9362, pp. 102–110, 2015.
DOI: 10.1007/978-3-319-25207-0_9

label of the {a/an, the, ϵ } representing the correct article which should be used in the context (ϵ stands for no article).

The contributions of this paper can be summarized as follows:

- Instead of employing features relying on human ingenuity and prior NLP knowledge, our approach simply takes contexts words surrounding the articles as features.
- Our approach could be trained on both an error annotated corpus and an error non-annotated corpus making it possible to learn a robust statistical model on sufficient examples of an error type.

2   Related Work

According to the approaches used, researches on GEC could be divided into five categories [2]: machine learning-based classifier approach [1,3], machine translation approach [4–6], hybrid classifier approach [7], language modeling-based approach [8] and rule-based [9] approach.

As deep learning approaches have achieved remarkable results in computer vision [10] and speech recognition [11], lots of researches have been done to explore how deep learning could be used to tasks of NLP. Word embeddings [12–14] have been one of the most successful research achievements of deep learning on NLP. In most work of deep learning on NLP, word embeddings features have been fed to a Convolutional Neural Network (CNN) to solve different tasks including semantic parsing [15], search query retrieval [16], sentence classification [17], and other traditional NLP tasks [18].

3   Model

Same as most researches on GEC, we treat the article error as a problem of multi-class classification with three labels: a/an, the, and ϵ. A convolutional neural networks (CNN) based method was proposed for this multi-class classification task. The idea behind the proposed method is that given the context of an article, the proper article may be chosen, just as a human would.

Instead of using much complicated syntactic or semantic features which may lead to the propagation of errors in the existing tools of NLP, the proposed method only takes words surrounding the article (not including the article) as features. In this way, the CNN model can be trained on both an error annotated corpus and an error non-annotated corpus. Through looking up the table of word embeddings, words are transformed into vectors. The deep feature representation of the contexts of articles are learned through CNN. Finally, the features learned are fed into a softmax classifier to compute the confidence of the each label that may occur in the given context.

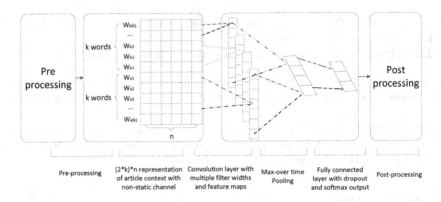

Fig. 1. Model Architecture for Article Error

3.1   Model Architecture

Figure 1 describes the architecture of the proposed article error correction method. The architecture includes the following three parts: preprocessing module, CNN module with one layer of convolution and pooling, and postprocessing module.

The preprocessing module produces the inputs for CNN module. It extracts the words surrounding the articles or the spaces at the beginning of a noun phrase (if there are no articles). The CNN module is the same as [17] except that only non-static channel is used. The postprocessing module is to discriminate when to use the article a or an by rules and to produce the output text.

3.2   Preprocessing Module

This module aims at extracting surrounding context of an article including ϵ representing not using an article. Firstly, we extract the surrounding context of a/an and the from the data. For ϵ, we get the word before the beginning of a noun phrase. If the word is not an article, we treat the surrounding context of space at the beginning of a noun phrase as ϵ's surrounding context.

A sentence can be denoted as the following format:

$\dots w_{b4}\ w_{b3}\ w_{b2}\ w_{b1}\ Art\ w_{a1}\ w_{a2}\ w_{a3}\ w_{a4}\dots$ where $Art \in \{a/an, the, \epsilon\}$, w_{bi} and w_{ai}, $i \in \{1, 2, 3, 4, \dots\}$ represent the ith word before and after Art. Let w_b^k refer to $w_{b(k)}, w_{b(k-1)}, \dots, w_{b1}$ and w_a^k refer to $w_{a1}, w_{a1}, \dots, w_{a(k)}$. The inputs of CNN module are M words $(w_b^{M/2}, w_a^{M/2})$ surrounding the article Art.

For input whose article have been annotated in the corpus, the correct one is its label, otherwise, the label is the Art.

3.3   CNN Module

This module first extracts contexts words surrounding the article. Then these words are mapped to vectors by looking up word-embeddings tables. After that,

CNN module takes the vectors of contexts words as input. New features are produced by applying many convolution operations to a window of h words in CNN module. A max over time pooling operation is applied to each feature map [18]. The idea behind this is to capture the most important feature by the maximum value. Finally, a fully connected layer with dropout is used to compute the confidence of each possible output class.

The CNN module uses non-static channel and word vectors are also taken as parameters [17]. In this way, article specific vectors could be learned through training. The optimal parameters are achieved through Stochastic Gradient Descent (SGD). And the implement of CNN is most based on the code[1] provided by [17].

3.4   Postprocessing Module

In English, there are rules to determine when to use *a* or *an* by considering the phonetic properties of word immediately after the a/an. Those rules implemented through CMU pronouncing dictionary[2], are employed to revise the output of our CNN module.

4   Dataset and Evaluation Metrics

To evaluate the performance of our approach, we use the data provided by CoNLL-2013. The training data use the NUCLE Corpus [19] and has been annotated with error-tag and correction-labels. In the shared task, 25 non-native speakers of English from NUS were recruited to write new essays to be used as blind test data. The statistics of the NUCLE corpus are shown in Table 1.

Table 1. Statistics of training and test data

	Train(NUCLE)	Test
#Essays	1,397	50
#Sentences	57,151	1,381
#Word Tokens	1,161,567	29,207
#ArtOrDet* Error	6,642	690

* ArtOrDet is short for articles and determiners.

The performance of a grammatical error correction system is evaluated by how well its proposed corrections or edits match the gold-standard edits. Precision, recall and F-score are often chosen as the evaluation criteria. The test data and official scorer (M^2 scorer [20]) provided by CoNLL-2013 are freely available[3].

---

[1] https://github.com/yoonkim/CNN_sentence
[2] In this paper, we use the interface provided by Natural Language Took Kit (NLTK).
[3] http://www.comp.nus.edu.sg/~nlp/conll13st.html

5   Experiments

In this section, two sets of experiment are conducted. One is to understand how the choice of hyperparameters affects the performance on development data set which was obtained by holding out 20% of the training data. The other evaluates the final performance of our approach on the test data of CoNLL-2013.

5.1   Pre-trained Word Embeddings

Researchers [18,21] have reported that initializing word embeddings with those learned from significant amounts of unlabeled data are far more satisfactory than the randomly initialized.

In this paper, we do not conduct a comparison of the available word embeddings, for it's beyond the scope of this paper. Embeddings provided by [22] are utilized in our experiments to initialize word embeddings table. Words not present in the set of pre-trained words are initialized randomly, which is same as [17] does.

5.2   Parameter Settings

We experimentally study the effects of the two parameters in our model: window size of error contexts and the number of feature map.

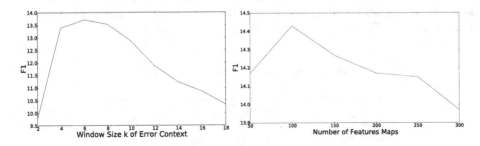

Fig. 2. Effects of Hyperparameters

In Figure 2, parameters about window size of error contexts and number of feature map are respectively varied. As shown in Figure 2 , the performance does not improve when the window size k of error contexts is larger than 6. Since the training data is limited, the model is prone to overfitting especially when the number of feature map (f) exceeds 100. The dimensionality of word embeddings (n) is the same as in [18]. The filter windows (h), dropout rate (p), l_2 constraint (s), and mini-batch size (m) are following [17]'s experiment settings. Table 2 lists the parameters used in the following experiments.

Table 2. Hyperparameter Settings

Parameter	k	n	h	p	s	m	f
Value	6	50	3,4,5	0.5	3	50	100

5.3 Experiment Results

Table 3 shows the results of article error correction of the top three systems in CoNLL-2013 [2] open evaluation and our approach. Besides, the linguistic features used in each systems are also given.

Table 3. The Article Result of Top3 in CoNLL-2013 and Our Approach

Team	Precision	Recall	F1	Linguistic Features
UIUC	47.84	25.65	33.40	lexical, POS, shallow parse
HIT	42.82	24.20	30.93	lexical, POS, constituency parse, dependency parse, semantic
NTHU	35.80	21.01	26.48	lexical, POS, constituency parse, dependency parse
Ours	30.15	51.74	38.10	lexical

Team UIUC employed a multi-class averaged perception for article error correction. Maximum entropy with confidence tuning was used for article error correction in team HIT. In team NTHU, N-gram-based and dependency-based language model was employed.

Our approach performs better than UIUC, HIT and NTHU in recall and F1, though only taking lexical as features. However, the precision of our approach is much lower. One possible reason is that source words are not used in our CNN model. Previous works have showed that the authors' word choices (source word) obey certain regularities [23, 24] and systems without employing the source word have a very poor precision [25]. When the source word is directly utilized as feature, model tends to have a low recall due to the error sparsity. [25] proposed an error inflation method to avoid this problem by adding artificial errors in train data based on the error distribution in the train set. We try to explain the low precision by adopting error inflation method to our model.

Figure 3 shows the result of effects of source word on development data set. The results show that source word could balance the recall the precision. When the inflation constant (C) is set to 0.7, F1 get the best result on the development set. The result on test is shown in Table 4. We fail to find the best trade-off between recall and precision on test set due to different distribution of error between train data and test data, but shed some light on why CNN model have a low precision.

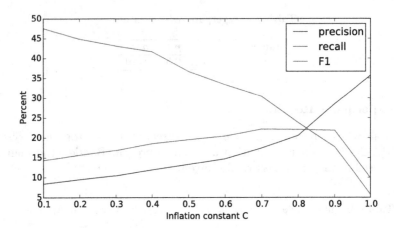

Fig. 3. Effects of Source Word

Table 4. CNN with Source Word

Precision	Recall	F1	C
45.10	22.03	29.60	0.7

6   Conclusion and Feature Work

In this paper, we exploit a convolutional neural network for English article error correction. Instead of employing features relying on human ingenuity and prior NLP knowledge, our approach simply takes words surrounding the contexts of articles as features. Though simple features are employed, experimental results conducted on CoNLL-2013's data set demonstrate the effectiveness of our approach. Besides, our approach could be trained on both an error annotated corpus and an error non-annotated corpus.

In the future, an effective post processing module determining whether to accept the correction will be explored to solve the drawback of low precision. The effectiveness of our approach on other error types will also be explored.

Acknowledgments. We thank reviewers for their helpful comments on an earlier version of this work. This work is supported by National Natural Science Foundation of China (61300114 and 61272383) and Natural Science Foundation of Heilongjiang Province(F201132).

References

1. Xiang, Y., Yuan, B., Zhang, Y., Wang, X., Zheng, W., Wei, C.: A hybrid model for grammatical error correction. In: Proceedings of the Seventeenth Conference on Computational Natural Language Learning: Shared Task, pp. 115–122 (2013)

2. Ng, H.T., Wu, S.M., Wu, Y., Hadiwinoto, C., Tetreault, J.: The conll-2013 shared task on grammatical error correction. In: Proceedings of the Seventeenth Conference on Computational Natural Language Learning: Shared Task, pp. 1–12 (2013)
3. Rozovskaya, A., Chang, K.W., Sammons, M., Roth, D.: The university of illinois system in the conll-2013 shared task. In: Proceedings of the Seventeenth Conference on Computational Natural Language Learning: Shared Task, pp. 13–19 (2013)
4. Yuan, Z., Felice, M.: Constrained grammatical error correction using statistical machine translation. CoNLL-2013, pp. 52–61 (2013)
5. Buys, J., van der Merwe, B.: A tree transducer model for grammatical error correction. CoNLL-2013, pp. 43–51 (2013)
6. Wilcox-OHearn, L.A.: A noisy channel model framework for grammatical correction. CoNLL-2013, pp. 109–114 (2013)
7. Xing, J., Wang, L., Wong, D.F., Chao, L.S., Zeng, X.: Um-checker: A hybrid system for english grammatical error correction. CoNLL-2013, 34 (2013)
8. Kao, T.H., Chang, Y.W., Chiu, H.W., Yen, T.H., Boisson, J., Wu, J.c., Chang, J.: Conll-2013 shared task: Grammatical error correction nthu system description. CoNLL-2013, 20 (2013)
9. Sidorov, G., Gupta, A., Tozer, M., Catala, D., Catena, A., Fuentes, S.: Rule-based system for automatic grammar correction using syntactic n-grams for english language learning (l2). CoNLL-2013, pp. 96–101 (2013)
10. Krizhevsky, A., Sutskever, I., Hinton, G.E.: Imagenet classification with deep convolutional neural networks. In: Advances in Neural Information Processing Systems, pp. 1097–1105 (2012)
11. Graves, A., Mohamed, A.R., Hinton, G.: Speech recognition with deep recurrent neural networks. In: 2013 IEEE International Conference on Acoustics, Speech and Signal Processing (ICASSP), pp. 6645–6649. IEEE (2013)
12. Bengio, Y., Ducharme, R., Vincent, P., Janvin, C.: A neural probabilistic language model. The Journal of Machine Learning Research 3, 1137–1155 (2003)
13. Yih, W.T., Toutanova, K., Platt, J.C., Meek, C.: Learning discriminative projections for text similarity measures. In: Proceedings of the Fifteenth Conference on Computational Natural Language Learning, Association for Computational Linguistics, pp. 247–256 (2011)
14. Mikolov, T., Sutskever, I., Chen, K., Corrado, G.S., Dean, J.: Distributed representations of words and phrases and their compositionality. In: Advances in Neural Information Processing Systems, pp. 3111–3119 (2013)
15. Yih, W.t., He, X., Meek, C.: Semantic parsing for single-relation question answering. In: Proceedings of ACL (2014)
16. Shen, Y., He, X., Gao, J., Deng, L., Mesnil, G.: Learning semantic representations using convolutional neural networks for web search. In: Proceedings of the Companion Publication of the 23rd International Conference on World Wide Web Companion, International World Wide Web Conferences Steering Committee, pp. 373–374 (2014)
17. Kim, Y.: Convolutional neural networks for sentence classification (2014). arXiv preprint, arXiv:1408.5882
18. Collobert, R., Weston, J., Bottou, L., Karlen, M., Kavukcuoglu, K., Kuksa, P.: Natural language processing (almost) from scratch. The Journal of Machine Learning Research 12, 2493–2537 (2011)
19. Dahlmeier, D., Ng, H.T., Wu, S.M.: Building a large annotated corpus of learner english: The nus corpus of learner english. In: Proceedings of the Eighth Workshop on Innovative Use of NLP for Building Educational Applications, pp. 22–31 (2013)

20. Dahlmeier, D., Ng, H.T.: Better evaluation for grammatical error correction. In: Proceedings of the 2012 Conference of the North American Chapter of the Association for Computational Linguistics: Human Language Technologies, Association for Computational Linguistics, pp. 568–572 (2012)

21. Socher, R., Pennington, J., Huang, E.H., Ng, A.Y., Manning, C.D.: Semi-supervised recursive autoencoders for predicting sentiment distributions. In: Proceedings of the Conference on Empirical Methods in Natural Language Processing, Association for Computational Linguistics, pp. 151–161 (2011)

22. Turian, J., Ratinov, L., Bengio, Y.: Word representations: a simple and general method for semi-supervised learning. In: Proceedings of the 48th Annual Meeting of the Association for Computational Linguistics, Association for Computational Linguistics, pp. 384–394 (2010)

23. Rozovskaya, A., Roth, D.: Annotating esl errors: Challenges and rewards. In: Proceedings of the NAACL HLT 2010 Fifth Workshop on Innovative Use of NLP for Building Educational Applications, Association for Computational Linguistics, pp. 28–36 (2010)

24. Lee, J., Seneff, S.: An analysis of grammatical errors in non-native speech in english. In: Spoken Language Technology Workshop, 2008. SLT 2008, pp. 89–92. IEEE (2008)

25. Rozovskaya, A., Sammons, M., Roth, D.: The UI system in the hoo 2012 shared task on error correction. In: Proceedings of the Seventh Workshop on Building Educational Applications Using NLP, Association for Computational Linguistics, pp. 272–280 (2012)

NLP for Social Media

Automatic Detection
of Rumor on Social Network

Qiao Zhang[1,2], Shuiyuan Zhang[1,2], Jian Dong[3],
Jinhua Xiong[2]([✉]), and Xueqi Cheng[2]

[1] University of Chinese Academy of Sciences, Beijing, China
[2] Institute of Computing Technology, Chinese Academy of Sciences, Beijing, China
xjh@ict.ac.on
[3] The Third Research Institute of Ministry of Public Security, Shanghai, China

Abstract. The rumor detection problem on social network has
attracted considerable attention in recent years. Most previous works
focused on detecting rumors by shallow features of messages, including content and blogger features. But such shallow features cannot distinguish between rumor messages and normal messages in many cases.
Therefore, in this paper we propose an automatic rumor detection
method based on the combination of new proposed implicit features
and shallow features of the messages. The proposed implicit features
include popularity orientation, internal and external consistency, sentiment polarity and opinion of comments, social influence, opinion retweet
influence, and match degree of messages. Experiments illustrate that our
rumor detection method obtain significant improvement compared with
the state-of-the-art approaches. The proposed implicit features are effective in rumor detection on social network.

1 Introduction

With the development of social network, the amount of information has been
growing explosively. However, the quality of information does not become better. All kinds of false information, especially rumor information, have permeated
almost every corner of social networks. Therefore, automatic assessment of information credibility has received considerable attention in recent years.

Rumor detection is one of the critical research topics of information credibility. It is often viewed as a tall tale of explanations of event circulating from
person to person and pertaining to an object, event, or issue in public concern
[6]. The diffusion of rumor is harmful to people's lives and the stability of the
society, and it has become a serious concern of social network. Rumor detection
is usually modeled as a classification problem based on shallow features of messages, including content and blogger features. But such shallow features cannot
distinguish between rumor messages and normal messages in many cases.

In this paper, we also formulate rumor detection as a binary classification
problem. and propose an automatic rumor detection classification method based
on the combination of new proposed implicit features and shallow features of the

© Springer International Publishing Switzerland 2015
J. Li et al. (Eds.): NLPCC 2015, LNAI 9362, pp. 113–122, 2015.
DOI: 10.1007/978-3-319-25207-0_10

messages. Shallow features are usually extracted from basic attributes of user or content, while implicit features are generated by mining the deep information of user or content. The implicit features are the most innovative part of the paper. They are obtained by analyzing the popularity, sentiment or viewpoint of message contents and user historical information, including popularity orientation, internal and external consistency, sentiment polarity and opinion of comments, social influence, opinion retweet influence, and match degree of messages. Experiments illustrate that our rumor detection method obtain significant improvement, compared with the state-of-the-art approaches. The proposed implicit features are effective, and make more contribution to rumor detection on social network compared to shallow features.

The rest of this paper is organized as follows: Section 2 gives an overview of related works. Section 3 describes our method to rumor detection, especially the process of analyzing and extracting shallow and implicit features. Section 4 presents our experiments. The last section draws a conclusion.

2   Related Works

There are a large number of related studies on rumor detection. Recently, rumor detection method is mainly based on supervised learning. In other words, it formulates the problem of rumor detection as a classification problem. One key factor of classification model is features, for determine the upper bound of rumor detection performance. Therefore, feature extraction is a critical step of detecting rumors accurately.

Currently, related studies focus on extracting useful and efficiency features for rumor detection. Generally speaking, features for rumor detection can be divided into four types: (1) content-based features; (2) user-based features; (3) propagation-based features; (4) other-based features.

For content-based features, Ratkiewicz et al. (2010) [8] created the Truthy system, identifying misleading political memes on Twitter using content-based features, including hashtags, links and mention.Qazvinian et al. (2011) [7] applied unigrams, bigrams and pos-tagging results to detect rumors. Takahashi et al. (2012) [10] found that vocabulary distribution of rumor messages are different from non-rumor messages, so they computed the ratio of the number of rumor and non-rumor messages vocabulary words, as one of the features to detect rumors.

For the user-based features, Castillo et al. (2011) [2] used registration age, number of user posted messages, number of followers, number of friends and other attributes of users to detect rumors. Al-Khalifa et al. (2011) [1] and Gupta et al. (2012) [3] also used some attributes of users as features.

As to the propagation-base features, Men-doza et al. (2010) [5] analyzed the retweet network topology and found that the diffusion patterns of rumors are different from news, and found that rumors tend to be questioned more than news by the Twitter community. Kwon et al. (2013) [4] discovered that rumor tweets had more cycle volatility, compared with non-rumor tweets. And they proposed PES (Periodic External Shocks) model to detect rumors. Wu et al. (2015) [11] introduced the propagation tree, and used random walk kernel algorithm to build rumor detection classifiers.

Finally, some other-based features are put forward. Yang et al. (2012) [12] proposed the client program that user has used to post a microblog and the actual place where the event mentioned by the messages has happened. Sun et al. (2013) [9] used the multimedia features of pictures in messages to detect event rumors on Sina microblog.

However, none of these work considered the implicit features of contents and users. In this paper, We propose some innovative features by analyzing the popularity, sentiment or viewpoint of message contents and user historical information, and build a effective classifier to detect rumors.

3 Proposed Method

We formulate rumor detection as a classification problem. For a given message, features are extracted first from different aspect of view, then we will use a classifier to determine whether this message is a rumor. Features are critical in our method, and we focus on implicit features of contents and users. In this section, we will introduce our general process of rumor detection and some key features that contribute to rumor detection a lot.

3.1 Rumor Detection Flow

Our proposed rumor detection method is a typical classification problems, and it mainly contains 3 parts which are data cleaning, feature extraction and model training.

There are a lot of spam message, these message will cause interference to our approach. In the data cleaning process, we filter out some spam message such as message which only contains URL or punctuation.

Feature extraction is a key step in our method, and we focus on features extracted from message contents and users. Contents and users are two key factors of a message, patterns of these two factors for rumors are obviously different from that of normal messages. We identify a set of implicit features based on contents and users, and these features make a great contribution to detect rumors. Also, we combine some features which have been studied in previous work, including shallow text features of contents and basic attribute features of users.

After feature extraction, a classifier model will be trained using the extracted features. A large amount of supervised model can be used such as Support Vector Machine, Random Forest.

3.2 Content-Based Implicit Features

Popularity Orientation refers to the relevance of message content and the current social hot topics or events. Since many contents of rumors are associated with the current hot topics, popularity is a valuable feature for rumor detection. Popularity orientation of content is defined as Equation 1.

$$Popularity\_Orientation = max\left(simi(W, T_1), simi(W, T_2), ..., simi(W, T_m)\right) \quad (1)$$

Where W means the keywords set of message, and T_i means a certain category of popular topic words. $simi(W, T_i)$ means the Jaccard similarity between W and T_i .

Internal and External Consistency refers to the correlation between the message content and the content of the corresponding external page. The more relevant they are, the less likely the message is rumor. Internal and external consistency is defined as Equation 2.

$$Internal\_External\_Consistency =$$
$$\begin{cases} 0, T\,notcontainURL \\ max(Rel(T, title), Rel(T, description), Rel(T, keywords), T\,containURL) \end{cases} \quad (2)$$

Where $Rel(T, title)$ means the Jaccard similarity between message T and title of external page.

Sentiment Polarity refers to the sentiment polarity of messages. Subjective information is a key factor of one message, contents of rumors are usually exaggerated, extreme words such as "disfigure", "poisonous" often appear in rumors. In order to get the sentiment polarity of messages, we use classification method together with some dictionary corpus to classify the messages.

In the process of classifying sentiment polarity of messages, we modify the traditional TF-IDF, and propose the TF-FW to compute item weight based on different types of dictionary. FW is defined as Equation 3.

$$FW_w = log_2(level_w + 1) \quad (3)$$

Where $level_w$ means the level of item w and its definition can be found in Table 1.

Table 1. Item level distribution table

Item	Level
Items in Sentiment Dictionary	5
Items in Emotional Dictionary	4
Items in Sensitive Dictionary	3
Items in Punctuation Dictionary	2
Other Items	1

The weight of k-th term can be calculated by Equation 4.

$$weight_k = \frac{(\log(f_k) + 1.0) \times \log_2(level_k + 1)}{\sqrt{\sum_{k=1}^{l} [(\log(f_k) + 1.0) \times \log_2(level_k + 1)]^2}} \tag{4}$$

Where f_k means the frequency of current term in message, l means the number of terms in message, and $level_k$ means the level of current term, which can be computed by Equation 3.

Then the sentiment polarity can be obtained by text classification, we classify the sentiment polarity of a message into three types including positive, negative and neutral.

Opinion of Comments refers to the degree of acceptance of the comments in message. A large number of doubtful and inquiring comments, such as "Gab", "Fake", always appear in rumor messages. Therefore, opinion analysis carried out on the comments can be used to obtain the credibility of the message. In this paper, we first get the opinion polarity of comment via classification method, the classification process is the same with that of sentiment polarity. Then the opinion of comments is defined as Equation 5.

$$Opinion\_Of\_Comments = \log \frac{N_{pos}}{N_{neg}} \tag{5}$$

Where N_{pos} means the number of comments supported, and N_{neg} means the number of comments nonsupport.

3.3 User-Based Implicit Features

Users are the core of social network. In the process of rumor propagation, user is not only the producer of rumors, but also is the disseminator of rumors. We use social influence to measure the impact of one user in the information diffusion process.

Social Influence refers to the communicative influence of one user on social network. It has a great relationship with the number of followers and friends of this and is defined as Equation 6.

$$Social\_Influence = \log \left(\frac{fol\_num - bi\_fol\_num}{fri\_num + 1} \right) \tag{6}$$

Where fol_num means the number of followers and fri_num means the number of friends, and bi_fol_num means the number of both followers.

Opinion Retweet Influence refers to the degree of acceptance by other users of the user's opinion. Generally speaking, the users with higher degree of acceptance by others, the lower the probability of publishing rumors. It can be calculated by Equation 7.

$$Opinion\_Retweet\_Influence = \frac{retweets\_num}{statuses\_num} \tag{7}$$

Where $retweets\_num$ means the total retweet number of user's all messages, and $statuses\_num$ means the total number of user's messages.

Match Degree of Messages refers to the match degree of the user's professional orientation and his message contents. The subject of rumors often serious discrepancies with the theme of the users historical content. We use topic model to get the distribution of user's historical contents and current message content. These distributions are viewed as a mixture of various topics, we use the match degree of historical and current messages' topic distribution to measure the the feature. It is defined by Equation 8 as follows.

$$
\begin{aligned}
Match\_Degree\_of\_Messages &= cosin\_simi(his\_topic, cur\_topic) \\
&= \frac{his\_topic \times cur\_topic}{|his\_topic| \times |cur\_topic|}
\end{aligned} \tag{8}
$$

Where his_topic and cur_topic are the topic distributions of the user's historical and current messages respectively.

3.4   Feature Fusion

In this part, we also use some shallow features that have been proposed in previous work to strengthen our model, and merge these features with implicit features proposed in this paper.

For content-based feature,features we extract are listed in Table 2.

For user-based feature, we choose some attributes of user profile, these are listed in Table 3.

4   Experiments

4.1   Dataset

We use the method proposed by Yang et al. (2012) [12] to collect Sina Weibo rumor messages from Sina Weibo Community Management Center. The crawled microblogs are used as out dataset, and it contains 3229 rumor microblogs, and 12534 non-rumor microblogs from the same time span. Two-thirds of them are used as training set, and the rest as test set.

Table 2. Fusion Features of Contents

Category	Features	Description
Shallow Text Features	Time_Span	The time interval between the time of posting and user registration
	Has_URLs	Whether the message includes a URL pointing to an external source
	Has_Multimedia	Whether the messages contains picture, videos, or audios
	Has_Refer	Whether the messages include '@' referring to others
implicit Content Features	Popularity Orientation	The popularity orientation of the message
	Internal and External Consistency	The internal and external consistency of the message
	Sentiment Polarity	The sentiment polarity of the message
	Opinion of Comments	The opinion of comments of the message

Table 3. Fusion Features of Users

Category	Features	Description
Basic Attribute Features	Verify_Type	The verify type of the user
	Gender	The gender of the user
	Has_Description	Whether the user has personal description
	Has_Profile_URL	Whether the user has profile URL
	Has_Domain	Whether the has domain
	Post_Num	The number of messages posted by the user
	User_Activity	The activity of the user
	Favorite_Num	The number of favorite messages
	Fans_Rate	The number of user's followers
	Friends_Rate	The number of user's friends
	Bi_Follow_Num	The number of users' binary followers
implicit User Features	Social_Influence	The user's social influence
	Opinion Retweet Influence	The user's opinion retweet influence
	Match Degree of Messages	The user's match degree of messages

4.2 Evaluation

To evaluate the performance of our methods, we use the standard information retrieval metrics of precision, recall and F1 [13]. The precision is the ratio of the number of rumors classified corrected to the total number of microblogs predicted as rumors. The recall is the ratio of the number of rumors classified correctly to the total number of true rumors. The F1 is a comprehensive assessment of precision and recall rate, and it is defined as Equation 9.

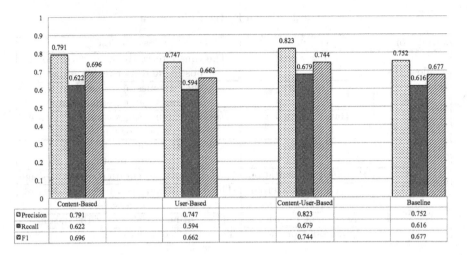

	Content-Based	User-Based	Content-User-Based	Baseline
Precision	0.791	0.747	0.823	0.752
Recall	0.622	0.594	0.679	0.616
F1	0.696	0.662	0.744	0.677

Fig. 1. Rumor detection results with SVM

$$F1 = \frac{2 \times precision \times recall}{precision + recall} \tag{9}$$

We use the method of Yang et al. (2012) [12] as baseline, and train a Support Vector Machine classifier with our proposed features. We conduct three sets of experiments to better understand the impact of different classification method and implicit features on rumor detection. We use Content-Based, User-Based, Content-User-Based and Baseline to indicate different features used in rumor detection.

The experimental results in Figure 1 show that the method combined with content and user features is better than others, with 7.1% improvement in precision and 6.3% improvement in recall rate combined with baseline. The reason is that in our method, we merge the implicit content and user features to improve the effectiveness on detecting rumors.

In order to further assess the effectiveness of implicit features proposed by this paper, we use shallow text features, implicit content features, user attribute features and implicit user features alone to detect rumors, and denote as Shallow-Content-Based, Implicit-Content-Based, Shallow-User-Based and implicit-User-Based respectively. Figure 2 is the result of using different types of features above.

The experimental results show that Implicit-Content-Based method have significant improvement compared with Shallow-Content-Based method, with 10.5% improvement in precision and 4.7% in recall rate. The main reason is that the implicit content features have better identification of rumor detection. As to the implicit-User-Based method, it also achieves better performance than Shallow-User-Based, and fully verified the effectiveness of the implicit features we proposed.

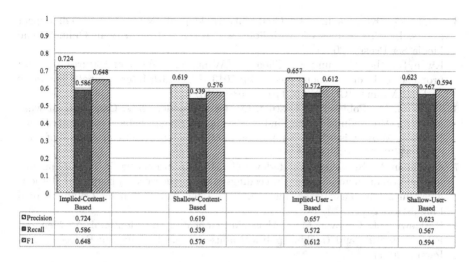

Fig. 2. RumorRumor Detection with Different Types of Features

5 Conclusion

In this paper we focus on detecting rumor on social network. To distinguish rumors from normal messages, we propose a rumor detection method based on implicit features of contents and users.

In the feature engineering process, we introduce some implicit features based on the characteristic of rumors. These features focus on popularity orientation, internal and external consistency, sentiment polarity and opinion of comments, social influence, opinion retweet influence, and match degree of messages.

User credibility is an important factor that impact information credibility, so the analysis of user credibility help detect message credibility. In the future, we would like to do some work on user credibility, and use this to improve the performance of rumor detection.

Acknowledgments. This research was supported by the National High Technology Research and Development Program of China (Grant No. 2014AA015204), the National Basic Research Program of China (Grant No. 2014CB340406), the NSFC for the Youth (Grant No. 61402442) and the Technology Innovation and Transformation Program of Shandong (Grant No.2014CGZH1103).

References

1. AlKhalifa, H.S., AlEidan, R.M.: An experimental system for measuring the credibility of news content in Twitter. International Journal of Web Information Systems 7(2), 130–151 (2011)
2. Castillo, C., Mendoza, M., Poblete, B.: Information credibility on twitter. In: Proceedings of the 20th, pp. 675–684 (2011)

3. Gupta, A., Kumaraguru, P.: Credibility ranking of tweets during high impact events. In: Proceedings of Workshop on Privacy and Security in Online Social Media Ser Psosm (2012)
4. Kwon, S., Cha, M., Jung, K., Chen, W., Wang, Y.: Prominent features of rumor propagation in online social media. In: 2013 IEEE 13th International Conference on Data Mining (ICDM), pp. 1103–1108 (2013)
5. Mendoza, M., Poblete, B., Castillo, C.: Twitter under crisis: Can we trust what we rt? In: Proceedings of the First Workshop on Social Media Analytics (2010)
6. Peterson, W.A., Gist, N.P.: Rumor and public opinion. American Journal of Sociology **57**(2), 159–167 (1951)
7. Qazvinian, V., Rosengren, E., Radev, D.R., Mei, Q.: Rumor has it: Identifying misinformation in microblogs. In: Proceedings of the Conference on Empirical Methods in Natural Language Processing, pp. 1589–1599. Association for Computational Linguistics (2011)
8. Ratkiewicz, J., Conover, M.D., Meiss, M., Goncalves, B., Flammini, A., Menczer, F.M.: Detecting and tracking political abuse in social media. In: Proceedings of Icwsm (2011)
9. Sun, S., Liu, H., He, J., Du, X.: Detecting event rumors on sina weibo automatically. In: Ishikawa, Y., Li, J., Wang, W., Zhang, R., Zhang, W. (eds.) APWeb 2013. LNCS, vol. 7808, pp. 120–131. Springer, Heidelberg (2013)
10. Takahashi, T., Igata, N.: Rumor detection on twitter. In: 13th International Symposium on Advanced Intelligent Systems (ISIS), 2012 Joint 6th International Conference on Soft Computing and Intelligent Systems (SCIS), pp. 452–457 (2012)
11. Wu, K., Yang, S., Zhu, K.Q.: False rumors detection on sina weibo by propagation structures. In: IEEE International Conference on Data Engineering, ICDE (2015)
12. Yang, F., Liu, Y., Yu, X., Yang, M.: Automatic detection of rumor on sina weibo. In: Proceedings of the ACM SIGKDD Workshop on Mining Data Semantics (2012)
13. Yang, Y.: An evaluation of statistical approaches to text categorization. Information Retrieval **1**(1–2), 69–90 (1999)

Multimodal Learning Based Approaches for Link Prediction in Social Networks

Feng Liu[✉], Bingquan Liu, Chengjie Sun, Ming Liu, and Xiaolong Wang

School of Computer Science and Technology,
Harbin Institute of Technology, Harbin, China
{fengliu,liubq,cjsun,mliu,wangxl}@insun.hit.edu.cn

Abstract. The link prediction problem in social networks is to estimate the value of the link that can represent relationship between social
members. Researchers have proposed several methods for solving link
prediction and a number of features have been used. Most of these models are learned with only considering the features from one kind of data.
In this paper, by considering the data from link network structure and
user comment, both of which could imply the concept of link value, we
propose multimodal learning based approaches to predict the link values. The experiment results done on dataset from typical social networks
show that our model could learn the joint representation of these datas
properly, and the method MDBN outperforms other state-of-art link prediction methods.

Keywords: Link prediction · Social networks · Multimodal learning ·
Deep Belief Networks · Support vector classifier

1 Introduction

With the explosive growth of SNS(Social Network Services) websites, there are
large scale data of social media [3]. The mass data includes the interactions
among social members, such as comments and links. The comment is always a
short paragraph with only one or a few sentences, which are sent from one user
to another. The link is usually a label with sign value that represent one user's
certain kind of opinion to another, such as expressing support or oppose.

Taking social members as vertexes and links as directed edges between them,
the link network can be represented as a graph. The classical link prediction
is the problem of predicting the existence of a link between two entities, based
on attributes of the objects and other observed links [6,13]. In this paper, the
predicting task based on link network structure, as shown in Fig. 1(a), is to
predict the relation of one user toward another from the evidence provided by
their relations with other members from the surrounding social network.

Many state-of-art link prediction methods are based on machine learning
models [20]. The machine learning based methods treat the link prediction problem as a classification task. Logistical Regression model is used to classify the

© Springer International Publishing Switzerland 2015
J. Li et al. (Eds.): NLPCC 2015, LNAI 9362, pp. 123–133, 2015.
DOI: 10.1007/978-3-319-25207-0_11

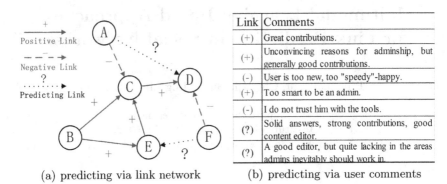

(a) predicting via link network   (b) predicting via user comments

Fig. 1. The Link Prediction Problem.

link values in [11,12]. Support vector machine is used to analyse how link network structure features effect link's values in [17]. Deep belief network based approaches for link prediction are introduced in [15,16]. However, these methods only used the data from link network structure.

The sentiment analysis or opinion mining is the computational study of people's opinions and attitudes toward entities and individuals [14]. Taking the 'link' as one user's opinion or attitude to another user, the link prediction task is some kind like the sentiment analysis or opinion mining as shown in Fig. 1(b). User comments are used in [2,5,21] to improve the performance of link prediction. It shows that the link prediction problem could be solved by using user comments with opinion mining methods.

As introduced in [19], a good multimodal learning model must satisfy certain properties. The joint representation must be such that similarity in the representation space implies similarity of the corresponding concepts. It is also desirable that the joint representation be easy to obtain even in the absence of some modalities. It should also be possible to fill-in missing modalities given the observed ones. In addition, it is also desirable that extracted representation be useful for discriminative tasks. Multimodal learning model focus on learning representations for speech audio which are coupled with videos of the lips is introduced in [18]. And multimodal models for image and text described the image are introduced in [10,22].

Assuming that one user have given good comments to the other, that user would be more likely to give a positive link. While the one who got bad comment are more likely to get a negative link. As a result, by learning the joint representation between the comments and links properly, we could predict the sign value of user links with high accuracy. At that same time, this joint representation could help to classify the polarity of user comment. This research would benefit link prediction and sentiment analysis, both of which are hot topic in study of social computing.

For the link prediction problem in this paper, the predicting via link network shown in Fig. 1(a), and predicting via user comments shown in Fig. 1(b), they imply similarity of the corresponding 'concept' as link value. This inspires us to use multimodal learning model to get the joint representation of link network features and user comments. And we assume that joint representation could improve the performance of link prediction based on only one kind of data. The multimodal learning based approaches for link prediction are proposed in this paper. The experiment results show that our method works well and the performance of link prediction is improved.

2 Background

In this section, we review the basic principles of RBM and DBN, which is used as a layer-wise building block for our models.

2.1 Restricted Boltzmann Machine

A Restricted Boltzmann Machine(RBM) is a neural network that contains two layers. It has a single layer of hidden units h that are not connected with each other. And the hidden units have undirected, symmetrical connections w to a layer of visible units v. As shown in Fig. 2(a), $\{V_1, W_1, H_1\}$ constructs a RBM. The model defines a probability distribution over v, h as

$$- \log P(v,h) \propto E(v,h) = - \sum_i v_i a_i - \sum_j h_j b_j - \sum_{i,j} v_i w_{ij} h_j \qquad (1)$$

where a_i is the bias of visible unit i, and b_j is the bias of hidden unit j. When input a vector v $(v_1, v_2...v_i...)$ to the visible layer, the binary state h_j of each hidden unit is set to 1 with probability by

$$p(h_j = 1|v) = \varphi(b_j + \sum_i v_i w_{ij}) \qquad (2)$$

where $\varphi(x) = 1/(1 + e^{-x})$. When input a vector h $(h_1, h_2, ...h_j...)$ to the hidden layer, the binary state v_i of each visible unit is set to 1 with probability by

$$p(v_i = 1|h) = \varphi(a_i + \sum_j h_j w_{ij}) \qquad (3)$$

The parameters $\{w, a, b\}$ are usually trained by using the Contrastive Divergence(CD) learning procedure, which is introduced in [1,7].

2.2 Deep Belief Network

The Deep Belief Network(DBN) is a multilayer, stochastic generative model that is created by learning a stack of RBMs, as shown in Fig. 2. It can be understood as only one RBM may not have enough abstracting ability to solve some complex problems for there are only two layers in one RBM. And the two layers can be though as transforming the input(visible layer) into another space(hidden layer)

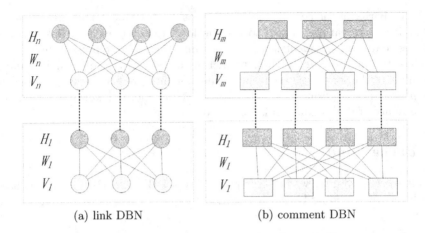

(a) link DBN   (b) comment DBN

Fig. 2. The Structure of DBN.

only once, so one RBM's ability is limited. One DBN is built up with a stack of RBMs that could have more abstracting ability, because each layer of RBM in DBN can make a space transformation [9].

When learning a DBN, only the first RBM is trained on the original samples by CD learning procedure. Then the second RBM is trained by the first RBM's hidden activation vectors which are generated from the original samples. Do that iteratively until the top RBM is learned. This greedy, layer-by-layer learning can be repeated as many times as desired [8]. If a sample vector is imputed to the first RBM of that DBN, the highly abstracted vector of that sample would be gotten from the top RBM's hidden layer.

3   Link Prediction Problem

In this paper, the link prediction problem is defined as follows. Taking the whole network as a directed graph $G = (V, E)$, V is the set of users and E is the set of edges. Each edge linking two nodes has a sign value(either positive or negative). Supposing there are two nodes u and v and an edge linking from u to v. Denote that edge as $e(u, v)$ and assume the sign value of $e(u, v)$ is "lost". Supposing there is a sub graph G', whose edges have the same assumption as $e(u, v)$. Meanwhile, the sign values of edges in $G - G'$ are known. We infer the sign value of edges in G' by using the information from the structure of G and the patterns of link values from $G - G'$. For illustration, in Fig.1(a), a small part of the whole SSN is shown to illustrate the link prediction problem.

4   Methodology

In this section, we describe how to extract features from link networks and user comments. Then we introduce how to learn discriminative DBN for predicting link values and the method of MDBN.

4.1   Link Network Structure Features

The link network could be represented as a directed graph $G = (V, E)$, where V is the set of users and E is the set of edges, and each edge e has a link value (either positive or negative) as shown in Fig. 1(a). The edge directly linking from user u to v is denoted as $e(u, v)$. And denote $Ne(u)$ as the set of u's neighbour nodes and $CNe(u, v)$ as the common neighbours shared with u and v. There are total 26 link network structure features, including node features and neighbour features.

The node features for u is counting the in-degrees and out-degrees with sign values. Denote $d_{in}^+(u)$ for positive in-degree, $d_{in}^-(u)$ for negative in-degree, $d_{out}^+(u)$ for positive out-degrees and $d_{out}^-(u)$ for negative out-degrees. There are 8 node features.

The neighbour features includes statistical information from $CNe(u, v)$, such as $c_{Ne}^N(u, v)$ is the number of nodes w in $CNe(u, v)$ and $c_{Ne}^E(u, v)$ is the number of edges between w and u, v. Then select any node w from $CNe(u, v)$, whose edges could have any direction with any sign value connected with u and v, denote $c(u_{\rightarrow}^+ w_{\leftarrow}^+ v)$ as the number of nodes who get positive links from both u and v. There are 2 directions and 2 kinds of sign values, so the relationships of u, v and w can be divided into 16 kinds. There are total 18 neighbour features.

4.2   User Comment Features

In some online social networks, users are allowed to make some comments when they tag a link to others, such as Wikipedia[1]. A small part of Wikipedia contributors are administrators, who are users with access to additional technical features that aid in maintenance. In order for a user to become an administrator a Request for Adminship (RfA) is issued and the Wikipedia community via a public discussion or a vote decides who to promote to adminship. Some examples of the RfA votes are shown in Fig. 1(b). The vote can be represented as a link with sign value, and users can also make some comments in their votes.

The text, such as comment, could be represented by the Bag of Words model(BOW). It treats each comment as the bag of its words, and represents text as a vector of words via the word dictionary. The word dictionary Dic contains all the appeared words, and the dictionary size is $lenDic$. The set of words appeared in comment from u to v is denoted as $W(u, v)$. Then build a word vector $w(u, v)$ with dimension $lenDic$, and set the ith position to '1' if the Dic's $ith\ word \in W(u, v)$, while all other positions are set to '0'.

The $lenDic$ is always very large for there are so many words appeared in comments. But most of these words are only used few times, and it makes $w(u, v)$ very sparse. Because the number of first layer RBM's visible units should be equal to $lenDic$. We select $number$ of top frequency words to build another Dic named as $top\ number$, and the new $lenDic$ is equals to $number$.

---

[1] https://www.wikipedia.org

4.3   Discriminative Deep Belief Networks

In order to use the unsupervised learned DBN to solve link prediction problem , we added a layer of linear output units for class labels at the top of link DBN and comment DBN. Each of the output unit stands for a class label, and the sample's label should be the output unit with the largest value. This output layer works as a linear Softmax classifier, when input a sample vector v to the bottom of DBN, the value of the jth output unit is

$$O_j(v) = \sum_{h_i \in H_{top}} w_{ij} h_i \tag{4}$$

where H_{top} is the hidden layer activations of the top RBM in DBN, and can be calculated by Eq.2 though all the layers by inputting v into the bottom RBM. And w is the weights between H_{top} and output layer.

Then the possibility of the class label of input vector v is

$$p(label = o_i|v) = \frac{e^{O_i}}{\sum_j e^{O_j}} \tag{5}$$

The Softmax classifier is learned with minimizing the cross-entropy loss error as $-\sum_i y_i \log p_i$ where y_i is the class label and p_i is the predicted link value. In order to get a better classification performance, the unsupervised DBN is fine-tuned when we update the output layer's weights.

4.4   Multimodal Deep Belief Networks

There are two approaches to learn a Multimodal Deep Belief Networks(MDBN) model, a shallow one as shown in Fig. 3 and a deep one as shown in Fig. 4. The circle neural 'Networks Unit' means this neural unit is learned with link network structure data, the rectangle neural 'Comment Unit' means this neural unit is learned with user comment data, and the diamond neural 'Multimodal Unit' means this unit is learned with both kinds of data. The main difference between the two approaches is that the shallow one is learned by original data, while the deep one is based on well learned DBNs, which are used for discrimination in Sect.4.3.

The shallow MDBN is a direct approach that trains a DBN over the concatenated link network data and user comment data. We joint the vector of link structure features and word vector as the input for the first layer RBM, and train the shallow MDBN with the greedy layer-by-layer learning, as shown in Fig. 3. As the RBM can model the distribution of input samples, the 1st RBM could represent the joint distribution of the link network and user comments. Then the joint input is abstracted by above RBMs. However, the correlations between the link network and user comments are highly non-linear. We found it is difficult for the 1st RBM to represent their relations properly, so we designed deep MDBN by taking the learned DBNs into account.

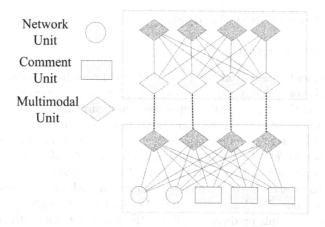

Fig. 3. Structure of Shallow Multimodal Deep Belief Networks.

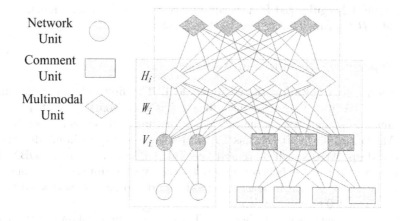

Fig. 4. Structure of Deep Multimodal Deep Belief Networks.

As introduction in Sect.2.2, well learned DBN could highly abstracted the inputs, and the results of discriminative DBN show that the represented inputs become linear classifiable. It inspires us to learn a multimodal model on higher RBMs. As shown in Fig. 4, we reuse the DBNs learned for link prediction introduced in Sect.4.3 as bottom RBMs. The link structure features is represented by the link DBN(in Fig. 2(a)) firstly, and the user comment word vector is represented by the comment DBN (in Fig. 2(b)). Then we joint the highly abstracted features together, shown as V_i in Fig. 4, and use it to learn higher RBMs. It can be easier for the model to learn higher-order correlations between link network data and user comments data.

5 Experiments and Analysis

5.1 Experiment Setup

In our experiments, the social media data of Wikipedia RfA prepared by West [21] is used. Before building the comment word vector, we removed the words such as 'support', 'supporting', 'supported' and similar words for 'oppose'. Removing such words is because that these words could imply the sentiment ploarity directly, and the model may nelgect all the other words. To avoid the samples imbalance effects, we randomly selected 50000 balanced samples. We use 40000 samples for training and the other 10000 for testing.

In order to make a more completely comparison with other state-of-art link prediction methods. We learned Support Vector based Classifier(SVC), which is widely used for link prediction [20], on the same dataset and features. We extracted features by programming in Python, train DBNs and MDBNs in Matlab, and the toolkit 'LIBLINEAR' by [4] is used for learning SVCs. The parameters of LIBLINEAR is set as $(' - s\ 1\ -c\ 1\ -B\ 1')$. The parameter $-s\ 1$ means using L2-regularized L2-loss support vector classification model, c is the cost and $-B\ 1$ means adding bias to the samples.

5.2 Experiment Results and Analysis

The experiment results are shown in Table 1. The first column is the name of method. We use 2 rows to introduce a DBN or MDBN, the first row is name and the second row is the network structure as dimension of each layer in RBM. The shallow MDBN has $26(link) + 2000(word)$ as input dimension. And the deep MDBN, which reused the link DBN and comment DBN, has $100(represented\ link) + 500(represented\ word)$ visible units. The second column is the features used to learn the model, and top 2000 means word vector built with top 2000 frequency words.

The experiment results is shown in Table 1. The first column is the name of method. SVC it a multi-class classifier based on support vectors. We use 2 rows to introducing a DBN or MDBN, the first row is name and the second row is the network structure as dimension of each layer in a RBM. For example, (26-25-50-100) means the first RBM has a visible layer of 26 units and 25 hidden units. Then the second RBM has 25 visible units and 50 hidden units, and the top RBM has 50 visible units and 100 for hidden. The first RBM always has a visible layer with the same dimension as the feature vector's. As a result, the shallow MDBN has $26(link) + 2000(word)$ as input dimension. And the deep MDBN, which reused the link DBN and comment DBN, has $100(representedlink) + 500(representedword)$ visible units. The second column is the features used to learn the model, including link structure features, word vector features, word vector built with top 2000 frequency words, and the combine of them.

First of all, the models learned on multiple source of data outperform the ones learned on single source of data. Both the results of SVC with concatenated features and MDBN have better accuracy than using only one kind of features.

Table 1. Experiment Results

Methods	Features	Accuracy %
SVC	link structure	81.17
SVC	word vector (all words)	84.81
SVC	word vector (top 2000)	82.65
SVC	link + all words	85.96
discriminative DBN (26-25-50-100)	link structure	82.53
discriminative DBN (2000-2000-1000-500)	word vector (top 2000)	85.22
shallow MDBN (2026-2000-1000-500)	link + top 2000	87.75
deep MDBN (600-600-800-1000)	link + top 2000	88.50

It shows that use the joint representation from different data spaces, which imply similarity concepts, could improve models' performance.

Secondly, by learning on the same data, discriminative DBN outperforms SVC. What is more, the DBN with top 2000 frequency words has better performance than SVC learned with all words. It means that the highly abstracted features are more suitable for discrimination. As the discriminative DBN's output layer is linear unit in our experiment, it means that the original features become more linear classifiable after transformation by DBN.

Thirdly, the MDBN's performance is best of all methods, it shows that our multimodal learning method is effective for link prediction problem. Both the shallow and deep MDBN have good performance, and the deep one has about 0.8% higher accuracy. This result shows it is better to learn a multimodal model on higher RBMs. Because the correlations between data from different space became easier for learning after they abstracted by DBN.

Another aspect need to care is the computing cost. Learning RBMs is a time and space costly process for a large number of weights need to be adjusted. We run the experiments on a 4 core 3.5 Ghz CPU and 16GB RAM, and learning the deep MDBN needs about 40 hours. So the SVC learned with concatenated features maybe a choice when fast training is required. It cost less than 10 minutes with an acceptable performance.

6   Conclusion

In this paper, we proposed multimodal learning approaches for predicting link values in social networks. By taking both the link network structure and user comment data into account, the MDBN outperforms state-of-art link prediction methods, such as support vector based classifier and discriminative deep belief network. From the analysis of the different network structure of MDBNs, shallow and deep, we found that it is better to learn a multimodal model on data represented by RBMs.

Our further work includes try to find some other data that imply link values. And we would try some other method for multimodal learning.

Acknowledgement. This work is supported by the National Natural Science Foundation of China (61272383 and 61300114), Specialized Research Fund for the Doctoral Program of Higher Education (No. 20132302120047), the Special Financial Grant from the China Postdoctoral Science Foundation (No.2014T70340), China Postdoctoral Science Foundation (No.2013M530156), and Natural Science Foundation of Heilongjiang Province(F201132).

References

1. Carreira-Perpinan, M.A., Hinton, G.E.: On contrastive divergence learning. In: Proceedings of the Tenth International Workshop on Artificial Intelligence and Statistics. vol. 2005, pp. 17–24. Society for Artificial Intelligence and Statistics, Savannah Hotel, Barbados (2005)
2. Chelmis, C., Prasanna, V.K.: Social link prediction in online social tagging systems. ACM Transactions on Information Systems (TOIS) **31**(4), 20–46 (2013)
3. Demailly, C.: The business impacts of social networking. AT&T White Papers, pp. 1–8 (2008)
4. Fan, R.E., Chang, K.W., Hsieh, C.J., Wang, X.R., Lin, C.J.: Liblinear: A library for large linear classification. The Journal of Machine Learning Research **9**, 1871–1874 (2008)
5. Freno, A., Garriga, G., Keller, M.: Learning to recommend links using graph structure and node content. In: Neural Information Processing Systems Workshop on Choice Models and Preference Learning, pp. 1–7. NIPS (2011)
6. Getoor, L., Diehl, C.P.: Link mining: a survey. ACM SIGKDD Explorations Newsletter **7**(2), 3–12 (2005)
7. Hinton, G.E.: Training products of experts by minimizing contrastive divergence. Neural Computation **14**(8), 1771–1800 (2002)
8. Hinton, G.E.: To recognize shapes, first learn to generate images. Progress in Brain Research **165**, 535–547 (2007)
9. Hinton, G.E., Salakhutdinov, R.R.: Reducing the dimensionality of data with neural networks. Science **313**(5786), 504–507 (2006)
10. Kiros, R., Salakhutdinov, R., Zemel, R.: Multimodal neural language models. In: Proceedings of the 31st International Conference on Machine Learning (ICML 2014), pp. 595–603 (2014)
11. Leskovec, J., Huttenlocher, D., Kleinberg, J.: Predicting positive and negative links in online social networks. In: Proceedings of the 19th International Conference on World Wide Web, pp. 641–650. ACM (2010)
12. Leskovec, J., Huttenlocher, D., Kleinberg, J.: Signed networks in social media. In: Proceedings of the SIGCHI Conference on Human Factors in Computing Systems, pp. 1361–1370. ACM (2010)

13. Liben-Nowell, D., Kleinberg, J.: The link-prediction problem for social networks. Journal of the American Society for Information Science and Technology 58(7), 1019–1031 (2007)
14. Liu, B., Zhang, L.: A survey of opinion mining and sentiment analysis. In: Mining Text Data, pp. 415–463. Springer (2012)
15. Liu, F., Liu, B., Sun, C., Liu, M., Wang, X.: Deep learning approaches for link prediction in social network services. In: Neural Information Processing, pp. 425–432. Springer (2013)
16. Liu, F., Liu, B., Sun, C., Liu, M., Wang, X.: Deep belief network-based approaches for link prediction in signed social networks. Entropy 17(4), 2140–2169 (2015)
17. Liu, F., Liu, B., Wang, X., Liu, M., Wang, B.: Features for link prediction in social networks: A comprehensive study. In: 2012 IEEE International Conference on Systems, Man, and Cybernetics (SMC), pp. 1706–1711. IEEE (2012)
18. Ngiam, J., Khosla, A., Kim, M., Nam, J., Lee, H., Ng, A.Y.: Multimodal deep learning. In: Proceedings of the 28th International Conference on Machine Learning (ICML-11), pp. 689–696 (2011)
19. Srivastava, N., Salakhutdinov, R.R.: Multimodal learning with deep Boltzmann machines. In: Advances in neural information processing systems, pp. 2222–2230 (2012)
20. Wang, P., Xu, B., Wu, Y., Zhou, X.: Link prediction in social networks: the state-of-the-art. Science China Information Sciences, pp. 1–38 (2014)
21. West, R., Paskov, S.H., Leskovec, J., Potts, C.: Exploiting social network structure for person-to-person sentiment analysis. Transactions of the Association of Computational Linguistics 2(1), 297–310 (2014). http://aclweb.org/anthology/Q14-1024
22. Xing, E.P., Yan, R., Hauptmann, A.G.: Mining associated text and images with dual-wing harmoniums, pp. 1–9 (2012). CoRR, arxiv:abs/1207.1423

Sentiment Analysis Based on User Tags
for Traditional Chinese Medicine in Weibo

Junhui Shen[1], Peiyan Zhu[2], Rui Fan[2], Wei Tan[3], and Xueyan Zhan[4(✉)]

[1] Information Center, Beijing University of Chinese Medicine, Beijing 100029, China
[2] State Key Lab of Software Development Environment, Beihang University,
Beijing 100191, China
[3] School of Management, Beijing University of Chinese Medicine,
Beijing 100029, China
[4] School of Chinese Materia Medica,
Beijing University of Chinese Medicine, Beijing 100029, China
zhan_xueyan@163.com

Abstract. With Western culture and science been widely accepted in China, Traditional Chinese Medicine (TCM) has become a controversial issue. So, it is important to study the public's sentiment and opinions on TCM. The rapid development of online social network, such as twitter, make it convenient and efficient to sample hundreds of millions of people for the aforementioned sentiment study. To the best of our knowledge, the present work is the first attempt that applies sentiment analysis to the fields of TCM on Sina Weibo (a twitter-like microblogging service in China). In our work, firstly, we collected tweets topics about TCM from Sina Weibo, and labelled the tweets as supporting TCM or opposing TCM automatically based on user tags. Then, a Support Vector Machine classifier was built to predict the sentiment of TCM tweets without tags. Finally, we presented a method to adjust the classifier results. The performance of F-measure attained by our method is 97%.

Keywords: Sentiment analysis · Machine learning · Support Vector Machine · Traditional Chinese Medicine (TCM) · Weibo · User tag

1 Introduction

At the beginning of the 20th century, western culture and science was introduced in China and soon was accepted. Traditional Chinese Medicine (TCM) was seriously thrown into doubts in terms of its scientific foundation. When such kind of debates are reviewed in respective of debaters' sentiment towards TCM, two types of sentiment are dominating: one school thinks that TCM cannot be proved by scientific experiments, so it is pseudo-science and should be abolished, while the other school believes TCM is effective in treating many diseases, therefore TCM is essentially a kind of science.

Microblogging today has become a very popular communication tool among internet users. In China, Sina Weibo (http://www.weibo.com), a Twitter-like

© Springer International Publishing Switzerland 2015
J. Li et al. (Eds.): NLPCC 2015, LNAI 9362, pp. 134–145, 2015.
DOI: 10.1007/978-3-319-25207-0_12

microblogging service launched in 2009, has accumulated more than 500 million users in less than four years, leading to its most important role in the social media marketing platform. Every second, approximately more than 1000 Chinese tweets are posted in Weibo. It is imaginable that the debates about TCM spread into cyber-space in an unbelievable speed.

So far, although many researches have been conducted on sentiment classification [1], there is little such work about Traditional Chinese Medicine. To the best of our knowledge, the present work is the first attempt that applies sentiment analysis to the fields of TCM on Sina Weibo. In our work, main contents are: collecting corpus and dictionary resources, labelling data automatically based on user tags, building a Support Vector Machine (SVM) classifier to predict the sentiment of TCM tweets and presenting a method to adjust the classifier results. The performance of F-measure attained by our method is 97%.

2   Data Collecting and Labelling

In this section, we discuss the collection and pre-processing of tweets which topics are about TCM. For each tweet in our corpus, we converted it into a sequence of words.

2.1   Corpus Collection Based on User Tags

In China, Sina Weibo is one of the most important social networking channels, and is the Chinese counterpart to Twitter. As with Twitter, Weibo users are allowed to post real-time messages, called tweets. Tweets are short messages, restricted to 140 characters in length.

There are some prominent differences between Twitter and Weibo. For example, users can freely tag himself/herself to indicate his/her interests and characteristics in Weibo. Of course, tagging is not mandatory in Weibo where users can tag up to 10 keywords.

In January 2014, we searched Weibo users interested in TCM by user tags. If someone has more than one user tags included in our search keywords list, he/she would be duplicated in our dataset. After filtering the duplicated users, we constructed a dataset including 48861 Weibo users, denoted as C. The user tags and the corresponding number of Weibo users are listed in Table 1. Among all tags, "Traditional Chinese Medicine" is used by 42608 users and occupies the dominating share of 87%, "Medicine Material", "Acupuncture and Moxibustion" and "Massage" follow but none of them takes the share of more than 8%. It is not surprised to find that "Traditional Chinese Medicine" is the main tag used because it is a wide concept, which often refers to not only TCM therapy but also includes "Medicine Material", "Acupuncture and Moxibustion" and "Massage". Using the Application Programming Interfaces (API) provided by Weibo, we collected the tweets which were posted by the users in C. Due to the limit of API, only the most recent 2000 tweets each user posted can be obtained, we gathered 21,242,370 tweets totally.

Table 1. The user tags and the corresponding number of Weibo users.

User Tag(Original Text)	User Tag(Translation)	the Counts of Weibo Users
中医	Traditional Chinese Medicine	42608
中药	Medicine Material	3827
针灸	Acupuncture and Moxibustion	3236
推拿	Massage	2198
艾灸	Moxa-moxibustion	763
中草药	Chinese Herb Medicine	417
针刺	Acupuncture	73
针推	Acupuncture and Massage	67
中成药	Chinese Patent Drug	50

The sentiment of a retweet is not always consistent with the tweet, especially when debating. For this reason, we split each tweet which followed with retweet and inserted each retweet into our corpus. Sometimes, one post had more than one re-posting, so we had much more tweets after splitting. Totally, we collected 43,012,068 tweets in our corpus, more than twice of original tweets amount.

2.2 Two Dictionary Resources

In this paper, we introduced two new resources for the pre-processing of Weibo data topics on TCM: custom dictionary and TCM terminology dictionary. We collected western medicine terminology, TCM terminology and popular vocabulary on the internet, totally 5307 words in the custom dictionary. It can be used as a helpful complement of built-in dictionary of general tool for Chinese Word Segmentation. The TCM terminology dictionary contained 2715 TCM terminology words including Traditional Chinese Medicine, Chinese Patent Medicine, Chinese Herb Medicine and acupuncture point etc. It can be used to filter the Weibo which topic is about TCM.

2.3 Pre-processing of Data

We pre-processed all the tweets as follows:

1) Translating the tweet to Chinese Simplified if it is written by Chinese Traditional;

2) Filtering URL links (e.g. http:// example.com), Weibo user names (e.g. @shenjh-with symbol @ indicating a user name), Weibo special words (e.g. reply), and emoticons from tweets;

3) Segmenting Chinese Word (with the ICTCLAS tool [2] and the custom dictionary as introduced in section 2.2) to generate a sequence of words;

4) Removing stop words (such as "oh") from the bag of words;

5) Filtering advertisements by key words (such as "sale").

2.4   Filtering Chinese Medicine Tweets

However, the topics of tweets posted by the users interested in TCM were diverse and not only about TCM. Therefore, in our study, we should screen out the tweets in which the real topic was not about TCM.

During our process, we filtered the tweets topics on TCM with the TCM terminology dictionary (introduced in section 2.2). Usually, a tweet on TCM contains a few key words which are about TCM, so we filtered the tweets including at least two different key words of TCM strictly. After filtering, 1,650,497 tweets remained in our corpus in which the real topic is about TCM.

2.5   Labelling the Data

When we were labelling the sentiment of tweets, our approach based on the basic principle: the user prone to have consistent opinions for a certain topic due to the principle of consistency [3]. It means that if the user's opinion is for TCM, the sentiment of all the tweets he/she posted is for TCM. In contrast, if the user's opinion is against TCM, the sentiment of all the tweets he/she posted is against TCM.

In our analysis, we acquired user's opinions about TCM by the user tags. The key words used as user tags are defined by the user. Consequently, the user tags could be different even if the sentiment to TCM is same. The user tags used to label the sentiment are listed in Table 2. Only the user tags which had been quoted by more than 10 users are included in the table. As a result, 1866 Weibo users were labelled as supporting TCM, while 290 Weibo users were labelled as opposing TCM. The rest were not labelled because we couldn't obtain obvious sentiment orientation from his/her user tags. Based on our basic principle, we labelled the sentiment of tweets according to the user's opinions on TCM. Finally, 40888 tweets were labelled as supporting TCM, and 6975 tweets were labelled as opposing TCM. Obviously, there was an imbalance but it is consistent with the reality. The tweets labelled would be used as the training dataset in the next step of our research.

3   Methodology

This section presents the methodology of sentiment classification system we used. First, feature selection method was used to pick out discriminating terms for training and classification. Then we used the machine learning method to build a sentiment classifier. Finally, we adjusted the classification results based on the basic principle that a user keeps consistent opinions for a certain topic.

3.1   Feature Selection

A number of feature selection metrics had been explored in text categorization, i.e. chi-square (CHI), information gain (IG), correlation coefficient (CC) and

Table 2. The user tags and corresponding Weibo user counts.

Sentiment	User Tag(Original Text)	User Tag(Translation)	User Counts
Supporting TCM	中医爱好	Love TCM	972
	爱中医	Love TCM	239
	中医师	Doctor of TCM	230
	喜欢中医	Love TCM	85
	中医粉	TCM Follower	55
	中医控	TCM Follower	52
	中药师	Pharmacist of TCM	51
	针灸师	Acupuncturist	42
	中医养生爱好	Regimen of TCM	29
	推拿师	Masseur	28
	中医达人	TCM Master	12
Opposing TCM	反中医	Oppose TCM	191
	中医黑	Abominate TCM	55
	反对中医	Oppose TCM	28

odds ratios (OR). All these methods computed a score for each individual feature and then picked out a predefined size of a feature set. In our approach, we used chi-square feature selection method, one of the most effective methods in text categorization [4]. Chi-square measures the lack of independence between a term t and a category c_i and can be compared to the chi-square distribution with one degree of freedom to judge extremeness.

3.2 Machine Learning Method

So far, most of the researches on sentiment classification focused on training machine learning algorithms to classify reviews [5],[6]. Support Vector Machine has been shown to be highly effective for traditional text categorization [7]. Based on the structural risk minimization principle from the computational learning theory, SVM seeks a decision surface to separate the training data points into two classes and makes decisions based on the support vectors that are selected as the only effective elements in the training set.

Here we limit our discussion to linear SVM due to its popularity and high performance in text categorization [8].

3.3 Adjusting Sentiment Classification Results

Based on the basic principle that the same user should have consistent opinions for a certain topic [3], we adjusted the sentiment classification results: assigned majority sentiment label to all the tweets the same user posted.

Based on the sentiment classification result, the number of tweets which are judged as supporting TCM posted by one user can be obtained as C_s, and the number of tweets posted by the same user which are judged as opposing TCM can be obtained as C_o. Then we define γ as

$$\gamma = \frac{max\{C_s, C_o\}}{C_s + C_o} \tag{1}$$

where $0.5 \leq \gamma \leq 1$. If $\gamma = 1$, it means the sentiment of the user is consistent absolutely. If $\gamma = 0.5$, it means Co is equal to Cs, then we don't need to adjust the sentiment classification result. When $0.5 < \gamma < 1$, we can adjust the classification result.

4   Experiments and Results

In our dataset, there were 1,650,497 tweets in which the topic focused on TCM, including 40,888 tweets labelled as supporting TCM, and 6,975 tweets labelled as opposing TCM (introduced in Section 2.5). Since it was imbalanced, we focused on not only the global performance, but also the performance of each class. Therefore, we chose F1 to evaluate the classification system.

After applying CHI feature selection to tweets, for all our experiments, we used Support Vector Machine and reported 5-fold cross-validation test results.

Pang et al. [7] argued that feature presence binary value is more useful than feature frequency for the SVM classier. Therefore, we used binary value for each feature instead of feature frequency.

4.1   The Performance Measure

To evaluate the imbalanced classification system, we used the F1 measure. This measure combines recall and precision in the following way:

$$Precision = \frac{number\ of\ correct\ positive\ cpredictions}{number\ of\ positivec\ predictions} \tag{2}$$

$$Recall = \frac{number\ of\ correct\ positive\ predictions}{number\ of\ positive\ examples} \tag{3}$$

$$F1 = \frac{2 * Precision * Recall}{Recall + Precision} \tag{4}$$

4.2   Feature Selection Results

The top 10 key words of each class selected by the CHI method were listed in Table 3. Among the proponents of TCM, it is not surprising that "Medicine Material", "Health Preservation", "Traditional Chinese Medicine" and "Body" were often used. The frequency of "State" and "China" could be due to that

Chinese government employed clear policy to support TCM. Among the opponents of TCM, "Aristolochia acid", "Cinnabar", "Longdan Xiegan Wan" and "Injection" were popular words. This could be attributed to that all these terms are related to untoward effects so the opponents wanted to shake the scientific foundation of TCM.

Table 3. The top 10 key words of each class selected by the CHI method

Supporting TCM (Original Text)	Supporting TCM (Translation)	Opposing TCM (Original Text)	Opposing TCM (Translation)
中药	Medicine Material	中成药	Chinese Patent Medicine
养生	Health Preservation	马兜铃酸	Aristolochic acid
国家	State	注射	injection
科学	Science	注射液	injection
中医药	TCM	方舟子	Zhouzi Fang
中国	China	朱砂	Cinnabar
身体	Body	事件	Events
医生	Doctor	反对	Oppose
健康	Health	马兜铃	Aristolochic
治疗	Cure	龙胆泻肝丸	Longdan Xiegan Wan

Figure 1 shows the classification performance curves using the CHI feature selection method vs. feature number. The performance of classifier is above 90% and the performance increases as the number of features increases. It is found that the performance of TCM proponent classifier is slightly better than the performance of the total classifier. It is notable that the performance of TCM opponent classifier increases significantly when the number of features increases. The performance of each class is relatively stable when the number of features exceed 3000. So, we fixed the number of features at 3000 in the following experiments.

4.3   Classification Results

Because the dataset was imbalanced, we tuned the wi parameter for SVM, where 0=wi=1. Figure 2 shows the performance of F1, precision and recall by varying the parameter wi from 0.1 to 1.0. When wi increases from 0 to 1, precision, recall and F1 all increase significantly and reach plateau. Figure 3 shows precision and recall separately with each class by varying the parameter wi. It is interesting that precision shows a reverse trend of that of recall. When wi increases from 0 to 1, the precision of supporting TCM gradually decreases while the precision of

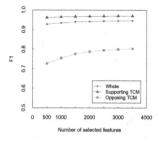

Fig. 1. The classification performance curves using the CHI feature selection method vs. feature number.

opposing TCM rapidly increases. During the same process, recall of supporting TCM increases while recall of opposing TCM significantly decreases. Figure 4 shows the performance of each class separately.

From these figures we can see that it is better to set wi to 0.9. It summarizes the performance of the classifier of supporting TCM and the classifier of opposing TCM. When wi gradually increases, for TCM proponents, Precision decreases from 98% to 96%, Recall increases from 91% to 98%, and F1 increases gradually to a plateau phase. For TCM opponents, Precision increases 62% to 86%, Recall decreases from 89% to 75%, and F1 increases gradually to a plateau phase.

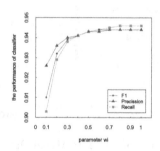

Fig. 2. The performance curves by varying the parameter wi

We viewed both the whole and the individual class: when wi increases from 0.1 to 1, F1 value increases gradually to a plateau phase. F1 value reaches the optimal when wi equals to 0.9.

4.4 Adjusted Classification Results

As introduced in Section 4.3, we can adjust the classification results based on the principle that the same user should have consistent opinions for a certain topic. Figure 5 shows the performance by varying the parameter γ from 0.5 to 1(and fixing wi=0.9). There is a notable decline of F1. When γ is set to 0.5, our model achieves the best performance of F1, which is 97%.

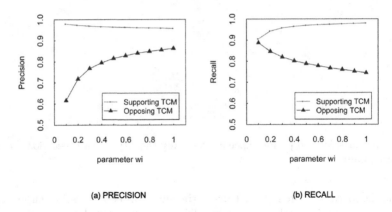

(a) PRECISION   (b) RECALL

Fig. 3. The precision and recall separately with each class by varying the parameter wi.

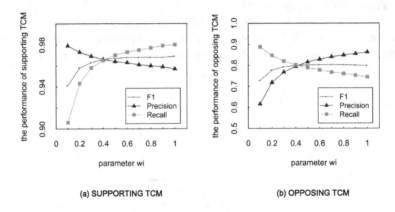

(a) SUPPORTING TCM   (b) OPPOSING TCM

Fig. 4. The Performance of Supporting TCM and Opposing TCM Separately.

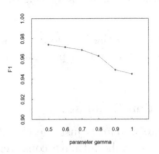

Fig. 5. The performance curve of sentiment classification by varying the parameter γ from 0.5 to 1.

4.5   Prediction

Besides the labelled tweets, there are 1,602,634 unlabelled tweets which the topic is about TCM. We can predict their sentiment with our trained classifier. Figure 6 shows the curves for the amount of tweets which, respectively, support TCM (a) and oppose TCM (b). The amount of tweets supporting TCM far exceeds the number of tweets opposing TCM. For the simple comparison, the tweets amount of both opposing and supporting TCM are converted to their log forms, as shown in (c). This result coincides with the real world. In china, most people support TCM, especially the regimen of TCM. There are only a small number of people opposing TCM. In addition, the tweets amount before 2010 was very small, due to the limit of Weibo where only the most recent 2000 tweets of each user can be obtained.

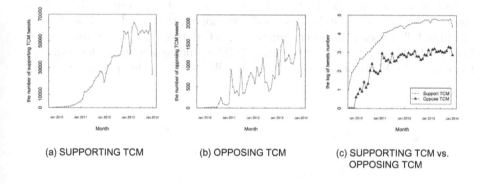

(a) SUPPORTING TCM   (b) OPPOSING TCM   (c) SUPPORTING TCM vs.
                                                        OPPOSING TCM

Fig. 6. The curves for the amount of tweets which, respectively, support TCM (a) and oppose TCM(b).

After the sentiment classification of tweets concerning TCM, we can monitor the sentiment fluctuation of TCM in Weibo. As shown in Figure 6, the number of tweets supporting TCM decreases significantly during January of 2012, 2013 and 2014. Because the three periods conflicted with Chinese New Year. The decrease could be due to that people did not log on Weibo during these holiday seasons. On the contrary, the number of tweets opposing TCM shows no clear trend. The erupt of the tweets opposing TCM could be caused by incidents related to TCM, which could be an interesting research topic in the future.

It is also found from Figure 6 that the tweet counts of both class reached the peak in Nov. 2013. We show the details of the curve in that month in Figure 7. In November 2013, the number of tweets supporting TCM is relatively stable while the number of tweets opposing TCM fluctuates drastically. This is in line with the overall trend of the number of each class.

Moreover, the top 50 key words of each class in Nov. 2013 is shown separately in Figure 8. "Traditional Chinese Medicine", "Health Preservation", "Food" etc.

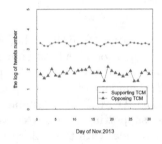

Fig. 7. The curves for the amount of tweets which, respectively, support TCM (a) and oppose TCM in Nov.2013.

often appears in tweets supporting TCM, while "Chinese Patent Medicine", "injection", "toxicity" etc. frequently appears in tweets opposing TCM. This is conformed with the Table 3. It is worth mentioning that words such as "toxic" or "harmful" appears in tweets supporting TCM too. This is not unexpected because TCM theory admits that a few TCM medicine is toxic so the dosage of these toxic TCM medicines should be controlled and be paid special attention.

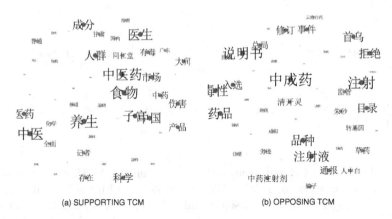

Fig. 8. The top 50 key words of each class(supporting TCM and opposing TCM) in Nov.2013.

5   Conclusion and Future Work

Traditional Chinese Medicine is an ancient but thriving and somewhat controversial discipline. Meanwhile, it is important to study the public's sentiment and opinions on TCM. To the best of our knowledge, the present work is the first attempt to study sentiment analysis for TCM based on user tags in Weibo. We classify the opinions on TCM into two categories: supporting TCM and opposing TCM. The F1 measure value of our method is 97%.

Moreover, we collect 48861 Weibo users who are interested in TCM and 1,650,497 tweets concerning about TCM. And we construct two dictionary resources for processing Chinese tweets topic on TCM. Based on the aforementioned corpora and resources, we build an effective classifier with SVM to analyse the sentiment opinions on TCM using Weibo tweets automatically.

In future work, we will explore more linguistic techniques to study sentiment analysis for TCM, such as parsing, semantic analysis and topic modelling.

Acknowledgments. This work is supported by the MOE (Ministry of Education in China) Project of Humanities and Social Sciences (Project Name:Public opinion analysis of Traditional Chinese Medicine based on sentiment analysis, Project No: 15YJCZH137).

References

1. Zhao, J., Dong, L., Wu, J. Xu, K.: Moodlens: an emoticon-based sentiment analysis system for Chinese tweets. In: Proceedings of the 18th ACM SIGKDD International Conference on Knowledge Discovery and Data Mining, pp. 1528–1531. ACM Press, Jeju island (2012)
2. Zhang, H., Yu, H., Xiong, D. Liu, Q.: HHMM-based Chinese lexical analyzer ICTCLAS. In: Proc of SIGHAN Workshop on Chinese Language Processing, pp. 758–759. ACM Press, Sapporo (2003)
3. Deng, H., Han, J., Li, H., Ji, H., Wang, H., Lu, Y.: Exploring and inferring user - user pseudo-friendship for sentiment analysis with heterogeneous networks. Statistical Analysis and Data Mining **7**, 308–321 (2014)
4. Yang, Y., Pedersen, J.: A comparative study on feature selection in text categorization. In: 14th Int'l Conf. Machine Learning, pp. 412–420. ACM Press, Nashville (1997)
5. Liu, B.: Sentiment Analysis and Opinion Mining. International Journal **5**, 1–167 (2012)
6. Agarwal, A., Xie, B., Vovsha, I., Rambow, O., Passonneau, R.: Sentiment analysis of Twitter data.In: Proceedings of the Workshop on Languages in Social Media, pp. 620–622. ACL Press, Portland (2011)
7. Pang, B., Lee, L. Vaithyanathan, S.: Thumbs up? Sentiment classification using machine learning techniques. In: Proceedings of Emnlp, pp. 79–86. ACL Press, Stroudsburg (2002)
8. Fan, R.E., Chang, K.W., Hsieh, C.J., Wang, X.R., Lin, C.J.: Liblinear: a library for large linear classification. Journal of Machine Learning Research **9**, 1871–1874 (2008)

Predicting User Mention Behavior
in Social Networks

Bo Jiang, Ying Sha[⊠], and Lihong Wang

Institute of Information Engineering,
Chinese Academy of Sciences, Beijing 10093, China
{jiangbo,shaying,wanglihong}@iie.ac.cn

Abstract. Mention is an important interactive behavior used to explicitly refer to target users for specific information in social networks. Understanding user mention behavior can provide important insights into questions of human social behavior and improve design of social network platforms. However, most previous works mainly focus on mentioning for the effect of information diffusion, few researches consider the problem of mention behavior prediction. In this paper, we propose an intuitive approach to predict user mention behavior using link prediction method. Specifically, we first formulate user mention prediction problem as a classification task, and then extract new features including semantic interest match, social tie, mention momentum and interaction strength to improve the performance of prediction. To evaluate the proposed approach, we conduct extensive experiments on Twitter dataset. The experimental results clearly show that our approach has 15% increase in precision compared with the best baseline method.

Keywords: Social network · Mention behavior · Link prediction · Classification

1 Introduction

In recent years, social network platforms such as Sina Weibo, Twitter and Facebook have become more and more popular, because they allow users to freely post a short message named tweet or status for sharing viewpoints and acquiring knowledge in real time. According to the latest statistics, approximately 500 million tweets are generated per day in Twitter[1]. As a result, the rich information in social networks not only expands our horizon, but also has wide applications in public opinions supervision, natural disaster prediction and political upheaval detection.

In order to better exchange ideas among users, Twitter provide mention function as a new feature to encourage interaction and conversation in social activities. Mention is enabled in a tweet by using "@username". It is placed anywhere in the body of the tweet and contains one or more target users. If a tweet

---

[1] https://about.twitter.com/company

© Springer International Publishing Switzerland 2015
J. Li et al. (Eds.): NLPCC 2015, LNAI 9362, pp. 146–158, 2015.
DOI: 10.1007/978-3-319-25207-0_13

contains more than one username, these people will all see the tweet in their own Mentions tabs. By using mention ensures that these tweets which usually have a higher priority to the target user do not get lost by the overwhelming stream of other tweets in the user timeline. Most studies on Twitter's mention function have focused on constructing mention network to explore the pattern of information diffusion and recommending the mentioned users to improve the visibility of the tweets. Although exist research works has been obtained some significant progress, these works neglect the most important fact: *the purpose of mentioning between users is not always to expand information diffusion, sometimes more to let a user know about a tweet or build a good relationship that a mentioned user is interested in information.*

In the paper, we try to predict who will be mentioned for a new coming tweet, it is important to predict future mention links since these links are a bridge for future spread of influence. To this end, we first formulate the user mention prediction problem as a classification task, and then extract a set of feature for the prediction. We would expect that exploiting these features for user's mention prediction would help improve the performance of prediction.

Our contributions are summarized as follows:

- We model mention behavior prediction problem as a supervised classification task combine user, textual, social tie and temporal information features.
- We use semantic enrichment technique to measure interest match between users and topic interest between tweet and user. We also propose a notion of mention momentum to quantify the mention behavior from the pairwise angle.
- We construct a large collection of dataset from Twitter service. The experimental study shows that these features can effectively improve the performance of user mention prediction comparing with other baseline methods.

The rest of the paper is organized as follows. Section 2 surveys the previous work on analysis of user mention in social networks. In Section 3, we formulate the user mention prediction problem and investigate the properties of the problem. Section 4 proposes our model. The empirical experiment results are reported in Section 5. At last, we conclude our work and give the future research directions in Section 6.

2   Related Work

There has been a number of efforts that study the properties of mention mechanism in social networks. We can roughly divide these works into two categories of models in this scope: (i) information diffusion models and (ii) recommend models. In the following, we will briefly summarize some representative works as follow.

The goal of information diffusion models is to study on analyzing how mention effect information diffusion. Yang et al. [13] construct a diffusion network based on @username mentioning to analyze the speed, scale and range of tweets on the same topic spread in Twitter. Itakura et al. [7] characterize the structural differences of the retweet graph, the mention graph and the reply graph. They employee PageRank and HITS algorithms to measure each graph, the result suggest that using the mention function is the most efficient method of reaching the mass audience in Twitter. Yu et al. [14] proposes a directed tree model based on user interaction that considering the history, type and frequency of interaction to describe the process of information diffusion.

Beside, mention function has a widely used in recommender systems. Tang et al. [10] present a context-aware recommendation framework and employ ranking support vector machine model to locate target users for posting promotion-oriented messages. Pramanik et al. [8] develop a Twitter app to recommend the best set of users to be mentioned in a tweet in order to maximise the spread of an information. Wang et al. [12] propose a new recommendation scheme to expand the diffusion of tweets by recommending proper users to mention. They formulate the problem as a ranking problem and use all the new features, including user interest match, user social ties and user influence, to achieve the best performance of the algorithm. Those recommendation models aim to find the right users to mention in a tweet for expanding the diffusion of tweets.

However, as far as I know, few research work predict whom to mention. To fill this gap, we now present an approach to predict user mention behavior through link prediction method.

3   Problem Statement

Mention is an important interactive behavior in social networks. The intuition behind is that the purpose of mention is to attract the attention of other people for specific information and to form interactions each other.

Mention forms a "tweet-to-user" interaction link between mentionee and mentioner, thus we formulate the problem of mention behavior prediction as a link prediction task. The link prediction problem can also be regarded as a classification problem.

For the convenience and simplicity of description, we formally define mention behavior prediction as follows.

Definition 1 Mention Prediction. We use a triple (u, t, v) to represent that user u mentioned v in tweet t. For user u, we denote all mentioned users as the mention candidate C. Given a tweet t that is to post by user u, the goal is to find whether u mention $v(\in C)$ in t or not.

Note that publisher u may mention multiple users within one mention tweet t, we will consider as multiple mention instances.

As above discussed, the problem of user mention prediction is modeled as a classification task and exist lots of the popular classification methods can be employed. But, the key challenge in this approach is to extract a set of features for the classification task. Next we will discuss the set of features that have been used successfully for mention prediction task.

4   User Mention Prediction Model

In this section, we present the process of extracting features used in our model and propose user mention behavior prediction model for task. These features are extracted from the content of tweets, the structure of network topology and interaction knowledge to predict user potentially mention behaviors in the future.

4.1   Feature Extraction

User-to-User Interest Match. This feature assumes that a user is more likely to mention these users who share similar topics of interests. However, profiling user interest is a challenge task due to the informal nature and ungrammatical language of tweets in Twitter [9]. For instance, given a tweet such as "#Gravity is beautiful http://fb.me/3SkiI5DND" it is difficult to understand to talk about movie or space based on topic modeling technics like LDA. Hong et al. [6] show that topic model techniques like LDA can not accurately depict the topic of tweet due to the short-length, ambiguous, noisy data feature in Twitter. Therefore, we use a semantic enrichment technique to construct user interest profile.

[5] have been shown some web services providing semantic enrichment such as Dbpedia Spotlight[2], TextRazor[3] and Zemanta[4], we opt for OpenCalais[5] due to its state-of-the-art semantic functionality and a high rate limit of 50,000 document per day. We use OpenCalais API to assign each tweet to known topics which now have 18 categorizations, such as sports, education, environment and politics. Then, for each user u, we measure the distribution of his/her interests, $D(u)$, represented as a vector over the set of topic categories:

$$D(u) = (\frac{N_{c_1}}{N}, \frac{N_{c_2}}{N}, \cdots, \frac{N_{c_n}}{N}) \quad where \quad N = \sum_i N_{c_i} \tag{1}$$

N_{c_i} is the number of tweets classified into category c_i published by user u in sampled dataset. N is the total number of tweets posted by user u. Thus, $D(u)$ represent the proportion of tweets made by the user u publishing about each topic category and also reflect the importance of each of interests.

As discussed in previous studies [1] for topics of interests of Twitter's user, we also propose to use Cosine Similarity (CS) to measure *interest match degree* between mentionee u and mentioner v through their interests distributions. That is,

$$UUIM(u, v) = \frac{D(u) \cdot D(v)}{\| D(u) \| \| D(v) \|} \tag{2}$$

Tweet-to-User Topic Match. In order to expand the visibility of tweets, publisher always prefers to mention these users who are interested in the new

---

[2] http://spotlight.dbpedia.org/
[3] https://www.textrazor.com/
[4] http://www.zemanta.com/
[5] http://www.opencalais.com/

coming tweet and more likely to retweet it. Therefore, in this paper we also calculate topic match degree between mentioner v and the new coming tweet t.

As is mentioned above, we also exploit OpenCalais API to represent the topic of tweet t as follow:

$$D(t) = (p_{c_1}, p_{c_2}, \cdots, p_{c_n}) \tag{3}$$

where p_{c_i} is the probability of tweet t belong to the i-th topic. Similarity, *topic match degree* between the new coming tweet t and mentioner v can be computed as Cosine Similarity as follows:

$$TUTM(t, v) = \frac{D(t) \cdot D(v)}{\| D(t) \| \| D(v) \|} \tag{4}$$

User Social Tie. Intuitively, the more common neighbors, the more they are familiar with each other. They are more likely to mention each other compared with a total stranger. In simple words, if user u follow w and v follow w, then there is a high probability that between u and v form a social tie. So, the number of common neighbors can be measure as the chance that u and u will have a mention between them. However, the common neighbors metric is not normalized. Therefore, we use Adamic/Adar [2] as a metric of the strength of *social tie* between two users. For a set of features z, it is defined as below.

$$UST(u, v) = \sum_{z \in \Gamma(u) \cap \Gamma(v)} \frac{1}{\log |\Gamma(z)|} \tag{5}$$

$UST(u, v)$ gives each common neighbor of user u and v a weight, $\frac{1}{\log|\Gamma(z)|}$, to denote its importance. According to exist works on link prediction, Adamic/Adar works better than common neighbors metrics.

Mentioner Influence Score. In social marketing, the goal of mentioning users in a tweet mainly hope to attract more people for discussing or retweeting it. Consequently, the influence of mentioners is an important consideration factor of information diffusion with publisher. Therefore, we can define *influence* for mentioner v who is to mentioned in a tweet t as follow:

$$MIS(t, v) = Num.Follower_v \tag{6}$$

where $Num.Follower_v$ is the number of mentioner v's followers.

Mention Momentum Based Time. We assume both of users often mentioned each other in the past, this tends to point out that they have been having a conversation and therefore share a good relation in social networks. Furthermore, they are more likely to mention each other in the near future. Specifically, given mentionee u and mentioner v, we denote the timestamp of u first mention v as $m_f(u, v)$, and the timestamp of u last mention v as $m_l(u, v)$. We define *mention time span* from u to v as below:

$$MTS(u,v) = m_l(u,v) - m_f(u,v) \tag{7}$$

Correspondingly, we also define *mention frequency* as the average mention interval from u to v as follow:

$$MF(u,v) = \frac{N(u,v)}{MTS(u,v)} \tag{8}$$

where $N(u,v)$ is the mention number of from u to v in the given time interval.

Furthermore, we also calculate *recent mention interval*, which is defined as the interval between $m_l(u,v)$ and the posted timestamp t_t of tweet t:

$$RMI(u,t,v) = t_t - m_l(u,v) \tag{9}$$

Finally, we formulate *mention momentum* as below:

$$MMT(u,t,v) = \frac{MF(u,v)}{RMI(u,t,v)} \tag{10}$$

Interaction Score. A lot of studies [7,13,14] show that interaction network such as retweet network, reply network and mention network are much more important than follow network in terms of link prediction. Therefore, we use *interaction score* to measure the strength of user interaction as below:

$$IS(u,v) = \sum_{i=1}^{K} \alpha_i N_i(u,v) \tag{11}$$

where K is the number of types for interaction behaviors. α_i and $N_i(u,v)$ is the weight of i-th interaction behavior and the number of times, respectively.

4.2 Mention Behavior Prediction

Now we have extracted all features of predicting user mention behavior, and then in this section we will discuss how to model the user mention prediction model (named as UMPM) using these features. As discussed in Section 3, the mention behavior prediction can be regarded as a classification problem: given a tweet t that is to posted by user u and a set of candidate C at a specific time point T, the goal is to find whether u mention $v(\in C)$ in t. We denote the classification label as $m_{u,t,v}$. $m_{u,t,v} = 1$ indicates that u will mention v in t, and $m_{u,t,v} = 0$ otherwise. The classification model is very flexible, thus we can integrate different combinations of the features into the model conveniently.

To solve the classification problem, many classification models for supervised learning can be used. In this paper, we opt to choose SVM, because SVM has a strong theoretical foundations and practical advantages [11], and [15] result shows that SVM is better than others, including Naive Bayes, decision trees, etc. in link prediction.

$$f(m_{u,t,v}|X) = sign(w^T X + b) \tag{12}$$

where X is the feature vector introduced above, and w are weights of the features and b is a bias.

5 Experiments and Analysis

In this section, we design the experiment with three goals: (1) show two sampling methods to collect Twitter dataset and then preprocess the dataset; (2) compare link prediction methods for evaluating our proposed approach; (3) leverage common metrics to evaluate the performance of our model.

5.1 Data Collection

For this experiment, our objective is to construct a dataset that reflects a comprehensive history of user mention over an extended period. Thus we devise two different sampling schemes - a snowball sample and an activity sample in order to obtain a more comprehensive dataset.

Snowball Sample. We first randomly select 10 seed users to perform a snowball sampling. For each seed user, we crawl all tweets on which that Twitter user has been published, and all followers list and all followees list. We then crawl all users appearing in the "list of lists". We repeat these last two steps. In total, our obtain 5,140 Twitter users. Our dataset also contains all tweets ever posted by the collected users, which consists of 11,104,955 tweets.

Activity Sample. To avoid potentially biased by the snowball sampling method, we also run a sample of users based on their activity. Specifically, we carefully choose 10 fields of celebrity users that those are verified, and crawl their all tweets and follow networks. Table 1 shows a more detailed information.

Table 1. Data Description of Activity Sample

Users' Category	Num.User	Num.Tweet	Example
Athletes	16	34645	TheRock, JohnCena, KAKA
Brands	15	45086	McDonalds, YouTube, CocaCola
Celebrities	17	37492	Justinbieber, katyperry
Games	19	49896	PlayStation, AngryBirds
Movies	17	33169	Starwars, SpiderManMovie
News	19	61367	CNN, espn, FoxNews, nytimes
Organizations	18	58019	NASA,Harvard,PBS,RED
Politicians	17	39928	BarackObama, MittRomney
Sport Teams	17	54890	Realmadrid, LFC, NBA
TV Shows	16	39857	BigBang_CBS, Discovery

In order to our prediction task, we need to run some steps to extract the data requested. From the tweets we only consider these tweets that user directly publish with @ symbol. To improve data quality, strict filtering is employed:

1. Only keep tweets that reference them using "@username".
2. Only keep alphanumeric characters and #. Remove URLs from tweet.
3. Remove all tweets containing RT as a stand-alone term.
4. Remove all tweets that reply other user.
5. Remove all tweets that are determined not to be English.

Figure 1 lists the results of this filtering where we only keep the roughly 21% tweets that does not match any of the filter criteria.

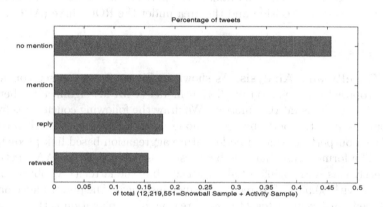

Fig. 1. Results of filtering tweet data to improve data quality

5.2 Comparison Methods and Evaluation Metrics

In [3], the authors survey a variety of techniques for link prediction. Here, we choose some typical link prediction methods for evaluating our proposed approach are described as follows:

- **Common Neighbors(CN).** $CN(u, v)$ uses the number of shared neighbor as the proximity score of user u and v. The proximity score can be formally defined as:

$$CN(u, v) = |\Gamma(u) \cap \Gamma(v)| \tag{13}$$

  The more $CN(u, v)$ is, the closer user u and v are in the network.
- **Jaccard Coeficient (JC).** $JC(u, v)$ normalizes the size of common neighbors in their neighbors as the proximity score of user u and v as below:

$$JC(u, v) = \frac{|\Gamma(u) \cap \Gamma(v)|}{|\Gamma(u) \cup \Gamma(v)|} \tag{14}$$

- **Adamic/Adar (AA).** $AA(u, v)$ also employees the common neighbors, and weighs the common neighbors with smaller degree of user u and v as below:

$$AA(u, v) = \sum_{z \in \Gamma(u) \cap \Gamma(v)} \frac{1}{log \, |\Gamma(z)|} \tag{15}$$

- **Preferential Attachment (PA).** $PA(u, v)$ gives higher scores to pairs of nodes for user u and v as below:

$$PA(u, v) = \Gamma(u) \cdot \Gamma(v) \tag{16}$$

We divide the constructed data set into training and testing data, and perform 10-fold cross validation. We evaluate the performance of mention behavior prediction in terms of Precision and the area under the ROC curve (AUC).

5.3 Results and Discussion

Overall Results and Analysis. As shown in Figure 2, we can see conclude that our proposed user mention prediction model (UMPM) significantly better other baseline methods in AUC metrics. We draw the following conclusions from these results. First, the performance of node neighborhood based link prediction methods outperform that of node feature aggregation based link prediction methods. The former show that the best results is Adamic/Adar (AA) method that correctly classifies 75.06% of all instances, but the latter (PA) obtain only the precision with 60.95% of correctly classified instances. The performance has a relatively well improvement for 23.15%. A reasonable explanation is the number of common neighborhood of both users can better measure user's relationship than node's attribute, they are more likely to mention each other in a new coming tweet in the future. In addition, the performance of AA is better than that of Common Neighbors (CN), this is in accordance with the conclude in [3]. Second, the performance of our proposed method (UMPM) outperforms all the comparison algorithms in the experiment. Even comparing the best result with AA, it shows 15% increase in precision. This indicates that the features that we had selected have good discriminating ability. Finally, the result also indicates that the proposed model can improve the performance of mention behavior prediction as we increase the observation period.

Feature Analysis. As discussed in Section 4.1, we extract a set of features for the mention prediction task, namely User-to-User Interest Match, Tweet-to-User Topic Match, User Social Tie, Mentioner Influence Score, Mention Momentum based Time and Interaction Score. To analyze how effectiveness of each feature used in our proposed algorithm contribute to the learned model, we design this contrast experiment by eliminating one feature at a time and observe how the performance of our model changes. Specifically, we mark without User-to-User Interest Match feature as No_UUIM, without Tweet-to-User Topic Match feature as No_TUTM, without User Social Tie feature as No_UST, without Mentioner Influence Score feature as No_MIS, without Mention Momentum based Time

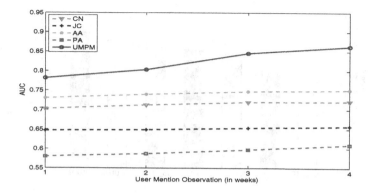

Fig. 2. AUC on link prediction with baseline methods

feature as No_MMT and without Interaction Score feature as No_IS. Similarly, we summarize performance for feature evaluation using the Precision metric as mentioned in Section 5.2.

All the results are shown in Figure 3, where the higher the precision is, the less important the feature is. Note that, when we remove User-to-User Interest Match feature (No_UUIM) and Tweet-to-User Topic Match feature (No_TUTM), the precision suffers from a 5.8% decline and 10.4% decline, respectively. The result shows the prediction importance of both features is lowest. It is in accordance with our instinct that both users who have the similarity interests or tastes are not necessarily mention each other. Moreover, when we eliminate Mention Momentum based Time feature (No_MMT) and Interaction Score feature (No_IS), the model suffers a 35.8% decline of precision and 27.7% decline of precision, respectively. We can conclude that although user interest match and user influence help to improve the prediction result, user interactions play a much more significant role in the mention prediction task. As discussed earlier, past user mentions has a high correlation with the user mention data in the future. This is as we expected because the users who often mention each other are more likely to have a mention relationship in the future than the users who have not mentioned each other in the past.

Discussion. In total, we aim to predict user mention behavior using machine learning method, but, as shown in the picture above, the accuracy is not more higher. One of the reasons being that the tweet feature is sparse and user profile information contributes very little to the classifier performance. In addition, our observation for Twitter is that the neighborhood does not completely identify the area of influence. For example, in our experimental dataset, we define four types of user relationship, namely stranger (both of users no follow each other), follower (one is following another user), followee (one is followed by another user), friend (both of users mutual follow each other). We extract the link relationship between mentionee u and montioner v. The result is shown in Figure 4. From

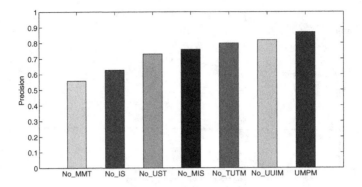

Fig. 3. Comparison on how different features affect the performance of UMPM

the figure, we can clearly see that most of mention behaviors are more likely to occur between friends. Meanwhile, we also find an interest fact that in second place is mentions between strangers, where the lowest is 13% in SportTeams and the highest is up to 44% in Brands. We can conclude that the purpose of a user mentions others in a tweet is very different in social networks. The set of our mentioned candidate have no consider the strangers, this is one of reasons that why the precision is not high in the above mentioned.

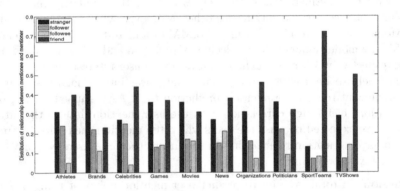

Fig. 4. Distribution of tweets per user relationship in activity sample

6   Conclusions and Future Work

In this paper we model the problem of user mention prediction as a classification task. Our analysis focuses on whom to be mentioned for a new coming tweet. To this end, we extract a series of prediction features and carefully choose prediction model. The experiment result shows that our proposed method outperform the baseline link prediction methods which only consider network topology or node attribute. Moreover, we find that novel users who never mentioned for the publisher in the past are refer to in a tweet. Although it is more difficult for predicting these novel users and the prediction accuracy may be lower, we will try to address this issue in the future due to the need of information diffusion monitoring. Meanwhile, we also will consider location-based feature in our mention prediction model.

Acknowledgments. We thank the three anonymous reviewers for their helpful comments and suggestions. This research is supported by National Key Technology R&D Program(No.2012BAH46B03), and the Strategic Leading Science and Technology Projects of Chinese Academy of Sciences(No.XDA06030200).

References

1. Abel, F., Gao, Q., Houben, G.J., Tao, K.: Analyzing temporal dynamics in twitter profiles for personalized recommendations in the social web. In: Proceedings of the 3rd International Web Science Conference, p. 2. ACM (2011)
2. Adamic, L.A., Adar, E.: Friends and neighbors on the web. Social networks **25**(3), 211–230 (2003)
3. Al Hasan, M., Zaki, M.J.: A survey of link prediction in social networks. In: Aggarwal, C.C. (ed.) Social Network Data Analytics, pp. 243–275. Springer, Heidelberg (2011)
4. Batuwita, R., Palade, V.: Class imbalance learning methods for support vector machines. In: Imbalanced Learning: Foundations, Algorithms, and Applications, pp. 83–99 (2013)
5. Derczynski, L., Maynard, D., Aswani, N., Bontcheva, K.: Microblog-genre noise and impact on semantic annotation accuracy. In: Proceedings of the 24th ACM Conference on Hypertext and Social Media, pp. 21–30 (2013)
6. Hong, L., Davison, B.D.: Empirical study of topic modeling in twitter. In: Proceedings of the First Workshop on Social Media Analytics, pp. 80–88 (2010)
7. Itakura, K.Y., Sonehara, N.: Using twitter's mentions for efficient emergency message propagation. In: 2013 Eighth International Conference on Availability, Reliability and Security (ARES), pp. 530–537 (2013)
8. Pramanik, S., Danisch, M., Wang, Q., Mitra, B.: An empirical approach towards an efficient "whom to mention?" Twitter app
9. Ritter, A., Clark, S., Etzioni, O.: Named entity recognition in tweets: an experimental study. In: Proceedings of the Conference on Empirical Methods in Natural Language Processing, pp. 1524–1534 (2011)
10. Tang, L., Ni, Z., Xiong, H., Zhu, H.: Locating targets through mention in Twitter. World Wide Web, pp. 1–31 (2014)
11. Vapnik, V.: The nature of statistical learning theory. Springer Science & Business Media (2013)
12. Wang, B., Wang, C., Bu, J., Chen, C., Zhang, W.V., Cai, D., He, X.: Whom to mention: expand the diffusion of tweets by@ recommendation on micro-blogging systems. In: Proceedings of the 22nd International Conference on World Wide Web, pp. 1331–1340 (2013)

13. Yang, J., Counts, S.: Predicting the speed, scale, and range of information diffusion in twitter. In: ICWSM, pp. 355–358 (2010)
14. Yu, M., Yang, W., Wang, W., Shen, G., Dong, G.: Information diffusion and influence measurement based on interaction in microblogging. In: Huang, H., Liu, T., Zhang, H.-P., Tang, J. (eds.) SMP 2014. CCIS, vol. 489, pp. 129–140. Springer, Heidelberg (2014)
15. Al Hasan, M., Chaoji, V., Salem, S., Zaki, M.: Link prediction using supervised learning. In: SDM (2006)

Convolutional Neural Networks for Multimedia Sentiment Analysis

Guoyong Cai[✉] and Binbin Xia

Guangxi Key Lab of Trusted Software, Guilin University of Electronic Technology,
Guilin 541004, Guangxi, China
{ccgycai,beybinxia}@gmail.com

Abstract. Recently, user generated multimedia contents (e.g. text, image, speech and video) on social media are increasingly used to share their experiences and emotions, for example, a tweet usually contains both texts and images. Compared to sentiment analysis of texts and images separately, the combination of text and image may reveal tweet sentiment more adequately. Motivated by this rationale, we propose a method based on convolutional neural networks (CNN) for multimedia (tweets consist of text and image) sentiment analysis. Two individual CNN architectures are used for learning textual features and visual features, which can be combined as input of another CNN architecture for exploiting the internal relation between text and image. Experimental results on two real-world datasets demonstrate that the proposed method achieves effective performance on multimedia sentiment analysis by capturing the combined information of texts and images.

Keywords: Multimedia · Sentiment analysis · Convolutional Neural Networks · Deep learning

1 Introduction

Online social networks are providing multiple forms of access to their users, for instance, people can post a tweet attached with images or videos. Social networks sites play an important role in people's live for requiring information and sharing experiences. Meanwhile, online users love to express their opinions on subjects they interested in, because of the free expression of speech on social networks. Sentiment analysis of online user generated data on social networks can be helpful to understand user behavior and improve applications aimed at online users. Among the large amount of data, we are particularly interested in analyzing sentiment of tweets containing both texts and images towards specific events and topics.

Deep neural networks have achieved remarkable performance in many fields, especially in compute vision [1, 2, 3, 4] and speech recognition [5] in recent years. In the field of natural language processing (NLP), works with deep learning methods were also widely used. As a challenging task NLP, sentiment analysis has been studied in various ways. Inspired by the enormous successes of deep learning, much research work on sentiment analysis has applied deep learning algorithms. However, most of

© Springer International Publishing Switzerland 2015
J. Li et al. (Eds.): NLPCC 2015, LNAI 9362, pp. 159–167, 2015.
DOI: 10.1007/978-3-319-25207-0_14

them are mainly focused on one single form of user content (such as text, image or video) separately instead of the combined representation. In fact, a great number of images posted do not contain any sentiment words in text at all, or the text sentiment is obvious but the image sentiment is unconspicuous. Figure 1 shows two tweets consisting of both text and image, their emotions can only be classified obviously through the combined text and image.

It's May 17th and it is SNOWING near Devils Lake, ND. Photo via NDDOT #NDwx

Final result of the #climate negotiation, they almost reached the goals! Well done! Now lunch #careplanet2015

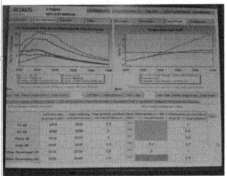

Fig. 1. Examples of two image tweets, text sentiment of the left one and image sentiment of the right one are not obvious. But image sentiment of the left one and text sentiment of the right one are relatively obvious.

In this paper, we focus on the problem of sentiment prediction based on the joint textual and visual information within an image post. Convolutional Neural Networks employed in prior works [6, 7, 8, 9] have been proved very powerful in solving image or text sentiment analysis tasks. Thus, to solving the challenging problem mentioned above, a novel deep learning architecture based on CNN was proposed. We intend to find out whether applying CNN to the joint information of text and image provides better performance than classifiers using only single form of information (either text or image).

The rest of the paper is organized as follows. In Section 2, we review research work focusing on sentiment analysis. Next we describe the proposed sentiment prediction based on CNN architecture in Section 3. Then we present datasets and experimental results in Section 4. Finally, some conclusions and future work are given in Section 5.

2   Related Work

In recent years, compared to traditional sentiment analysis on text, sentiment analysis of visual content has also attracted much attention, especially prominent performance has been witnessed on image classification based on deep learning algorithms [3, 23, 25]. In this section, we review research work closely related to our study focusing on textual and visual sentiment analysis.

2.1   Textual Sentiment Analysis

Sentiment analysis of text has been a challenging and fascinating task since it is pro-posed, and researchers have developed different approaches to solve this problem. Generally, two main approaches can be distinguished: dictionary based method and machine learning method.

Dictionary based method for sentiment analysis usually depends on the pre-defined sentiment dictionaries. Turney [10] presented a simple unsupervised learning algorithm for classifying users' reviews by leveraging the average semantic orientation score of the phrases, which is calculated by mutual information measures.

For machine learning approaches, Pang et al. [11] took n-gram and POS as features to classify movie reviews with Naive Bayes, maximum entropy classification, and support vector machines. As a sub-field of machine learning, deep learning methods achieved tremendous success, which motivate researchers to employ different kinds of deep learning methods for textual sentiment analysis. Socher et al. [12] proposed a semi-supervised approach based on recursive autoencoders for predicting sentiment distributions. Kim [6] and dos Santos et al. [7] developed deep convolutional neural network built on top of word2vec for performing textual sentiment analysis.

2.2   Visual Sentiment Analysis

In contrast to textual sentiment analysis, research work focusing on sentiment prediction of visual content falls far behind. Previous researches on visual sentiment analysis have mostly been conducted by utilizing low-level image features [13, 14, 26] or mid-level image attributes [15, 16]. Jia et al. [13] developed a semi-supervised framework based on factor graph model, which takes advantage of color features and social correlation among images. Yang et al. [14] proposed a novel emotion learning method to exploit social effect correlate with the emotion of images. The method jointly modeled images posted by social users and comments added by friends. Yuan et al. [15] proposed an image sentiment prediction framework based on mid-level attributes which were generated from four general scene descriptors. Borth et al. [16] constructed a large-scale Visual Sentiment Ontology and a novel visual concept detector library to visual sentiment prediction.

In addition, several researches employed deep learning methods [6, 7] for visual sentiment analysis. Xu et al. [6] proposed a novel visual sentiment prediction frame-work with CNN. The framework performs transfer learning from a CNN [4] with millions of parameters, which is pre-trained on large-scale data for object recognition. In order to solve challenging problem of image sentiment, You et al. [7] proposed a progressive CNN, which a probabilistic sampling algorithm was employed to select the new training subset, namely removing instances with similar sentiment scores for both classes with a high probability.

2.3   Multimedia Sentiment Analysis

To our best knowledge, there are few works focusing on sentiment analysis of combined textual and visual content. Borth et al. [16] compare the performance on twitter dataset using text only (SentiStrength), visual only (SentiBank), and their combination. The experimental results show that visual content predicted by the SentiBank-based classifier plays a much more important role in predicting the overall sentiment of the tweet, and the combined classifier achieve best performance. Wang et al. [17] propose a novel Cross-media bag-of-words Model (CBM) for Microblog sentiment analysis. The model represent the text and image of a tweet as a unified bag-of-words features, which are taken as input of machine learning methods (i.e., NB, SVM and Logistic Regression).

3   Textual and Visual Sentiment Analysis with CNN

Previous works have proven the powerful performance of CNN for textual [6, 7] and visual [8, 9] sentiment analysis. In this section, we introduce a comprehensive framework for joint sentiment analysis with CNN. As shown in Figure 2, the overall architecture of the proposed framework consists of three components: text CNN, image CNN and multi CNN, and each of the three components is a CNN architecture. Multi CNN takes joint text-level and image-level representation as input, and two kinds of representation are respectively extracted by vectorizing the features in the penultimate layer of text CNN and image CNN.

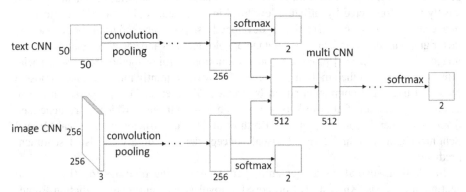

Fig. 2. The overall architecture of the proposed multimedia sentiment prediction framework

3.1   Textual Sentiment Analysis with CNN

We develop the text CNN for textual sentiment analysis to generate text-level representation. Pre-trained word vectors are used to initialize the word representations, which are taken as input of the text CNN. Detailed process of learning pre-trained word vectors will be discussed in Section 4. The overall architecture of the text CNN consists of three convolutional layers, two full connected layers and one softmax layer. Each convolutional layer is connected to a max pooling layer. Detailed information of text CNN is described as follows.

The first convolutional layer filters the word representations with 16 kernels of size 5*5, the second convolutional layer takes the pooled output of the first convolutional layer as input with 32 kernels of size 4*4. Pooled output of the second convolutional layer is connected to the third convolutional layer with 64 kernels of size 3*3. The last max pooling layer is followed by two full connected layers, and each of them has the same amount of neurons. The last softmax layer is used to classify the output of last full connected layer over two class labels. The text-level representation v_{text} is computed as follows:

$$v_{text} = f(\mathbf{w} \cdot (CNN_{text}(T)) + b) \tag{1}$$

Where f denotes the activation function, CNN_{text} is the text CNN, $\mathbf{w}$ is the weight matrix and b is a bias term. Thus, each text T can be represented as a fixed dimension vector v_{text}.

3.2   Visual Sentiment Analysis with CNN

Similar to the text CNN, the image CNN is developed for visual sentiment analysis and generating image-level representation. The image CNN is composed of five convolutional layers, three full connected layers and one softmax layer. The input images for the first convolutional layer is resized to the same size (256*256*3). Details of image CNN is discussed as follows.

The number of kernels of each convolutional layer is same with [2], and the corresponding size of kernels is respectively 17*17, 13*13, 7*7, 5*5 and 3*3. The output of the last full connected layer is taken as input for softmax layer. The formula of computing image-level representation v_{image} is similar with Equation (1).

3.3   Multimedia Sentiment Analysis with CNN

Aiming at solving the problem of multimedia sentiment analysis, we develop the multi CNN to take the joint text-level and image-level representation as input. The multi CNN does not contain any convolutional layer and max pooling layer at all, it just consists of four full connected layer and one softmax layer. The multi CNN is described in detail as follows.

The input features are mapped by four full connected layer, and the output features are passed to a softmax layer, which produces a distribution over two class (positive or negative) labels.

3.4   Classification

As described above, text-level and image-level representation are both in vector form, which can be taken as features for linear classifiers, such as Naïve Bayes, SVM and Logistic Regression. The experiment results of Borth et al. [16] show that Logistic Regression achieves better performance than SVM for visual sentiment prediction, and Logistic Regression is also employed in Xu et al. [8]. In Section 4, we employ Logistic

Regression as classifier with the vectorized features in the penultimate layer of text CNN, image CNN and multi CNN.

4   Experimental Setup and Results

4.1   Datasets

In our work, training dataset is constructed with randomly chosen 20K image posts (one image post consists of one image and corresponding description) from SentiBank [16], which consists of collected image posts on Flickr. The SentiBank are weakly labeled by 1200 adjective noun pairs (ANPs), which are based on psychological theory, Plutchik's Wheel of Emotions [24]. Similar to the work [9], we employ a probabilistic sampling algorithm to generate the new training dataset.

We evaluate the performance of proposed CNN architecture on two real-world twitter datasets, which have respectively been used in prior work [16, 9]. Both of two datasets are collected from image tweets, each of which contains text and corresponding image. The first twitter datasets (TD1) includes 470 positive tweets and 133 negative tweets, and the second one (TD2) includes 769 positive tweets and 500 negative tweets.

4.2   Pre-trained Word Vectors

In this work, word vectors initialized by skip-gram model [19], which has shown powerful performance in previous works. Word vectors are trained with word2vec tool on the latest English Wikipedia corpus, which is processed by removing paragraphs are not in English and sentences are less than 20 characters. The dimension of word vectors is set to 50 with a context window of size 5.

4.3   CNN Training

In our experiments, training is processed by stochastic gradient descent (SGD) with mini-batch size of 128 for optimization. Early-stopping [20] and dropout [21, 22] (with probability of 0.5) are employed for avoiding over-fitting. ReLU [21, 23] is adopted as activation function for text CNN, image CNN and multi CNN. Words are not in pre-trained word vectors are initialized randomly and the randomly initialized vectors are taken as parameters of networks, which will be fine-tuned in training process. In order to handle sentences of variable length, the maximum length of sentence is fixed to 50 for text CNN, zero vectors are padded if length is less than 50. The dimension of text-level and image-level representation are both set to 256. We implement our experiments for the proposed CNN architecture on Keras, which is an effective deep learning framework implementation.

4.4   Results

We compare the text CNN with Naïve Bayes, SVM and Logistic Regression for textual sentiment analysis. As for visual sentiment analysis, the image CNN is compared with low-level features [26], SentiBank [16] and Sentribute [15]. Since little works focus on multimedia sentiment analysis, we just take text CNN, image CNN, the combination of SentiStrength and SentiBank, SVM and Logistic Regression as comparative methods against multi CNN. Results of the proposed method on two twitter datasets can be respectively seen in table 1, table 2 and table 3. The experimental results show that the proposed multi CNN lead to better performance than other methods for multimedia sentiment analysis.

Table 1. Accuracy of algorithms on twitter datasets of text

Algorithms	TD1	TD2
NB	0.70	0.72
SVM	0.72	0.74
LR	0.73	0.76
Text CNN	0.74	0.77

Table 2. Accuracy of algorithms on twitter datasets of image

Algorithms	TD1	TD2
Low-level	0.710	0.664
SentiBank	0.709	0.662
Sentribute	0.738	0.696
Image CNN	0.773	0.723

Table 3. Accuracy of algorithms on twitter datasets

Algorithms	TD1	TD2
SentiStrength +SentiBank	0.72	0.723
SVM	0.76	0.781
LR	0.77	0.783
Multi CNN	0.78	0.796

5   Conclusions

In this paper, we propose a new CNN architecture that fully uses joint text-level and image-level representation to perform multimedia sentiment analysis. Based on idea of the complementary effect of the two representations as sentiment features, the proposed method exploits the internal relation between text and image in image tweets and achieves better performance in sentiment prediction. In future work, we would like to explore multimedia sentiment analysis with much more combination among text, image and other type of social media.

References

1. LeCun, Y., Boser, B., Denker, J.S., et al.: Backpropagation applied to handwritten zip code recognition. Neural computation 1(4), 541–551 (1989)
2. Hinton, G.E., Osindero, S., Teh, Y.W.: A fast learning algorithm for deep belief nets. Neural computation 18(7), 1527–1554 (2006)
3. Ciresan, D.C., Meier, U., Masci, J., et al.: Flexible, high performance convolutional neural networks for image classification. In: IJCAI Proceedings-International Joint Conference on Artificial Intelligence 22(1), p. 1237 (2011)
4. Krizhevsky, A., Sutskever, I., Hinton, G.E.: Imagenet classification with deep convolutional neural networks. In: Advances in neural information processing systems, pp. 1097–1105 (2012)
5. Graves, A., Mohamed, A., Hinton, G.: Speech recognition with deep recurrent neural networks. In: 2013 IEEE International Conference on Acoustics, Speech and Signal Processing (ICASSP), pp. 6645–6649. IEEE (2013)
6. Kim, Y.: Convolutional neural networks for sentence classification (2014). arXiv preprint arXiv:1408.5882
7. dos Santos, C.N., Gatti, M.: Deep convolutional neural networks for sentiment analysis of short texts. In: Proceedings of the 25th International Conference on Computational Linguistics (COLING), Dublin, Ireland (2014)
8. Xu, C., Cetintas, S., Lee, K.C., et al.: Visual Sentiment Prediction with Deep Convolutional Neural Networks (2014). arXiv preprint arXiv:1411.5731
9. You, Q., Luo, J., Jin, H., et al.: Robust image sentiment analysis using progressively trained and domain transferred deep networks. In: The Twenty-Ninth AAAI Conference on Artificial Intelligence (AAAI) (2015)
10. Turney, P.D.: Thumbs up or thumbs down?: semantic orientation applied to unsupervised classification of reviews. In: Proceedings of the 40th Annual Meeting on Association for Computational Linguistics. Association for Computational Linguistics, pp. 417–424 (2002)
11. Pang, B., Lee, L., Vaithyanathan, S.: Thumbs up?: sentiment classification using machine learning techniques. In: Proceedings of the ACL 2002 Conference on Empirical Methods in Natural Language Processing, vol. 10. Association for Computational Linguistics, pp. 79–86 (2002)
12. Socher, R., Pennington, J., Huang, E.H., et al.: Semi-supervised recursive autoencoders for predicting sentiment distributions. In: Proceedings of the Conference on Empirical Methods in Natural Language Processing, pp. 151–161. Association for Computational Linguistics (2011)
13. Jia, J., Wu, S., Wang, X., et al.: Can we understand van gogh's mood?: learning to infer affects from images in social networks. In: Proceedings of the 20th ACM International Conference on Multimedia, pp. 857–860. ACM (2012)
14. Yang, Y., Jia, J., Zhang, S., et al.: How Do Your Friends on Social Media Disclose Your Emotions. In: Proc. AAAI, 14, pp. 1–7 (2014)
15. Yuan, J., Mcdonough, S., You, Q., et al.: Sentribute: image sentiment analysis from a mid-level perspective. In: Proceedings of the Second International Workshop on Issues of Sentiment Discovery and Opinion Mining, p. 10. ACM (2013)
16. Borth, D., Ji, R., Chen, T., et al.: Large-scale visual sentiment ontology and detectors using adjective noun pairs. In: Proceedings of the 21st ACM International Conference on Multimedia, pp. 223–232. ACM (2013)

17. Wang, M., Cao, D., Li, L., et al.: Microblog sentiment analysis based on cross-media bag-of-words model. In: Proceedings of International Conference on Internet Multimedia Computing and Service, p. 76 ACM (2014)

18. Go, A., Bhayani, R., Huang, L.: Twitter sentiment classification using distant supervision. CS224N Project Report, Stanford, pp. 1–12 (2009)

19. Mikolov, T., Chen, K., Corrado, G., et al.: Efficient estimation of word representations in vector space (2013). arXiv preprint arXiv:1301.3781

20. Caruana, R., Lawrence, S., Giles, C.L.: Overfitting in neural nets: backpropagation, conjugate gradient, and early stopping. In: Advances in Neural Information Processing Systems 13, Proceedings of the 2000 Conference, p. 402. MIT Press (2001)

21. Hinton, G.E., Srivastava, N., Krizhevsky, A., et al.: Improving neural networks by preventing co-adaptation of feature detectors (2012). arXiv preprint arXiv:1207.0580

22. Dahl, G.E., Sainath, T.N., Hinton, G.E.: Improving deep neural networks for LVCSR using rectified linear units and dropout. In: 2013 IEEE International Conference on Acoustics, Speech and Signal Processing (ICASSP), pp. 8609–8613. IEEE (2013)

23. Krizhevsky, A., Sutskever, I., Hinton, G.E.: Imagenet classification with deep convolutional neural networks. In: Advances in neural information processing systems, pp. 1097–1105 (2012)

24. Plutchik, R.: Emotion: A psychoevolutionary synthesis. Harpercollins College Division (1980)

25. Simonyan, K., Zisserman, A.: Very deep convolutional networks for large-scale image recognition (2014). arXiv preprint arXiv:1409.1556

26. Siersdorfer, S., Minack, E., Deng, F., et al.: Analyzing and predicting sentiment of images on the social web. In: Proceedings of the International Conference on Multimedia, pp. 715–718. ACM (2010)

Applications on Language Computing

An Adaptive Approach to Extract Characters from Digital Ink Text in Chinese Based on Extracted Errors

Hao Bai[✉]

Beijing Language and Culture University, Beijing, China
baihao@blcu.edu.cn

Abstract. Extracting characters from digital ink text is an essential step which leads to more reliable recognition of text and also a prerequisite for structured editing. Casualness and diversity of handwriting input result in unsatisfied accuracy of extracted characters. Reprocessing the initial extracted characters based on context makes some considerable improvement. Therefore, this paper proposes an approach to adaptively extracting characters from digital ink text in Chinese based on extracted errors. The approach firstly classified the extracted errors in the primary extraction. According to different types of extracted errors, the approach gives different operations. Experimental data shows that the approach is effective.

Keywords: Digital ink · Character extraction · Error classification

1 Introduction

Comparing to keyboard, handwriting meets the needs of input conventionally and ergonomically, which provides the naturalness of writing and the rich expression of ink. With the development of such devices like digital pen, tablet PC and different pads, vast digital ink texts are accumulated and needed to be recognized. Extracting characters from text leads to more reliable recognition and availability of structured editing.

Self-adaptability of existing character extraction method is unsatisfactory in the results of processing. According to detailed analysis of the errors, many errors can be avoided in secondary extraction by rule-making. As a result, this paper proposes a method of self-adaptively extracting character from Chinese digital ink text based on error classification, firstly classifying and then processing the primary wrong extraction results with corresponding methods.

This paper is supported by Science Foundation of Beijing Language and Culture University(supported by "the Fundamental Research Funds for the Central Universities") (no.14YJ160202), and the National Natural Science Foundation of China (no. 61202249).

J. Li et al. (Eds.): NLPCC 2015, LNAI 9362, pp. 171–181, 2015.
DOI: 10.1007/978-3-319-25207-0_15

2   Related Works

In Chinese digital ink text, characters occupy a large proportion, including word, punctuation, number, letter, English word, etc. Based on the information used, character extraction methods can be divided into four types:

1. Method based on time and space distance of adjacent strokes

Subrahmonia et al [1] extracted words from English digital ink text based on time and spatial threshold, but did not specify the method for threshold determination; in addition, threshold ranges of different digital ink texts differ greatly. Zhang Xiwen et al [2] extracted characters from Chinese digital ink texts based on multi-level information. Zhang Shilong et al [3] extracted Chinese characters based on classification of stroke spacing. Han Yong et al [4] extracted characters according to minimum spanning tree (MST) of stroke, structural arrangement of Chinese characters and spatial arrangement of strokes. Su Rui et al [5] extracted characters based on histogram projection.

2. Method based on shape and structure

Tseng et al [6] calculated candidate character spacing with the minimum bounding box, firstly merged strokes according to structure of Chinese characters and finally used dynamic programming method to further merge candidate characters. Zhao Yuming et al [7] also calculated candidate character spacing with the minimum bounding rectangle, and gradually merged strokes, to extract separate Chinese character. Shilman et al [8] combined strokes with similar size and direction, and extracted words from English digital ink text following a bottom-to-top direction.

3. Method based on probability

Artieres [9] extracted words based on the probabilistic features grammars with genetic algorithm, considering the context information. Blanchard et al [10] put forward a method based on probability for English digital ink text, which takes the context and probabilistic features grammars into account.

4. Method based on the results of recognition

Subrahmonia et al [1] extracted letters from English digital ink text, considering all segmentation hypotheses, and took segmentations achieving the best effect as the satisfactory result. Chen et al. [11] extracted characters according to spacing of candidate characters, built grid based on the results of recognition, studied the best path according to scores of recognition and language model, to get results of Chinese characters extraction.

The above methods mostly extract words relying on information of the text. Due to the complexity of structure of Chinese characters and randomness of Chinese handwriting, the methods stated above have unsatisfactory adaptability for different languages and poorly structured data and need to integrate multi-level information besides the data, such as the information of primary extraction, etc. Therefore, this paper presents a re-extraction method, which raises extraction accuracy on the basis of existing technical level.

3   Classification of the Extraction Errors

Figure 1 shows the results after extracting character from a digital ink text, and the circular circles mark the wrong characters extracted.

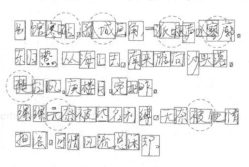

Fig. 1. Results after character extraction

Errors existing in character extraction can be divided into three categories: deficient extraction, beyond extraction and false extraction [12].

3.1   Deficient Extraction

Deficient extraction refers to the situation in which adjacent characters are deemed as one single character. The common errors of deficient extraction after primary extraction contain two types:

1. Adjacent characters are deemed as one single character due to the careless writing, undersize space between characters or low aspect ratio of rectangular bounding box. As shown in Fig. 2, the characters are extracted as one character because of the small bounding box and undersize space between characters.
2. Exaction error as a result of mixing punctuation with characters, as shown in Fig. 3, the comma after the character is taken as a stroke of the character.

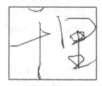

Fig. 2. Deficient extraction of type 1   **Fig. 3.** Deficient extraction of type 2

3.2   Beyond Extraction

Beyond extraction refers to the situation in which a component or partial strokes are taken as another character, this kind of error covers two types after primary extraction:

1. A character in up-down structure is divided into two characters as shown in Fig. 4.
2. A character in left-right structure, as a result of the oversize spacing between strokes, is divided into two characters, as shown in Fig. 5.

Fig. 4. Beyond extraction of type 1   **Fig. 5.** Beyond extraction of type 2

3.3   False Extraction

The reasons causing false extraction errors after primary extraction include: undersize spacing between adjacent characters or oversize spacing between the left and the right components, such as "被" and "他" as shown in Fig. 6, the left component of "被" is taken as a character while the right component "皮" and the left component of its adjacent "他" are extracted as the second character while the right component "也" is taken as the third character.

Fig. 6. False extraction

3.4   Classifying Rules

In view of structure of Chinese characters [13], even handwritten Chinese characters have some common features [14], such as aspect ratio of rectangular volume as the bounding box of a character, number of strokes of a character, and bounding box width of a character, etc.

1. Aspect ratio of rectangular bounding volume Rw/h=W/H, through analyzing 10 Chinese digital ink texts with an average of 300 characters, it was found that average aspect ratio of common Chinese characters ranges between 0.9187292 and 1.036495, thus the range of aspect ratio can be used to preliminarily judge whether a character extraction is correct or not, and the secondary processing will be conducted for false extraction.

2. Number of strokes (Ns): stroke of commonly used Chinese characters is averaged by 9.17 [14], and deficient extraction may occur if strokes of character extracted exceeds the average.
3. Width (W) of the bounding box: different from western words, Chinese characters are "square" shapes, and their width and line height (except for punctuations) show linear change, thus false extraction may occur if W exceeds the normal range of line height.
4. When one single extraction hypotheses is recognized as more than one characters, deficient extraction may exist in word segmentation.

Specifically, step 1, to recognize extraction hypotheses, if more than one character is recognized, deficient extraction occurs. Step 2, check Rw/h of rectangular volume and W of rectangular volume, if they exceed the normal range, false extraction may occur; Step 3, check the number of strokes (Ns) of single character, if Ns exceeds the normal range, deficient extraction may occur.

Based on analyzing extraction errors of the above types, it was found that secondary extraction of deficient extraction errors may result in beyond extraction errors; therefore, with the error classification-based method, the deficient errors should be firstly processed and then the beyond extraction errors.

4   Method for Deficient Extraction Errors

Unlike printed words, digital ink text always has poorly structures varying in size, spacing and internal distance, thus adjacent characters may easily be extracted as one due to the small spacing or undersize aspect ratio of bounding box. Punctuation marks may also be taken as a stroke of a character. In testing 10 Chinese digital ink texts with an average of 300 characters, the accuracy in primary extraction only reached an average of 80%. It was noticed that though handwriting is relatively arbitrary, basic positional arrangement of interior strokes is fixed; therefore, this paper suggests taking relative position of strokes of a character as the feature and conducting clustering analysis to process deficient errors, specifically,

1. Calculating central point of the strokes and take its horizontal projection value as value of the sample point;
2. Processing the data object with condensed hierarchical clustering algorithm [15], to obtain the number of clusters K;
3. Conducting clustering analysis against central points with K-means clustering algorithm [16] and the number of clusters K, to obtain each cluster as extraction.

4.1   Eigenvectors Extraction

Considering the left-right structure of character writing, obvious spacing generally exists between characters, which makes extraction of horizontal projection value from central points of strokes feasible. Taking the deficient extraction as shown in Fig. 7, projection values of abscissa of stroke centers are shown in dashed lines, specifically,

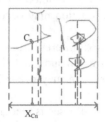

Fig. 7. Projection diagram of stroke center

1. Calculating central point C_n of the bounding boxes;
2. Calculating horizontal projection value of the central point X_{Cn};
3. Get date set $Cluster_0$ as the objects of X_{Cn}.

4.2   Hierarchical Agglomeration Clustering Algorithm

In this paper, K-means algorithm is used in clustering analysis; however, as the number of clusters K is required to be given in advance when this algorithm is used and inappropriate K can produce non-ideal clustering results, thus K should be knownbefore using K-means clustering algorithm. This research adopts hierarchical agglomeration to determine the number of clusters, obtain an initial clustering and improve it

Table 1. Similarities among Clusters (top 40 characters)

No.	Characters	Similarities	No.	Characters	Similarities	No.	Characters	Similarities
1	0	0	15	正	0.12	29	江	0.2
2	4	0.07	16	致	0.32	30	似	0.41
3	可	0	17	国	0.22	31	冻	0.33
4	桂	0.16	18	港	0.3	32	,	0
5	枝	0.31	19	秋	0.37	33	革	0.15
6	香	0.15	20	了	0	34	绎	0.3
7	王	0.01	21	天	0.31	35	如	0.2
8	安	0.12	22	气	0.19	36	预	0.39
9	石	0.29	23	初	0.31	37	0	0
10	登	0.24	24	卵	0.29	38	归	0.35
11	临	0.25	25	0	0	39	虹	0.34
12	送	0.23	26	于	0.13	40	支,	0.25
13	目	0.17	27	里	0.38			
14	,	0	28	澄	0.28			

with K-means. Defining the similarity of clusters as a termination condition is the key to the algorithm. This research takes internal and external distance ratio to represent similarity of clusters, the higher the value, the higher the degree of polymerization of data objects in the clusters. Table 1 shows the top 40 characters out of 134 from a Chinese digital ink text and whose extracted errors was manually corrected. According to the results of testing text data of 10 Chinese ink texts averaged by 300 characters, if the similarity was below 0.5, the target cluster was obtained. Internal distance of a cluster adopts average spacing of projection values (X_{Cn}) of stroke center in a cluster, while cluster spacing adopts spacing of projection values of stroke center between clusters.

4.3   K-means Clustering Algorithm

With the value of K obtained in hierarchical agglomeration algorithm and the initial data set $Cluster_0$ in calculating the eigenvector, K-means algorithm can be used to iterate and improve $Cluster_0$, in order to obtain the optimal results of extraction, as shown in Fig. 8. Fig. 9 shows comparison of a sample before and after application of deficient extraction algorithm, and dotted line marks the characters with varying results in extraction.

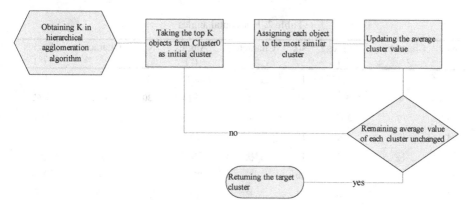

Fig. 8. K-means algorithm flow chart

Fig. 9. Results comparison

5   Method for Beyond Extraction

According to the habit of Chinese handwriting, centers of characters in a line form a relatively stable straight line, as shown in Fig. 10. For two adjacent characters in a row, their horizontal angle of the center point connection is small (excluding punctuations). Table 2 lists horizontal angles of the center point connection of adjacent characters in a 134-character text extraction of which is corrected. Considering the structure of characters, the algorithm can be proceeded by,

1. Calculating center of each character;
2. Calculating horizontal angle of the center point connection;
3. If the angle exceeds the threshold (above 30°), take them as a character.

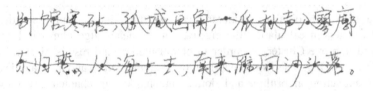

Fig. 10. Center line of characters

Table 2. Horizontal angles

No.	Characters	Angles (°)	No.	Characters	Angles (°)	No.	Characters	Angles (°)
1	0	12.25	15	正	3.02	29	江	1.58
2	4	5.1	16	致	5.14	30	似	6.62
3	可	3.13	17	国	0.75	31	冻	13.45
4	桂	3.32	18	港	3.97	32	,	0
5	枝	1.5	19	秋	28.51	33	革	0
6	香	0	20	了	22.27	34	绎	9.65
7	王	3.31	21	天	2.93	35	如	0.94
8	安	6.52	22	气	0.87	36	预	21.44
9	石	0	23	初	12.08	37	0	21.86
10	登	1.84	24	峥	12.78	38	归	0.79
11	临	2.51	25	0	0	39	虹	1.68
12	送	8.73	26	于	1.02	40	支,	12.78
13	目	20.58	27	里	0.76			
14	,	14.5	28	澄	3.49			

Through testing 10 Chinese digital ink texts averaged by 300 characters, it was found that apart from left-right structure, this algorithm effectively avoided extraction error for characters with up-down structure and half-investing structure. Punctuation identification (period, comma, etc.) is the key point of the algorithm, to avoid taking punctuation as beyond extraction. In this research, structural feature of punctuation and its positional feature in the text were combined: firstly, number of strokes of punctuation is less than 2; secondly, width and height of punctuation are far smaller than the height of the line. Fig. 11 shows the comparison of results before and after using the beyond extraction respectively, and the dotted lines mark the characters with varying results of extraction.

Fig. 11. Examples of beyond extraction

6   Performance Test

Based on the proposed approach, in this research, a prototype system was developed using C# programming language and development platform of Microsoft visual studio 2005. This system operates on PC with Windows XP SP3. In the following part, quantitative analysis of results of a large number of Chinese digital ink texts is conducted to determine performance of the method proposed in this paper. Writings of six undergraduates, collected with digital pen produced by Swedish Anoto [17], were taken as Chinese digital ink test data of this research, and the ink data was extracted and rendered by prototype system developed with MS Tablet PC SDK [18]. Based on the experimental data, three performance indicators are put forward:

1. Extraction efficiency: the ratio between total time consumption and the total number of characters;
2. Initial accuracy: the ratio between the initial number of extraction hypotheses and the total number of characters;
3. Classification accuracy: the ratio between the number of extraction hypotheses and the total number of characters;

Partial statistical results of the algorithm proposed in this paper are shown in Table 3, and the experimental data shows that (1) the method for initial results of extraction could effectively improve accuracy; (2) for data with lower initial extracting accuracy, the method reached obviously higher accuracy; while for data with higher accuracy, the method is less effective; (3) as the method is a secondary extraction based on primary extraction, with the increase in number of text characters, the time consumption will be multiplied.

Table 3. Extraction algorithms based on error classification

Data sample	001	007
Total number of characters	114	309
Extracting time (sec)	12.81	48.8
Extraction efficiency (sec/words)	0.11	0.16
Primary accuracy	78.95%	93.20%
Accuracy achieved in this research	87.72%	94.50%

7   Conclusions

Due to the subjectivity and differentiation of handwriting text, single-pass character extraction always fails to reach satisfactory accuracy; through reprocessing the results of single-pass extraction, considerable improvement can be made. For this reason, this paper proposed an approach of extracting character from Chinese digital ink text based on error classification: firstly classify different types of errors and then processing the primary wrong extraction results with corresponding methods. As the method is a secondary extraction based on primary extraction, with the increase in number of text characters, the time consumption will be multiplied. However, considering the short time of extraction and the improvement of hardware performance, the extraction accuracy for texts in different length can be maintained in an acceptable range, thus the method proposed in this paper is effective in improving accuracy of single-pass extraction.

Reference

1. Subrahmonia, J., Zimmerman, T.: Pen computing: challenges and applications, vol. 2, pp. 60–66 (2000)
2. Xiwen, Z., Xiujuan, G., Guozhong, D.: Adaptive Character Extraction from Continuous Handwriting Chinese Text. Computing Technology and Automatics **3**, 73–77 (2003)
3. Shilong, Z., Xiwen, Z.: Adaptive character extraction from continuous handwriting Chinese textbased on classifying between - stroke gaps. Information Technology **8**, 80–82 (2005)
4. Yong, H., De, X., Guo-Zhong, D.: Using MST in Handwritten Chinese Characters Segmentation. Journal of Software **3**, 403–409 (2006)
5. Rui, S., Xi-wen, Z., Yong-quan, L., Guo-zhong, D.: Intelligent Editing of Handwriting Based on Structure Understanding. Journal of System Simulation **z1**, 371–373 (2006)
6. Yu, L., Rung, T., Chen, C.: Segmenting handwritten Chinese characters based on heuristic merging of stroke bounding boxes and dynamic programming. Pattern Recognition Letters **8**, 963–973 (1998)
7. Yuming, Z., Xingzhi, J., Pengfei, S.: Algorithm for off-line handwritten Chinese character segmentation based on extracting and knowledge-based merging of stroke bounding boxes. Infrared and Laser Engineering **1**, 23–27 (2002)
8. Shilman, M., Wei, Z., Raghupathy, S., Simard, P., Jones, D.: Discerning structure from freeform handwritten notes, pp. 60–65 (2003)

9. Artières, T.: Poorly structured handwritten documents segmentation using continuous probabilistic feature grammars, pp. 5–8 (2003)

10. Blanchard, J., Artieres, T.: On-line handwritten documents segmentation.: frontiers in handwriting recognition. In: Ninth International Workshop on IWFHR-9, pp. 148–153 (2004)

11. Chen, H., Loudon, G., Yimin, W., Zitserman, R.: Segmentation and recognition of continuous handwriting Chinese text. International Journal of Pattern Recognition and Artificial Intelligence **1998**, 223–232 (1998)

12. Kun, Z., Xi-wen, Z.: Comparison of features and classifiers for detailedly classifying handwriting characters in Chinese ink text. Application Research of Computer, pp. 3486–3489 (2008)

13. Donghan, L.: Character Structure and Evolution. Shanghai Education Publishing House, pp. 73–75(1959)

14. Lian, Z., Chenxiao, W., Jicang, H.E., et al.: The effects of font, stroke and contrast on the reading speed of Chinese characters. Chinese Journal of Optometry & Ophthalmology, pp. 96–99 (2008)

15. Dunham, M.H.: Data Mining Introductory and Advanced Topics. Prentice Hall, pp. 112–115 (2003)

16. Han, J., Kamber, M.: Data Ming: Concepts and Techniques. Elsevier Inc., pp. 263–269 (2006)

17. Anoto Inc. http://www.anoto.com

18. Microsoft Windows XP Tablet PC Edition Software Development Kit 1.7. http://www.microsoft.com/downloads/details.aspx?familyid=b46d4b83-a821-40bc-aa85-c9ee3d6e9699&displaylang=en

Context-Dependent Metaphor Interpretation Based on Semantic Relatedness

Chang Su[1]([✉]), Shuman Huang[1], and Yijiang Chen[2]

[1] Cognitive Science Department, Xiamen University, Xiamen 361000, Fujian, China
suchang@xmu.edu.cn
[2] Computer Science Department, Xiamen University, Xiamen 361000, Fujian, China

Abstract. The previous work of metaphor interpretation mostly focused on single-word verbal metaphors and ignored the influence of contextual information, leading to some limitations(e.g. ignore the polysemy of metaphor). In this paper, we creatively propose the aspect-based semantic relatedness, and we present a novel metaphor interpretation method based on semantic relatedness for context-dependent nominal metaphors. First, we obtain the possible comprehension aspects according to the properties of source domain. Then, combined with contextual information, we calculate the degree of relatedness between the target and source domains from different aspects. Finally, we select the aspect which makes the relatedness between target and source domains maximum as comprehension aspect, and the metaphor explanation is formed with corresponding property of source domain. The results show that our method has higher accuracy. In particular, when the information of target domain is insufficient in corpus, our method still exhibits the good performance.

Keywords: Metaphor interpretation · Semantic relatedness · Contextual information · Comprehension aspect

1 Introduction

Metaphor, as a kind of common phenomenon in natural language, is not only a rhetorical means, but also a cognitive style of human beings. In recent years, the research of metaphor interpretation has attracted attention of people gradually. The existing methods of metaphor interpretation paid more attention to single-word metaphors expressed by verbs, such as "stir-excitement". Instead, the interpretation mechanism of nominal metaphors is more complex than verbal metaphors. It not only needs to find the relevance between target and source domains, but also requires the consideration about contextual background of metaphors. Hobbs [1] regarded metaphor interpretation as a part of the general discourse processing problem. Only in context the metaphorical expression can be properly interpreted. To avoid misunderstandings, speaker will guide people toward the desired direction to understand metaphor through the adjustment and supplement of discourse. Consider the following examples:

© Springer International Publishing Switzerland 2015
J. Li et al. (Eds.): NLPCC 2015, LNAI 9362, pp. 182–193, 2015.
DOI: 10.1007/978-3-319-25207-0_16

(1) 他是猫。(He is a cat.)

(2) 他是只晒太阳的慵懒的猫。(He is a lazy cat and enjoys the sunshine.)

When the contextual information is insufficient, metaphor interpretation is hard to determine. The source domain "猫(cat)" of examples has the properties, such as "顽皮(naughty)", "敏捷(quick)", "慵懒(lazy)" and so on. According to these properties of source domain, we can find different comprehension aspects to make target and source domain relevant. In the example (1), we might have multiple explanations, such as "他是顽皮的(He is naughty)", "他是敏捷的(He is quick)" and so on. In the example (2), based on the contextual information, we can make sure that the explanation "他是慵懒的(He is lazy)" is much better.

In nominal metaphors, a seemingly unrelated concept is usually associated with another concept. And metaphor interpretation is finding the appropriate aspect in line with knowledge background to make the relatedness between target and source domains maximum. In this paper, we present a metaphor interpretation method based on semantic relatedness, aiming at the context-dependent nominal metaphors. First, we extract the properties of source domain, and select possible comprehension aspects according to these properties. Then, we apply vector representations of words and integrate the relatedness between context and source domain to calculate the relatedness between target and source domains from different aspects. Finally, we choose the aspect which makes the relatedness of target and source domains maximum as the comprehension aspect, and metaphor explanation is formed with corresponding property of source domain.

The contributions of this paper are as follows: 1. We creatively propose the aspect-based semantic relatedness computational method. 2. We consider the influence of contextual information to comprehend metaphors, and apply the aspect-based semantic relatedness to context-dependent nominal metaphor interpretation. Combined with the contextual information, we calculate the relatedness between target and source domains, and obtain the metaphor explanation. The experiment results show that our method is effective.

The remainder of this paper is organized as follows. In Section 2, we provide an overview of related work. In Section 3, we show the theoretical basis at first. Then we introduce the aspect-based semantic relatedness computational method and the context-dependent nominal metaphor interpretation method. Section 4 shows the experimental results and evaluation. Our conclusions and future work are given in Section 5.

2 Related Work

2.1 Metaphor Interpretation

Shutova [2] proposed a method which first extracted a set of potential substitutes by selecting all words that appeared in a particular syntactic relation with the metaphorical verb in the BNC. Then it narrowed down the list of candidates by selecting the verbs that shared a hypernym with the metaphorical verb in WordNet. She used automatically induced selectional preferences to discriminate between figurative and literal paraphrases. Shutova et al.[3] presented a novel approach to metaphor interpretation with a vector space model using a non-negative matrix factorization to compute the meaning list of target verbs. Bollegala and Shutova [4] presented an unsupervised metaphor interpretation method that used the Web to find literal paraphrases for metaphorical expressions. Ekaterina et al. [5] presented a metaphor interpretation approach based on abduction. They mapped linguistic metaphors to conceptual metaphors, and interpreted conceptual metaphors in terms of both logical predicates and natural language expressions.

Comparing to other work that focused on single-word verbal metaphors, we utilize contextual information to deal with nominal metaphor interpretation through aspect-based semantic relatedness.

2.2 Semantic Relatedness

There are two kinds of semantic relatedness calculation model, knowledge-based and corpus-based approaches.

Knowledge-based methods employ information extracted from manually constructed lexical taxonomies, e.g., WordNet. Previous studies have focused on gloss [6] and the structure of the lexicon [7], such as the semantic path and depth. Siblini and Kosseim [8] used all 26 semantic relations found in WordNet in addition to information found in glosses.

Corpus-based approaches mainly use context information and structural pattern of corpus, such as using paths in the Wikipedia category structure, using the contents of the articles, or using the hyperlinks between articles. For example, Explicit Semantic Analysis (ESA) [9] as well as Salient Semantic Analysis (SSA) [10] was proposed to incorporate large amounts of human knowledge such as Wikipedia into word relatedness computation. They both represented a word as a concept vector, where each dimension corresponds to a Wikipedia concept. Yazdani and Popescu-Belis [11] used the contents and links for computing text semantic relatedness. Agirre et al. [12] studied the different types of links in Wikipedia, and applied random walk algorithm on the full graph based on those links for word relatedness and named-entity disambiguation. Mikolov et al. [13] used machine learning techniques to compute continuous vector representations of words from large datasets, and then calculated the word vector distance to measure semantic relatedness. They observed large improvements in accuracy at much lower computational cost.

In this paper, we adopt the corpus-based method. Compared with knowledge-based methods, our method has the advantage that information content is much richer. However, the existing corpus-based methods regard a concept as a whole. For example, compare all attributes of concepts, the more their common attributes the higher relatedness they are. Thus, they ignored that the semantic relatedness between concepts will change with comprehension aspect. And aspect-based semantic understanding is a noticeable phenomenon in natural language; aspect-based semantic relatedness computation can help understand natural language. Therefore, based on the previous researches, we propose an aspect-based semantic relatedness computational method.

3   Our Method

Our method consists of the following steps. Given the nominal metaphor with context, which has marked target and source domains, we extract the properties of source domain. Then, we obtain the possible comprehension aspects according to these properties. And from different aspects, we calculate the degree of relatedness between target and source domains combined with contextual information. Finally, we select the aspect which makes relatedness maximum as the comprehension aspect, and the corresponding property of source domain forms the metaphor explanation. Section 3.1 illustrates the theoretical basis of our methods. Section 3.2 presents the aspect-based semantic relatedness computation method and our method for the context-dependent nominal metaphor interpretation.

3.1   Theoretical Basis

Davidson [14] indicated that, generally speaking, the literal meaning of metaphor is clear error or absurd. In other words, the target domain and the source domain are not related literally. But, according to relevance theory, any words of discourse are related on the semantics, and the process of discourse understanding is looking for the relevance of discourse to support the "contextual effect" of discourse. Thus, although the target and source domains are not related in literal meaning. From some comprehension aspects, the semantic relatedness will be found, which we call aspect-based semantic relatedness. The process of metaphor interpretation is to find the appropriate aspect from which the relevance between target and source domains is constructed.

Semantic relatedness is based on the aspect, the related concepts from a certain aspect may be irrelevant from another aspect. For example, from the aspects "色彩(color)" "智慧(intelligence)" "行为(behavior)", "狐狸(fox)" has the properties"红色(red)" "聪明(smart)" " 敏捷(quick)", respectively. Based on the aspect "色彩(color)", target domain "律师(lawyer)" and source domain "狐狸(fox)" are irrelevant. But they are relevant from the aspect "智慧(intelligence)", because both of them have the property "聪明(smart)". Thus, the key to metaphor interpretation is the comprehension aspects.

What is more, because of the openness and uncertainty of metaphorical meaning [14], people might find multi-aspects to produce different explanations. For example, "张三是狼(Zhangsan is a wolf)" can be interpreted as "张三很凶狠(Zhangsan is cruel)" "张三很狡猾(Zhangsan is crafty)" "张三很多疑(Zhangsan is suspicious)". To avoid misunderstanding, speaker will guide listener to the correct understanding direction through the selection and adjustment of context, such as emphasizing on the typical properties of target domain, extending and stating the non-significant properties of source domain. When the contextual information is sufficient, the optimal explanation can be determined. According to relevance theory, the optimal metaphor comprehension should have the greatest relatedness with intent of discourse, and the context will make certain properties of source domain salient [15]. In this sense, the relatedness between context and source domain will help determine the best explanation.

Searle's metaphor theory [16] also supported our idea to some extent. It mainly explained how to interpret metaphor. For simple example, "S is P" which means "S is R". He pointed out that metaphor interpretation is utilizing context or existing knowledge base to obtain the relations among S,P and R. For this kind of metaphor, "S is P", he proposed six principles to get R through P. 1. R is in the definition of P. For example, because of the definition of "giant" is "big", the metaphor "Sam is a giant" can be interpreted as "Sam is big". 2. Under certain conditions, P is R, R is an important and well known feature of P. For example, "Sam is a pig" can be interpreted as "Sam is filthy, gluttonous, and sloppy, etc". 3. P is often said to be R, although P does not have the feature R. For example, the feature of P in a familiar myth. 4. As a result of natural or cultural reasons, we feel that there is a relation between P and R. For example, "I am in a black mood" can be interpreted as "I am angry and depressed". 5. P and R are not similar, but the situations which they are in are similar. For example, the metaphor "You have become an aristocrat" means the living conditions is similar to an aristocrat. 6. In some cases, P and R are same or similar in the sense, but the applied range of R is limited, then people use P. For example "His brain is addled", although "addled" only applies to "egg". These principles are to obtain the related features R of P from different aspects, then construct the semantic relatedness between S and P based on these aspects.

3.2   Metaphor Interpretation Based on Semantic Relatedness

Given the metaphor with context, we first extract the notional words(nouns, verbs, adjectives and adverbs), which are denoted as $w_1, w_2, ..., w_N$ (except the marked target domain $Target$ and source domain $Source$).

The knowledge of source domain is crucial to metaphor interpretation. In this paper, we select possible comprehension aspects according to the properties of source domain, and calculate the aspect-based semantic relatedness between the target and source domains. We extract the properties of the source domain by using $Attribute\ Database$[1] and $Sardonicus$[2], which are denoted as $p_1, p_2, ..., p_M$, the corresponding aspects are $r_1, r_2, ..., r_M$, respectively.

Then, metaphor interpretation can be regarded as a problem to find the comprehension aspect $\mathbf{r} = r_i$, which makes the semantic relatedness between target and source domains $Rel(Target, Source, \mathbf{r})$ maximum, and the property p_i of source domain which is corresponding to the aspect r_i will be salient. Thus, we express the metaphor explanation briefly as "$Target\ Be\ p_i$".

---

[1] A database developed by NLP Lab of Xiamen University.

[2] An adjective classification retrieval. http://afflatus.ucd.ie/sardonicus/tree.jsp.

In order to measure the semantic relatedness between words, a word w is represented by a vector $\vec{w}$, as follows:

$$\vec{w} = < c_1, c_2, ..., c_q > \tag{1}$$

where, q is the dimension of vector, $c_i (1 \leq i \leq q)$ is the value of dimension i.

Mikolov et al. [13] proposed two model architectures for computing continuous vector representations of words from large data sets, Continuous Bag-of-Words (CBOW) model and distributed Skip-gram model. They measured the quality of these representations in a word similarity task and compared their methods with different types of neural networks. The results revealed their methods had large improvements in accuracy at much lower computational cost. In this paper, we apply CBOW model[3] to obtain the vector representations of words.

Assuming source domain $Source$ has the property p from the aspect r, then the semantic relatedness between word w and word $Source$ based on aspect r, $Rel(w, Source, r)$, could be computed using cosine distance measure as follows:

$$Rel(w, Source, r) = dis_{cos}(\vec{w}, \vec{p}) \tag{2}$$

$$dis_{cos}(\vec{w}, \vec{p}) = \frac{\sum_{j=1}^{q} c_j e_j}{\sqrt{\sum_{j=1}^{q} c_j^2} \sqrt{\sum_{j=1}^{q} e_j^2}} \tag{3}$$

where, w is represented as $\vec{w} = < c_1, c_2, ..., c_q >$, p is represented as $\vec{p} = < e_1, e_2, ..., e_q >$.

Combined with the relatedness between context and source domain, we can obtain the semantic relatedness between target domain $Target$ and source domain $Source$ based on aspect r, $Rel(Target, Source, r)$, as follows:

$$
\begin{aligned}
Rel(Target, Source, r) &= dis_{cos}(\overrightarrow{Target}, \vec{p}) + \frac{1}{N} \sum_{i=1}^{N} Rel(w_i, Source, r) \\
&= dis_{cos}(\overrightarrow{Target}, \vec{p}) + \frac{1}{N} \sum_{i=1}^{N} dis_{cos}(\vec{w_i}, \vec{p})
\end{aligned}
\tag{4}
$$

where, N is the number of extracted notional words(except the marked target domain $Target$ and source domain $Source$). If $Target$(or w_i) absents in the corpus, let $dis_{cos}(\overrightarrow{Target}, \vec{p}) = 0$(or $dis_{cos}(\vec{w_i}, \vec{p}) = 0$).

We obtain the semantic relatedness between target and source domains from all possible aspects. Then we choose the aspect r which makes the relatedness between target and source domains maximum as the comprehension aspect, as shown in Eq(5).

---

[3] https://code.google.com/p/word2vec/.

$$\mathbf{r} = \underset{r_i}{\operatorname{argmax}}\{Rel(Target, Source, r_i)\}$$

$$= \underset{r_i}{\operatorname{argmax}}\{dis_{cos}(\overrightarrow{Target}, \overrightarrow{p_i}) + \frac{1}{N}\sum_{j=1}^{N} Rel(w_j, Source, r_i)\} \qquad (5)$$

$$= \underset{r_i}{\operatorname{argmax}}\{dis_{cos}(\overrightarrow{Target}, \overrightarrow{p_i}) + \frac{1}{N}\sum_{j=1}^{N} dis_{cos}(\overrightarrow{w_j}, \overrightarrow{p_i})\}$$

where $r_i \in \{r_1, r_2, ..., r_M\}$.

The corresponding property $\mathbf{p}=p_i$ of source domain from the aspect $\mathbf{r}=r_i$ will become salient. The property p_i forms the metaphor explanation which is expressed briefly as "$Target$ Be p_i".

4   Experiment and Evaluation

In our experiments, we use *Reader Corpus*[4] as the corpus, and use *Segtag*[5] to support CBOW model in computing the vector representations of words. Considering the data sparse problem in corpus, we extend the synonyms of the property word p using the *Tongyi Cilin (Extended)*[6]. The synonyms set of p is represented as $S = \{v_1, v_2, ..., v_{|S|}\}$, where $|S|$ is the number of synonyms. If r is the corresponding aspect of property p, the semantic relatedness between word w and word *Source* based on aspect r, $Rel(w, Source, r)$, could be computed as follows:

$$Rel(w, Source, r) = dis_{cos}(\overrightarrow{w}, \overrightarrow{p}) = \frac{1}{|S|}\sum_{i=1}^{|S|} dis_{cos}(\overrightarrow{w}, \overrightarrow{v_i}) \qquad (6)$$

Specially, if w and p is synonym, let $Rel(w, Source, r) = 1$.

For the metaphor "如今的华为已经成了狮子，成为电信行业当之无愧的王者(Now Huawei has become a lion, who is the king of the telecommunications industry)", the marked target domain is "华为(Huawei)" and source domain is "狮子(lion)". We extract the properties of "狮子(lion)" and select possible comprehension aspects according to the properties. Table 1 contains the semantic relatedness between target and source domains from different aspects(the corresponding aspects are given by *Adjectives Database*[7]).

As shown in Table (1), we see the relatedness between word "王者(king)" and source domain "狮子(lion)" are higher from the aspects "心情(mood)" and "神情(manner)", thus the corresponding properties "恼怒(irritated)" and "威严(august)" become more salient.

---

[4] A Chinese Corpus. URL: www.duzhe.com.

[5] A word segmentation tool of NLP Lab of Xiamen University.

[6] A Chinese Thesaurus, http://ir.hit.edu.cn/.

[7] An adjective classification database developed by NLP Lab of Xiamen University.

Table 1. The relatedness between word "王者(king)" and source domain "狮子(lion)" from different aspects

w	Source	p	r	$\mathrm{Rel}(w, Source, r)$
		饥	衣食	0.1135108
		hungry	subsistence	
		凶猛	行为	0.1261268
王者	狮子	fierce	behavior	
king	lion	恼怒	心情	0.1703408
		irritated	mood	
		威严	神情	0.1515766
		august	manner	

In order to better illustrate the performance of our method, we compare our method with the following three simple methods:(1)RT: Ignoring the contextual information, it just calculates the semantic relatedness between the target and source domains from different aspects, which represents as $RT(Target, Source, r)$; (2)RS: Select the most salient property of source domain and we use aspect-based semantic relatedness to measure the degree of saliency, which represents as $RS(Source, Source, r)$; (3)RTS: Combined RT with RS, it not only utilizes the semantic relatedness between the target and source domains, but also the saliency of properties, which represents as $RTS(Target, Source, r)$. The computational formulas are as follows:

$$RT(Target, Source, r) = dis_{cos}(\overrightarrow{Target}, \overrightarrow{p}) \tag{7}$$

$$RS(Source, Source, r) = dis_{cos}(\overrightarrow{Source}, \overrightarrow{p}) \tag{8}$$

$$RTS(Target, Source, r) = RT(Target, Source, r) + RS(Source, Source, r) \tag{9}$$

where, p is the corresponding property of source domain from the aspect r.

Considering the following examples, the words with underline are the extracted notional words of context:

(1) "高龄 八十九岁的萧乾，仍然是一匹生气勃勃的野马。"
   "XiaoQian with the advanced age of eighty-nine is still a vibrant wild horse."
   The marked target domain is "萧乾(XiaoQian)", source domain is "野马(wild horse)".

(2) "不用 上班，不用 上学，就躺着晒太阳，多 好啊！我是一只没有 忧虑 烦恼的猫。"
   "How wonderful to just lay in the sun without work and school ! I am a no-worry cat."
   The marked target domain is "我(I)", source domain is "猫(cat)".

(3) "这个塞维利亚球员长着一张娃娃脸，但在球场上却是个不折不扣的魔鬼。"
   "The baby-faced Sevilla player is a real devil on the pitch."
   The marked target domain is "球员(player)", source domain is "魔鬼(devil)".

(4) "古老的济南，城内那么狭窄，城外又那么宽敞，山坡上卧着些小 村庄，小村庄的房顶上卧着点雪，这是张小 水墨画，也许是唐代的名手 画的吧。"

"Old Jinan is so narrow in the city, but it is so spacious outside the city, small villages lay on the hillside, little snow lies on the roof of the villages. It is a little wash painting, and perhaps drawn by a famous artist of Tang Dynasty."

The marked target domain is "这(it)"(refer to the scenery), source domain is "水墨画(wash painting)".

The results of four methods are demonstrated in Table 2.

As the example (1) shows, the source domain "萧乾(XiaoQian)" does not appear in the corpus, thus we do not have access to relevant information of source domain. It conforms to the common phenomenon in the metaphor that listeners are not familiar with the target domain, thus they are guided by the source domain and the current context to understand the metaphor. RT can not get a result. RS and RTS choose the most salient property "魁梧(strapping)" as the result. Our method considers the relevance between context and source domain, which get the best interpretation and the chosen property is "矫健(strong and vigorous)". The example shows that our method still has a good effect in the absence of the target domain information, and proves the guidance of the context information for metaphor interpretation.

As the example (2) shows, the property "顽皮(naughty)" is obviously more salient than other properties of source domain "猫(cat)". Thus, RS and RTS choose the "顽皮(naughty)" as the results. The effect of context makes the property "慵懒(lazy)" become more salient in our method. It reveals that contextual information emphasizes the properties of the target domain.

As the example (3) shows, the results of RT reveal the semantic relatedness between target domain "球员(player)" and source domain "魔鬼(devil)" from the aspect "感觉(sense)" is greatest. Then, RT chooses the corresponding property "可怕(horrible)". But, the most salient property of source domain "魔鬼(devil)" is "邪恶(evil)" according to RS and the saliency is obviously higher than others, which results in the most salient property is still "邪恶(evil)" in RTS. And in our method, combined the relatedness between context and source domain, the property "可怕(horrible)" becomes the most salient property.

As the example (4) shows, the marked target domain "这(it)" refers to the scenery of Jinan. In other words, the marked target domain does not provide available information. The results of RT show there is not significant difference among the relatedness from various aspects. From the results of RS, we see the property "简洁(concise)" of source domain "水墨画(wash painting)" is more salient than others obviously. Our method obtains the better result than RS and RTS. It reveals our method still work well when the information of target domain is absent. It also shows the computation of aspect-based semantic relatedness between context and source domain utilizes the contextual information validly.

We evaluate the experiment with the help of human annotators that annotate 80 instances of nominal metaphors with context from the Web, Blogs, and the Books. The data contains various genres: news/journal articles, politics, finance,

Table 2. The results of four methods

Target	Source	p	RT	RS	RTS	Our Method
萧乾 XiaoQian	野马 wild horse	高大 tall	NIL	0.09607011	0.09607011	0.092578129
		魁梧 strapping	NIL	0.18563326	0.18563326	0.110308271
		强壮 strong	NIL	0.12608944	0.12608944	0.127122275
		矫健 strong and vigorous	NIL	0.11942509	0.11942509	0.134411583
我 I	猫 cat	贴心 intimate	0.01600922	0.01757664	0.03358586	0.070006938
		灵活 flexible	0.03114080	0.07590316	0.10704396	0.075094045
		敏捷 quick	-0.0261091	0.02083970	-0.0052694	-0.00909032
		顽皮 naughty	0.09468020	0.21980882	0.31448899	0.174238095
		慵懒 lazy	0.09193315	0.0590001	0.15093327	0.197266790
球员 player	魔鬼 devil	可怕 horrible	0.029382963	0.06316606	0.092549023	0.091602731
		残忍 cruel	0.009653102	0.14321677	0.152869872	0.059492234
		邪恶 evil	0.008303388	0.17595066	0.184254048	0.077411169
这 it	水墨画 wash painting	名贵 valuable	0.043466200	0.08059552	0.08059552	0.045615660
		简洁 concise	0.056759820	0.12486820	0.12486820	-0.00535744
		美丽 beautiful	0.088287376	0.06179154	0.06179154	0.061062775
		相同 same	0.057834150	0.06957777	0.06957777	0.026469391
		简单 simple	0.058563670	0.05398252	0.05398252	0.023385722

essays, fiction and speech. We have 5 volunteer annotators who are all native speakers of Chinese and their agreement on the preliminary test was $0.66(\kappa)$ [17], which is considered reliable. Then we evaluated our method against their judgments in terms of accuracy and compared with other methods. We divided the acceptability into five levels instead of simple binary decision(accept/decline), because the five-level method makes the evaluation finer-grained. We asked annotators to score the acceptability of each result and took average acceptability below three as the incorrect results. The results are demonstrated in Table 3.

Table 3. The results of four methods

	RT	RS	RTS	Our Method
Accuracy	0.59	0.55	0.61	**0.84**

The results show our method achieves higher accuracy and it has obvious improvements compared with other three methods, which reveals our method has the good performance. Then, we analyze the errors and propose the solutions to improve our approach in the future.

1. We use the *AttributeDatabase* and *Sardonicus* to obtain the properties of source domain, but some properties are still difficult to be extracted, causing that the method is unable to obtain the appropriate comprehension aspect. In the future, we will improve the extraction mechanism of properties to solve this problem.
2. Context information provides the cues of proper comprehension aspect, at the same time, it may also bring some noise. In the future, we can introduce the weights to reduce the influence of noise.

5   Conclusions

The previous researches of metaphor interpretation mostly focus on single-word metaphors expressed by verbs and are lacking in the effective use of context information. In this paper, we creatively propose the aspect-based semantic relatedness and present a novel metaphor interpretation method based on semantic relatedness for context-dependent nominal metaphors. We obtain the possible comprehension aspects according to the properties of source domain. Then, combined with the relatedness between context and source domain, we calculate the relatedness between target and source domain from different aspects. Finally, we select the aspect which makes the relatedness between target and source domains maximum as the comprehension aspect, and the corresponding property of source domain will be salient, forming the metaphor explanation. The experimental results show that the aspect-based semantic relatedness computation is reasonable and the context information can effectively guide to the appropriate understanding of metaphor. We evaluate our method and compare it with other methods, the results show that our method has good performance. In future work, we will improve our approach and apply the method to other NLP tasks such as word sense disambiguation and text clustering.

Acknowledgments. Funding was provided by the National Natural Science Foundation of China under Grant(No. 61075058).

References

1. Hobbs, J.R.: Metaphor and abduction. In: Ortony, A., Slack, J., Stock, O. (eds.) Communication from an Artificial Intelligence Perspective: Theoretical and Applied Issues. NATO ASI Series, pp. 35–58. Springer, Heidelberg (1992)
2. Shutova, E.: Automatic metaphor interpretation as a paraphrasing task. In: Human Language Technologies: The 2010 Annual Conference of the North American Chapter of the Association for Computational Linguistics, pp. 1029–1037. Association for Computational Linguistics (2010)
3. Shutova, E., Van de Cruys, T., Korhonen, A.: Unsupervised metaphor paraphrasing using a vector space model. In: COLING (Posters), pp. 1121–1130 (2012)
4. Bollegala, D., Shutova, E.: Metaphor interpretation using paraphrases extracted from the web. PloS one **8**(9), e74304 (2013)
5. Ovchinnikova, E., Israel, R., Wertheim, S., Zaytsev, V., Montazeri, N., Hobbs, J.: Abductive inference for interpretation of metaphors. In: ACL 2014, pp. 33 (2014)
6. Zesch, T., Gurevych, I.: Wisdom of crowds versus wisdom of linguists - measuring the semantic relatedness of words. Natural Language Engineering **16**(01), 25–59 (2010)
7. Liu, B., Feng, J., Liu, M., Liu, F., Wang, X., Li, P.: Computing semantic relatedness using a word-text mutual guidance model. In: Zong, C., Nie, J.-Y., Zhao, D., Feng, Y. (eds.) NLPCC 2014. CCIS, vol. 496, pp. 67–78. Springer, Heidelberg (2014)
8. Siblini, R., Kosseim, L.: Using a weighted semantic network for lexical semantic relatedness. In: RANLP, pp. 610–618 (2013)
9. Gabrilovich, E., Markovitch, S.: Computing semantic relatedness using wikipedia-based explicit semantic analysis. In: IJCAI, vol. 7, pp. 1606–1611 (2007)
10. Hassan, S., Mihalcea, R.: Semantic relatedness using salient semantic analysis. In: AAAI (2011)
11. Yazdani, M., Popescu-Belis, A.: Computing text semantic relatedness using the contents and links of a hypertext encyclopedia. In: Proceedings of the Twenty-Third International Joint Conference on Artificial Intelligence, pp. 3185–3189. AAAI Press (2013)
12. Agirre, E., Barrena, A., Soroa, A.: Studying the wikipedia hyperlink graph for relatedness and disambiguation (2015). arXiv:1503.0165
13. Mikolov, T., Chen, K., Corrado, G., Dean, J.: Efficient estimation of word representations in vector space (2013). arXiv preprint arXiv:1301.3781
14. Davidson, D.: What metaphors mean. Critical inquiry, pp. 31–47 (1978)
15. Macagno, F., Zavatta, B.: Reconstructing metaphorical meaning. Social Science Electronic Publishing **28**(4), 453–488 (2014)
16. Searle, J.: Metaphor (1978)
17. Sidney, S.: Nonparametric statistics for the behavioral sciences. The Journal of Nervous and Mental Disease **125**(3), 497 (1957)

Context Vector Model for Document Representation: A Computational Study

Yang Wei[1,2], Jinmao Wei[1,2(✉)], and Hengpeng Xu[1,2]

[1] College of Computer and Control Engineering,
Nankai University,Weijin Rd. 94, Tianjin 300071, China
weiyang_tj@outlook.com, weijm@nankai.edu.cn,
xuhengpeng@mail.nankai.edu.cn
[2] College of Software, Nankai University, Weijin Rd. 94, Tianjin 300071, China

Abstract. To tackle the sparse data problem of the bag-of-words model for document representation, the Context Vector Model (CVM) has been proposed to enrich a document with the relatedness of all the words in a corpus to the document. The nature of CVM is the combination of word vectors, wherefore the representation method for words is essential for CVM. A computational study is performed in this paper to compare the effects of the newly proposed word representation methods embedded in CVM. The experimental results demonstrate that some of the newly proposed word representation methods significantly improve the performance of CVM, for they estimate the relatedness between words better.

Keywords: Document representation · Word vector · Relatedness

1 Introduction

Since representing documents in a feature space [21] is a pre-requisite work for many machine learning algorithms, e.g., text classification and clustering, converting a raw text to a fixed-length vector has long been studied. Perhaps the most common vector representation for texts is the bag-of-words (BOW) model due to its simplicity, comprehensibility and acceptable accuracy. However, words are assumed to be independent of each other in BOW, where relatedness actually exists. The neglect of word relatedness incurs the sparse data problem. Specifically, BOW cannot reveal the similarities between documents composed of different words. In other words, BOW has little sense about the semantic meanings of documents.

Some dimensionality reducing methods [9,11,14,23] have successfully constructed compact feature spaces. Probabilistic generative algorithms [1] and the neural probabilistic language model [15] are the outstanding artifacts in this branch. Significant improvements have been achieved with these methods. Nevertheless, the parameters, especially dimension of the space, are often difficult to be decided.

© Springer International Publishing Switzerland 2015
J. Li et al. (Eds.): NLPCC 2015, LNAI 9362, pp. 194–206, 2015.
DOI: 10.1007/978-3-319-25207-0_17

The Context Vector Model (CVM) [3,8,13,20], which represents documents in the same feature space of BOW, tackles the sparse data problem by considering the relatedness of all the words in a corpus to a document. This mechanism is achieved by the combination of word vectors. More precisely, the document is represented as the weighted sum of word vectors, where the weights of the word vectors are estimated based on the frequencies of the words in the document.

Representing words as vectors, the relatedness of the dimensions to a word is evaluated by their co-occurrences with the word [5,22]. Since the relatedness between words cannot be obtained directly from the dimensionality reduction methods of word representation, e.g., LDA (Latent Dirichlet Allocation) [4] and SVD (Singular Value Decomposition) [7], or the distributed representation for words, e.g., word2vec [17], the distributional representation in the feature space of BOW [6,12] is used as the statistical foundations. With the weighted sum of the distributional word vectors, the relatedness of a word to a document is obtained by the weighted sum of the relatedness of the word to the original words in the document. In this sense, the relatedness between words plays an important role for CVM. Since several word representation methods [6,12,19] have been proposed in recent years, it's meaningful to test their effects embedded in CVM for document representation. Hence a computational study on these methods is performed in this paper.

Besides, the weighted sum of word vectors has already been proved to hinder its usage on representing the semantic meanings of phrases [18]. Intuitively, the relatedness of a word to several original words in a document, which describe the same topic, may contain repeating information. Then the sum of the relatedness overestimates the relatedness of the word to the document. We propose to keep the maximum values of the quantified scores on each dimension of the word vectors to guarantee that there is no repeating information. This combination scheme of word vectors is a variance of CVM. The word representation methods embedded in different combination schemes are also compared in this paper.

The remainder of this paper is organized as follows: Sect. 2 provides the preliminaries for BOW, CVM and the word representation methods. The detailed experimental setup is described in Sect. 3. The experimental results are presented in Sect. 4. Finally, in Sect. 5, the conclusion and the direction of future work is provided.

2   Preliminaries

2.1   The BOW Model

According to BOW, the raw collection of n documents, D, must be preprocessed for vector representation. The necessary pre-operations include tokenization, to split sentences into individual tokens; stemming, a process of reducing words to their basic forms; and stopword removal. The derived words by preprocessing constitute the collection's vocabulary V. If there are m words in the vocabulary, a feature space with m-dimensions are generated. Hence a document d could be represented as:

$$\Phi_{bow} : \mathbf{d} = (c_{v_1|d}, c_{v_2|d}, \cdots, c_{v_m|d}) \in \mathbb{R}^m, \tag{1}$$

where $c_{v_x|d}$ is the occurrence times of the word v_x in d. $c_{v_x|d}$ could be the raw occurrence times of v_x in d, while it is usually re-weighted by the popular tf·idf weighting scheme:

$$c_{v_x|d} = \text{tf}_{v_x|d} \cdot \text{idf}_{v_x} = \frac{c_{v_x|d}}{\sum_{y=1}^{m} c_{v_y|d}} \cdot (1 + \log_2(\frac{n}{n_{v_x}})), \tag{2}$$

where n_{v_x} is the number of documents in which v_x occurs. $\text{tf}_{v_x|d}$ is called the Term Frequency of v_x in d, and idf_{v_x} is the Inverse Document Frequency of v_x in the whole corpus.

2.2 Context Vector Model

Since BOW cannot figure out similar documents composed of different words, the Context Vector Model (CVM) tries to reveal the meanings of documents with a set of weighted word vectors [3,8,13,20]. $\forall v_x \in V$, its word vector is usually defined as [5,22]:

$$\mathbf{v}_x = (\frac{c_{v_x,v_1|D}}{c_{v_x|D}}, \frac{c_{v_x,v_2|D}}{c_{v_x|D}}, ..., \frac{c_{v_x,v_m|D}}{c_{v_x|D}}), \tag{3}$$

where $c_{v_x,v_y|D}$ is the co-occurrence times between v_x and v_y in the whole corpus, and $c_{v_x|D}$ is the total times v_x occurs in D. The basic assumption of word vectors is that words that occur in similar contexts tend to have similar meanings [10]. Generally, the meanings of words should be independent of the corpus size, so $c_{v_x|D}$ is introduced to give the basic context of a word [6]. The values in a word vector measure the relatedness of the dimensions to the word.

Together with all the word vectors generated by (3), an $m \times m$ matrix is obtained, which is called *context matrix*:

$$\mathbf{V} = (\mathbf{v}_1, \mathbf{v}_2, ..., \mathbf{v}_m) . \tag{4}$$

Then the new document vector generated by CVM is:

$$\Phi_{cvm} : \mathbf{d}' = \mathbf{dV} = \left(\sum_{x=1}^{m} c_{v_x|d} \frac{c_{v_x,v_1|D}}{c_{v_x|D}}, \sum_{x=1}^{m} c_{v_x|d} \frac{c_{v_x,v_2|D}}{c_{v_x|D}}, ..., \sum_{x=1}^{m} c_{v_x|d} \frac{c_{v_x,v_m|D}}{c_{v_x|D}}\right). \tag{5}$$

Instead of tf·idf, $c_{v_x|d}$ will be evaluated by the simple tf scheme in this paper, because the global importance of a word in D indicated by idf has already been described in detail by the context matrix. CVM is a method which combines the local term weights (tf) and the global word relatedness together. As a result, the generated document vector is the weighted sum of the word vectors in the context matrix according to their tf weights in d. In fact, the values for each dimension are re-estimated according to their relatedness to the documents, where the relatedness of a word to the document is estimated by the weighted sum of its relatedness to the original words in the document. The nature of CVM incurs the reasonability of the weighted sum of word vectors strategy, for the simple sum of the relatedness of a word to the original words may contain duplicate counts. Intuitively, if two terms express the same meaning in some perspective, the more informative one will cover the other one in most cases. In particular, supposing a word is related to two original words in a document, the accumulation of the relatedness of the word to the two original words will contain duplicate counts if the original words express the same meaning. So CVM is an aggressive strategy which cannot hold unless all the words in a document are independent to each other.

On the contrary, we give a conservative strategy, which takes the strongest relatedness of a word to the original words in a document as the final relatedness of the word to the document. This strategy corresponds to the following representation for a document:

$$\Phi_{crm} : \mathbf{d}' = \left(\max_{x=1}^{m} c_{v_x|d} \frac{c_{v_x,v_1|D}}{c_{v_x|D}}, \max_{x=1}^{m} c_{v_x|d} \frac{c_{v_x,v_2|D}}{c_{v_x|D}}, \ldots, \max_{x=1}^{m} c_{v_x|d} \frac{c_{v_x,v_m|D}}{c_{v_x|D}} \right).$$
(6)

The new representation is named as Context Vector Model with the Maximum-value-aware strategy (CVMM in short), as it reserves the maximum values on each dimension of the word vectors. CVMM will hold when all the words in a document are related to each other. The weighted sum strategy used by CVM is renamed as Context Vector Model with the Accumulation strategy (CVMA). CVMA and CVMM only differ in the combination schemes of word vectors. In practice, neither of the assumptions required by CVMA or CVMM is true. CVMA will overestimate the relatedness of a word to a document, for the duplicate counts exist. While CVMM will underestimate the relatedness of a word to a document, for the non-maximum relatedness of the word to the original words is overlooked. The practical effects of CVMA and CVMM will be compared in our experiments. In the following, CVM refers to both CVMA and CVMM.

2.3 Generating the Word Vectors

According to (5) and (6), the relatedness between words plays a central role for CVM. As there are several methods to generate the word vectors in the feature space of BOW, our motivation is to embed these methods in CVM to find the best algorithm for document representation. The first word representation method to be compared is the one proposed along with CVMA [3]:

$$\frac{c_{v_x,v_y|D}}{c_{v_x|D}} = \frac{\sum_{a=1}^{n} \frac{c_{v_x|d_a}}{\sum_{z=1}^{m} c_{v_z|d_a}} \cdot \frac{c_{v_y|d_a}}{\sum_{z=1}^{m} c_{v_z|d_a}}}{\sum_{a=1}^{n} \left(\frac{c_{v_x|d_a}}{\sum_{z=1}^{m} c_{v_z|d_a}} \sum_{b=1,b\neq x}^{m} \frac{c_{v_b|d_a}}{\sum_{z=1}^{m} c_{v_z|d_a}} \right)}.$$
(7)

The other word representation methods are listed in Table 1, which are proposed for word representation independently [6,12,19]. The co-occurrence times and the total occurrence times used in Table 1 are usually counted with the context window method [16], where a "window", representing a span of words,

Table 1. Methods for Generating Word Vectors

Method Name	Weighting Scheme
Binary (BNR)	0 or 1
Term Frequency (TF)	$c_{v_x,v_y|l,D} / \sum_{y=1}^{m} c_{v_x,v_y|l,D}$
Log of TF (LTF)	$\log_2 c_{v_x,v_y|l,D} / \log_2 \sum_{y=1}^{m} c_{v_x,v_y|l,D}$
Log of TF-IDF (IDF)	$\log_2 c_{v_x,v_y|l,D} / log_2 \sum_{y=1}^{m} c_{v_x,v_y|l,D} \cdot \log_2 \frac{n}{n_{v_x}}$
Mutual Information (MI)	$\log_2 \dfrac{c_{v_x,v_y|l,D}/c_D}{(\sum_{x=1}^{m} c_{v_x,v_y|l,D}/c_D)(\sum_{y=1}^{m} c_{v_x,v_y|l,D}/c_D)}$

is passed over the corpus being analyzed, and words within this window are recorded as co-occurring. For instance, by setting the window length to two, the fragment "the key to success, the success to $\cdots$" can be decomposed to (the key), (key to), (to success), (success the), (the success), and (success to). The co-occurrence times between "success" and "the" are two due to the appearances of (success the) and (the success). Whereas the co-occurrence times between "success" and "key" is zero for neither (success key) nor (key success) appears. In our experiments, the window length l will be set to two, which gives the strictest definition of "co-occurrence", and is the way to involve least non-related words for a target word.

The symbol $c_{v_x,v_y|l,D}$ denotes the co-occurrence times between v_x and v_y in the corpus D with the window length l, and $c_D = \sum_{x=1}^{m} \sum_{y=1}^{m} c_{v_x,v_y|l,D}$. The BNR method sets $\frac{c_{v_x,v_y|D}}{c_{v_x|D}}$ to one if $c_{v_x,v_y|l,D}$ is bigger than zero; otherwise $c_{v_x,v_y|l,D}$ is set to zero.

3   Experimental Setup

The performance of the pairwise similarity evaluation is an important index to verify the qualities of the representations for documents. Generally, with good representation, the similarities between semantically related documents should obtain high scores, while the similarities between unrelated documents should obtain low scores. This is consistent with the purpose of the clustering task that similar documents are organized into the same group, while dissimilar documents are organized into different groups. Therefore, we evaluate our algorithms on document clustering problem with the Group-average Agglomerative Hierarchical Clustering (GAHC) algorithm. The evaluation of the similarities between documents directly affects the results of GAHC, thus can reflect the qualities of the representation methods for documents.

Table 2. Characteristics of the Reuters Subsets

NO.	Topics	m	l_d	$\bar{O}$	NO.	Topics	m	l_d	$\bar{O}$
D_1	earn, money-supply	1452	101.54	0.126	D_2	coffee, sugar	2586	214.28	0.118
D_3	money-supply, sugar	1987	155.62	0.121	D_4	sugar, interest	2272	161.84	0.095
D_5	crude, money-supply	2460	169.64	0.107	D_6	coffee, interest	2518	186.76	0.102
D_7	money-supply, interest	1766	128.00	0.112	D_8	crude, interest	2685	175.86	0.094
D_9	trade, interest		2727	196.76	0.096	D_{10}	coffee, money-fx	2759	213.54	0.112
D_{11}	ship, sugar		2801	180.82	0.104	D_{12}	crude, sugar	2804	203.48	0.106
D_{13}	trade, sugar		2856	224.38	0.105	D_{14}	crude, money-fx	2930	202.64	0.102
D_{15}	ship, money-fx		2950	179.98	0.094	D_{16}	acq, crude	2956	176.50	0.095
D_{17}	acq, trade		3009	197.40	0.094	D_{18}	coffee, ship	3021	205.74	0.111
D_{19}	crude, coffee		3044	228.40	0.113	D_{20}	crude, trade	3229	238.40	0.102
D_{21}	all topics		3907	126.82	0.101					

3.1 Dataset

Two document collections are used in our experiments. The first is the NSF research award abstracts[1]. One hundred documents are selected randomly from the category Materials Research (MR) and Industrial Technology (IT), fifty documents per category.

The second is the twenty-one subsets extracted from Reuters[2]. The characteristics of these subsets are described in Table 2. Column *Topics* in Table 2 states the predefined categories of each subset (*all topics* of subset D_{21} means the categories acq, coffee, crude, earn, interest, money-fx, money-supply, ship, sugar, and trade are all included in this subset). Column m states the vocabulary sizes of each subset. Column l_d states the average lengths of documents in each subset. And Column $\bar{O}$ states the average overlap ratios defined in (10). There are 200 documents in $D_1 - D_{20}$, 100 per category; and five hundred documents in D_{21}, where the ratios of the number of documents in each category reserves the original ratios in the full dataset of Reuters.

All the datasets are preprocessed by removing tags, tokenizing, stemming and stopword removal. A word is considered to be stopword if its frequency in the dataset is bigger than 0.5.

3.2 Methods to Be Compared

Thirteen methods will be tested in the following experiments, namely:

1. The BOW model with the tf · idf weighting scheme, which is used as the baseline;
2. CVMA incorporated with ACP, TF, LTF, IDF, MI and BNR, respectively;
3. CVMM incorporated with ACP, TF, LTF, IDF, MI and BNR, respectively.

---

[1] http://archive.ics.uci.edu/ml/datasets/NSF+Research+Award+Abstracts+
1990-2003

[2] http://archive.ics.uci.edu/ml/datasets/Reuters-21578+Text+
Categorization+Collection

3.3 Similarity and Distance Measure

Cosine is used to quantify the pairwise similarity between two documents. For document vectors, d_i and d_j, their cosine similarity is given by:

$$\text{sim}_{\cos}(\mathbf{d}_i, \mathbf{d}_j) = \frac{\mathbf{d}_i \mathbf{d}_j^\top}{\|\mathbf{d}_i\|_2 \|\mathbf{d}_j\|_2} . \tag{8}$$

The distance between two documents is defined as the following accordingly:

$$\text{Dist}(\mathbf{d}_i, \mathbf{d}_j) = 1 - \text{sim}_{\cos}(\mathbf{d}_i, \mathbf{d}_j) . \tag{9}$$

3.4 Evaluation Metrics

Overlap Ratio. The overlap ratio is a metric to evaluate the discrimination between two vectors, which is defined as:

$$O(\mathbf{d}_i, \mathbf{d}_j) = \frac{2 * |\mathbf{d}_i \cap \mathbf{d}_j|}{|\mathbf{d}_i| + |\mathbf{d}_j|}, \tag{10}$$

where $|\mathbf{d}_i|$ is the number of dimensions with non-zero values in $\mathbf{d}_i$, and $|\mathbf{d}_i \cap \mathbf{d}_j|$ is the number of dimensions with non-zero values in both $\mathbf{d}_i$ and $\mathbf{d}_j$.

Standard Deviation. The standard deviation of the values in a word vector is defined as:

$$\text{std}(\mathbf{v}_x) = \sqrt{\sum_{y=1}^{m} \left(\frac{c_{v_x,v_y|D}}{c_{v_x|D}} - \frac{1}{m} \sum_{z=1}^{m} \frac{c_{v_x,v_z|D}}{c_{v_x|D}} \right)^2} . \tag{11}$$

F$_1$-Score. With the index i to denote the i-th class and j to denote cluster j,

$$F_1 = \sum_i \frac{n_i}{n} \max_j \frac{2RP}{P+R}, \tag{12}$$

where n_i is the number of documents in class i, $R = \frac{n_{i,j}}{n_i}$ which is called recall in the field of Information Retrieval (IR), and $P = \frac{n_{i,j}}{n_j}$ which is called precision in IR. $n_{i,j}$ is the number of documents in both class i and cluster j [2].

Normalized Mutual Information

$$\text{NMI} = \frac{\sum_{i,j} n_{i,j} \log \frac{n n_{i,j}}{n_i n_j}}{\sqrt{(\sum_i n_i \log \frac{n_i}{n})(\sum_j n_j \log \frac{n_j}{n})}} . \tag{13}$$

The range of NMI is $[0, 1]$, where a value of one denotes a perfect match between clusters and reference classes [2].

3.5 Clustering Method

GAHC considers each document as a unique cluster initially and selects a pair of clusters to merge repeatedly in the merging procedure. In each turn, the pair of the most similar clusters is selected to be merged. The similarity of two clusters is calculated as the average pairwise similarities between the documents in the two clusters. The stop of the merging procedure for GAHC is achieved by predefining the target number of clusters. Specifically, the target numbers for the subsets D_1 to D_{20} range from 2 to 20. Then the number corresponding to the best F_1-score is reserved as the final number of clusters. Similarly, for dataset D_{21}, the target numbers of clusters are set in a range from 10 to 26.

4 Experimental Results

4.1 Discrimination of Document Vectors

The discrimination of document vectors generated by different models was evaluated on the NSF dataset. CVMA and CVMM were performed with ACP. Each point in the scatter diagrams of Fig. 1 represents the results of a comparison between two document vectors. The circles in the first column are the results produced by BOW; the stars in the second column represents the results of CVMA; and the plus signs in the last column represents the results of CVMM. The horizontal axis stands for the overlap ratio of two vectors, and the vertical axis stands for their similarity. The first row in Fig. 1 exhibits the results of which both documents to be compared were extracted from MR, the second row exhibits the results of which both documents were extracted from IT, and the third row illustrates the comparing results of which one document was selected from MR and the other was selected from IT.

It's shown that the overlap ratios between the document vectors generated by BOW are very low, and the document similarities wander around a small value.

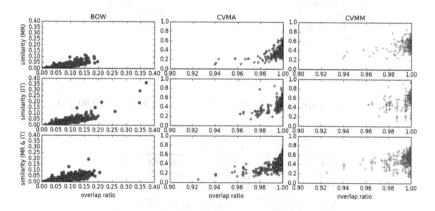

Fig. 1. Discrimination of document vectors using different representing methods.

On the contrary, the overlap ratios with CVMA or CVMM are quite high, and the document similarities range in a wide scope. This contrast reveals the advantages of using word relatedness: overcoming data sparse and magnifying the differences between document vectors. With CVM, the words which have not occurred in a document will be involved if they are related to any original words in the document. Hence the generated document vectors will have more words in common. Therefore, CVM has more smoothing power than BOW. This lead to the result that the similarities between documents composed of different words will be revealed according to the newly discovered common related words. Then similar documents are distinguished from dissimilar ones.

4.2   Performance on Document Clustering

In this experiment, the subsets of Reuters listed in Table 2 were used. On the subsets D_1 to D_{20}, the performances of the document representation methods on the variation of the data properties caused by topic changing were observed by restricting the number of categories and the number of documents, specifically, two categories and two hundred documents per subset. Since the experiment

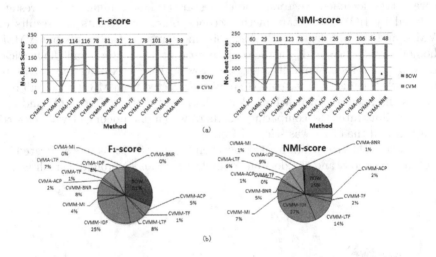

Fig. 2. Comparison between BOW and CVM on 200 datasets.

Table 3. Average Scores of the Versions of CVM on 200 Datasets

	ACP	TF	LTF	IDF	MI	BNR
			F_1-score			
CVMM	0.846	0.782	0.879	0.886	0.839	0.852
CVMA	0.775	0.757	0.835	0.857	0.773	0.782
			NMI-score			
CVMM	0.509	0.343	0.613	0.627	0.520	0.551
CVMA	0.341	0.275	0.486	0.561	0.334	0.358

was repeated ten times where all the documents for each subset were reselected randomly, document clustering on two hundred (20×10) subsets was performed actually. The summarized results of the document clustering are shown in Fig. 2 and Table 3. Figure 2(a) shows the numbers of the best scores each method has achieved compared with BOW, and in Fig. 2(b), the overall winning frequencies of each method on the datasets are shown. Table 3 gives the average scores of the versions of CVM on the 200 subsets. The NMI scores are the corresponding results when each method achieved their best F_1-scores.

As shown in Fig. 2, both CVMA and CVMM got competitive results compared with BOW, especially when incorporated with LTF or IDF. This agrees with our analysis that CVM obtains more smoothing power than BOW by utilizing word relatedness. However, BOW does have its particular advantages in some conditions, for BOW won 63 of 200 times in the overall competition. On subset D_{21}, the performances of BOW, CVMA and CVMM became undistinguishable. BOW got the average F_1-score of 0.625, CVMA got the average of 0.624 incorporated with ACP, and CVMM got the average of 0.636 incorporated with BNR. Since all the topics were contained in D_{21}, none of the three methods could handle the rich documents together well.

In addition, according to the relative comparison with BOW and the average scores, CVMM is superior to CVMA. This demonstrates that duplicate counts actually exist with the weighted sum strategy, and CVMM seems more plausible in practice. While the overall comparison shown in Fig. 2(b) illustrates that CVMM cannot beat CVMA all the times. It's reasonable to switch the weighted sum strategy and the maximum-value-aware strategy according to the particular dataset.

4.3   Discussion about Word Representation Methods

Six word representation methods are applied in this paper, namely, ACP, TF, LTF, IDF, MI and BNR. The experimental results in Fig. 2 and Table 3 demonstrate that the word representation methods affect the performances of CVM apparently. Instead of the different ranges of the quantified relatedness scores between words estimated by these methods, it's the relative differences among

Table 4. Correlations between Standard Deviations

	ACP	TF	LTF	IDF	MI	BNR
ACP	1.00	0.97	0.95	0.96	-0.26	0.36
TF	0.97	1.00	0.99	0.97	-0.14	0.47
LTF	0.95	0.99	1.00	0.94	-0.14	0.55
IDF	0.96	0.97	0.94	1.00	-0.13	0.31
MI	-0.26	-0.14	-0.14	-0.13	1.00	0.39
BNR	0.36	0.47	0.55	0.31	0.39	1.00

the values of the dimensions in the same vector that really matters. The standard deviation was used to evaluate the relative differences in the word vectors of the subsets D_1 to D_{20} generated with different word representation methods, respectively. The results in the same subset were averaged. The correlations between the average standard deviations corresponding to each word representation method were calculated by the Pearson Correlation[3], and the results are shown in Table 4.

It's shown that the average standard deviations with ACP, TF, LTF, and IDF have high correlations. The common input of TF, LTF and IDF is the word co-occurrence times. For a word v_x, its co-occurrence times with any dimension are smoothed by its total occurrence times with TF. The nature of ACP is the same as TF, but ACP treats words as co-occurrence as long as they appear in the same document. In LTF, the co-occurrence times are further smoothed by taking logarithm, and the co-occurrence times are smoothed one step further in IDF by introducing the inverse document frequency. Similarly, MI smooths the word co-occurrence times by taking into account the occurrence times of both the target word and the dimensions. BNR takes an extreme smoothing approach by assigning dimensions the binary scores. According to the average scores shown in Table 3, the word representation methods LTF and IDF performed consistently better than the other methods embedded in both CVMA and CVMM, which demonstrates that ACP and TF are under-smoothing policies, while MI and BNR are over-smoothing policies.

5   Conclusions

In this paper, we have compared the performance of the Context Vector Model (CVM) with the classical Bag-of-Words model (BOW) for document representation. The experimental results demonstrate that CVM has more smoothing power than BOW by considering the relatedness between words. Six representation methods for words have been embedded into CVM in our experiments; the corresponding results show that CVM severely relies on the representation methods for words. The methods incorporated with the log of term frequency and the inverse document frequency are proved to get the overall superiorities.

Besides, the combination scheme of word vectors in CVM still remains uncertain. Both the traditional weighted sum of word vectors and the proposed maximum-value-aware strategy achieve competitive results. Further study is expected to explore the inherent difference between the two combination schemes.

Acknowledgments. This work was supported by the National Natural Science Foundation of China under grant 61070089, the Science Foundation of TianJin under grant 14JCYBJC15700.

---

[3] http://en.wikipedia.org/wiki/Pearson_correlation_coefficient

References

1. Anastasiu, D.C., Tagarelli, A., Karypis, G.: Document clustering: The next frontier. Tech. rep., Technical Report. University of Minnesota (2013)
2. Andrews, N.O., Fox, E.A.: Recent developments in document clustering. Computer Science, Virginia Tech, Tech Rep (2007)
3. Billhardt, H., Borrajo, D., Maojo, V.: A context vector model for information retrieval. Journal of the American Society for Information Science and Technology **53**(3), 236–249 (2002)
4. Blei, D.M., Ng, A.Y., Jordan, M.I.: Latent dirichlet allocation. Journal of Machine Learning Research **3**, 993–1022 (2003)
5. Blunsom, P., Grefenstette, E., Hermann, K.M., et al.: New directions in vector space models of meaning. In: Proceedings of the 52nd Annual Meeting of the Association for Computational Linguistics (2014)
6. Bullinaria, J.A., Levy, J.P.: Extracting semantic representations from word co-occurrence statistics: A computational study. Behavior Research Methods **39**(3), 510–526 (2007)
7. Bullinaria, J.A., Levy, J.P.: Extracting semantic representations from word co-occurrence statistics: stop-lists, stemming, and SVD. Behavior Research Methods **44**(3), 890–907 (2012)
8. Cheng, X., Miao, D., Wang, C., Cao, L.: Coupled term-term relation analysis for document clustering. In: The 2013 International Joint Conference on Neural Networks (IJCNN), pp. 1–8. IEEE (2013)
9. Deerwester, S.C., Dumais, S.T., Landauer, T.K., Furnas, G.W., Harshman, R.A.: Indexing by latent semantic analysis. JASIS **41**(6), 391–407 (1990)
10. Harris, Z.S.: Distributional structure. Word (1954)
11. Hofmann, T.: Probabilistic latent semantic indexing. In: Proceedings of the 22nd Annual International ACM SIGIR Conference on Research and Development in Information Retrieval, pp. 50–57. ACM (1999)
12. Iosif, E., Potamianos, A.: Unsupervised semantic similarity computation between terms using web documents. IEEE Transactions on Knowledge and Data Engineering **22**(11), 1637–1647 (2010)
13. Kalogeratos, A., Likas, A.: Text document clustering using global term context vectors. Knowledge and Information Systems **31**(3), 455–474 (2012)
14. Karypis, G., Han, E.: Concept indexing: A fast dimensionality reduction algorithm with applications to document retrieval and categorization. Tech. rep, DTIC Document (2000)
15. Le, Q.V., Mikolov, T.: Distributed representations of sentences and documents. In: Proceedings of the 31st International Conference on Machine Learning, vol. 32, JMLR W&CP (2014)
16. Lund, K., Burgess, C.: Producing high-dimensional semantic spaces from lexical co-occurrence. Behavior Research Methods, Instruments, & Computers **28**(2), 203–208 (1996)
17. Mikolov, T., Chen, K., Corrado, G., Dean, J.: Efficient estimation of word representations in vector space. arXiv preprint arXiv:1301.3781 (2013)
18. Mitchell, J., Lapata, M.: Composition in distributional models of semantics. Cognitive Science **34**(8), 1388–1429 (2010)
19. Pangos, A., Iosif, E., Potamianos, A., Fosler-Lussier, E.: Combining statistical similarity measures for automatic induction of semantic classes. In: 2005 IEEE Workshop on Automatic Speech Recognition and Understanding, pp. 278–283. IEEE (2005)

20. Rungsawang, A.: Dsir: The first trec-7 attempt. In: TREC, pp. 366–372. Citeseer (1998)
21. Salton, G., Wong, A., Yang, C.S.: A vector space model for automatic indexing. Communications of the ACM **18**(11), 613–620 (1975)
22. Turney, P.D., Pantel, P., et al.: From frequency to meaning: Vector space models of semantics. Journal of Artificial Intelligence Research **37**(1), 141–188 (2010)
23. Wong, S.K.M., Ziarko, W., Raghavan, V.V., Wong, P.: On modeling of information retrieval concepts in vector spaces. ACM Transactions on Database Systems (TODS) **12**(2), 299–321 (1987)

NLP for Search Technology and Ads

Refine Search Results Based on Desktop Context

Xiaoyun Li$^{(\boxtimes)}$, Ying Yu, and Chunping Ouyang

Department of Computer Science and Technology,
University of South China, Hengyang, Hunan, China
lxy.yy@yeah.net, yyingu@sina.com, ouyangcp@gmail.com

Abstract. During a search task, a user's search intention is possible inaccurate. Even with clear information need, it is probable that the search query cannot precisely describe the user's need. And besides, the user is utterly impossible browse all the returned results. Thus, a selected and valuable returned search list is quite important for a search system. Actually, there are lots of reliable and highly relevant personal documents existing in a user's personal computer. Based on the desktop documents, it is relevantly easy to understand the user's current knowledge level about the present search subject, which is useful to predict a user's need. An approach was proposed to exploit the potential of desktop context to refine the search returned list. Firstly, to attain a comprehensive long-term user model, the operational history and a series of time-related information were analyzed to achieve the attention degree that a user paid to a document. And the keywords and user tags were focused on to understand the content. Secondly, working scenario was regarded as the most valuable information to construct a short-term user model, which directly suggested what exactly a user was working on. Experiment results showed that desktop context could effectively help refine the search returned results, and only the effectively combination of the long-term user model and the short-term user model could offer more relevant items to satisfy the user.

Keywords: Desktop context · Information retrieval · User model · Personal information space

1 Motivations

In big data era, the increasing growth of various kinds of digital resources leads to the difficulty and inefficiency to find accurate information that user need. Considering the whole process of information retrieval, the reasons can be concluded into three defects. Firstly, people do not know what they need, that is the search intention is unclear. Secondly, the search intention probably cannot be described accurately and adequately by the short search query. And thirdly, it is totally impossible to check numerous returned result pages.

In order to satisfy the user need in a better way, an approach was proposed to exploit the desktop context information, and two user models were constructed. One is the long-term user model that is responsible for collecting all the data related to a user's interest preference. This model is relatively reliable and comprehensive.

© Springer International Publishing Switzerland 2015
J. Li et al. (Eds.): NLPCC 2015, LNAI 9362, pp. 209–218, 2015.
DOI: 10.1007/978-3-319-25207-0_18

The other one is a short-term user model that is focusing on find out what exactly the user is working on. But this model is temporary and the data is relatively sparse. To achieve a better understanding of the user' search need for a specific task, the short-term model is not sufficient. The experiment results showed that combining the two models helps predict the user need and refine the search returned results.

2   Related Works

To overcome those inherent defects mentioned above, many research works contributed to improve the performance of retrieval systems. Most works were based on collecting implicit feedback information to gather user personal information. For example, Mouse dynamics were regarded relatively unique from person to person by Zheng [1], so they used the point-by-point angle-based metrics of mouse movements for user authentication. They proved that an ordinal regression model for user feedback could greatly improve the accuracy of a recommender system. And a user clicking on an item has been proven as implicit evidence that the user was interested on it [2,3,4]. Lee [5] even found that a user's first search result click could provide valuble insight into this user's subsequent interaction with the returned result list. Radinsky [6] used time-series models to represent the dynamics of search behavior over time and the results showed that it could effectively improve ranking and query suggestions. These works, to some extent, improved search performance but did not solve the intrinsic problems like data sparsity and cold start.

Lots of research works focused on other factors to improve personalized information service. Social annotation [7] was explored by Lin as the expansion term resource, and used the term co-occurrence method to demonstrate that the expansion terms extracted from social annotation were better than those from feedback documents. Otsuka [8] used the seasonal and topical facets on the interfaces to provide appropriate terms in the systems of Community Question Answering. White [9] focused on all user's search histories, not only one user's. They analyzed a user's current search task, and mined other users' historical behavior who had performed similar tasks to leverage the current user's on-task search behavior. Lu [10] built a probabilistic model to identify implicit local intent queries, and leverage user's physical location to improve Web search results for these queries. Considered different people usually have different perceptions about the same document, for each document, Xu [11] got a personalized document profile for each individual user to better summarize his/her perception about this document, then constructed user profile as the sum of all of the user's personalized document profiles to better characterize a user's preferences. This method was proved effective to achieve better personalized ranking on the Social Web. Liu [12] proposed methods for analyzing and modeling user search behavior in search sessions, and generated prediction models of document usefulness from behavior data collected in a controlled lab experiment. And the documents predicted useful and not useful by the models were used to modify the queries in each search session. Their results showed that these models could lead to consistently improved performance.

In a personalized system, one of the most difficult works is to understand a user's need. Kotov [13] focused on cross-session search tasks, predicted if a user would return to the present task, and next time, identified queries from earlier sessions on the same task. By this method, a search engine could recommend queries to re-find helpful past results. Actually, there are lots of reliable and highly relevant personal documents precisely existing in a user's personal computer, which are very useful to predict a user's need. Indeed, lots of researchers had paid much attention on desktop information to raise the quality of information retrieval [6, 14] and personalization [6, 15]. Work task and interaction context also have been proven effective to predict information need. Work tasks and search tasks play different roles when a user interacted with an information system, and different work tasks would lead to different types of search tasks [16].

However, these works still ignore lots of precious information on a user's personal computer.

3   Desktop Context Model

Although there is plentiful precious personal information in a user's computer, in a particular time, the user generally works on a specific task, such as writing an academic paper, searching some specific information, shopping online or exploring some news. Under this special consideration, the system model was separated into two main parts, as shown in following figure.

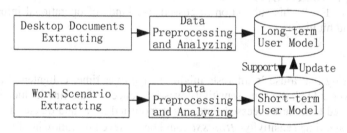

Fig. 1. Desktop Context Model

These two parts were dependent but connected. Nowadays, there are dramatically increasing volume of digital files in a user's computer, such as various documents, images, and multimedia files. These files were regarded as the desktop documents in this paper. For one part, desktop context extracting helped analyze the user's interest in the digital world, and helped establish a relatively complete profile of the user's personal information space. For another part, work scenario extracting focused on understanding what exactly a user was working on. This was very useful to predict the user's current need. Here the user's long-term model and the short-term model were mutually supportive to assure the best information retrieval service.

3.1   Analyzing Document Information

A user saved those desktop documents in his or her personal computer, so it is reasonable to assume that these documents were quite valuable for this user. However, in the long run, these documents did not own the same importance. To understand a user, a traditional way was to analyze the documents' content the user explored. Actually, there was some pretty important information hidden in the operational history, which, to some extent, reflected the real requirement degree for the document. Therefore, two kinds of information were employed to find those documents that owned highly values, as shown in figure 2.

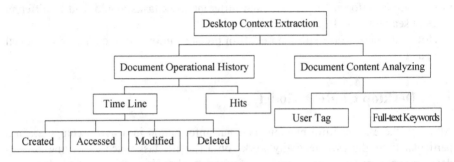

Fig. 2. Desktop Context Extraction

Extracting Document Operational History

According to Fig 2, this subsection discussed two kinds of operational information: time line and hit history.

- Time line

Considering that a user's interests might change over time, enlightened by Jaime Teevan's works [15], the time sensitivity was introduced. Time line included a document's created time, last accessed time, latest modified time, and deleted time. With these raw data, time sensitivity *timesst* could be achieved by following formula.

$$timesst = del \cdot (\alpha \cdot crttime + \beta \cdot accsstime + \gamma \cdot mdftime) \qquad (1)$$

where α, β and γ were parameters. *crttime*, *accsstime* and *mdftime* separately represented the time sensitivity of a document being created, viewed and modified. Time information was processed by fuzzy linguistic [14].

Besides, the deletion information was quite valuable. Once a document was deleted, it was sensible to believe that the user regarded this document as irrelevant or no longer useful. So the deletion time was also recorded here and it was denoted by *del*. But this record was only reserved for a period of time. Once the time exceeded a predefined threshold, then the document record would be deleted. Experiments showed that users rarely clicked those links that they had deleted. And excluding the corresponding item links from the returned results' list could effectively improve the user experience.

- Hit history

People own different behavioral habits like periodically cleanup disks or not. In a personal computer, there are various documents, fresh but useless or relatively old but frequently used. Thus, *hit* was introduced to record a document's hit history. And then combined with the time sensitivity, the importance of a document could be achieved by formula (2).

$$doc \; (timesst, hit) = A \cdot timesst + B \cdot hit + C \qquad (2)$$

where A, B, C were constants that were achieved and adjusted in the experiments.

Extracting Document Content
The notion of relevance is the key of a personalized system. It is determined by the document content. We focused on user tags and keywords. User tags were generally marked by the authors, editors or peers. In our experiments, not all documents owned this kind of tags. And the keywords were extracted through full-text segmentation and preprocessing. In this paper, the document segmentation relied on the Institute of Computing Technology Chinese Lexical Analysis System1. The document weight was gained by TF-IDF, as shown in formula (3).

$$weight_i = \frac{n_{i,j}}{\sum n_{k,j}} \cdot \log\left(\frac{N}{n}\right) \qquad (3)$$

where $weight_i$ represented the weight of word i; $N_{i,j}$ was the occurrence number of the word i in document j; the denominator was the occurrence number of all words in document; N was the total number of documents in the corpus, and n was the number of documents where the word i appeared.

Thus, the weight of word i could be achieved by formula (4).

$$word_i = doc_j \cdot \left(\sigma \cdot weight_i + \lambda \cdot tag_i\right) \qquad (4)$$

where σ and λ were constants that were achieved and adjusted in experiments.

3.2   Analyzing Work Scenario

A user generally focuses on some particular task in a certain time. Consider an example scenario: while a user concentrated on topic A, like writing a paper, there might be in a particular time, the user switched to another topic B that might be generating a list to listen to the music. We called this task switch as the task context switch. No matter what the task was, there was a corresponding short-term context in the personal computer. We regarded current opened documents in the same work scenario as a

---

[1] http://www.nlp.org.cn/project/project.php?proj_id=6

work corpus, which would take more responsibilities than other documents in the personal computer to personalize information retrieval. Task identifying was not the research issue in this paper, so in our experiments, many work scenario were identified artificially.

Working corpus included local documents and online Web pages. It was more than a kind of real-time desktop resources analysis. Three kinds of information were employed to build a short-term user model:

1. Extracting the document content in the work corpus could display what kind of task the user was working on.
2. Analyzing the corresponding exploring time for each document might show how valuable the document was.
3. Several special actions, including print, collect, and save, implied that the user regarded these documents or pages as valuable.

No matter explicit or implicit, all these information, to some extent, reflected a user's interest preference. They could be used to build the short-term model.

4   Experimental Evaluation

There were 23 participants joined our experiments, who were asked to search pdf format documents separately through Bing, Baidu and the experimental system. The Discounted Cumulative Gain (DCG) [15] was introduced to measure the quality of the search engine result sets. DCG was a measure to evaluate the effectiveness of a search engine algorithm or related applications, which owned two assumptions:

1. Highly relevant documents were more useful when appearing earlier in a search engine result list (have higher ranks).
2. Highly relevant documents were more useful than marginally relevant documents, which are in turn more useful than irrelevant documents.

The DCG was defined as formula (5).

$$DCG\,(i) = \begin{cases} G(1) & ,if \quad i = 1 \\ DCG\,(i-1) + \dfrac{G(i)}{\log_2(i)}, otherwise \end{cases} \qquad (5)$$

where $G(i)$ was the graded relevance of the result at position i. In order to determine whether a returned document was relevant to a user's search intention, each participant was asked to give a gain value to each returned document: 2 (if highly relevant), 1 (if relevant), or 0 (if not relevant).

For each query, DCG was cumulated for all ranks and offers us a simple method to measure the quality of a results set. However queries that had more relevant documents should have a higher DCG, so the DCG was normalized to a value between 0 (the worst possible DCG given the ratings) and 1 (the best possible DCG given the ratings).

The normalized DCG values could be averaged to measure the average performance of a search engine's ranking algorithm.

To evaluate the potential of desktop context, we compared the top-10 search results between the experimental system and two general search engines, *Bing* and *Baidu*, for the query *"Mercedes Benz"* for user A, who majored in automotive engineering and there were lots of documents about design, manufacture and operation of vehicle in his personal computer. According to the short-term model achieved from work scenario analysis, *"car design"* was taken into account. And the search results were filtered and re-ranked according to the long-term document model. The relevance of each result document was graded by the user, and the results are shown in following table.

The results in above Table 1 clearly showed that the experimental system used the keywords gained from work scenario analysis as the expansion query could provide more relevant items for the user than the other two general search engines and the sequence of gain value for each returned documents in refined results lists showed that those documents with more relevance were ranked better than its original lists.

Table 1. Comparison of top-10 search results between the experimental system and two general search engines

Bing Original Result	Gain	Baidu Original Result	Gain	Experimental Result	Gain
ec.europa.eu/competition	0	worldauto.com.cn/html/sj	0	bric.com/archive/pdf/	2
mbusi.com/pages/MBUS	0	mcgraw-hill.cn/pdf/20070	0	bridgestone.eu/Englis	2
mbusi.com/images/MB%	0	mercedes-benz.co.jp/e/ne	0	me.cycu.edu.tw/csme	2
daimler.com/Projects/c2	1	autechk.com/upfile/20092	1	abc-shanghai.com/en/	1
daimler.com/Projects/c2	1	total-lub.jp/lub/content/N1	1	cd1.edb.hkedcity.net/	1
ac-rerefined.com/Endors	0	businesscenter.jdpower.c	0	abc-shanghai.com/en/	0
fire.state.mn.us/Respons	2	img1.toocle.com/uppic/a/	1	bharatbook.com/Sam	0
epa.gov/compliance/res	0	westfalia-van.de/fileadmin	0	web.mit.edu/aeroastr	2
iatfglobaloversight.org/d	0	english.people.com.cn/90	0	mercedes-benz.co.jp/	0
eaton.com/ecm/groups/p	0	mercedes-benz.com.tw/c	0	mercedes-benz.co.jp/	0
DCG	1.64	DCG	1.29	DCG	6.86

To evaluate the performance of the system for the same query to different users with different background and information need, we randomly selected User B, who paid much attention to vehicle market information. According to the short-term model of User B, *"market report"* was taken into account when user B was searching. Compare the best ranking results of *"Mercedes Benz"* for user A and user B. Baidu served as the meta-search engine.

Table 2. Best ranking results and experimental results for two users for the same query

Best Ranking of the Result				Experimental Re-ranked Result			
User A	Gain	User B	Gain	User A	Gain	User B	Gain
ieeexplore.ieee.org/Xpl	2	mercedes-benz-financia	2	wzb.eu/gwd/wpa/pdf/blo	1	caranddriver.com/asset	1
mercedes-benz.com.au/d	2	mitsubishifuso.lv/Proje	2	ieeexplore.ieee.org/Xpl	2	mercedes-benz-financia	2
oliverwyman.com/ow/pd	2	aldrichquaihoi.com/asse	2	autonews.com/assets/PI	1	osmanauction.com/Nex	1
mercedes-benz.com/flee	2	caranddriver.com/assets	1	mercedes-benz.com.au/c	2	bishkek.usembassy.gov	0
mercedes-benz.com/flee	2	osmanauction.com/Nex	1	oliverwyman.com/ow/pc	2	nrel.gov/docs/fy09osti	1
wzb.eu/gwd/wpa/pdf/blo	1	smmt.co.uk/downloads	1	mercedes-benz.com/flee	2	mitsubishifuso.lv/Proje	2
autonews.com/assets/PI	1	nrel.gov/docs/fy09osti	1	mercedes-benz.com/flee	2	aldrichquaihoi.com/ass	2
mercedes-benz-classic.c	0	ec.europa.eu/competiti	1	wcoty.com/files/2010_1	0	smmt.co.uk/downloads	1
oldtimer-doctor.com/lin	0	bishkek.usembassy.gov	0	mercedes-benz-classic.c	0	ec.europa.eu/competiti	1
mercedes-benz-classic.c	0	us-cdn.creamermedia.c	0	oldtimer-doctor.com/lir	0	us-cdn.creamermedia.c	0
Normalized DCG	1	Normalized DCG	1	Normalized DCG	0.89	Normalized DCG	0.85

The best ranking of the results was the best possible ranking for a query a search engine could do for a user, and its normalized DCG was 1.00. The results showed that experimental refined list owned a pretty high normalized DCG, which meant the refined ranking was closed to the best ranking results. Take all 23 participants into consideration, we compared the variation of the average normalized DCG in three situations, including searched only with long-term model, only with short-term model, and combined those two models. The results showed in figure 3.

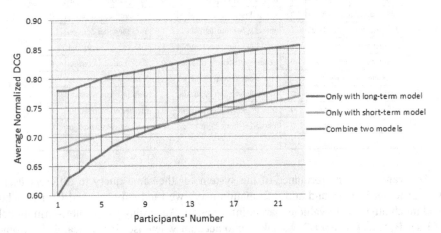

Fig. 3. Comparison of the average normalized DCG according to the participants' number

This figure showed that the short-term model served better for a specific search task than the long-term model. But in the long run, the long-term model showed more potential. These two models had their own merits and demerits. The combination of these two models could overcome the shortcomings of each models and support each other. The above figure displayed obvious advantage of this combination. The results also showed that with the increasing number of the participants, in all cases, the average value of the normalized DCG was gradually stabilized at a relatively high level, which indicated the potential of desktop context information that did help predict a user's current need and improve the personalized information retrieval.

5   Conclusion

This paper addressed the potential of desktop information, and proposed an approach to extract the desktop context to construct a relatively comprehensive personal information space. With the support of the long-term user model achieved by local resources extraction, we further concentrated on extracting the information of the work scenario that helped to understand what exactly a user need in a particular time. A series of experimental results reflected the potential of the desktop information that could improve traditional query expansion and effectively refine returned search results for the user. However, the present experiments were conducted in a limited field and we will continue to study to improve the information service in the future.

Acknowledgements. This work is supported by the National Natural Science Foundation of China (No.61402220), Hunan Provincial Natural Science Foundation of China (No.13JJ4076), the Scientific Research Fund of Hunan Provincial Education Department for excellent talents (No.13B101), Foundation of University of South China (No.2012XQD28), the Construct Program for the Key Discipline in University of South China (No.NHxk02), the Construct Program for Innovative Research Team in University of South China.

References

1. Zheng, N., Paloski, A., Wang, H.N.: An efficient user verification system via mouse movements. In: Proc. 18th ACM Conference on Computer and Communications Security, Chicago, Illinois, USA, October 17–21, 2011
2. Stern, D., Herbrich, R., Graepel, T.: Matchbox: large scale online bayesian recommendations. In: Proc. 18th WWW Conference, Madrid, Spain, pp. 111–120. April 20–24, 2009
3. Das, A.S., Datar, M., Garg, A., et al.: Google news personalization: scalable online collaborative filtering. In: Proc. 16th International Conference on World Wide Web, New York, NY, USA, pp. 271–280 (2007)
4. Lee, C.-J., Teevan, J., de la Chica, S.: Characterizing multi-click behavior and the risks and opportunities of changing results during use. In: Proc. 37th Annual International ACM SIGIR Conference on Research and Development in Information Retrieval (SIGIR 2014), Gold Coast, Australia, July 2014
5. Radinsky, K., Svore, K., Dumais, S.T., Shokouhi, M., Teevan, J., Horvitz, E.: Behavioral Dynamics on the Web: Learning, Modeling and Prediction. Proc. ACM Transactions on Information Systems (TOIS) **31**(3) (2013)
6. Teevan, J., Dumais, S.T., Horvitz, E.: Potential for Personalization. Proc. ACM Transactions on Computer-Human Interaction (TOCHI) special issue on Data Mining for Understanding User Needs, March 2010
7. Lin, Y., Lin, H.F., Jin, S., et al.: Social annotation in query expansion a machine learning approach. In: Proc. the 34th Annual ACM SIGIR Conference, Beijing, China, pp. 405–414, July 2011

8. Otsuka, A., Seki, Y., Kando, N., et al.: QAque: faceted query expansion techniques for exploratory search using community QA resources. In: Proc. the 21st International Conference Companion on World Wide Web. Lyon, France, pp. 799–806, April 2012
9. White, R.W., Chu, W., Hassan, A., et al.: Enhancing Personalized Search by Mining and Modeling Task Behavior. In: WWW 2013, Rio de Janiero, Brazil, May 13–17, 2013
10. Lu, Y.M., Peng, F.C., Wei, X., et al.: Personalize web search results with user's location. In: Proc. 33rd International ACM SIGIR Conference on Research and Development in Information Retrieval, SIGIR 2010, Geneva, Switzerland, pp. 763–764, July 19–23, 2010
11. Xu, Z., Lukasiewicz, T., Tifrea-Marciuska, O.: Improving personalized search on the social web based on similarities between users. In: Straccia, U., Calì, A. (eds.) SUM 2014. LNCS, vol. 8720, pp. 306–319. Springer, Heidelberg (2014)
12. Liu, C., Belkin, N.J., Cole, M.J.: Personalization of search results using interaction behaviors in search sessions. In: Proc. 35th International ACM SIGIR Conference on Research and Development In Information Retrieval. Portland Oregon, USA, pp. 205–214, August 12–16, 2012
13. Kotov, A., Bennett, P.N., White, R.W., et al.: Modeling and analysis of cross-session search tasks. In: Proc. 34th SIGIR. Beijing, China (2011)
14. Findlater, L., Moffatt, K., McGrenere, J.: Ephemeral adaptation: the use of gradual onset to improve menu selection performance. In: Proc. 27th International Conference on Human Factors in Computing Systems, Boston, MA, USA, pp. 1655–1664, April 4–9, 2009
15. Teevan, J., Dumais, S.T., Horvitz, E.: Personalizing search via automated analysis of interests and activities. In: Proc. 28th Annual ACM Conference on Research and Development in Information Retrieval (SIGIR 2005), Salvador, Brazil, August 2005
16. Li, Y.L., Belkin, N.J.: An exploration of the relationships between work task and interactive information search behavior. Journal of the American Society for Information Science and Technology (ASIS & T) **61**(9), 1771–1789 (2010)

Incorporating Semantic Knowledge with MRF Term Dependency Model in Medical Document Retrieval

Zhongda Xie[✉], Yunqing Xia, and Qiang Zhou

Department of Computer Science, TNList,
Tsinghua University, Beijing 100084, China
{xzd13,yqxia,zq-lxd}@tsinghua.edu.cn

Abstract. Term dependency models are generally better than bag-of-word models, because complete concepts are often represented by multiple terms. However, without semantic knowledge, such models may introduce many false dependencies among terms, especially when the document collection is small and homogeneous(e.g. newswire documents, medical documents). The main contribution of this work is to incorporate semantic knowledge with term dependency models, so that more accurate dependency relations will be assigned to terms in the query. In this paper, experiments will be made on CLEF2013 eHealth Lab medical information retrieval data set, and the baseline term dependency model will be the popular MRF(Markov Random Field) model [1], which proves to be better than traditional independent models in general domain search. Experiment results show that, in medical document retrieval, full dependency MRF model is worse than independent model, it can be significantly improved by incorporating semantic knowledge.

Keywords: Semantic knowledge · Term dependency model · Markov random field · Medical document retrieval

1 Introduction

Nowadays, term dependency retrieval models [1–3] generally outperform the traditional bag-of-word models. This is because a complete concept in a query is often represented by multiple terms; so, the occurrence of multiple dependent terms in a document can provide a stronger evidence that the document is relevant with the query. In general domain search, these models are significantly better. MRF model [1] proposed by Metzler and Croft is among the best, which indicates that full dependency model is better than sequential dependency and full independency models. However, in some specific areas, like medical document retrieval, the data collection is homogeneous, the documents are filled with formalized terminologies, and these term dependency models will introduce too many false dependencies.

Take the following two queries as examples:

© Springer International Publishing Switzerland 2015
J. Li et al. (Eds.): NLPCC 2015, LNAI 9362, pp. 219–228, 2015.
DOI: 10.1007/978-3-319-25207-0_19

(Query1): coronary artery disease and myocardial infarction.

(Query2): abdominal pain and helicobacter pylori and cancer.

In Query1, *coronary artery disease* is a kind of disease, while *myocardial infarction* is another, and the patient wants to know how to distinguish the two. The main difference between medical queries and general domain queries is that, the concepts in medical domain are represented by formalized terminologies. That is to say, if a medical concept is to appear in a document, the words are most likely to be in the same order as the query, and tend to be close to each other, with no other word inserted inside. So, the good dependency detected for this query is *coronary artery disease* and *myocardial infarction*; while traditional MRF model will generate other dependency relations like *coronary infarction, myocadial disease, artery myocardial* and so on, which are all false dependency relations. The main reason of such difference is that, in general domain queries, most single word is meaningful enough, so arbitrary combination of the words is likely to represent a complete concept, which can be related to the query; But in medical domain, the multiple words belonging to a terminology must get together to represent the right concept, while a single word can hardly do.

In Query2, *helicobacter pylori* is a kind of bacteria, and it CAUSES *cancer*, now the patient wants to know whether *abdominal pain* is the SYMPTOM of such *cancer*. Now that we know there exist a CAUSES relation between *helicobacter pylori* and *cancer*, and a SYMPTOM_OF relation between *abdominal pain* and *cancer*, we can assign accurate dependencies to these concept pairs. And we will leave *abdominal pain* and *helicobacter pylori* independent, for they don't have direct relations to each other. Such kind of detection reply highly on semantic knowledge web, this is why we will incorporate it with state-of-art term dependency models.

In our work, we use UMLS(Unified Medical Language System) as our knowledge base, and SemMedDB(built on the same concepts in UMLS) as our semantic web. We first use MetaMap program to extract medical concepts in the query, and add the concept as matching unit. We make the comparison between two ways of matching the concept: using the exact sequence of the concept words; and detecting unordered concept words in a window of document text, as the traditional MRF model will do. Then, we use the SemMedDB database to detect related concept pairs, and take such pairs as additional matching units(e.g. [*helicobacter pylori, cancer*]). Finally, we extract the words indicating relationship between concept pairs from the SemMedDB, and put the related concept pairs together with their relation word as matching units(e.g. [*helicobacter pylori, CAUSES, cancer*]). The experiments are conducted on CLEF eHealth 2013 data set, results show that our model is better than BM25 and the three variants of traditional MRF model(full independency, sequential dependency, full dependency).

The rest of this paper is organized as follows. In section 2 we will summarize the related works. The details of our methods are in section 3, while the experimental results and analysis will be in section 4. We make the conclusion in section 5.

2   Related Work

Term dependency models generally have advantage over bag-of-word models, because the co-occurrence of dependent words in a document has a stronger sign that the document is relevant. So, much work has been done in this direction.

Xu et al. [2] introduced a extend BM25 model. It extends the traditional BM25 equation from single word based to n-gram based, and sums up the scores of the n-grams, ranging from all lengths and all start positions. This model is very simple, but will always outperform the basic IR systems. However, this model cannot accurately catch the dependency among separated words, due to its n-gram based nature. Take the Query2 from section 1 as an example, it cannot precisely get the unit *abdominal pain cancer*, which has a SYMPTOM_OF relation between the two concepts, and it has to get the whole query as a 7-gram in order to capture this relation.

Later, Many works [3,4] rely on text parse trees to extract more accurate dependencies. However, these models still only consider about pairs of words. Park et al. [4] uses a language model to make use of the parse tree, in which a document generates a parsing path in a certain probability, and that path will generate the query. They treat all the paths equally important, which will introduce many noisy word dependencies, and the parsing process on the whole document is rather computation intensive.

Metzler and Croft [1] then propose a MRF model to represent the term dependencies. MRF model is very flexible, by defining different kinds of feature functions, they can put various features(single word; neighboring, ordered dependent words; separate, unordered words and so on) together in one universal model. This model has three basic variants: full independency model, sequential dependency model, full dependency model. They differ on what combination of words is dependent. The model works out very well in general domain search, and the three variants are in ascending order of precision as listed previously. However, the full dependency model is bad in small, homogeneous collections(like medical report collection), as has been explained in section 1. And it needs appropriate knowledge to extract precise dependency relations.

In medical document retrieval, much work has shown that concept detection and concept pair relation can be helpful. Qi et al. [5] uses a concept-based model, and experiment with many popular methods for word based models, and find out that vector space model together with pseudo relevance-feedback works well. Later, researchers go further to concept level relations. Khoo et al. [6] make an attempt to use the cause-effect relation between concepts, which they think is the most helpful relation. Lee et al. [7] use ontology relations between concepts, like *IS_A* and *CO-OCCURS_WITH*. However, Vintar et al. [8] find that those are coarse-grained relations, which have no real meaning, so they want to use fine-grained relations extracted from semantic web, like *TREAT* and *SYMP-TOM_OF*. They use such relations as a document filter in a boolean manner, which only improves traditional models slightly. So, Xia et al. [9] introduce a way to represent the query and documents in relation level, and compute their relevance in both word level and relation level, the final score is a combination of the scores in two levels.

Inspired by above promising works, we want to incorporate medical semantic web and its corresponding techniques with MRF model, which is flexible and rather powerful in general domain. We want to assign accurate dependencies to the words, thus improving precision of the system.

3   Methodology

In order to incorporate the medical semantic knowledge with MRF model. First we need to detect the medical concepts and the find-grained relations among them. After that, a core problem is how we will reflect such knowledge in the model, and the flexible feature functions of MRF provides a way. In the following sections, we will discuss about the details of our modules.

3.1   Concept Detection

The knowledge base we use is UMLS(Unified Medical Language System), developed by NLM(National Library of Medicine) of U.S.. It integrates medical concepts from different authoritative resources, merges the medical terminologies with the same meaning into one concept, and gives each concept a unified, unique identifier.

In order to extract the concepts, we use a program called SemRep, which is developed based on the UMLS ontology. It will split the text into separated parts, and map each part into a concept(actually a unique concept identifier, called CUI) in UMLS. For example, in Query2 of section 1, we will get the result as following(we have changed the original output format for the sake of easy understanding):

[abdominal pain: C0000737] and [helicobacter pylori: C0079488] and [cancer: C0004382]

The brackets indicate split parts of the text, and the sequence after the terminology is the CUI, the unique identifier of the concept.

The output of the program may have several ways of splitting the text and various mappings from a terminology to the concept in the UMLS database. We only choose the best parsing result(indicated by the program in terms of accurate probability), in order to make our detection process precise enough.

3.2   Concept Relation Extraction

The concept relation in our work does'n mean coarse-grained relations, which includes the common *IS_a* and *CO-OCCURRES_WITH* relations. They only indicate the ontology hierarchical structures, but not the real semantic relations. What we need is the find-grained relations like *TREATS* and *DIAGNOSES*, which represent meaningful relations in medical domain.

The semantic web we use is SemMedDB, it is built on top of UMLS ontology, so the medical concept extracted by SemRep can find its match in SemMedDB, using the CUI assigned to it. SemMedDB integrates around 70 million labeled medical domain sentences as its resource. For each sentence, there will be several pairs of related concepts, each called a *predication*. For each predication, we can follow the information in it to extract its corresponding two concepts, as well as the relation type between them. Take the following sentence as an example:

It is said that the helicobacter pylori often causes cancer, so, is my recent abdominal pain a possible symptom of my cancer?

From SemMedDB, we can know that there is a predication [*helicobacter pylori, cancer*], and the relation type is *CAUSE*; we can also find another predication [*abdominal pain, cancer*], and the relation type is *SYMPTOM_OF*.

So, we have shown how to get the related concept pairs in a sentence in the database. But we are faced with is different: we have a query, we extract all its concepts in the way described in section 3.1, enumerate all the possible pairs, and we want to find out whether each concept pair is related(or belong to a predication). And the solution goes in the opposite direction against the way we get predictions from a labeled sentence in database. We get the CUIs of the two concepts and find their entries in SemMedDB, then search the predication table to find whether they appear in the same predication; we can use the predication ID to further get the labeled sentence that the predication comes from. In this way, we can start from two concepts in the query, and find all the possible relations that have been labeled in the SemMedDB database.

There is other information that we can extract. For a related concept pair, we can extend it into a triple, containing the keyword that indicates the relation between them. For instance, for the pair [*helicobacter pylori, cancer*], it can be extended to [*helicobacter pylori, causes, cancer*], [*helicobacter pylori, caused, cancer*], [*helicobacter pylori, lead to, cancer*] and so on, all these extensions reflect the *CAUSE* relation between the concepts. We will use these triples as matching unit in our last module, the occurrence of the keyword *causes, caused* and *lead to* in the document can give a stronger evidence that it is talking about the two concepts, specifically with the wanted relation *CAUSE*.

In order to obtain the keywords indicating a certain relation, we need to preprocess the SemMedDB database. We traverse the SENTENCE_PREDICTION table, which contains all the predications. For each predication, we follow the link in it to extract the relation type corresponding with it; we also look for the sentence that produce the predication, use the position information from SENTENCE_PREDICATION table to find the exact keyword that indicates the relation.

Finally, we obtain fifty-seven relation types from the database, and remain forty-eight of them(the fine-grained relations only), with half positive and half negative. The twenty-four positive relations are: *ADMIN-ISTERED_TO, AFFECTS, ASSOCIATED_WITH, AUGMENTS, CAUSES, COEXISTS_WITH, COMPLICATES, CONVERT_TO, DIAGNOSES, DIS-*

RUPTS, INHIBITS, INTERACTS_WITH, LOCATION_OF, MANIFESTA-TION_OF, METHOD_OF, OCCURS_IN, PART_OF, PRECEDES, PRO-CESS_OF, PRODUCES, PREDESPOSES, STIMULATES, TREATS, USES, PREVENTS. And the negative relations are ones that begin with *NEG_*, for example, the negative relation against *CAUSES* is *NEG_CAUSES*, which means something is not the cause of a disease.

The negative category of relations is rather useful. In case a patient wants to find out whether a kind of bacteria is the cause of a illness, while actually it isn't. If we only have the positive relations, the bacteria and the illness will obviously not appear in the *CAUSE* keyword list, and we will ignore the patient's intent. Only by remaining the *NEG_CAUSE* relation, can we understand what the patient wants.

3.3   Feature Function

In term dependency models, the three main characters will make the difference:

1. The number of words together as a matching unit.
2. Whether the words in the document should appear in the same order as the query or not.
3. The text window in the document that we detect related multiple words.

We will talk about the choice of traditional MRF model briefly, and explain our strategy when concept and relations have been detected.

In MRF model, there are three types of combination of query words: single word, sequential words, separated words. Different type leads to different strategies, and it is reflected by using different feature functions. Feature function is a very flexible character of the MRF model, we can merge the above various types in a universal way, and finally sum up the score of all the functions. The work in [1] uses Indri as search engine, which provides convenient syntax to represent all the feature functions.

The table below is the feature functions used in [1] and their corresponding Indri search queries. In the column of *Indri Query*, the #1() means the words inside needs to appear in order and consecutively, the #uwN() means the words inside only need to appear in a text window of size N, with no strict order required.

Table 1. Feature functions and associated Indri queries used in traditional MRF model

Type	Feature	Indri Query
Single Word	$f_T(q_i, D)$	q_i
Ordered Phrase	$f_O(q_i, q_{i+1}, ..., q_{i+k}, D)$	$\#1(q_i, q_{i+1}, ..., q_{i+k})$
Unordered Phrase	$f_U(q_i, ..., q_j, D)$	$\#uwN(q_i, ..., q_j)$

In tradition MRF model, q_i indicates the ith word in the query. Consecutive sequential words will use two feature functions, the ordered phrase type(which appear ordered and consecutively in the document), and the unordered phrase type(which only needs to appear in a text window). On the other hand, Separate words in the query, like (q_1, q_2, q_4), can only be matched through the unordered phrase function.

In our model, we have different strategies for single concept and concept pairs, which will be explained separately in the following sections.

(1)Single Concept Feature
The number of words is not fixed, because the basic unit is concept, so the number will be up to the length of the current concept. The concept words appear consecutively in the query. If it is in the traditional MRF model, two features will be computed, both as ordered and unordered phrase. But for a medical concept, we only compute its score in a ordered way. This is mainly due to the formalization feature of the medical terminologies, which has been explained in Section 1. The feature functions are listed in table 2, and we only choose *Ordered Phrase* feature.

Table 2. Feature functions and Indri queries used in our model

Type	Feature	Indri Query
Single Word	$f_T(q_i, D)$	q_i
Ordered Phrase	$f_O(c_1, c_2, ..., c_{|c|}, D)$	$\#1(c_1, c_2, ..., c_{|c|})$
Combined Phrase	$f_C(p_1, p_2, D)$	$\#uwN(\#1(p_1)\#1(p_2))$

p_i represents the ith phrase, which is also $c_{1,1}, c_{1,2}, ..., c_{1,|c_1|}$.

(2)Concept Pair Feature
Concept pair is made up of two concepts. In order to keep up with the formalization of medical terminologies, the single concept alone is still computed in an ordered way. However, in concept level, the two concepts can be unordered. For example, the pair *[some_disease, some_illness]* may appear in the document as *some_disease causes my some_illness*, or *my some_illness is caused by some_disease*. So, we define a *Combined Phrase* type for such kind of feature.

Later, we extract keywords for each relation type, and this is easily realized by adding the keyword in the *Combined Phrase* type.

4   Experiments

We use CLEF 2013 eHealth Lab Medical Retrieval data set as our collection. 50 queries are provided officially as test set, and they cover a wide range of health topics.

The evaluation metrics used are the popular ones in information retrieval: (1) p@10: precision considering only the top 10 returned documents. (2)nDCG@10: normalized Discounted Cumulative Gain, also assessed at top 10 documents returned.

4.1 Experiment 1: Feature Function for Single Medical Concept

In algorithms described in section 3, we only use ordered phrase feature for single medical concept. Actually, before we go on to the next steps, we make the comparison between ordered phrase feature and unordered phrase feature for single concept. And the experimental result is as follows:

Table 3. Results using different features for single medical concept

Features	P@10	nDCG@10
Ordered Phrase	0.4960	0.5043
Unordered Phrase	0.4840	0.4967
Ordered and Unordered Phrase	0.4880	0.4963

Result shows that only using ordered phrase feature is the best, this looks up to our analysis of the formalized feature of medical concepts. In fact, we find some examples indicating unordered phrase feature is not fit for medical concepts. In the returned documents, the model matches *coronary heart disease and coronary artery revascularization* as relevant to query *coronary artery disease*. It shows matching a medical concept in strict order is very important.

4.2 Experiment 2: Results of Different Models

We compare our revised models against the three variants of MRF model. The abbreviation of each model is listed below:

-**MRF(I)**: full independency model of traditional MRF.
-**MRF(S)**: sequential dependency model of traditional MRF.
-**MRF(F)**: full dependency model of traditional MRF.
-**MRF(C)**: our model using features for single medical concept(only ordered phrase feature).
-**MRF(CR)**: our model using both features for single medical concept and concept pairs.
-**MRF(CR-EX)**: extended model(using relation keyword) of **MRF(CR)**.

The results is as table 4.

The first three models are the variants of the traditional MRF model, they introduce more and more word dependencies. The full dependency model(**MRF(S)**) works out best in general domain, but worst among the three in

Table 4. Results using different models

Model	P@10	nDCG@10
MRF(I)	0.4940	0.5087
MRF(S)	0.4780	0.5028
MRF(F)	0.4580	0.4762
MRF(C)	0.4960	0.5100
MRF(CR)	0.5060	0.5193
MRF(CR-EX)	0.5100	0.5203

medical domain. Without careful assigned dependencies, term dependency model cannot work well in small, homogenous collections like medical report collection. **MRF(C)** introduces medical semantic knowledge, only uses the extracted medical concepts, and only in a ordered phrase form. The accurately detected dependency relation significantly regains the precision of the system, especially against **MRF(F)** model. Then, **MRF(CR)** model uses related concept pair as a matching unit, which can give a stronger evidence that the document is talking about both of them. The extended model also improves a little, better ways of using the extended keyword should be proposed, and the gain will be larger.

5 Conclusion

In this work, we find out that, the main reason for the failure of traditional MRF term dependency model in medical domain, is the formalized feature of the medical terminologies. Thus, we make use of medical semantic knowledge to extract medical concepts, and only remain dependency relations of these concepts in a ordered form. The accurate dependencies in our model provides significant gain in precision of our system against the full dependency MRF model(**MRF(F)**). We also process the semantic web database to detect relations between concept pairs, and define a combined way of using ordered concept phrase and unordered concept pair. The related concept pair can give stronger evidence that the document is about both the concepts and the relation between them, thus again improves our system.

The extended model is not so good, one reason is that the keyword list need to be cleaned. Currently, the multiple relations existing between a concept pair are treated as equally important, this can be another reason that the extended model doesn't perform as we want. So, in the future, we will develop algorithms to assign different weights to the relations.

Acknowledgement. This work is supported by National Science Foundation of China(NSFC: 61272233). We thank the anonymous reviewers for the valuable comments.

References

1. Metzler, D., Croft, W.B.: A markov random field model for term dependencies. In: Proceedings of the 28th Annual International ACM SIGIR Conference on Research and Development in Infromation Retrieval, SIGIR 2005, pp. 472–479. ACM, New York (2005)
2. Xu, Jun, Li, Hang, Zhong, Chaoliang: Relevance ranking using kernels. In: Cheng, Pu-Jen, Kan, Min-Yen, Lam, Wai, Nakov, Preslav (eds.) AIRS 2010. LNCS, vol. 6458, pp. 1–12. Springer, Heidelberg (2010)
3. Gao, J., Nie, J.-Y., Wu, G., Cao, G.: Dependency language model for information retrieval. In: Proc. 27th Ann. Intl. ACM SIGIR Conf. on Research and Development in Information Retrieval, pp. 170–177 (2004)
4. Park, J.H., Croft, W.B., Smith, D.A.: A quasi-synchronous dependency model for information retrieval. In: Proceedings of the 20th ACM International Conference on Information and Knowledge management. CIKM 2011, pp. 17–26. NY, USA, New York (2011)
5. Qi, Y., Laquerre, P.F.: Retrieving medical records: NEC Labs America at TREC 2012 medical track. In: TREC 2012, Gaithersburg, Maryland, NIST(2012)
6. Khoo, C.S.G., Myaeng, S.H., Oddy, R.N.: Using cause-effect relations in text to improve information retrieval precision. Inf. process. Manage. **37**(1), 119–145 (2001)
7. Lee, J., Min, J.K., Oh, A., Chung, C.W.: Effective ranking and search techniques for web resources considering semantic relationships. Inf. Process. Manage. **50**(1), 132–155 (2014)
8. Vintar, S., Buitelaar, P., Volk, M.: Semantic relations in concept-based cross-language medical information retrieval. In: Proceedings of ECML/PKDD workshop on Adaptive Text Extraction and Mining. ATEM (2003)
9. Xia, Y., Xie, Z., Zhang Q., et al.: Cannabis_TREATS_cancer: Incorporating Fine-Grained Ontological Relations in Medical Document Ranking. Communications in Computer & Information Science (2014)

A Full-Text Retrieval Algorithm for Encrypted Data in Cloud Storage Applications

Wei Song[1,2], Yihui Cui[2], and Zhiyong Peng[1,2(✉)]

[1] State Key Laboratory of Software Engineering, Wuhan University, Wuhan, China
{songwei,peng}@whu.edu.cn
[2] School of Computer, Wuhan University, Wuhan, China
cuiyihui@whu.edu.cn

Abstract. Nowadays, more and more Internet users use the cloud storage services to store their personal data, especially when the mobile devices which have limited storage capacity popularize. With the cloud storage services, the users can access their personal data at any time and anywhere without storing the data at local. However, the cloud storage service provider is not completely trusted. Therefore, the first concern of using cloud storage services is the data security. A straightforward method to address the security problem is to encrypt the data before uploading to the cloud server. The encryption method is able to keep the data secret from the cloud server, but cloud server also can not manipulate the data after encryption. It will greatly undermine the advantage of the cloud storage. For example, a user encrypts his personal data before uploading them to the cloud. When he wants to access some data at the cloud, he has to download all the data and decrypt them. Obviously, this service mode will incur the huge overheads of communication and computation. Several related works have been proposed to enable the search over the encrypted data, but all of them only support the encrypted keyword search. In this paper, we propose a new full-text retrieval algorithm over the encrypted data for the scenario of cloud storage, in which all the words in a document have been extracted and built a privacy-preserved full-text retrieval index. Based on the privacy-preserved full-text retrieval index, cloud server can execute full-text retrieval over the large scale encrypted documents. The numerical analysis and experimental results further validate the high efficiency and scalability of the proposed algorithm.

1 Introduction

Cloud computing attracts considerable attentions and interests from both industry and academia because of its scalability, flexibility, and cost-effective features. With cloud storage services, users can rent the cloud storage space to store their personal data and does not have to buy the storage hardware. This mechanism is particularly suitable for the mobile devices which have the limited storage spaces. However, due to the nature of the public, users usually prefer not to store their sensitive private information into the cloud in the plaintext form even if the data privacy is enforced by law.

J. Li et al. (Eds.): NLPCC 2015, LNAI 9362, pp. 229–241, 2015.
DOI: 10.1007/978-3-319-25207-0_20

A straightforward solution for users to protect their data privacy is to encrypt the data before outsourcing. This service mode has been adopted by Amazon IS service. By encryption, the data privacy is preserved, but the data utilization, i.e., search, becomes difficult over the encrypted data. A naive way requires the user to download and decrypt the data and execute the search over the plaintext. Obviously, such approach incurs tremendous overheads of communication and computation and greatly undermines the advantages of using cloud storage applications.

Searchable encryption schemes [1–6] have been developed in recent years for balancing the search efficiency and data privacy. However, the existing approaches mainly focus on the keyword-based search which is difficult to meet the requirements of the large-scale cloud storage systems. Full-text retrieval is a successful information retrieval technology for content search over the large scale data. Its effectiveness and efficiency have been verified by the success of Internet search engine systems. However, the full-text retrieval technology needs to extract all the words in the contents of documents, which makes the scale of index words is much larger than that in the keyword-based search. Therefore, it is far from practical to provide the full-text retrieval services for the cloud storage applications using the existing searchable encryption schemes.

To achieve the full-text retrieval over the large scale encrypted documents, we design a privacy-preserved full-text retrieval index based on which a full-text retrieval algorithm have been proposed. In our scheme, all the documents are encrypted, the search processes do not need to decrypt the data. We analyze the efficiency of the proposed scheme and prove its security. The main contributions of this paper can be summarized as follows:

- To the best of our knowledge, this is the first work that identifies the problem of privacy-preserved full-text retrieval over the encrypted data for a large-scale cloud storage system.
- To address this problem, we propose a secure index structure, based on which an efficient and secure full-text retrieval scheme over the encrypted data has been proposed.
- We analyze the security and efficiency of the proposed scheme. Moreover, we demonstrate the effectiveness and efficiency of the proposed scheme through extensive experimental evaluation.

The rest of our paper is organized as follows. In the next section, we discuss some related work. Then we introduce our full-text retrieval algorithm in Section 3. In Section 4, we analysis the efficiency of our scheme and prove the security. We evaluate the performance of our scheme by the experiments in Section 5. Finally, our paper is concluded in Section 6.

2   Related Work

Our work mainly focuses on addressing the problem of the secure and efficient full-text retrieval over the encrypted data at the cloud. To the best of our knowledge, no existing research has addressed this issue. The existing researches that are similar to ours can be found in the areas of keyword searchable encryption, rich functional encrypted data search, and ranked search over encrypted data.

2.1   Keyword Searchable Encryption

The existing keyword searchable encryption schemes [1–6] usually build an encrypted searchable index such that content of sensitive documents is hidden to the cloud server. The client gives appreciate query trapdoors through the secret key to retrieve the interested documents. It is first studied by Song et al. [4] with the symmetric surroundings. Some improvements and advanced security definitions, including Goh [1], Chang et al. [2], Curtmola et al. [5], and Kamara et al. [8] are given. In the public key setting, Boneh et al. [6] propose the first public key searchable encryption construction (PEKS), where anyone encrypts the data using public key but only authorized users with private key can create query trapdoors to search. Public key solutions usually have a high computation overhead. The Bloom filter is an effectively searchable index structure by which some researches [2,3,6] have been given to implement encrypted data query. However, these researches put keywords of a document into a Bloom filter, this document-based index pattern needs to match Bloom filter one by one during query, so the query efficiency will be reduced with document scale increasing. Our work, which builds the hierarchical Bloom filter tree index based on the word, can keep the high query efficiency with a large scale of documents.

2.2   Rich Functional Encrypted Data Search

To achieve authorized encrypted data search, the public-key searchable encryption schemes [9–12] have been proposed. A common approach of them is to use a paring based encryption to construct the searchable indexes. However, these solutions are inefficient to support the cloud storage application because paring operations are expensive. To enrich the search functionalities, conjunctive keyword search, fuzzy keyword search, and subset query [9,13–16] over encrypted data have been proposed. These schemes result in large overhead for their functional computations, such as bilinear computation cost in [13], communication cost for secret sharing in [14]. And a more general search schemes, predicate encryption schemes [17–19] are recently proposed to support both conjunctive and disjunctive search. The disjunctive search is similar to our full-text retrieval which returns every document that contains a subset of the query keywords. Moreover, our full-text retrieval scheme designs the effective index structure contains all words of a document, so it can describe document contents more accurately and comprehensively and make the cloud with high availability.

2.3   Ranked Search Over Encrypted Data

Ranked searchable encryption enables users to retrieve the most relevant documents from the cloud in the case that both the user queries and data are in the encrypted form to protect user privacy. The work in [20] is the first research for ranked search encryption. It only supports single-keyword search, and encrypts documents and queries with a one-to-many OPSE (Order Preserving Symmetric Encryption) scheme [21] and utilizes keyword frequency to rank query results. Their following work [22], which supports multi-keyword searches, uses the secure KNN scheme [23] to rank the results based on inner product value.

3   Full-Text Retrieval Algorithm Over Encrypted Data

3.1   Full-Text Retrieval Model and Its Security Problem

The full-text retrieval model is a successful technology in the scenario of information retrieval applications, its efficiency has been verified in the popular search engine applications. The straightforward method to achieve the full-text retrieval over encrypted data is to improve the index structure in the existing full-text retrieval model. To protect the privacy of user's personal data, we should encrypt the full-text retrieval index. According to the encryption granularity, the encrypted index can be divided into two main categories: '*index level*' and '*token level*'. The '*index level*' mode is to encrypt the whole index. During the search processes, cloud server has to decrypt the entire index or partial index and execute the query over the plaintext index. The searches in the '*index level*' mode will lead to a great deal of encryption/decryption operations, so it is not suitable for the large-scale cloud data. Moreover, this service mode makes the cloud server be able to decrypt the index on the cloud, which makes it be not able to resist to the internal threatens.

The '*token level*' mode encrypts the tokens and builds the secure full-text retrieval index to enable the searches over the encrypted data. But, the existing various full-text retrieval algorithms are based on the token offset position and the token frequency, which will leak the user's privacy. During the search processes, the server needs to compute the offset position and the token frequency, so we can not directly encrypt the offset position and the token frequency. We design a novel full-text retrieval algorithm over the encrypted data without using the offset position and the token frequency.

To introduce our scheme, we first define the full-text retrieval model.

Definition 1: A full-text retrieval system R can be defined as $\{D, Q, F\}$, where D is the set of documents in R, and Q is the expression of the user's queries, and F denotes the framework of document expression and content extraction.

Based on the above definition, we introduce our privacy-preserved full-text retrieval algorithm as below. For a document d_i, it is mapped into a set of tokens $T_i((t_1, p_1), (t_2, p_2), \ldots, (t_k, p_k))$ under the framework F, in which t_j represents a token in d_i's contents, and p_j represents the t_j's offset position in d_i. After extracting all the tokens in d_i, client uses a one-way hash function H to process every plaintext token t_j by its private key key_{pri} and outputs the encrypted token e_j as Equation 1.

$$e_j = H(t_j | key_{pri}) \tag{1}$$

Once the client gets the set of encrypted tokens $ET_i = (e_i, p_i)_{1 \leq i \leq k}$ as in Equation 2, it builds the encrypted full-text retrieval index $EIndex$ for d_i and uploads to the cloud server.

$$d_i \xrightarrow{F} T_i((t_1, p_1), \ldots, (t_k, p_k)) \xrightarrow{Encrypt} ET_i((e_1, p_1), \ldots, (e_k, p_k)) \xrightarrow{Index} EIndex \quad (2)$$

3.2 System Model

We first present the overview of the proposed privacy-preserved full-text retrieval framework in this subsection, which is shown in Fig. 1.

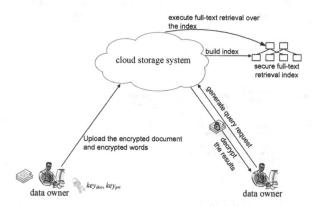

Fig. 1. The full-text retrieval framework of our scheme

The processing flow of the full-text retrieval over the cloud encrypted data is as follows. While a new user joins the cloud storage system, he first chooses a private key key_{pri} and a document encryption key key_{doc} and stores them at local. And the cloud server initializes a hash function H for the new user. H represents a one-way hash function which maps arbitrary string to an integer between 1 and 2^m. m is a system global parameter decided by the cloud server. To protect data privacy, the data owner utilizes the symmetric encryption algorithm to encrypt the documents before outsourcing. Besides encrypting the documents, the data owner extracts all the words from the contents of the document and encrypts these words by the hash function H with the key key_{pri}. Finally, the data owner uploads the encrypted documents and the encrypted index words to the cloud. Once the cloud server receives the files uploaded from the data owner, it inserts them into the full-text retrieval index to provide the secure and efficient full-text retrieval services.

When the user wants to query the data in the cloud with certain query words, he generates encrypted query words using the key key_{pri} and the hash function H. After the cloud server receives the query request, it executes the full-text retrieval over the index. Finally, the authorized user decrypts the results returned from cloud with key_{doc} to finish the query processes.

3.3   Privacy-Preserved Full-Text Retrieval Index Based on B+ Tree

But the token's offset position will leak the user's privacy, so we design a privacy-preserved full-text retrieval index based on B+ tree without token's offset position. The index structure is shown in Figure 2.

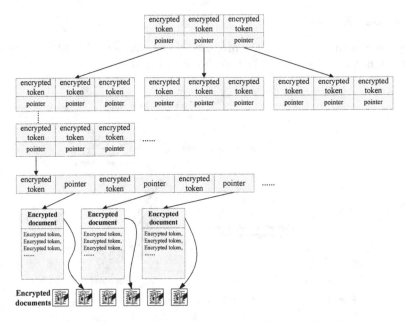

Fig. 2. The encrypted full-text retrieval index based on B+ tree

First, the client calls the document processing algorithm which we introduce in the next section to extract all the tokens $T_i(t_1, t_2, \ldots, t_k)$ for a document d_i. To protect user's privacy, the index in our scheme does not include the tokens offset position information. Then, we improve the index generation in Equation 2 as Equation 3

$$d_i \xrightarrow{F} T_i(t_1, t_2, \ldots, t_k) \xrightarrow{Encrypt} ET_i(e_1, e_2, \ldots, e_k) \xrightarrow{Index} EIndex \qquad (3)$$

In our scheme, the search processes does not have the decryption operations. After the cloud server receives the query request from the client, it converts the request to the query condition formed by the encrypted query words.

3.4   Document Pre-processing

In the traditional plaintext full-text retrieval algorithm, the server determines whether a long compound word in a document's contents based on the word's offset position. For example, if a word *cloud*'s offset position in the document

d is x and a word *computing*'s offset position in the document d is $x + 1$, then the server can tell that the compound word 'cloud computing' must be in d. To protect user's privacy, we do not store the tokens' offset position information in the full-text retrieval index, so we propose the privacy-preserved full-text retrieval algorithm without offset position supporting.

If we put all the possible compound words into the index, then we can achieve the full-text retrieval over the encrypted data without offset position. But, putting all the compound words of a document into the index will damage the search efficiency. Moreover, the compound word which has a long size has a small possibility for query hit. So, we design the maximal word length k extraction algorithm in Algorithm 1 to collect all the words for a document, which will extract all the single words and compound words, the length of which are no more than k.

Algorithm 1. Extract the index words for a document

Input: d, a document to be uploaded; k, maximal number of single words in a compound index word; SW, stop words list

Output: *words*, the index words for d

1   extract all the single words from d and put them into d_{words};
2   **for** $(i = 1; i \leq d_{words}.size; i + +)$ **do**
3      $t = d_{words}[i]$;
4   **if** $(t\ in\ SW)$ **then**
5   continue;

6   **else**
7   add t into *words*;

8   **for** $(j = 1; j < k; j + +)$ **do**
9   **if** *(the jth word after t is in SW)* **then**
10   break;

11   **else**
12   add the compound word from t to the (j)th word after t into *words*;

13   remove reduplicative compound words from *words*;
14   **return** *words*;

After the data owner collects all the index words for an uploading document d, he uses his private key key_{pri} and the hash function H to encrypt these index words as Equation 1. Then, the data owner uploads these encrypted index words with the encrypted document to the cloud server.

While the cloud server receives an encrypted document d and its corresponding encrypted index words $e_i(1 \leq i \leq k)$, it searches the full-text retrieval index at cloud and finds all the nodes which are equal to one encrypted index word e_i. At last, the cloud server creates the pointer links from these nodes to the encrypted document to finish the document insertion.

3.5 Full-Text Retrieval Algorithm Over the Encrypted Data

While a user wants to search his interested files at cloud, he first gives some query words like using the web search engine applications. The search processes are described by Equation 4. A user's query Q is composed by several query words $qw_i(1 \leq i \leq n)$. The user uses the same hash function H and his private key key_{pri} to encrypt these query words by $ew_i = H(qw_i|key_{pri})(1 \leq i \leq n)$ and submits them to the cloud server.

$$Q \rightarrow qw_1(\wedge, \vee)qw_2, \ldots, qw_n \xrightarrow{Encrypt} EQ(ew_1, ew_2, \ldots, ew_k) \xrightarrow{Search} Results \quad (4)$$

Once the cloud server receives the encrypted query words ew_i from the user, it executes the query based on the index in Fig. 2. Based on the characters of B+ tree, the cloud server finds all the nodes which equal to a encrypted word ew_i from the root node. Afterwards, the cloud server returns the documents which have pointer links to these nodes as the result of the query to the user.

4 Performance and Security Analysis

In this paper, we propose a privacy-preserved full-text retrieval algorithm over the encrypted data. During the services, our scheme does not need the decryption operations. In this section, we analyze the efficiency and the security of our scheme.

4.1 Query Precision of Our Scheme

The query precision rate indicates the ratio of the exactly relevant documents to all the returned documents as illustrated in Equation 5, in which D_{match} represents the total documents which correctly match a query, and D_{return} represents the total documents returned from the cloud server.

$$Precision = \frac{D_{match}}{D_{return}} \times 100\% \quad (5)$$

The index of our scheme is built based on the Hash function H, so the query false positive rate brought by the confliction of Hash function has to be considered. We assume that the width of the index node is m, then the confliction of a node is $\frac{1}{2^m}$. Consider a document d' with k index words uploaded by a user u' whose private key is key'_{pri}, then the encrypted document d' has been mapped into k index nodes. While a user u whose private key is key_{pri} launches a query Q, the probability of a query word qw equal to an index node linked to d' but this index node is not the word qw is $1 - (1 - \frac{1}{2^m})^k$. For example, when $m = 16, k = 1000$ the false positive rate is 1.51%, and the false positive rate is 0.47% when $m = 20, k = 5000$.

4.2   Query Efficiency of Our Scheme

The query processes in our scheme include three main steps: 1) the user encrypts the query words and submits them to the cloud server. 2) the cloud server executes the query over the full-text retrieval index. 3) the user decrypts the encrypted documents returned from the cloud server by the key key_{doc}.

In the 1st query step, the cloud user first selects several original query words, then encrypts the query words by H. The computation cost of the first step is $c \times T_{hash}$, where c represents the number of original query words given by the user, and T_{hash} represents the computation cost of one hash operation.

For a privacy-preserved full-text retrieval index which depth is l, the cloud server needs to route to the nodes which are equal to the encrypted query word from the root node. So, the computation cost of the 2nd step is $c \times l$.

In the 3rd step, the user decrypts the encrypted documents returned from the cloud by his private key key_{doc}. Assuming that the cloud server returned d documents for a query, and then the computation cost for this step is $d \times T_{dec}$, in which T_{dec} represents the time cost of decrypting a document.

4.3   Security Analysis of Our Scheme

In our scheme, the cloud server stores and processes three types of data including the encrypted documents E, the full-text retrieval index I, and the query requests R from the user. In these information, E is encrypted by the data owner. Meanwhile, the key key_{doc} is grasped by the data owner. We assume that it is impossible for an adversary to stole key_{doc} from the data owner. Therefore, based on the security of encryption algorithms, the attacker is unable to break the data privacy through attacking E without key_{doc}.

For an external attacker, i.e., the hacker or the internal attacker, he masters the full-text retrieval index in the cloud. We assume that an adversary $\mathcal{A}$ tries to break the security of our scheme. First, $\mathcal{A}$ attempts to guess the word in the full-text retrieval index node and to guess the information of encrypted documents. To guess the word in an index node, $\mathcal{A}$ has to answer the one-way Hash function H. Based on the index structure, $\mathcal{A}$ can output the word in an index node with probability roughly $q_H/2^m$, where q_H is the total number of the queries on H. Take the index structure ($m = 24$) as the example, the probability that $\mathcal{A}$ can exactly guess the word in an index node is no more than $1/16,777,216$ in one time query. Therefore, through the analysis, we can think of the data privacy in our scheme is guaranteed.

5   Experiments

5.1   Experimental Setup

We use JAVA language to implement the proposed full-text retrieval scheme in this paper. We carry out the experiments on a PC machine running Windows 7 with a 64-bits, 3.0 GHz CPU and 4GB main memory. The parameters of our scheme in the experiments are shown in Table 1. In the experiments, we use ICTCLAS[1] to extract the words from the contents of documents. Moreover, we run the experiments over the dataset: Chinese laws and regulations ceremony which size is 60,000.

Table 1. Experimental Parameters in the Experiments

Parameters	Parameter Descriptions	Values
m	the bit length of index node	24
k	the maximal length of the compound index word	5, 6, 7

5.2   Storage Overhead of Our Scheme

Usually, a cloud storage server stores large scale of documents. So, the storage overhead of secure full-text retrieval index is significant important for the performance of cloud storage system. We design the experiments to measure the storage space of the index in our scheme. We compare our scheme with 2-MCIS [7] which is multi-keyword search over the encrypted data. The experimental results are shown in Fig. 3.

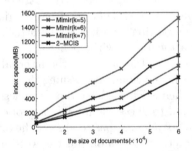

Fig. 3. Storage overhead of our scheme

From the experimental results, we can find that the storage overhead of our scheme is comparable with that in 2-MCIS. When $k = 5$ and $d = 60,000$, the index is no more than 830MB. So, our scheme achieves a satisfied storage performance which makes our scheme is able to efficiently support a large scale cloud storage system.

---

[1] NLPIR. http://ictclas.nlpir.org/

5.3 Query Efficiency of Our Scheme

For a cloud storage system, the query response time is one of the most important indicators. We design the experiments to evaluate the mean query response time with different k value. The experimental results are shown in Fig. 4.

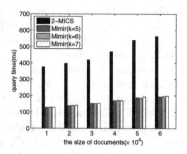

Fig. 4. query efficiency with various scale of documents

We can find from the experimental results in Fig. 4 that our scheme is more efficient than 2-MICS. For a query, our algorithm is able to make the cloud server execute the query in 200 ms. Moreover, the query efficiency will not reduce greatly with the increasing of the document scale. This interesting feature makes our scheme be able to support large scale cloud storage system.

6 Conclusion

In this paper, for the first time we define and address the problem of supporting efficient yet privacy-preserved full-text retrieval services to enrich the query function over the cloud encrypted data. We design a privacy-preserved full-text retrieval index structure to allow the authorized user to execute the full-text retrieval over the encrypted documents at the cloud. Through the rigorous security and performance analysis, we demonstrate that the proposed solution is secure and privacy preserving. Extensive experimental results further validate the effectiveness and efficiency of our solution.

Acknowledgments. This work is supported by National Natural Science Foundation of China No. 61202034 and 61232002, CCF Opening Project of Chinese Information Processing No. CCF2014-01-02, and the Program for Innovative Research Team of Wuhan No. 2014070504020237.

References

1. Goh, E.-J.: Secure indexes. IACR Cryptology ePrint Archive (2003)
2. Chang, Y.-C., Mitzenmacher, M.: Privacy preserving keyword searches on remote encrypted data. In: Ioannidis, J., Keromytis, A.D., Yung, M. (eds.) ACNS 2005. LNCS, vol. 3531, pp. 442–455. Springer, Heidelberg (2005)
3. Watanabe, C., Arai, Y.: Privacy-preserving queries for a DAS model using encrypted bloom filter. In: Zhou, X., Yokota, H., Deng, K., Liu, Q. (eds.) DASFAA 2009. LNCS, vol. 5463, pp. 491–495. Springer, Heidelberg (2009)
4. Song, D., Wagner, D., Perrig, A.: Practical Techniques for Searches on Encrypted Data. In: Proceedings of S&P, pp. 44–55 (2000)
5. Curtmola, R., Garay, J.A., Kamara, S., Ostrovsky, R.: Searchable symmertric encryption: improved definitions and efficient constructions. In: CCS (2006)
6. Boneh, D., Di Crescenzo, G., Ostrovsky, R., Persiano, G.: Public key encryption with keyword search. In: Cachin, C., Camenisch, J.L. (eds.) EUROCRYPT 2004. LNCS, vol. 3027, pp. 506–522. Springer, Heidelberg (2004)
7. Wang, N., Zhao, W., Liu, G., Zhao, C.: K-mapping chipher index scheme as to character data in outsourced database. Journal of Yanshan University **33**(5), 438–443 (2009)
8. Kamara, S., Papamanthou, C., Roeder, T.: Dynamic Searchable Symmetric Encryption. In: Proceedings of CCS, pp. 965–976 (2012)
9. Li, M., Yu, S., Cao, N., Lou, W.: Authorized Private Keyword Search over Encrypted Data in Cloud Computing. In: ICDCS, pp. 393–402 (2011)
10. Sun, W., Yu, S., Lou, W., Hou, Y., Li, H.: Protecting your Right: Attribute-based Keyword Search with Fine-grained Owner-enforced Search Authorization in the Cloud. In: Proceedings of INFOCOM, pp. 226–234 (2014)
11. Rhee, H.S., Park, J.H., Susilo, W., Lee, D.H.: Trapdoor Security in a Searchable Public-key Encryption Scheme with a Designated Tester. Journal of System and Software **83**(5), 763–771 (2010)
12. Boneh, D., Waters, B.: Conjunctive, subset, and range queries on encrypted data. In: Vadhan, S.P. (ed.) TCC 2007. LNCS, vol. 4392, pp. 535–554. Springer, Heidelberg (2007)
13. Li, J., Wang, Q., Wang, C., Cao, N., Ren, K., Lou, W.: Fuzzy keyword search over encrypted data in cloud computing. In: INFOCOM, pp. 441–445 (2010)
14. Ballard, L., Kamara, S., Monrose, F.: Achieving efficient conjunctive keyword searches over encrypted data. In: Qing, S., Mao, W., López, J., Wang, G. (eds.) ICICS 2005. LNCS, vol. 3783, pp. 414–426. Springer, Heidelberg (2005)
15. Bellare, M., Boldyreva, A., O'Neill, A.: Deterministic and efficiently searchable encryption. In: Menezes, A. (ed.) CRYPTO 2007. LNCS, vol. 4622, pp. 535–552. Springer, Heidelberg (2007)
16. Boneh, D., Kushilevitz, E., Ostrovsky, R., Skeith III, W.E.: Public key encryption that allows PIR queries. In: Menezes, A. (ed.) CRYPTO 2007. LNCS, vol. 4622, pp. 50–67. Springer, Heidelberg (2007)
17. Katz, J., Sahai, A., Waters, B.: Predicate encryption supporting disjunctions, polynomial equations, and inner products. In: Smart, N.P. (ed.) EUROCRYPT 2008. LNCS, vol. 4965, pp. 146–162. Springer, Heidelberg (2008)
18. Lewko, A., Okamoto, T., Sahai, A., Takashima, K., Waters, B.: Fully secure functional encryption: attribute-based encryption and (Hierarchical) inner product encryption. In: Gilbert, H. (ed.) EUROCRYPT 2010. LNCS, vol. 6110, pp. 62–91. Springer, Heidelberg (2010)

19. Shen, E., Shi, E., Waters, B.: Predicate privacy in encryption systems. In: Reingold, O. (ed.) TCC 2009. LNCS, vol. 5444, pp. 457–473. Springer, Heidelberg (2009)
20. Wang, C., Cao, N., Li, J., Ren, K., Lou, W.: Secure ranked keyword search over encrypted cloud data. In: Proceedings of ICDCS, pp. 253–262 (2010)
21. Boldyreva, A., Chenette, N., Lee, Y., O'Neill, A.: Order-preserving symmetric encryption. In: Joux, A. (ed.) EUROCRYPT 2009. LNCS, vol. 5479, pp. 224–241. Springer, Heidelberg (2009)
22. Cao, N., Wang, C., Li, M., Ren, K., Lou, W.: Privacy-preserving multi-keyword ranked search over encrypted cloud data. In: Proceedings of INFOCOM (2011)
23. Wong, W.K., Cheung, D.W., Kao, B., Mamoulis, N.: Secure kNN computation on encrypted databases. In: Proceedings of SIGMOD, pp. 139–152 (2009)

How Different Features Contribute
to the Session Search?

Jingfei Li[1], Dawei Song[1,2], Peng Zhang[1(✉)], and Yuexian Hou[1]

[1] Tianjin Key Laboratory of Cognitive Computing and Application,
Tianjin University, Tianjin, P.R. China
jingfl@foxmail.com, {dwsong,pzhang,yxhou}@tju.edu.cn
[2] The Computing Department, The Open University, Milton Keynes, UK

Abstract. Session search aims to improve ranking effectiveness by
incorporating user interaction information, including short-term interactions within one session and global interactions from other sessions (or
other users). While various session search models have been developed
and a large number of interaction features have been used, there is a lack
of a systematic investigation on how different features would influence
the session search. In this paper, we propose to classify typical interaction features into four categories (current query, current session, query
change, and collective intelligence). Their impact on the session search
performance is investigated through a systematic empirical study, under
the widely used Learning-to-Rank framework. One of our key findings,
different from what have been reported in the literature, is: features based
on current query and collective intelligence have a more positive influence than features based on query change and current session. This would
provide insights for development of future session search techniques.

Keywords: Session features · Query change · Collective intelligence

1 Introduction

Session search aims to rank documents/web pages based on not only a current
query, but also single or collective user interactions such as query reformulations
and document clicks, in the current search session or longer-term search history
[4][5][11][13][14]. Various session search models have been proposed [5][13][14]
based on different interaction features extracted from specific resources.

However, it is still an open question how effectively different features contribute to improving the session search performance. This paper aims at a systematic investigation on this problem. We propose to classify typical interaction
features into 4 categories (see Figure 1): (i) current query features; (ii) query
change features; (iii) whole session features; and (iv) collective intelligence features. Each category is based on a specific assumption, namely *query relevance*,
search intent change, *search intent relatedness*, and *collective intelligence helpfulness*, respectively. These assumptions (detailed in Section 3), individually,

© Springer International Publishing Switzerland 2015
J. Li et al. (Eds.): NLPCC 2015, LNAI 9362, pp. 242–253, 2015.
DOI: 10.1007/978-3-319-25207-0_21

are largely implied in different personalized search (including session search) approaches, e.g., [1][4][5][11][13]. We investigate these four assumptions (and the corresponding features) in one unified framework, i.e., the widely used Learning to Rank (Learning2Rank) framework. The most related work to ours is Bennett et al. [1] that investigated different user profiles built from users' long-term and short-term search behaviors. Our work is intrinsically different, in that we investigate session search based on different assumptions which reflect users' current Information Need (IN), evolving IN and collective intelligence, while Bennett et al. [1] studied the personalization based on features from different temporal views, i.e. historic view, session view and aggregate view. Moreover, our work is the first comparative study of the query change features against the other three types of features in the Learning2Rank framework.

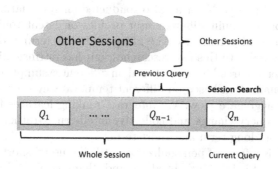

Fig. 1. An illustration of how we select features from current query, previous query, the whole session and other sessions with similar search goals to this session.

Specifically, after extracting different categories of features, we integrate them into the Learning2Rank model LambdaMART [2], then re-rank the original results returned by a baseline search engine. The re-ranking performance based on different features are compared. Experimental results on a real-world query log demonstrate that different categories of features have different impacts on re-ranking. One of our key findings is that the current query features and collective intelligence features are relatively more influential to the re-ranking performance. This provides new insights to the design of future session search models.

The remainder of this paper is organized as follows. Section 2 presents related work. Section 3 formalizes the background of this study. Section 4 introduces involved features in detail. Section 5 conducts extensive experiments to analyze how different features influence the session search. Conclusions and future work are discussed in Section 6.

2   Related Work

There are several lines of work related to ours including general personalized search and session search.

There have been many attempts to personalize the web search [4][11] and to investigate personalization-related IR problems (e.g., potential to personalization and personalization risks etc.) [1][9]. Dou et al. [4] proposed to re-rank the web search results with two personalization strategies (person-level and group-level), and found for the first time that personalization only works on a small subset of queries with larger click-entropy, this finding is also supported by many subsequent contributions [1][9][11]. Vu et al. [11] improved the search personalization with dynamic group of user profiles constructed in responds to the user's input query. Teevan [9] systematically studied the potential to personalization for queries with variations of features, and proposed that search personalization should be applied in different manners according to queries' characteristics. Bennett et al. investigates how different user profiles modeling user's long- or short- term search behaviors contribute to the re-ranking performance of web search results [1]. Similarly, Vu et al. also conducted an systematic investigation for personalization and built different user profiles on latent topic space from more fine grit temporal perspectives, i.e., long term, day term and session term [10]. Our work is similar to this two researches, but has intrinsic differences. We investigate the user's search behaviors based on different assumptions, while they studied the personalization based on different temporal views, i.e. historic view, session view and aggregate view. We integrate ranking features from different perspectives, e.g., query change and collective intelligence, rather than a single temporal angle, e.g., long-term and short-term.

Session search is a form of personalization utilizing users' short term interaction behaviors [5][7][13][14]. Guan et al. [5] and Zhang et al. [13] utilized query change information in session search which is based on the "search intent change" assumption introduced in Section 1. They listed some representative sessions from the TREC session tracks as examples to illustrate the phenomenon of user's query reformulation behaviors. They found that the current query of a session is composed of three parts, i.e. the common part, added part and removed part compared with previous query. Different weights are assigned to retrieved documents for three parts in ranking process. More specifically, the Markov Decision Process (MDP) was utilized to model the query change information in session search [5][13]. Luo et al. [7] and Zhang [14] also conducted session search considering the topic drifting and user's dynamics of information needs among queries within the same session. They model session search as a win-win game for the user and the search engine with the Partially Observable Markov Decision Process. Our work is inspired by them [5][7][13][14] which consider the dynamic information need of the user within a search session when retrieving documents. Differently, we focus on analyzing how effectively the query change features influence the overall retrieval performance within in a unified Learning-to-Rank framework, rather than developing a novel retrieval model.

3 Background

A session, formalized as $S = < q_1, ..., q_{n-1}, q_n >$, is a sequence of queries sorted by timestamp issued by one user. More general, a search session can be seen as a single search task or goal [5][14]. In session search, q_n is regarded as the current query whose original results are to be re-ordered. The re-ranking model may utilize the user's interactions with the search engine (recorded in previous queries $q_1, ..., q_{n-1}$), such as query issuing, query reformulation, results clicking, dwell time and paginating etc. [5][13]. To take advantage of the collective intelligence, some re-ranking models have incorporated other users' interaction information [4][10][11]. There exists so much information for session search and different information may interwine together, which makes session search a exceedingly challenging retrieval task.

LambdaMART [2] is a Learning2Rank model which has been proven to be effective for document ranking. For training and testing of the LambdaMART models, we design a semi-automated labeling algorithm for estimating the relevance of each document with regard to a query. The SAT-clicked documents (with dwell time more than 30 seconds [1][11]) are labeled as 2, the clicked documents with a dwell time less than 30 seconds are labeled as 1 and the other documents are labeled as 0. This dwell time based grading method is also used in [14].

4 A Classification of Interaction Features

In this paper, we classify the typical interaction features into four categories (i.e., current query features, query change features, whole session features, and collective intelligence features), based on four underlying assumptions. First, the "current query" features are underpinned by the *query relevance* assumption. It assumes that a user-issued query represents the user's Information Need (IN) directly. However, the representation is incomplete due to the limitation of user knowledge. Therefore, more information (features) needs to be imported to enrich the representation of user IN. Second, the *search intent change* assumption, which underpins the "query change" features, is based on the fact that user's IN evolves continuously. The query reformulation information between the current and previous queries provides some clues to capture these changes. Third, the *search intent relatedness* assumption, which is related to the "whole session" features, considers that queries in the same session have a similar search intent, and the previous queries and clicked documents in the same session can reveal the search intent of current query to some extent. Finally, *collective intelligence helpfulness* underpinning the "collective intelligence" features, assumes that the interactions of other people, especially the ones who have similar search interests or have submitted similar queries in the past, will provide useful clues for ranking documents for the current user.

In the rest of this section, we will describe each category of features in detail. An empirical comparative study is then reported in Section .

4.1 Current Query Features

This category of features only consider the similarity of a document d to the current query q. Three traditional scoring schemes which have been proven to be effective are included in this category. They are BM25 [8][1], Query Likelihood model [12][2] and $tf \cdot idf$ based ranking function [6]. Additionally, the ranks of documents in the original result list for the current query q is also considered. A score is estimated as $frank(q, d) = \frac{1}{log_2(1+rank_d)}$, where $rank_d$ is the rank of document d. The current query features are summarized as Table 1-(a). The computation equations for different features are formalized as follows.

One of the most prominent instantiation among the whole family of BM25 based ranking functions is formulated as follows:

$$bm25(d, q) = \sum_{w \in q} idf(w) \cdot \frac{c(w, d) \cdot (k_1 + 1)}{c(w, d) + k_1 \cdot (1 - b + b \cdot \frac{|d|}{avgdl})} \cdot \frac{(k_3 + 1) \cdot c(w, q)}{k_3 + c(w, q)} \quad (1)$$

where $c(w, d)$ and $c(w, q)$ are w's term frequencies in d and q respectively; $avgdl$ is the average document length of the text collection. $idf(w) = log\frac{N - df_w + 0.5}{df_w + 0.5}$, where N is the total document count in a collection, df_w is the number of documents containing word w; k_1, b and k_3 are three parameters which, in this paper, are empirically set as 1.2, 0.75 and 7 respectively.

The query likelihood model is a language model used in IR, and it can be interpreted as being the likelihood of a document being relevant to a query. The relevance score based on query likelihood is as follows:

$$QL(q|\theta_d) = \prod_{w \in q} p(w|d) \propto \sum_{w \in q} log\, p(w|d), \quad p(w|d) = \frac{c(w, d) + \mu p(w|C)}{|d| + \mu} \quad (2)$$

where θ_d is a unigram language model, i.e., $\theta_d = p(w|d)_{w \in V}$, V is the vocabulary, $p(w|d)$ is the probability of word w. In this paper we use Dirichlet prior smoothing method to estimate the probability of w, where $p(w|C)$ is a language model for the collection and the smoothing parameter $\mu = 2500$ here.

We follow the definition of $tf \cdot idf$ score of a document d given a query q as Liu et al.[6] does. $c(w, d)$ is the term frequency of w in document d, df_w is the document frequency of w, and $|C|$ is the total count of documents in collection.

$$tfidf(q, d) = \sum_{w \in q \cap d} (0.5 + \frac{0.5 \times c(w, d)}{max\{c(w, d) : w \in d\}}) log \frac{|C|}{df_w} \quad (3)$$

4.2   Query Change Features

Query change, also known as query reformulation, is an important type of user interaction with the search engine. Lacking satisfaction with the returned results for an initial query, the user may change the query to achieve her/his search target. How to reformulate the query reflects the change direction of the user's

---

[1] The well known parameters k_1, b and k_3 for Okapi BM25 are empirically set as 1.2, 0.75 and 7 respectively in this paper.

[2] The Dirichlet prior smoothing method is used to estimate the probability of a word, and the smoothing parameter $\mu = 2500$ here.

Table 1. Features and their description. (a) is the current query dependent features; (b) is the whole session features; (c) is the collective intelligence features. d_c and d_{no} denote the clicked documents and non-clicked documents respectively.

(a) Current Query Features		
Feature	Formulas	Descriptions
C1	$bm25(q,d)$	BM25 ranking function
C2	$QL(q|\theta_d)$	Query Likelihood function
C3	$tfidf(q,d)$	$tf \cdot idf$ relevance score
C4	$frank(q,d)$	Rank based feature

(b) Whole Session Features		
(C is the clicked set, NC is the non-clicked set in the whole session)		
feature	Formulas	Descriptions
W1	$\sum_{d_c \in C} sim(d,d_c) \cdot bm25(q,d_c)$	Weighted sum of BM25 ranking function for clicked documents given current query
W2	$\sum_{d_c \in C} sim(d,d_c) \cdot QL(q|\theta_{d_c})$	Weighted sum of Query Likelihood score for clicked documents given current query
W3	$\sum_{d_c \in C} sim(d,d_c) \cdot tfidf(q,d_c)$	Weighted sum of tfidf score for clicked documents given current query
W4	$\sum_{d_{no} \in NC} sim(d,d_{no}) \cdot bm25(q,d_{no})$	Weighted sum of BM25 ranking function for non-clicked documents given current query
W5	$\sum_{d_{no} \in NC} sim(d,d_{no}) \cdot QL(q|\theta_{d_{no}})$	Weighted sum of Query Likelihood score for non-clicked documents given current query
W6	$\sum_{d_{no} \in NC} sim(d,d_{no}) \cdot tfidf(q,d_{no})$	Weighted sum of tfidf score for non-clicked documents given current query
W7	$\sum_{d_c \in C} sameDomain(d^{url}, d_c^{url})$	The number of document whose domain name is the same as the current document

(c) Collective Intelligence Features		
(C is the clicked documents set in other sessions)		
Feature	Formulas	Descriptions
I1	$\sum_{d_c \in C} sim(d,d_c) \cdot bm25(q,d_c)$	Weighted sum of BM25 ranking function for clicked documents given current query
I2	$\sum_{d_c \in C} sim(d,d_c) \cdot QL(q|\theta_{d_c})$	Weighted sum of Query Likelihood score for clicked documents given current query
I3	$\sum_{d_c \in C} sim(d,d_c) \cdot tfidf(q,d_c)$	Weighted sum of tfidf score for clicked documents given current query

search intent. Existing researches [5][13][14] analyzed different strategies of query reformulation from two aspects, i.e., query formation and query semantic. In this paper, we extract query change features based on the change of query formation, including adding and removing query terms. To this end, we segment the current query into three parts, i.e., the common part (obtained by $q_{com} = wordset(q_n) \cap wordset(q_{n-1})$), added part ($q_{add} = wordset(q_n) - q_{com}$) and removed part ($q_{rmv} = wordset(q_{n-1}) - q_{com}$) compared with previous query. The query-document features are computed based on the weighed sum of

Table 2. Query change features and their description. d_c and d_{no} denote the clicked documents and non-clicked documents respectively.

Query Change Features (C is the clicked documents set, NC is the non-clicked set in previous query)		
Feature	Formulas	Descriptions
Q1	$\sum_{d_c \in C} sim(d, d_c) \cdot bm25(q_{add}, d_c)$	Weighted sum of BM25 ranking function for clicked documents given the added query part
Q2	$\sum_{d_c \in C} sim(d, d_c) \cdot bm25(q_{rmv}, d_c)$	Weighted sum of BM25 ranking function for clicked documents given the removed query part
Q3	$\sum_{d_c \in C} sim(d, d_c) \cdot bm25(q_{com}, d_c)$	Weighted sum of BM25 ranking function for clicked documents given the common query part
Q4	$\sum_{d_c \in C} sim(d, d_c) \cdot QL(q_{add}|\theta_{d_c})$	Weighted sum of Query Likelihood score for clicked documents given the added query part
Q5	$\sum_{d_c \in C} sim(d, d_c) \cdot QL(q_{rmv}|\theta_{d_c})$	Weighted sum of Query Likelihood score for clicked documents given the removed query part
Q6	$\sum_{d_c \in C} sim(d, d_c) \cdot QL(q_{com}|\theta_{d_c})$	Weighted sum of Query Likelihood score for clicked documents given the common query part
Q7	$\sum_{d_c \in C} sim(d, d_c) \cdot tfidf(q_{add}, d_c)$	Weighted sum of tfidf score for clicked documents given the added query part
Q8	$\sum_{d_c \in C} sim(d, d_c) \cdot tfidf(q_{rmv}, d_c)$	Weighted sum of tfidf score for clicked documents given the removed query part
Q9	$\sum_{d_c \in C} sim(d, d_c) \cdot tfidf(q_{com}, d_c)$	Weighted sum of tfidf score for clicked documents given the common query part
Q10	$\sum_{d_{no} \in NC} sim(d, d_{no}) \cdot bm25(q_{add}, d_{no})$	Weighted sum of BM25 ranking function for non-clicked documents given the added query part
Q11	$\sum_{d_{no} \in NC} sim(d, d_{no}) \cdot bm25(q_{rmv}, d_{no})$	Weighted sum of BM25 ranking function for non-clicked documents given the removed query part
Q12	$\sum_{d_{no} \in NC} sim(d, d_{no}) \cdot bm25(q_{com}, d_{no})$	Weighted sum of BM25 ranking function for non-clicked documents given the common query part
Q13	$\sum_{d_{no} \in NC} sim(d, d_{no}) \cdot QL(q_{add}|\theta_{d_{no}})$	Weighted sum of Query Likelihood score for non-clicked documents given the added query part
Q14	$\sum_{d_{no} \in NC} sim(d, d_{no}) \cdot QL(q_{rmv}|\theta_{d_{no}})$	Weighted sum of Query Likelihood score for non-clicked documents given the removed query part
Q15	$\sum_{d_{no} \in NC} sim(d, d_{no}) \cdot QL(q_{com}|\theta_{d_{no}})$	Weighted sum of Query Likelihood score for non-clicked documents given the common query part
Q16	$\sum_{d_{no} \in NC} sim(d, d_{no}) \cdot tfidf(q_{add}, d_{no})$	Weighted sum of tfidf score for non-clicked documents given the added query part
Q17	$\sum_{d_{no} \in NC} sim(d, d_{no}) \cdot tfidf(q_{rmv}, d_{no})$	Weighted sum of tfidf score for non-clicked documents given the removed query part
Q18	$\sum_{d_{no} \in NC} sim(d, d_{no}) \cdot tfidf(q_{com}, d_{no})$	Weighted sum of tfidf score for non-clicked documents given the common query part

basic ranking scores (e.g., BM25, Query Likelihood and tfidf scores) of clicked documents in previous query given the three query parts respectively. For example, the cumulative okapi BM25 score of a document given the common part is formalized as $cumulative\_bm25(q_{com}, d) = \sum_{d_c \in C} sim(d, d_c) \times bm25(q_{com}, d_c)$, where d_c is a clicked document in the clicked documents set C of previous query, $sim(d_1, d_2)$ is the Cosine similarity between two documents represented with $tf \cdot idf$ vectors. The query change features are summarized as Table 2.

4.3   Whole Session Features

The features of this category are similar to the short-term features in Bennett et al. [1]. Although the information needs for different queries in the same session vary, they are supposed to somehow relate to the current query. For instance, one may issue a query to ask about the basic information of one city, and then issue another related query about the representative historical figures or famous scenery spots. Utilizing the previous related queries for session search may help to disambiguate the current query. We obtain the whole session features by computing the cumulative basic scores based on all clicked documents in previous queries in the same session. The features are summarized in Table 1-(b).

4.4   Collective Intelligence Features

Massive attempts on search personalization have shown that the performance of IR models can be improved by enriching global information related to current user [4][11]. It is also a popular phenomenon that our searching problems have been solved by others. Our searching behaviors may also be inspired by some popular events. Therefore integrating the collective intelligence features in IR is a natural choice in our investigation. We extract features from the clicked documents in other sessions topically similar to the current session. To this end, we utilize the Latent Dirichlet Allocation (LDA) to learn topics from clicked documents in the query log. Let T, W and D be variables which represent a latent topic, a single word, and a document respectively. The session variable is denoted as S. The instances of T, W, D and S are denoted as t, w, d, and s. $P(W|T)$ corresponds to a distribution of words for each topic, which shows the relevant probability of a word to the topic. $P(T|D)$ corresponds to a distribution of these learned latent topics for each document, which shows the probability a topic is relevant to the document. Based on the trained topics, we define the probability of a session s being relevant to a learned latent topic t as a conditional probability [11]:

$$p(t|s) = \frac{1}{|C(s)|} \sum_{d \in C(s)} p(t|d) \tag{4}$$

where $C(s)$ is the set of clicked documents in session s, and $p(t|d)$ is the probability of topic t given the document d. In this way, a session s can be represented as a vector of topics, denoted VT_s. Formally, $VT_s = < p(t_1|s), p(t_2|s), ..., p(t_{|T|}|s) >$, where t_i is the i_{th} latent topic, $|T|$ is the number of latent topics. We select top K sessions to extract the collective intelligence features according to the Cosine similarity between the topic vectors of current session and other sessions. The features of this category are summarized in the Table 1-(c).

5   Empirical Comparison of Different Features

5.1   Experimental Setup

Our experiments have been conducted on a query log containing 489,384 queries[3]. Each query in the log comes with the issued timestamp, anonymous user-id, clicked URLs, dwell time on clicked URLs and a list of URLs returned by the search engine. We segmented the query log into 126,103 sessions according to some simple but widely accepted criteria, i.e., the time interval between two subsequent sessions of a user are more than 30 minutes and the queries within one session are sorted by their issued timestamps [1]. Figure 2 reports the distribution of query number and session number on dates, which shows that the distributions are relatively uniform over all active days. However, massive

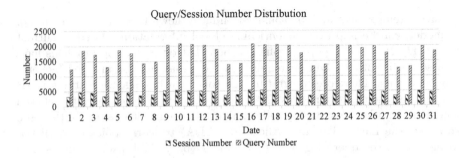

Fig. 2. Distribution of query (session) number on dates.

existing work has demonstrated that there is little potential to personalization for queries with click entropy[4] larger than some threshold [1][4][11]. Thus, in our experiments, we only selected a part of sessions, in which the current queries' click entropies are larger than 2.

The RankLib[5] is utilized to run the LambdaMART algorithm, in which "-norm" is set as "zscore", all LambdaMART-specific parameters (e.g., "-tree" and "-leaf") are set as default values. The selected sessions in the first 3 weeks are

---

[3] The query log is collected from the Bing search engine in 4 weeks (from July 1th 2012 to July 28th 2012) for 1166 users. All queries are from the US market, non-English queries are filtered out.

[4] Click Entropy[4] is a direct indication of query click variation, less click entropy means more focus of URLs on a query. It is defined as follows: $ClickEntropy(q) = \sum_{u \in U(q)} -P(u|q) \log_2 P(u|q)$, where $U(q)$ is the set of web pages (URLs) that are clicked with respect to the distinct query q, and $P(u|q)$ is the percentage of the clicks on URL u among all the clicks for the query q.

[5] https://sourceforge.net/p/lemur/wiki/ranklib/

randomly partitioned into a training set (2663 sessions) and a validation set (203 sessions). All selected sessions in the last week are set as the test set (851 sessions). The target metrics are respectively set as "ERR@10" and "NDCG@10" corresponding to two evaluation metrics used in this study.

We set the original results ("ORI") given by the search engine as the baseline model. Given that our aim is to investigate how different categories of features contribute to the session search performance, we design different strategies (considering different feature groups) to train and test the LambdaMART ranking models, which are list as follows.

1. CUR, only current query features are considered by the learner;
2. CHA, only query change features are considered by the learner;
3. WHO, only the whole session features are considered by the learner;
4. COL, only the collective intelligence features are considered;
5. ALL, all features are considered by the learner;
6. AECUR, all features except for current query features are considered;
7. AECHA, all features except for query change features are considered;
8. AEWHO, all features except for whole session features are considered;
9. AECOL, all features except for collective intelligence features;

Note that, it is important to determine the number K of sessions selected to extract the collective intelligence features. We conducted a series of pilot experiments and eventually selected $K = 2$, which gained the best performance.

5.2 Results and Analysis

We adopt ERR@10 [3] and NDCG@10 [8] as the evaluation metrics (and as target in training). Table 3 reports the evaluation results of ranking models considering different categories of features.

As illustrated in Table 3, when only one category of features is used (by learner), CUR and COL have better re-ranking performance than CHA and WHO. This shows that the current query features and collective intelligence features have more positive influence than the query change features and the whole session features. It is in accordance with our expectation that the current query features are important since the current query represents user's IN directly. We also find that models considering collective intelligence features outperform

Table 3. Experimental results evaluated with ERR@10 and NDCG@10. Rows Chg% report the change percentage of evaluation metrics compared with the baseline.

					Evaluation Results					
				(Chg% means change%, the symbol ‡ means $p < 0.01$ with paired t-test, † means $p < 0.05$)						
	ORI	CUR	CHA	WHO	COL	ALL	AECUR	AECHA	AEWHO	AECOL
ERR@10	0.247	0.267	0.243	0.225	0.270	0.297	0.281	0.291	0.293	0.270
Chg%	-	+8.177†	-1.453	-8.603	+9.591‡	+20.611‡	+14.014‡	+18.092‡	+18.936‡	+9.328†
NDCG@10	0.523	0.549	0.513	0.505	0.534	0.583	0.537	0.580	0.589	0.559
Chg%	-	+5.028	-1.882	-3.452	+2.239	+11.613‡	+2.842	+11.015‡	+12.673‡	+6.981‡

those considering the query change features and whole session features. We consider this finding meaningful, as it deviates from the observations in the existing work and may benefit the design of future session search algorithms. A possible interpretation of this phenomenon is that the topically similar sessions provide very useful information for current search task. To our best knowledge, we are the first to discover this phenomenon that other similar sessions' features have more positive influence than current session features including the query change features and the whole session features.

Ranking models considering multiple categories of features outperform all models that only consider a single category features. This illustrates that the combination of different categories features can improve the effectiveness of session search. Moreover, various feature combinations have different influences on re-ranking performance. With regard to ERR@10, the best feature combination is ALL, and removing any category of features will hurt the effectiveness of session search to different degrees. For NDCG@10, AEWHO is the best performing combination which outperforms ALL. This reflects that the whole session features may have some negative influences on session search. Removing current query features or collective intelligence features have more impact than query change features and the whole session features, with respect to both ERR@10 and NDCG@10.

6   Conclusions and Future work

In this paper, we have classified different interaction features for session search into four categories. We then trained and tested a series of session models considering different categories of features. Experimental results show that the current query features and collective intelligence features have more positive influence on re-ranking performance than query change features and whole session features. Our findings will potentially bring benefits for the design of future information retrieval models which can take full advantages of the collective intelligence besides the features extracted from the current query.

Although, in this paper, query change features did not gain a good performance, we consider this category of interactions very important to detect user's evolving IN in exploratory search, thus worth further investigating in the future. To our best knowledge, we are the first to integrate the query change information into the Learning2Rank framework, and we are the first to explicitly formalize these four categories of features together. In the future, we believe that the integration of multidimensional features which can reflect the dynamics of users' information need within a search session will be promising research topic. Additionally, in order to have a better understanding on how different features work on session search, the analysis of retrieval performances on different sessions (e.g., with different queries) could be conducted in the future.

Acknowledgments. The work presented in this paper is sponsored in part by the Chinese National Program on Key Basic Research Project (973 Program, grant No.

2013CB329304, 2014CB744604), the Chinese 863 Program (grant No. 2015AA015403), the Natural Science Foundation of China (grant No. 61402324, 61272265), and the Research Fund for the Doctoral Program of Higher Education of China (grant no. 20130032120044).

References

1. Bennett, P.N., White, R.W., Chu, W., Dumais, S.T., Bailey, P., Borisyuk, F., Cui, X.: Modeling the impact of short-and long-term behavior on search personalization. In: SIGIR, pp. 185–194. ACM (2012)
2. Burges, C.J.: From ranknet to lambdarank to lambdamart: An overview. Learning **11**, 23–581 (2010)
3. Chapelle, O., Metlzer, D., Zhang, Y., Grinspan, P.: Expected reciprocal rank for graded relevance. In: CIKM, pp. 621–630. ACM (2009)
4. Dou, Z., Song, R., Wen, J.-R.: A large-scale evaluation and analysis of personalized search strategies. In: WWW, pp. 581–590. ACM (2007)
5. Guan, D., Zhang, S., Yang, H.: Utilizing query change for session search. In: SIGIR, pp. 453–462. ACM (2013)
6. Liu, T.-Y., Xu, J., Qin, T., Xiong, W., Li, H.: Letor: benchmark dataset for research on learning to rank for information retrieval. In: Proceedings of SIGIR 2007 Workshop on Learning to Rank for Information Retrieval, pp. 3–10 (2007)
7. Luo, J., Zhang, S., Yang, H.: Win-win search: dual-agent stochastic game in session search. In: SIGIR, pp. 587–596. ACM (2014)
8. Manning, C.D., Raghavan, P., Schütze, H.: Introduction to information retrieval, vol. 1. Cambridge University Press Cambridge (2008)
9. Teevan, J., Dumais, S.T., Horvitz, E.: Potential for personalization. TOCHI **17**(1), 4 (2010)
10. Vu, T., Willis, A., Tran, S.N., Song, D.: Temporal latent topic user profiles for search personalisation. In: Hanbury, A., Kazai, G., Rauber, A., Fuhr, N. (eds.) ECIR 2015. LNCS, vol. 9022, pp. 605–616. Springer, Heidelberg (2015)
11. Vu, T., Song, D., Willis, A., Tran, S.N., Li, J.: Improving search personalisation with dynamic group formation (2014)
12. Zhai, C.: Statistical language models for information retrieval. Synthesis Lectures on Human Language Technologies **1**(1), 1–141 (2008)
13. Zhang, S., Guan, D., Yang, H.: Query change as relevance feedback in session search. In: SIGIR, pp. 821–824. ACM (2013)
14. Zhang, S., Luo, J., Yang, H.: A pomdp model for content-free document re-ranking. In: SIGIR, pp. 1139–1142. ACM (2014)

Web Mining

Beyond Your Interests: Exploring the Information Behind User Tags

Weizhi Ma(✉), Min Zhang, Yiqun Liu, Shaoping Ma, and Lingfeng Chen

State Key Laboratory of Intelligent Technology and Systems,
Tsinghua National Laboratory for Information Science and Technology,
Department of Computer Science and Technology,
Tsinghua University, Beijing 100084, China
mawz14@mails.tsinghua.edu.cn, {z-m,yiqunliu,msp}@mail.tsinghua.edu.cn,
clf0506@cs.duke.edu

Abstract. Tags have been used in different social medias, such as Delicious, Flickr, LinkedIn and Weibo. In previous work, considerable efforts have been made to make use of tags without identification of their different types. In this study, we argue that tags in user profile indicate three different types of information, say *the basics* (age, status, locality, etc), *interests* and *specialty* of a person. Based on this novel user tag taxonomy, we propose a tag classification approach in Weibo to conduct a clearer image of user profiles, which makes use of three categories of features: general statistics feature (including user links with followers and followings), content feature and syntax feature. Furthermore, different from many previous studies on tag which concentrate on user specialties, such as expert finding, we find that valuable information can be discovered with *the basics* and *interests* user tags. We show some interesting findings in two scenarios, including user profiling with people coming from different generations and area profiling with mass appeal, with large scale tag clustering and mining in over 6 million identical tags with 13 million users in Weibo data.

Keywords: Weibo · Tag classification · User group profiling

1 Introduction

Recent years, social tagging has become popular with the launch of sites like Delicious, Flickr, LinkedIn and Weibo. Since then, lots of social systems that support tagging of a variety of resources have been built. Tagging is a process in which a user assigns tags to an object. On Delicious, user can tag URL. On Flickr, user can assign tags to a photo. On LinkedIn and Weibo, user can add tags to themselves, named user tags. Take Weibo as an example, each user can add no more than ten tags to himself. The length of each tag is limited in 7 Chinese characters or 14 English letters, while the content of tag can be anything you want to describe yourself.

Due to the widely usage of tags, different techniques are employed to study various aspects of tagging [1]. The information behind tags is valuable in many

This work was supported by National Key Basic Research Program (2015CB358700) and Natural Science Foundation (61472206, 61073071) of China.

© Springer International Publishing Switzerland 2015
J. Li et al. (Eds.): NLPCC 2015, LNAI 9362, pp. 257–269, 2015.
DOI: 10.1007/978-3-319-25207-0_22

research areas. For example, Giannakidou et al. investigate to co-cluster tags with social data sources [2] and user interests discovering with tags [6]. Many researchers concentrate on using the information behind tags in Flickr to improve the performance of image retrieval [3–5]. Pennacchiotti et al. [7] propose a machine learning approach for twitter user classification based on hashtags. Many social tag prediction or recommendation work are conducted[8–10].

Tag studies based on Weibo come in many ways. Ghosh et al. [11] and Liang et al. [12] try to make use of user tags in Weibo to conduct expert finding studies. An automatic tag recommendation algorithm for Weibo is proposed in Wang et al.'s work [13]. To the best of our knowledge, in previous work, studies make use of different user tags indiscriminately. In fact, we find that user tags reflect user characteristics in three dimensions: 1) Tags show the attributes and status of a user, which are always ignored. 2) Tags indicate the topics that users are interested in. 3) Tags reveal the users's specialities. Therefore, we propose that user tags indicate three different kinds of information: *the basics*, *interests* and *specialty*. Moreover, we proposed a novel feature extraction method with the help of search engine. The analysis and classification will be introduced in Section 2.

As mentioned above, a lot of work in tag analysis concentrates on user specialties such as expert finding or user interests discovering, respectively. Several research efforts have been made for extracting profile information of a person [14,15]. Tang et al. [16] take user tags into account in user profiling. With the help of tag taxonomy, we find that valuable information can be discovered to profile user groups by considering both user *basics* and *interests*. We conduct user group profiling with different generations and area profiling with mass appeal by creating lists of keywords in Section 3 & 4.

Our main contributions are the following:

- Contrary to make use of different tags indiscriminately, we find that user tags indicate three different types of information, *the basics*, *interests* and *specialty*.
- We propose several novel tag feature extraction methods, which take features from tag links, user content and search engine for user tag classification.
- We find that valuable information can be explored in user tags with the help of tag taxonomy , such as user profiling with different generations and area profiling with mass appeal.

The remainder of this paper is organized as follows: In Section 2, we introduce work about user tag analysis and classification procedure. In Section 3, we present the result of user profiling with different generations. While in Section 4, we present our attempts in area profiling based on mass appeal and the method of characteristic tags extraction. We draw final conclusions and the outline of future work in Section 5.

2   User Tag Classification in Weibo

2.1   User Tag Taxonomy

In previous tag studies, researchers adopt identical data processing methods to different tags. In fact, tags may carry different types of information. For example,

Gloria Tang Tsz-kei, a famous Hong Kong singer, has a user tag list in Weibo which contains: "After 90's", "Musician", "Lively", "Singer", "Leo", "Like to amuse". We can see that "After 90's", "Leo" and "Lively" is *the basics* of hers, while user tag "Like to amuse" is her *interests*, "Singer" and "Musician" indicate her *specialty*. Based on analysis on large scale of user tag data, we propose user tag taxonomy as follows:

- *The basics*: Tags which indicate a user's age, state, locality, constellation, blood type and other user basics, like "After 90's", "Libra", etc.
- *Intersests*: Tags that show a user's interests. For instance: "singing", "sports", "traveling", etc.
- *Specialty*: Tags which reveal a user's specialty, like "doctor", "teacher", etc.

2.2 Feature Extraction for Classification

We attempt to use an automatic method to conduct tag classification. The first step is user tag's feature extraction. We design some features for tags:

Statistical Features. We designed five features based on statistic as follows:

1. **Popularity**: the usage percentage of tag t in all users in the dataset.

$$Popularity(t) = \frac{|\{u|t \in tag(u)\}|}{|u|}$$

2. **Absolute position**: the average of tag t's rank position in user's tag list.

$$Absolute\text{-}Position(t) = \frac{\sum_{u \in \{v|t \in tag(v)\}} rank(t,u)}{|\{v|t \in tag(v)\}|}$$

($rank(t,u)$ means tag t's rank position in user's tag list u.)

3. **Relative position**: the average relative occurrence position of tag t in user's tag list.

$$Relative\text{-}Position(t) = \frac{\sum_{u \in \{v|t \in tag(v)\}} \frac{rank(t,u)}{|tag(u)|}}{|\{v|t \in tag(v)\}|}$$

4. **Co-occurrence percentage in followers**: the usage percentage of this tag in the followers of the user who has this tag.

$$Followers\text{-}Co(t) = \frac{\sum_{u \in \{v|t \in tag(v)\}} \frac{|\{w|u\text{-}>w, t \in tag(w)\}|}{|\{w|u\text{-}>w\}|}}{|\{v|t \in tag(v)\}|}$$

(u->w means u follows w.)

5. **Co-occurrence percentage in followings**: the usage percentage of this tag in the following of the user who already has this tag.

$$Following\text{-}Co(t) = \frac{\sum_{u \in \{v|t \in tag(v)\}} \frac{|\{w|w\text{-}>u, t \in tag(w)\}|}{|\{w|w\text{-}>u\}|}}{|\{v|t \in tag(v)\}|}$$

Feature 1 is based on the popularity of user tags, feature 2 & 3 are related to the user tag's position in tag list. We suppose that different types of tags have different popularity in Weibo, and the position in tag list could be a useful feature for the reason that users may tend to tag similar tags together. Different from feature 1, 2, 3, feature 4 & 5 are features extracted from user links(with follower and following relationship).

Content Features. Content features are extracted based on tag vector representation. The Word2vec algorithm provides an implementation of the continuous bag-of-words and skip-gram architectures for computing vector representations of words. It takes a text corpus as input and word vectors as output. We use an open-source package of Word2vec[1]. The user tag dataset is regarded as the input to Word2vec, which is formated as follows:

$$t_{i,1} \quad t_{i,2} \quad ... \quad t_{i,n_i} \qquad (u_i\text{'s tag list.})$$
$$...$$
$$t_{j,1} \quad t_{j,2} \quad ... \quad t_{j,n_j} \qquad (u_j\text{'s tag list.})$$

As a result, each tag get a 200-dimension floating-point vectorized representation in the output of Word2vec.

Search Engine Based Syntax Features. We can extract syntax features by considering the co-occurrence frequency of the tags within certain sentences. We try to find the frequency with the help of a search engine. More specifically, we construct some sentences pattern using the user tags, put the sentences into Baidu[2] one by one in Chinese, and record the count of items and exact matching items returned by search engine. We propose 3 types of patterns including 9 instances, and the feature dimension we extracted from syntax is 18.

- *The basics* **related patterns:**
  I am __ (我是__, 我__.). E.g: I am 18 (我18岁). I am in after 90's (我是90 后).
  I am an Aries (我是白羊座).
  My __ (__的我). E.g: My passion (热情的我).
  I'm very __ (我很__). E.g: I'm very humorous (我很幽默).
- *Intersest* **related patterns:**
  I like __ (我喜欢__). E.g: I like traveling (我喜欢旅行).
  I love __ (我爱__). E.g: I love painting (我爱画画).
  (redundancy will be eliminated if there is existing "love" or "like" in the tag.)
- *Specialty* **related patterns:**
  I'm good at __ (我懂__). E.g: I'm good at design(我懂设计).
  __ expert (__专家, __家). E.g: political expert(政治家).

---

[1] https://code.google.com/p/word2vec/
[2] https://www.baidu.com, a popular search engine in China.

2.3   Dataset

The dataset we used in our work is a public Weibo dataset, provided by China Pameng[3]. The dataset is collected in Weibo from October 2012 to May 2013, which contains user's tag lists and user's follower & following relationships.

The number of user accounts in the dataset is 13,170,561. The sum of unique tags is 6,157,143. Each user has 123 followings and 6.21 tags on average.

It is necessary to preprocess the dataset due to the impact of noise brought by zombie users [17]. In this work, we filter out users whose followings or followers are fewer than 10. After filtering, we get 10,659,899 users with 6,156,993 unique tags.

2.4   Experiments and Results

In proposed taxonomy, a tag could belong to more than one types, for example: "Photography" and "music" can be classified into both *intersests* and *specialty*. At this time, the following labeling criteria can be adopted:

- If a user tag could be *the basics & interests*, or *the basics & specialty*, we tend to label it as *interests* or *specialty* but not *the basics* for the reason that we suppose *interests* and *specialty* tags convey more information.
- When it is really hard to judge whether the tag belongs to *interests* or *specialty*, we tent to classify it into a new type *interests & specialty*.
- If a user tag is meaningless, such as user tag "sser", "just", we take it as *noise*.

We choose the top 100 frequent tags and randomly select 1,086 tags from the whole tag set. We label these 1,186 tags according to the criteria manually. The distribution of tags is shown in Figure 1.

As we can see from Figure 1, more than 40% of user tags are classified into *interests*, which indicates that *interests* is a key component of user tags. The noise of user tags is about 20%, suggesting that a certain quantity of user tags is difficult to be understood.

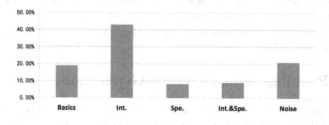

Fig. 1. Tag labeling result

---

[3] An organization which collects Weibo data. The official website is http://cnpameng.com.

As can be seen from the distribution of user tags, the number of user tags in different types varies widely. Thus, before conducting classification experiment, we carry out over sampling to make the count of tags in different types balanced. Specially, in our experiments, tags in *interests & specialty* are regard as both *interests* and *specialty* tag in classification experiments.

After feature normalization, we leveraged different classification algorithms, such as SVM, Multilayer Perceptron, Naive Bayes and Decision Tree. The performance in Decision Tree is the best. Hence we choose Decision Tree as the classification algorithm. We use the statistic, content and syntax features separately for classification experiments. In 10-fold cross validation, the classification results are listed in Table 1.

Table 1. Classification Results in Using Different Features

Feature	Basics			Interests			Specialty		
	Precision	Recall	F1	Precision	Recall	F1	Precision	Recall	F1
Statistics	81.6%	81.6%	81.6%	65.9%	65.4%	65.2%	83.7%	83.0%	82.9%
Content	87.4%	87.1%	87.1%	77.8%	77.2%	77.1%	89.8%	89.3%	89.3%
Syntax	83.2%	82.9%	82.8%	74.7%	74.6%	74.6%	86.7%	86.6%	86.6%
Statistics & Content	**88.4%**	**88.1%**	**88.0%**	79.9%	79.3%	79.2%	89.3%	88.7%	88.6%
Statistics & Syntax	85.9%	85.4%	85.3%	77.2%	77.0%	76.9%	89.7%	89.5%	89.4%
Content & Syntax	87.7%	87.3%	87.3%	78.1%	77.7%	77.6%	89.3%	88.8%	88.8%
All	87.9%	87.4%	87.3%	**81.4%**	**80.8%**	**80.7%**	**90.1%**	**89.7%**	**89.6%**

Table 1 indicates that the performance in *specialty* type classification is better than others and the performance in *interests* tags classification is the worst, which may be caused by the variety of *interests* tags. The performance of using content features individually is pretty good, which shows that Word2vec is really effective. Furthermore, comparing the classification results in using statistics & syntax features with content features, we find that the precision, recall and F1-Measure values are nearly the same, showing that the statistics & syntax features are effective.

Moreover, we find that the syntax features are not always helpful. In identifying *interests* and *specialty* tag tasks, the performance will be better when syntax features are applied, but worse in *the basics* identification task. It can be attributed to that tags in *the basics* are easier to be identified, the result is good enough without syntax features, while as *interests* or *specialty* tags, they are more difficult to be classified correctly with only statistic & content feature for their variety.

3   User Profiling with Different Generations

Researchers concentrate on expert finding and recommendation in some special domains by utilizing user tags in previous studies. However, we find that after user tag classification, extra valuable information can be discovered. In fact, we can divide users into different user groups according to their *basics* tags, for example, age, constellation, location, etc. In this section, we focus on user profiling with different generations based on user age. The result indicates that with the help of tag taxonomy, some extra information can be found in *interests* tags.

3.1 User Generation Distribution

In this section, our work focus on user group's interest distribution modeling and user group profiling. First, users are divided into different generation groups according to their tag lists. We set three generation groups: "After 70s", "After 80s" or "After 90s" (People who were born between 1970 and 1979 belong to "After 70s", and others are defined in a similar way.). For the reason that the number of "After 60s" and "After 00s" users is smaller than 1,000 based on user tags, we don't take them into consideration. If user's tags don't reveal his/her age, we will ignore this user.

The dataset has been introduced in Section 2.3, and the distributions of users in different generations are as follows: There are 19,821 users in user group "After 70's", 1,703,438 users in user group "After 80's", and 1,681,892 users in user group "After 90's". We find that many users don't have age tags, which can be our future work.

3.2 *Interests* Tag Clustering for User Profiling

Over 1.3 million different user tags are tagged in these users. The frequency of user tags follows a power law distribution. Considering that the quantity of user tags is extremely large, we filter out the tags whose frequency is less than 500. After filtering, the number of user tags has reduced to 1,733, while the frequency of these tags accounted for more than 80% of all tag's frequency.

It's hard to identify user's interests distribution in a large scale, so we try to cluster the tags into several tag sets. The method we chose to conduct a tag clustering experiment is K-means. We assigned the k value with 15 and put all the 1,733 user tags into tag clustering experiment. Figure 2 shows the percentages of users in each tag set. Users may have more than one tag, so the sum of percentages in each set is not 100%. We can find that the distributions of After 80's and After 90's are very similar. It's hard to get useful information from this clustering result.

Thus, before clustering the user tags, we try to filter out tags that do not belong to *interests* type. We conduct an automatic classification which is introduced in Section 2.4 and get 907 user tags that belong to *interests* type.

Clustering experiment is conducted on the 907 user tags. The result is that each tag set can be assigned with a keyword as its feature at this time, which has better performance than using all user tags. For example, design, comic, art, etc. But it's unsatisfactory in that one of the sets contains too many user tags, so we labeled this tag set and divided it into 5 tag sets manually. Moreover, famous stars from different countries are clustered into 3 sets, we combined the three sets into a big set. At last, the user tags are clustered into 17 sets. The keywords of each set are shown in Table 2. The keyword "Others" means that this set is mixed by a variety of *interests* tags without a keyword.

Table 2. Interests Clustering Results

Set	Key Word	Set	Key Word	Set	Key Word	Set	Key Word	Set	Key Word	Set	Key Word
0	Design	1	Art	2	Music	3	Stars	4	Social Science	5	Technology
6	Reading	7	Others	8	Housing	9	Geek	10	Comic	11	Fashion
12	Travel	13	e-commerce	14	Finance	15	Food	16	Sports		

3.3 Result and Analysis

We calculate the tag amount of each clustering set in user groups of different generations. The statistical results show the differences of interests in different generations, which are drawn in Figure 3. We find that the five types, Design, Music, Art, Stars and Technology, contain much more users than other groups. It indicates that these are social common interests. To see the results of other interest sets more clearly, we removed 6 tag sets and draw Figure 4.

In Figure 4, it is apparent that the popularity of different Interests varies in each group. Users in the group of "After 70's" are keen on Reading and Housing, which is reasonable as people in this age are more concerned about living a better life and personal finance. Young people, aged in 15-35, show more interest in finance than "After 70's". Moreover, we find that "After 70's" even show more interests in sports than "After 80's" and "After 90's". It indicates that young people's enthusiasm in sports is relatively low, which is an ominous sign.

We can find many other interesting information from the results. In fact, this work implies that valuable information can be mined in user tags after tag taxonomy. Moreover, these analyses will be helpful in tracing the transformation and evolution of social common interests.

4 Area Profiling with Mass Appeal

In Section 3, we introduced the findings in user profiling with different generations based on user tag taxonomy. In this section, we focus on area profiling with mass appeal, which is based on user groups from different provinces/cities.

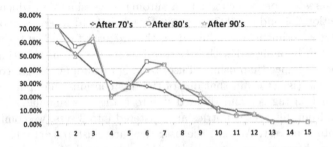

Fig. 2. User Interests Distribution without tag filtering

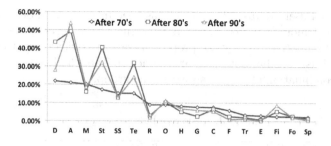

Fig. 3. User Interests Distribution

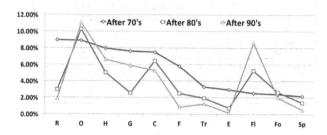

Fig. 4. User Interests Distribution in Eleven Sets

Users' area tags indicate the location of the users in usual cases. Thus we can infer the location information of users from their tag lists. According to this, users can be classified into different province/city user groups. Each group has a tag list which is unioned by the tags owned by the users who belong to the group. Then, we can perform area profiling experiments to find the characteristic tags in each user group to describe the province/city. Through comparing the differences in filtering out and not filtering out *the basics* user tag, we find that user tag taxonomy is helpful in characteristic tag finding. Moreover, the result is evaluated by a labeling task on Zhongbao, a Chinese crowdsourcing platform[4].

4.1   Area Characteristic Tags Extraction

For the reason that most of Weibo users are Chinese, many user tags about location are provinces or cities in China. Considering that if we choose the city tag appearing in a user's tag list to classify users into different groups directly, the number of user groups will be large and the amount of user in each city will be small. So we merged the user groups according to the affiliations of the areas. For example, Guangzhou is a city of Guangdong province in China, if a user tagged himself/herself with "Guangzhou", we put him/her into the user group of Guangdong province. Furthermore, people who have no location tags are filtered out in this experiment. For the reason that there are 34 provinces, autonomous regions, municipalities and special administrative regions in China,

---

[4] http://www.chinacrowds.com

we construct 34 user groups. The dataset has been introduced in Section 2.3. As a result, we get more than 360,000 users from the 34 provinces/cities.

Inspired by relative entropy, we propose a feature extraction method names tag entropy. Basic symbol notations are defined in Table 3. The tag list of each area is the combination of user's tag list in the group.

Table 3. Basic Symbols Notation

Symbol	Definition
A	The set of the 34 areas $\{a_1, a_2, ..., a_{34}\}$.
B	Complementary set of the 34 area $\{b_1, b_2, ..., b_{34}\}$, b_i represents the union set of $\{a_1, a_2, ..., a_{i-1}, a_{i+1}, ..., a_{34}\}$.
$|a_i|$, $|b_i|$	The number of users in set a_i, b_i.
$TagA(x, i)$	The frequency of tag x in a_i's tag list.
$TagB(x, i)$	The frequency of tag x in b_i's tag list.
$TE_{x,k}$	The tag entropy of tag x in a_k.

The formulations to calculate tag entropy of tag x in a_k are as following:

$$P_{x,i} = \begin{cases} \frac{TagA_{x,i}}{|a_i|} & x \text{ in } a_i \\ \frac{1}{|a_i|} & x \text{ not in } a_i \end{cases} \qquad (4-1)$$

$$Q_{x,i} = \begin{cases} \frac{TagB_{x,i}}{|b_i|} & x \text{ in } b_i \\ \frac{1}{|b_i|} & x \text{ not in } b_i \end{cases} \qquad (4-2)$$

$$TE_{x,k} = P_{x,k} * log(\frac{P_{x,k}}{Q_{x,k}}) \qquad (4-3)$$

We can get the tag entropy of each tag in different user groups. In our approach, we use the top N tags in $TE_{x,k}$ value as the characteristic tag of each province/city, and we ignore the order of the top N tags.

With the method introduced above, we get the characteristic tags of each province/city of China. For example, the top five characteristic tags of Qinghai province is: Qinghai Lake, Qinghai-Tibet Plateau, The Origin Of Three Rivers, Tibetan, Xia Du (an alias of Xining.); The result of Qinghai is consistent with generally acknowledged. However, in fact, it's challenging to evaluate the result of province/city characteristic tags extracted by the experiments directly, because the impression of an area is usually based on people's background. So we proposed a labeling task to evaluate the results.

4.2 Evaluation and Analysis

We design a labeling task to evaluate the results: We choose the top 20 tag entropy tags and 180 tags selected by using the method of multistage stratified sampling in the province's tag list. So that each province/city has 200 tags. Then, the 200 tags of each province/city will be evaluated whether the tag is a characteristic tag of the province/city with 3 level labeling: "Relevant", "partially relevant" or "irrelevant". Each tag is labeled by three users and the labeling of the tag is depended on the majority opinion. If the labels of a tag given by the three annotators are different with each other, this tag will be labeled as "partially relevant".

The labeling task is released on Zhongbao crowdsourcing platform. In order to compare the differences between raw province's tag list and province's tag list after *the basics* tag filtering, we calculate the precision, recall, F1-Measure on the two data sets by considering top k characteristic tag result as the right answer. In evaluation, we attempt two method: One is regarding the tags labeled with "relevant" as characteristic tags, the other one is regarding the tags labeled with "relevant" or "partially relevant" as characteristic tags. Table 4 shows the results of the evaluation.

Table 4. "relevant" and "relevant & partially relevant" label counts

Result		Raw data			Filtered data		
		Precision	Recall	F1	Precision	Recall	F1
Relevant	Top5	10.85%	44.12%	17.42%	11.43%	46.47%	18.35%
	Top10	23.44%	47.65%	31.42%	23.88%	48.53%	32.01%
	Top20	47.90%	48.68%	48.29%	**48.34%**	49.12%	48.73%
Relevant & Partially Relevant	Top5	11.79%	89.41%	20.83%	12.10%	**91.76%**	21.38%
	Top10	23.12%	87.65%	36.59%	23.27%	88.24%	36.83%
	Top20	45.31%	85.88%	59.32%	45.85%	86.91%	**60.03%**

We find that the performance on filtered data is better than raw data in Table 4, which indicates that after tag filtering, we can get better results. For the reason that only considering "relevant" tag is more strict than using both "relevant" and "partially relevant", the recall is obviously lower than the latter. The precision of the data increases with the increment of k, as the right answer set is expanded with the increment of k.

The highest precision, 48.34% , is achieved in "relevant" tags with top 20 on filtered data. The highest recall, 91.76%, is found in "relevant" and "partially relevant" tags with top 5 on filtered data. For the reason that the value of k restricts the precision and recall, the highest F1 only achieved in 60.03% with Top 20 "relevant" and "partially relevant" tags on filtered data.

In this part, we use tag entropy to conduct city profiling with mass appeal experiments. Our labeling task shows that tag taxonomy is useful in area profiling, which helps get better results.

5 Conclusion and Future Work

In this paper, we presented our work in user tags on Weibo. Firstly, we showed that tags in user profile indicate three different kinds of information: *basics*, *interests* and *specialty* of a person. We introduced our analysis about user tags in Section 2.1. With the help of search engine, we proposed a novel user tag feature extraction method. We conduct experiments to classify the user tags into different types with Decision Tree. The classification results show that the statistic and syntax features we extracted are effective.

Furthermore, we find that valuable information can be discovered with the *basics* and *interests* user tags. We present some interesting findings in Section 3 and Section 4: User profiling with different generations and area profiling with mass appeal. In user profiling with different generations, we take both user age and user interests into consideration. The result shows the interests distribution of users in different generations in using *interests* user tags. In area profiling with mass appeal, we proposed a method to extract area characteristic tags with tag entropy. Furthermore, we designed a labeling task to verify that if the tag classification is helpful. In fact, we find many valuable information behind the user tags by considering the *basics* and *interests* at the same time. We believe that valuable information can be found in other platforms. Not only user tags can be classified into different types, tags in other platforms indicate different information. But in other platforms, the taxonomy of tags may be different.

Future work include user *basics*, *interests* and *specialties* finding with the combination of the implicit user information in Weibo content, the explicit information behind hashtags and user tag information. Moreover, besides profiling users with generation and area basics, further profiling work in users with other basics, such as status and personality, will be conducted.

References

1. Gupta, M., Li, R., Yin, Z., et al.: Survey on Social Tagging Techniques. ACM Sigkdd Explorations Newsletter **12**(1), 58–72 (2010)
2. Giannakidou, E., Koutsonikola, V., Vakali, A., et al.: Co-clustering tags and social data sources. In: The Ninth International Conference on Web Age Information Management, WAIM 2008, pp. 317–324 (2008)
3. Li, X., Snoek, C.G.M., Worring, M.: Unsupervised multi-feature tag relevance learning for social image retrieval. In: Conference on Image and Video Retrieval, pp. 10–17 (2010)
4. Xiao, J., Zhou, W., Tian, Q.: Exploring tag relevance for image tag re-ranking. In: Proceedings of the 35th International ACM SIGIR Conference on Research and Development in Information Retrieval, pp. 1069–1070. ACM (2012)
5. Zhu, X., Nejdl, W., Georgescu, M.: An adaptive teleportation random walk model for learning social tag relevance. In: Proceedings of the 37th International ACM SIGIR Conference on Research & Development in Information retrieval, pp. 223–232. ACM (2014)
6. Giannakidou, E., Koutsonikola, V., Vakali, A., et al.: In & out zooming on time-aware user/tag clusters. Journal of Intelligent Information Systems **38**(3), 685–708 (2012)
7. Pennacchiotti, M., Popescu, A.M.: A Machine Learning Approach to Twitter User Classification. ICWSM **11**, 281–288 (2011)

8. Heymann, P., Ramage, D., Garcia-Molina, H.: Social tag prediction. In: Proceedings of the 31st Annual International ACM SIGIR Conference on Research and Development in Information Retrieval, pp. 531–538. ACM (2008)
9. Sigurbjörnsson, B., Zwol, R.V.: Flickr tag recommendation based on collective knowledge. In: Www 2008 Proc of International Conference on World Wide Web pp. 327–336 (2008)
10. Seitlinger, P., Kowald, D., Trattner, C., et al.: Recommending tags with a model of human categorization. In: Proceedings of the 22nd ACM International Conference on Information & Knowledge Management, pp. 2381–2386. ACM (2013)
11. Ghosh, S., Sharma, N., Benevenuto, F., et al.: Cognos: crowdsourcing search for topic experts in microblogs. In: Proceedings of the 35th International ACM SIGIR Conference on Research and Development in Information Retrieval, pp. 575–590. ACM (2012)
12. Liang, C., Liu, Z., Sun, M.: Expert finding for microblog misinformation identification. In: COLING (Posters), pp. 703–712 (2012)
13. Wang, X., Li, S., Zou, X., et al.: An automatic tag recommendation algorithm for micro-blogging users. In: 2013 International Conference on Computer Sciences and Applications (CSA), pp. 398–401. IEEE (2013)
14. Cunningham, H., Maynard, D., Bontcheva, K., et al.: GATE: an architecture for development of robust HLT applications. Proceedings of the 40th Annual Meeting on Association for Computational Linguistics. Association for Computational Linguistics, pp. 168–175 (2002)
15. Yu, K., Guan, G., Zhou, M.: Resume information extraction with cascaded hybrid model. In: Proceedings of the 43rd Annual Meeting on Association for Computational Linguistics. Association for Computational Linguistics, pp. 499–506 (2005)
16. Tang, J., Yao, L., Zhang, D., et al.: A Combination Approach to Web User Profiling. ACM Transactions on Knowledge Discovery from Data 5(1), 293–302 (2010)
17. Binlin, C., Jianming, F., Jingwei, H.: Detecting zombie followers in sina microblog based on the number of common friends. International Journal of Advancements in Computing Technology 5(2) (2013)

Nonparametric Symmetric Correspondence Topic Models for Multilingual Text Analysis

Rui Cai[✉], Miaohong Chen, and Houfeng Wang

Key Laboratory of Computational Linguistics,
Peking University, Ministry of Education, Beijing, China
{cairui,wanghf}@pku.edu.cn, chenmiaohongnlp@gmail.com

Abstract. Topic model aims to analyze collection of documents and has been widely used in the fields of machine learning and natural language processing. Recently, researchers proposed some topic models for multilingual parallel or comparable documents. The symmetric correspondence Latent Dirichlet Allocation (SymCorrLDA) is one such model. Despite its advantages over some other existing multilingual topic models, this model is a classic Bayesian parametric model, thus can't overcome the shortcoming of Bayesian parametric models. For example, the number of topics must be specified in advance. Based on this intuition, we extend this model and propose a Bayesian nonparametric model (NPSymCorrLDA). Experiments on Chinese-English datasets extracted from Wikipedia (https://zh.wikipedia.org/) show significant improvement over SymCorrLDA.

Keywords: Multilingual text analysis · Topic model · Bayesian nonparametric model

1 Introduction

Getting valuable information from a large number of materials has attracted increasing research interest in recent years. A topic model is a type of statistical model for discovering the abstract topics that occur in a large collection of documents. In topic modeling, each document is regarded as a mixture of topics, and Latent Dirichlet Allocation (LDA)[2] is perhaps the most common topic model currently in use. Generally other models are extensions of LDA, most of which are appropriate for processing monolingual texts while some can be used for multilingual parallel or comparable documents to capture the statistical dependencies between multiple representations. Both parallel documents and comparable documents are merged documents consisting of multiple language parts, and parts of the former are translations from one language to another while the latter just describe similar concepts and events.

Bilingual topic models for bilingual parallel documents that have word-to-word alignments have been developed ([9]) and are directed towards machine translation. In contrast, some topic models focus on analyzing dependencies

© Springer International Publishing Switzerland 2015
J. Li et al. (Eds.): NLPCC 2015, LNAI 9362, pp. 270–281, 2015.
DOI: 10.1007/978-3-319-25207-0_23

among languages by modeling multilingual comparable documents, each of which consists of multiple language parts that are not translations of each other but instead describe similar concepts and issues. Conditionally Independent LDA (CI-LDA)[5] and SwitchLDA[6] (extension of CI-LDA) are multilingual topic models, but they only share per-document multinomial distributions between different languages, which produces weak dependencies. Correspondence LDA (CorrLDA)[7] and Symmetric Correspondence LDA (SymCorrLDA)[8] can also handle multilingual documents. In modeling, a pivot language plays a key role, text of which is translated to other languages, and the pivot selection is an important process to ensure estimation quality. SymCorrLDA incorporates a hidden variable to control the pivot language while CorrLDA must specify it in advance. In general, CorrLDA outperforms CI-LDA and SwitchLDA in processing comparable documents, and SymCorrLDA works more effectively than CorrLDA.

As a classic Bayesian parametric model, SymCorrLDA also bears the shortcomings of parametric models and the number of topics must be specified in advance. Nonparametric models can determine automatically the parameters scales and the complexity of models according to observed data while parametric models hold a strong assumption of data distribution. Hierarchical Dirichlet process (HDP) is a nonparametirc Bayesian model for clustering problems involving multiple groups of data. Our Nonparametric Symmetric Correspondence LDA (NPSymCorrLDA) is based on Hierarchical Dirichlet Process. Different from SymCorrLDA, NPSymCorrLDA generates a topic for each word according to the Dirichlet Process. The number of topics is open-ended, whose value grows at rate logarithmic in the number of words. Experiments on Chinese-English datasets extracted from Wikipedia show significant improvement over SymCorrLDA.

2   Multilingual Topic Models

Some researchers explored multilingual topic models that based on the premise of using multilingual dictionaries or WordNet ([10], [11], [12]). In contrast, CorrLDA and SymCorrLDA only require multilingual comparable documents that can be easily obtained. Below we introduce LDA-style topic models that handle multiple classes and can be applied to multilingual comparable documents.

2.1   Correspondence LDA(CorrLDA)

CorrLDA is a topic model for multilingual comparable documents proposed in [7]. This model first generates topics for one language part of a document. This language is referred as pivot language. For the other languages, CorrLDA uses the topics that are already generated in pivot language. Figure 1(a) shows a graphical model of CorrLDA. The p in the graph is the pivot language that is specified in advance. Algorithm 1 shows the process of generating a document according to CorrLDA model.

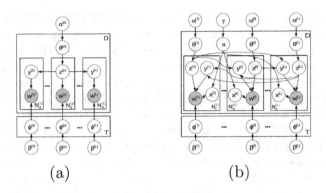

Fig. 1. Graphical model representations of (a) CorrLDA,(b) SymCorrLDA.

Algorithm 1.. CorrLDA

---

for all D documents' pivot language part **do**
    Sample $\theta_d^p \sim$ Dirichlet(α^p)
end for
for all T topics and all L languages **do**
    Sample $\phi_t^l \sim$ Dirichlet(β^l)
end for
for each of the N_d^p words w_i^p in language p of document d **do**
    Sample a topic $z_i^p \sim$ Multinomial(θ_d^p)
    Sample a word $w_i^p \sim$ Multinomial$(\phi_{z_i^p}^p)$
end for
for each of the N_d^l words w_i^l in language l of document d **do**
    Sample a topic $y_i^l \sim$ Uniform$(z_1^p, ..., z_{N_d^p}^p)$
    Sample a word $w_i^l \sim$ Multinomial$(\phi_{y_i^l}^l)$
end for

2.2 Symmetric Correspondence LDA(SymCorrLDA)

Compared with CI-LDA and SwitchLDA, CorrLDA can capture more direct depency between languages. However as discussed before, it has to select the pivot language in advance. Since the pivot language may differ based on the subject, such as the country a document is about, it is often difficult to appropriately select the pivot language. To address this problem, Symmetric Correspondence LDA(SymCorrLDA) are proposed in [8]. Different from CorrLDA, SymCorrLDA can generates a flag that specifies a pivot language for each word, adjusting the probability of being pivot languages in each language part of a document according to a binomial distribution for bilingual data or a multinomial distribution for data of more than three languages. Figure 1(b) shows a graphical model of SymCorrLDA. Algorithm 2 shows the process of generating a document according to CorrLDA model. The pivot language flag x_i^l for an arbitrary language l indicates that the pivot language for the word w_i^l is its own language l, and

$x_i^l = m$ indicates that the pivot language for w_i^l is another language m different from its own language l. The indicator function δ takes the value 1 when the designated event occurs and 0 if otherwise.

Algorithm 2.. SymCorrLDA

---

for all D documents' pivot language part **do**
    for all L languages **do**
        Sample $\theta_d^l \sim$ Dirichlet(α^l)
    end for
    Sample $\pi_d \sim$ Dirichlet(γ)
end for
for all T topics and all L languages **do**
    Sample $\phi_t^l \sim$ Dirichlet(β^l)
end for
for each of the N_d^p words w_i^p in language p of document d **do**
    Sample a pivot language flag $x_i^l \sim$ Multinomial(π_d)
    if $x_i^l == l$ **then**
        Sample a topic $z_i^l \sim$ Multinomial(θ_d^l)
    else if $x_i^l == m$ **then**
        Sample a topic $y_i^l \sim$ Uniform($z_1^m, ..., z_{M_d^m}^m$)
    end if
    Sample a word topic $w_i^l \sim$ Multinomial($\delta_{x_i^l=l}\phi_{z_i^l}^l + (1 - \delta_{x_i^l=l})\phi_{y_i^l}^l$)
end for

---

3 Nonparametric Symmetric Correspondence LDA(NPSymCorrLDA)

When the SymCorrLDA model is applied to parallel or comparable documents, the number of topics(hyper parameter T) must be specified in advance. In practice, the number of topics often differs based on the scale of corpus. To find the best assignment of T, one natural method is to try different values and pick the best one. However, this is often time-consuming due to the high time complexity of model inference. Since SymCorrLDA uses Dirichlet distribution(DD) as the prior of parameters, it is a Bayesian parametric model. In this paper, we propose a nonparametric Bayesian model in the hope of overcoming the disadvantages mentioned above as well as achieving improvement over SymCorrLDA. Our model is based on Hierarchical Dirichlet Process(HDP)[13], which is a nonparametric Bayesian approach to cluster grouped data in statistics and machine learning. Here, we give a brief introduction of DP and HDP.

3.1 Dirichlet Process(DP)

The Dirichlet Process is a stochastic process used in probability theory. Let G_0 be a probability distribution and α be a positive real number. A Dirichlet Process

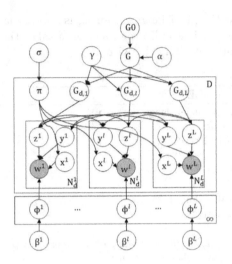

Fig. 2. Graphical model representations of NPSymCorrLDA.

G is defined as

$$G \sim DP(\gamma, G_0) \tag{1}$$

where γ is called the concentration parameter and G_0 the base measure. DP can be presented as the sticking-breaking process or the Chinese Restaurant Process(CRP). CRP is often used when we apply Gibbs sampling. In addition, Dirichlet Process is viewed as the infinite dimensional generalization of Dirichlet distribution. As with all Bayesian nonparametric models, DP's parameters grow as more data are observed. DP-based models have gained considerable popularity in the field of machine learning because of the above-mentioned flexibility([14,15]), especially in unsupervised learning. The flexibility also makes it an ideal candidate for clustering problems where the distinct number of clusters is unknown beforehand.

3.2   NPSymCorrLDA Based on HDP

The hierarchical Dirichlet process(HDP) is an extension to DP and often used to model grouped data. It uses Dirichlet process(DP) for each group of data, and all Dirichlet Process for grouped data share the same Dirichlet process as their base measure. For instance, for each group labeled j:

$$G_j \sim DP(\alpha, G) \tag{2}$$

where G itself is a DP

$$G \sim DP(\gamma, G_0) \tag{3}$$

Here α and γ are concentration parameters and H is a base measure. Like DP, HDP can be presented as sticking-breaking construction or Chinese Restaurant

Franchise(CRF)[13]. CRF is a generative process shown in Figure 3 and the metaphor is as follows: we have J restaurants, each with n_j customers(ϕ_{ji}'s), who sit at tables(ψ_{jt}'s). Now each table is served a dish (θ_k's) from a menu common to all restaurants. Suppose customer i sat at table t_{ji}. The conditional distributions according to $DP(\alpha, G)$ are:

$$t_{ji}|t_{j1},...,t_{ji-1},\alpha \sim \sum_t \frac{n_{jt}}{\sum_{t'} n_{jt'} + \alpha}\delta_t + \frac{\alpha}{\sum_{t'} n_{jt'} + \alpha}\delta_{t^{new}} \quad (4)$$

where n_{jt} is the number of customers currently at table t. Let m_k denote the number of tables enjoying dish k. The customer i has to select a dish from the global dish menu if he chooses a new table, which is again distributed according to $DP(\gamma, G_0)$. A dish θ_k that has already been shared in the global menu would be chosen with probability $m_k/(\sum_{k'} m_{k'} + \gamma)$ and a new dish with probability $\gamma/(\sum_{k'} m_{k'} + \gamma)$.

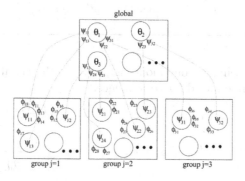

Fig. 3. An instantiation of CRF representation for the 3 group HDP. Each of the 3 restaurants has customers sitting around tables ,and each table is served a dish.

In our nonparameteric symmetric correspondence topic model(NPSymCorr LDA), the prior and posterior are not probability distributions any more, but stochastic process. This allows the model to determine the scale of parameters according to the data it observes. Like the SymCorrLDA, our model generates a flag that specifies a pivot language for each word, adjusting the probability of being pivot languages in each language part of a document according to a multinomial distribution over languages. This makes it possible for the model to estimates the best pivot language at the word level in each document. Unlike the SymCorrLDA, when the pivot flag is assigned with x, NPSymCorrLDA generates a topic for each word in document m according to the Dirichlet process Gm, x, and all the DPs share the same base measure G. Figure 2 shows a graphical model representation of NPSymCorrLDA. Algorithm 3 shows the process of generating a document according to NPSymCorrLDA model.

We use Gibbs sampling for model inference, and we sample from the posterior of HDF using CRF representation. The full conditional probability for

---

Algorithm 3.. NPSymCorrLDA

---

for all D documents' pivot language part **do**
    for all L languages **do**
        Sample $G_{d,l} \sim \mathrm{DP}(\gamma, G)$
    end for
    Sample $\pi_d \sim \mathrm{Dirichlet}(\sigma)$
end for
for all ∞ topics and all L languages **do**
    Sample $\phi_k^l \sim \mathrm{Dirichlet}(\beta^l)$
end for
for each of the N_d^l words w_i^l in language l of document d **do**
    Sample a pivot language flag $x_i^l \sim \mathrm{Multinomial}(\pi_d)$
    if $x_i^l == l$ **then**
        Sample a topic $z_i^l \sim G_{d,l}$
    else if $x_i^l == m$ **then**
        Sample a topic $y_i^l \sim \mathrm{Uniform}(z_1^m, ..., z_{M_d^m}^m)$
    end if
    Sample a word topic $w_i^l \sim \mathrm{Multinomial}(\delta_{x_i^l=l}\phi_{z_i^l}^l + (1 - \delta_{x_i^l=l})\phi_{y_i^l}^l)$
end for

---

collapsed Gibbs sampling of our model is given by the equations below. The hyper-parameters are omitted for convenience.

$$P(z_i^l = k, x_i^l = l | w^l, x_{-i}, z_{-i}^l) \propto (n_{d,-i}^l + \sigma) \cdot P(z_i^l = k | z_{-i}^l) \cdot f_k^l(w_i^l) \qquad (5)$$

$$P(y_i^l = k, x_i^l = m | w^l, x_{-i}, z^m) \propto (n_{d,-i}^l + \sigma) \cdot \frac{n_{d,k}^m}{\sum_{k=1}^T n_{d,k}^m} \cdot f_k^l(w_i^l) \qquad (6)$$

where

$$f_k^l(w_i^l) = \frac{n_{k,-i}^l + \beta_k^l}{\sum_{k=1}^T n_{k,-i}^m l + \beta_k^l} \qquad (7)$$

The conditional posterior probability $P(z_i^l = k | z_{-i}^l)$ is calculated according to the CRF representation of HDP.

4 Experiments

In this section, we demonstrate some examples with NPSymCorrLDA, and then we compare our model with baseline model using various evaluation methods. For evaluation, we use held-out log-likehood on English and Chinese dataset, the task of finding an English article that is on the same topic as that of a Chinese article, and a task with the languages reversed.

4.1 Settings

The dataset used in this work is a collection of Wikipedia articles: it is in English and Chinese, and articles in this collection are connected across languages via specific inter-language links. We extracted text content from original Wikipedia articles, removed link information, revision history information and other useless parts. We use Wikipedia Extractor[1] for this purpose. Chinese articles are edited by simplified and traditional type, we used Opencc[2] to convert traditional type to simplified type. For English articles, we removed over 1000 types of standard stop words. As for Chinese articles, we obtained words from Chinese sentences without stopping tags, using Chinese word segmentation system ICTCLAS[3]. The statistics of the datasets after preprocessing are shown in Table 1. We assumed each set of Wikipedia articles connected via inter-language links between two languages as a comparable document that consists of two language parts. To compute held-out log-likelihood, we randomly divided each of the training documents at the word level into 90% training set and 10% held-out set. To carry out the evaluation in the task of finding couterpart articles , we randomly divided the Wikipedia document collection at the document level into 90% training documents and 10% test documents.

Table 1. Summary of bilingual data.

	Chinese	English
No. of documents	5,029	5,029
No. of word types(vocab)	60,149	63,677
No. of word tokens	4,187,721	5,614,182

We estimated our model and SymCorrLDA as a baseline, using collapsed Gibbs sampling with the training set. In addition, we estimated a special implementation of NPSymCorrLDA, setting π_d in a sample way for comparison, where the pivot language flag for each word is randomly selected according to the proportion of the length of each language part. For all the models, we set symmetric Dirichlet hyperparameters $\alpha = 0.1$, $\beta = 0.01$ and $\gamma = 1$. We imposed the convergence condition of collapsed Gibbs sampling, such that the percentage change of held-out log-likelihood is less than 0.1%

4.2   Pivot Assignments

Figure 4 demonstrates how the frequency distribution of the pivot language-flag(binomial) parameter $\pi_{d,1}$ for the Chinese language with the bilingual dataset in NPSymCorrLDA changes while in iterations of collapsed Gibbs sampling. This figure shows that the piovt language flag is randomly assigned at the initial state, and then it converges to an appropriate bias for each document as the iterations proceed. We next demonstrate how the pivot language flags are assigned to each document. Figure 5 shows the title of eight documents and the corresponding $\pi_{d,1}$ when using the bilingual data. If $\pi_{d,1}$ is close to 1, the article can be consider

---

[1] http://medialab..di.unipi.it/wiki/
[2] https://code.google.com/p/opencc/
[3] http://ictclas.nlpir.org/

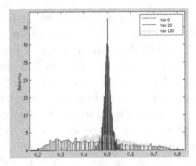

Fig. 4. Change of frequency distribution of $\pi_{d,1}$ according to number of iterations.

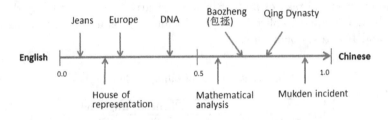

Fig. 5. Document titles and corresponding $\pi_{d,1}$

to be more related to a subject on Chinese or China. In contrast, if $\pi_{d,1}$ is close to 0 and therefore $\pi_{d,2} = 1 - \pi_{d,1}$ is close to 1, the article can be considered to be more related to a subject on English or English-speaking countries.

We further demonstrate the proportions of pivot assignments at the topic level. Figure 6 shows the content of 6 topics through 10 words with the highest probability for each language and for each topic when using the bilingual data, some of which are biased to Chinese or English while the others have almost no bias. It can be seen that the pivot bias to specific languages can be interpreted.

4.3   Held-Out Log-Likelihood

To estimate the performance of each topic model, we can calculate the held-out log-likelihood. The number of topics automatically converges to about 80 when training NPSymCorrLDA. We trained SymCorrLDA with setting topic numbers of 50 and 80 for comparsion. We know that the model with the higher held-out log-likelihood has greater predictive ability. The held-out set which we used in experiment is mentioned in Section 4.1.

The held-out log-likelihood of SymCorrLDA which shows superiority over other models is contained in [8]. We can see through Table 2 that our model NPSymCorrLDA is better than SymCorrLDA in both. Actually the improvements with NPSymCorrLDA are statistically significant compared to the previous model, according to the Wilcoxon signed-rank test at the 1% level in terms

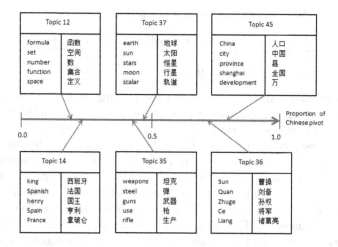

Fig. 6. Topic examples and corresponding proportion of pivots assigned to Chinese.

Table 2. Per-word held-out log-likelihood with bilingual data.

Methods	Chinese		English	
	T=50	T=80	T=50	T= 80
SymCorrLDA	-8.92	-8.83	-8.63	-8.55
NPSymCorrLDA	**-8.35**		**-8.27**	

of the word-by-word held-out log-likelihood. This indicates that our nonparametric model gets better estimation of the documents over SymCorrLDA. As for scalability, NPSymCorrLDA is as scalable as SymCorrLDA since the time complexity of these two models is the same.

4.4   Finding Counterpart Articles

Given an article, its hidden counterpart articles in other form of languages can be detected using a multilingual topic model. To fulfill this job, we conducted experiment with our bilingual dataset. We evaluated document-topic distribution of test documents for each language, using the topic-word distributions estimated by multilingual topic model with training documents. Using information mentioned above, we evaluated the performance of finding English counterpart articles using Chinese articles as queries, and vice versa. In our work, we utilized the technology of re-sampling to evaluate the document-topic distributions of test documents, and the process of re-sampling utilized the topic-word distribution estimated beforehand. Then we used the Jensen-Shannon(JS) divergence between a document-topic distribution of Chinese and that of English for each test document. JS divergence should be small if two counterpart articles have accurate latent topic evaluation. In light of this information, we first assumed each held-out Chinese article to be query and the corresponding English article

Table 3. MRR in Counterpart article finding task.

Methods	Chinese to English	English to Chinese
SymCorrLDA	0.3200	0.2948
NPSymCorrLDA	**0.3840**	**0.3814**

to be its counterpart, and then estimated the ranking of all the test articles of English in ascending order of the JS divergence, then we conducted the experiment with the languages reversed.

The mean reciprocal rank(MRR) are shown in Table 3. The definition of reciprocal rank is the multiplicative inverse of the rank of the counterpart article corresponding to each query article, and the MRR is the average of it over all the query articles. The topic number of SymCorrLDA is set to 80. And we can see through this table that our model shows its superiority over SymCorrLDA both with the Chinese and English queries. Statistically, the improvements with NPSymCorrLDA are significant according to the Wilcoxon signed-rank test at 1% level. Therefore, it is transparent that NPSymCorrLDA evaluates multilingual topics the most accurately in this experiment.

5 Conclusions

Many previous multilingual topic models connect assuming parallelism at the sentence level or word level, while works about document level is less than expected. SymCorrLDA incorporates a hidden variable to control a pivot language, in an extension of CorrLDA. However, SymCorrLDA has the same problem of how to assign the unknown parameters as other parameterized graph models. In this paper, we propose a nonparameterized symmetric correspondence LDA model(NPSymCorrLDA) that can be applied to multilingual documents, not using multilingual dictionaries. NPSymCorrLDA has an advantage that it does not require the number of topics specified in advance. We compared the performance of SymCorrLDA and NPSymCorrLDA in terms of held-out log-likelihood and in the task of cross-lingual link detection. Experiment results show that NPSymCorrLDA works significantly than SymCorrLDA. NPSymCorrLDA can be applied to other kinds of data that have multiple classes of representations, such as annotated image data. We plan to investigate this in future work.

Acknowledgments. Our work is supported by National High Technology Research and Development Program of China (863 Program) (No.2015AA015402), National Natural Science Foundation of China (No.61370117 & No.61433015) and Major National Social Science Fund of China (No.12&ZD227).

References

1. Hofmann, T.: Probabilistic latent semantic indexing. In: Proceedings of the 22nd Annual International ACM SIGIR Conference on Research and Development in Information Retrieval, pp. 50–57. ACM, 1999, August

2. Blei, D.M., Ng, A.Y., Jordan, M.I.: Latent dirichlet allocation. The Journal of machine Learning research **3**, 993–1022 (2003)
3. Mimno, D., Wallach, H.M., Naradowsky, J., Smith, D.A., McCallum, A.: Polylingual topic models. In: Proceedings of the 2009 Conference on Empirical Methods in Natural Language Processing: Volume 2-vol. 2, pp. 880–889. Association for Computational Linguistics, 2009, August
4. Ni, X., Sun, J.T., Hu, J., Chen, Z.: Mining multilingual topics from wikipedia. In: Proceedings of the 18th International Conference on World Wide Web, pp. 1155–1156. ACM, April 2009
5. Stephen, E.E., Fienberg, S., Lafferty, J.: Mixed membership models of scientific publications. In: Proceedings of the National Academy of Sciences (2004)
6. Newman, D., Chemudugunta, C., Smyth, P.: Statistical entity-topic models. In: Proceedings of the 12th ACM SIGKDD International Conference on Knowledge Discovery and Data Mining, pp. 680–686. ACM, August 2006
7. Blei, D.M., Jordan, M.I.: Modeling annotated data. In: Proceedings of the 26th Annual International ACM SIGIR Conference on Research and Development in Informaion Retrieval, pp. 127–134. ACM, July 2003
8. Fukumasu, K., Eguchi, K., Xing, E.P.: Symmetric correspondence topic models for multilingual text analysis. In: Advances in Neural Information Processing Systems, pp. 1286–1294 (2012)
9. Zhao, B., Xing, E.P.: BiTAM: Bilingual topic admixture models for word alignment. In Proceedings of the COLING/ACL on Main Conference Poster Sessions, pp. 969–976. Association for Computational Linguistics, July 2006
10. Boyd-Graber, J., Blei, D.M.: Multilingual topic models for unaligned text. In: Proceedings of the Twenty-Fifth Conference on Uncertainty in Artificial Intelligence, pp. 75–82. AUAI Press, June 2009
11. Jagarlamudi, J., Daumé III, H.: Extracting multilingual topics from unaligned comparable corpora. In: Gurrin, C., He, Y., Kazai, G., Kruschwitz, U., Little, S., Roelleke, T., Rüger, S., van Rijsbergen, K. (eds.) ECIR 2010. LNCS, vol. 5993, pp. 444–456. Springer, Heidelberg (2010)
12. Zhang, D., Mei, Q., Zhai, C.:. Cross-lingual latent topic extraction. In: Proceedings of the 48th Annual Meeting of the Association for Computational Linguistics, pp. 1128–1137. Association for Computational Linguistics, July 2010
13. Teh, Y.W., Jordan, M.I., Beal, M.J., Blei, D.M.: Hierarchical dirichlet processes. Journal of the american statistical association **101**(476), (2006)
14. Neal, R.M.: Markov chain sampling methods for Dirichlet process mixture models. Journal of computational and graphical statistics **9**(2), 249–265 (2000)
15. Rasmussen, C.E.: The infinite Gaussian mixture model. In: NIPS, vol. 12, pp. 554–560 (1999)

Knowledge Acquisition
and Information Extraction

Mining RDF from Tables in Chinese Encyclopedias

Weiming Lu[✉], Zhenyu Zhang, Renjie Lou, Hao Dai, Shansong Yang,
and Baogang Wei

College of Computer Science and Technology, Zhejiang University, Hangzhou, China
luwm@zju.edu.cn

Abstract. Web tables understanding has recently attracted a number of studies. However, many works focus on the tables in English, because they usually need the help of knowledge bases, while the existing knowledge bases such as DBpedia, YAGO, Freebase and Probase mainly contain knowledge in English.

In this paper, we focus on the RDF triples extraction from tables in Chinese encyclopedias. Firstly, we constructed a Chinese knowledge base through taxonomy mining and class attribute mining. Then, with the help of our knowledge base, we extracted triples from tables through column scoring, table classification and RDF extraction. In our experiments, we practically implemented our approach in 6,618,544 articles from *Hudong Baike* with 764,292 tables, and extracted about 1,053,407 unique and new RDF triples with an estimated accuracy of 90.2%, which outperforms other similar works.

1 Introduction

Nowadays, large-scale knowledge bases (KBs) are playing an increasing role in many intelligent applications. These knowledge bases contain millions of facts, such as information about people, locations, organizations, which are represented as RDF triples (subject-predicate-object triples).

Most of the knowledge bases are primarily builded by integrating the existing structured knowledge (e.g. Wikipedia's infoboxes for DBPedia [3]), or extracting knowledge from unstructured text such as NELL [5]. However, unstructured text can be very noisy, and the existing structured knowledge is quite limited. Web tables are content-rich, and relatively easier for knowledge extraction than the unstructured text. Therefore, many approaches have been tried to populate KBs by using Web tables.

The main challenge for RDF mining from Chinese tables is that the existing knowledge bases such as DBPedia, YAGO [18] and Freebase [4] contain very limited knowledge in Chinese, which makes the understanding of Chinese text very difficult. Fortunately, there are two large-scale collaboratively Chinese encyclopedias named Baidu Baike[1] and Hudong Baike[2], and they claimed they contain

---

[1] http://www.baike.baidu.com
[2] http://www.baike.com

© Springer International Publishing Switzerland 2015
J. Li et al. (Eds.): NLPCC 2015, LNAI 9362, pp. 285–298, 2015.
DOI: 10.1007/978-3-319-25207-0_24

more than 11.2 and 12.2 million articles respectively[3]. Therefore, in this paper, we will build a Chinese knowledge base from these encyclopedias, and then mine RDF from tables in encyclopedias for knowledge population.

Our paper has the following differences against other works. Firstly, we have to construct a Chinese knowledge base from scratch to help RDF mining. In addition to the *subclass-of*, *instance-of* and *class-attribute* relations, the probabilities for these relations should also be provided. Secondly, we directly classify tables into *Genuine Table with Header* and *Genuine Table without Header*, and detect object columns besides subject column in tables, which can improve the performance of the table understanding. This is because the attributes of an entity extracted from *infobox* could be incomplete or even missing, and information about different entities could be mixed in one table. For example, Figure 2a mixes the information about persons and organizations.

The rest of the paper is organized as follows. Section 2 provides the overview of our work. Then, we describe the Chinese knowledge base construction in Section 3, and the table understanding in Section 4. Experiments, related work and conclusions appear in Section 5, 6 and 7.

2   Approach Overview

The procedure of our work contains three major components: table extraction, table classification and RDF extraction, as shown in Figure 1. In order to help the RDF mining, we construct a Chinese knowledge base from the encyclopedias, which includes *Taxonomy Mining* and *Class Attribute Mining*.

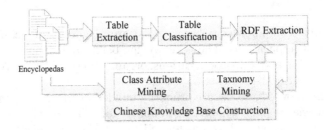

Fig. 1. The architecture of RDF mining from Chinese encyclopedias

Table Extraction. We mainly focus on the *wikitables* in encyclopedias as in [14]. At first, we locate the tables by looking for table-related HTML tags (e.g. table, tr and td), and then parse the table into a matrix by using the normalization technologies in [14]. In addition, many *wikitables* in encyclopedias always put table's abstract in the top row, which has only one column, so we delete this row in the tables.

---

[3] At Jan 10, 2015.

Table Classification. Although well-formed tables have been obtained, there are still several categories in the tables, which will lead to different subsequent operations. The categories include *Genuine Table with Header*, *Genuine Table without Header* and *Non-genuine Table* (entities in the same class are distributed in different columns), are shown in Figure 2. The aim of table classification is to classify the tables into categories, with the help of the Chinese knowledge base.

排名	财富(亿元)	姓名	公司	总部	行业	年龄	出生地
1	430	黄光裕	鹏润投资	北京	家电零售、房地产、投资	39	广东
2	360	杜双华	日照钢铁	山东	钢铁	43	河北
3	330	杨惠妍	碧桂园	广东	房地产	27	广东

(a) Genuine Table with Header

华北	北京	故宫博物院 · 天坛公园 · 颐和园 · 八达岭长城
	天津	天津古文化街旅游区 · 天津盘山风景名胜区
	河北	秦皇岛市山海关景区 · 保定市安新白洋淀景区 · 承德避暑山庄
	山西	大同市云冈石窟 · 忻州市五台山风景名胜区
东北	辽宁	沈阳市植物园 · 大连老虎滩海洋公园 · 老虎滩极地馆
	吉林	长春市伪满皇宫博物院 · 长白山景区

(b) Genuine Table without Header

中国旅游景点（一）

中国地域辽阔，美丽的自然风光与丰富的人文古迹比比皆是，美不胜收。

梨园剧场	和平寺	北京九龙游乐园	新文化运动纪念馆
颐和园	大杨山森林公园	青龙湖水上乐园	劳动人民文化宫
平谷老象峰景区	虎峪自然风景区	紫竹院公园	十渡景区
密云南山滑雪场	望姑坨风景区	李大钊烈士陵园	玉渊潭
房山十渡	平西抗日烈士陵园	中央电视塔	银山塔林

(c) Non-genuine Table

Fig. 2. Three categories of tables

RDF Extraction. When *Genuine Tables* are obtained, we extract RDF triples from these tables. The header could be used as the predicate in triples for *Genuine Table with Header*. For tables without header, we estimate the predicate with the help of our KB.

Class Attribute Mining and **Taxonomy Mining** are two components for building the Chinese knowledge base from encyclopedias. In **Class Attribute Mining**, we mine the proper attributes for each class in the knowledge base, and provide the probabilities of $p(attribute|class)$ and $p(class|attribute)$. **Taxonomy Mining** mines the probability of $p(class|entity)$ from encyclopedias.

We crawled 6,618,544 unique articles from *Hudong Baike*, where each article represents an entity, and it may include infobox, catalog, tags and innerlinks, which can be used to build the Chinese knowledge base. Finally, we extracted 4,897,722 *toc*s (table of content), 728,039 *infobox*s and 764,292 *wikitable*s, where the *wikitable*s are our focus in this paper for RDF mining.

3 Chinese Knowledge Base from Scratch

In this section, we mainly focus on the Chinese knowledge base construction from scratch, including *taxonomy mining* and *class attribute mining*.

3.1   Taxonomy Mining

Hudong Baike provides a category system for article navigation, but it does not form a real subsumption hierarchy, which includes *isa* and *notisa* relations between categories. Therefore, we have to distinguish *isa* and *notisa* relations, and use *isa* relation to form *subclass-of* relations. We treat this problem as a binary classification problem: given two categories c_i and c_j, we train a classifier to predicate whether c_i and c_j have the *isa* relation, where linguistic features and structural features [23] are used for the classifier.

Similarly, *instance-of* relation induction problem is also treated as a binary classification problem: given an article a and its category c, we train a classifier to predicate whether a is an instance of c. Here, we also use linguistic features and structural features as in [23] for the classifier.

In practice, we trained SVM classifiers and used precision, recall, and F1 to evaluate the performance of *subclass-of* and *instance-of* relations based on a manually labeled data set including randomly selected 1000 pairs of categories, and 1000 pairs of article and its category. We got precision=87.27%, recall=86.75%, F1=87.01% for *subclass-of* relation, and precision=80.33%, recall=67.58%, F1=73.41% for *instance-of* relation.

Finally, the induced taxonomy could be represented as $T = \{E, C, R_i, R_c, P_i, P_c\}$, where E and C is the set of entities and classes. When $r_i(e, c) \in R_i \subset E \times C$, it means entity e and class c has *instance-of* relation. Similarly, $r_c(c_i, c_j) \in R_c \subset C \times C$ indicates c_i is a subclass of c_j. In addition, we also consider the probability output of classifiers as the probability of relations. $p_i(e, c) \in P_i$ is the probability of *instance-of* relation for e and c, and $p_c(c_i, c_j) \in P_c$ is the probability of *subclass-of* relation for c_i and c_j.

If entity e can reach a class c along with *instance-of* and *subclass-of* relations, there is a path $path(e, c) = \langle e, c_1, c_2, ..., c_k \rangle$, where $c_k = c$, $r_i(e, c_1) \in R_i$, and $r_c(c_i, c_{i+1}) \in R_c$. Then, for each entity $e \in E$, we can get its classes $C(e) = \{c | \exists path(e, c), c \in C\}$. For each class c, we can get its entities $E(c) = \{e | \exists path(e, c), e \in E\}$.

Entity e and its classes $C(e)$ can form an graph $G = (V_G, E_G)$ as shown in Figure 3, where $V_G = \{e, c \in C(e)\}$ and $E_G = \{r_i(e, c \in C(e)) \in R_i, r_c(c_i \in C(e), c_j \in C(e)) \in R_c\}$.

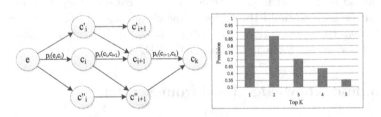

Fig. 3. Graph for an entity   **Fig. 4.** Performance of RWR

In this graph, we use RWR (random walk with restart)[19] to calculate the relevance score $p(e, c \in C(e))$ for each pair of $(e, c \in C(e))$. Formally, the walker starts from the node e, and then follows an edge to another node at each step. Additionally, at every step, the walker would return to the node e with a non-zero probability c. Let v_e be a vector of zeros with the element corresponding to node e set to 1 ($v_e(e) = 1$), and then the steady state probability vector $u_e = (u_e(e), u_e(c_1), u_e(c_2), ..., u_e(c_N))$ could be estimated by matrix multiplication. Let A be the adjacency matrix of the graph G, where $A(e, c \in C(e)) = p_i(e, c)$ or $A(c_i \in C(e), c_j \in C(e)) = p_c(c_i, c_j)$, then u could be calculated by $u_e = (1 - c)Au_e + v_e$. Finally, $p'(e, c \in C(e)) = u_e(c)$.

We randomly selected 85 entities with its corresponding ranked class list, and asked students to evaluate them. The precision of top K is shown in Figure 4, which shows the RWR can find proper classes for entities.

3.2 Class Attribute Mining

After taxonomy mining, we could use *infobox* to mine the attributes for each class with the help of the taxonomy.

Let E_b be the set of entities whose corresponding articles have *infobox*, and then the attribute set of each entity $e \in E_b$ would be represented as $A(e)$, which are extracted from its *infobox*. Then, given a class c, we can obtain its attribute set $A(c) = \cup_{e \in E(c)} A(e)$. Meanwhile, we can also obtain the class set $C'(a) = \{c | \exists e \in E, a \in A(e), c \in C(e)\}$ for attribute a. The co-occurrent frequency of attribute a and class c can be calculated by $f(a, c) = |\{e | a \in A(e) \cap e \in E(c)\}|$, and the occurrence of class c and attribute a are $f_{cls}(c) = |\{e | e \in E(c)\}|$ and $f_{attr}(a) = |\{e | a \in A(e), e \in E\}|$ respectively. However, even attribute a and class c occur frequently, they may not be the right attribute for class or right class for attribute. Figure 5 ranks the attribute set $A(Movie)$ and the class set $C'(Release\ time)$ respectively according to the occurrence frequency.

Attributes	Frequency
导演(Director)	58361
上映时间(Release time)	45688
主演(Actors)	44093
类型(Type)	41733
制片地区(Region)	34429
外文名(English name)	33411
编剧(Writer)	32787
对白语言(Language)	28776
片长(Length)	23502
色彩(Color)	20335
......	

Categories	Frequency
电影(Movie)	45688
影视(Film and television)	18514
艺术(Art)	17493
电视剧(Drama)	9463
剧情片(Feature Film)	6274
娱乐(Entertainment)	5737
导演(Director)	4204
年(Year)	4154
喜剧(Comedy)	3523
影片(Film)	3481
......	

(a) The attribute rank list for the class *Movie*   (b) The class rank list for the attribute *Release time*

Fig. 5. Attribute and class rank list according to the frequency

But we find that although *Art* and *Entertainment* are ranked in the front, they are not the proper classes for attribute *Release time*. So, we should filter some improper classes for a given attribute.

Here, we train a classifier to filter them, and the features for attribute-class pair (a, c) are list as follows.

1. $|C'(a)|$.
2. $R(C'(a), c)$. The rank of c in $C'(a)$ ordered by $f(a, c)$.
3. (2)/(1).
4. $f(a, c)/f(c)$.
5. $\frac{1}{|C(a)|} \sum_{c' \in C'(a)} \frac{f(a, c')}{f_{cls}(c')}$.
6. (4)/(5)
7. $|A(c)|$.
8. $R(A(c), a)$. The rank of a in $A(c)$ ordered by $f(a, c)$.
9. (8)/(7)
10. $\frac{1}{|A(c)|} \sum_{a' \in A(c)} \frac{f(a', c)}{f_{attr}(a')}$.

In order to train classifiers, we randomly select 408 attribute-class pairs from $(a, c \in C(a))$ or $(a \in A(c), c)$, and the pair (a, c) must satisfy $|C(a)| \geq 10, |A(c)| \geq 10$. Then, students are asked to label the pairs to form training data and testing data. Four classifiers in Weka: *Naive Bayes, Decision Tree J48, Logistic Classifier*, and *Random Forest* are trained, and the test results are shown in Table 1.

Table 1. The performance of (a, c) classification with different classifiers

Method	Prec	Rec	F1	Method	Prec	Rec	F1
Naive Bayes	0.828	0.811	0.819	Decision Tree J48	0.805	0.805	0.805
Logistic Classifier	0.811	0.811	0.811	Random Forest	0.812	0.811	0.811

We use Naive Bayes as our classifier to clean the $C(a)$ and $A(c)$, and then score $p''(a, c \in C(a)) = f(a, c)/f_{attr}(a)$.

4   Table Classification and Understanding

In this section, we extract RDF from *Genuine Tables* with the following steps: *Column Scoring, Table Classification*, and *RDF Extraction*.

4.1   Column Scoring

Wikitable is parsed into a matrix $T(m, n)$ with m rows and n columns in Section 2, so we can combine the information of row and column to compute the score for each column. Here, we don't detect the header at first as other related works, while compute the row score for first row directly, and then use the score with other features to classify tables, which will be described in Section 4.2.

For the ith column, we firstly obtain its candidate class set $C_i = \cup_{1 \leq j \neq i \leq n} C'(T(1, j))$. Then, the row score for i^{th} column and class $c \in C_i$ can be calculated by $s_{row}(i,c) = \sum_{1 \le j \ne i \le n} p''(T(1,j),c)$. The column score for i^{th} column and class $c \in C_i$ can be calculated by $s_{col}(i,c) = \sum_{1 \le j \le m} p'(T(j,i),c)$. Finally, the overall score for i^{th} column is calculated by $score(i) = \max_{c \in C_i} s_{row}(i,c) \cdot s_{col}(i,c)$.

Obviously, the column with larger score is more likely to be the subject column. Therefore, subject column can be determined by $sub_{col} = \arg \max_i score(i)$, and the corresponding class is $sub_{cls} = \arg \max_{c \in C_{sub_{col}}} [s_{row}(sub_{col},c) \cdot s_{col}(sub_{col},c)]$. The object columns are much related to the header column, so they can be filtered by comparing $p''(T(1,j), sub_{cls})$ with a threshold λ. That is, the columns with $p''(T(1,j), sub_{cls}) \ge \lambda, 1 \le j \ne sub_{cls} \le n$ can be considered as object columns, which are denoted as $objs_{col}$.

4.2 Table Classification

Subject column and object columns could be detected in *wikitables* in Section 4.1, but it may be not correct for some tables, especially for *Non-genuine Tables*. So in this section, we directly classify *wikitables* into three categories *Genuine Table with Header*, *Genuine Table without Header* and *Non-genuine Table* by combining the column scores and other features.

We classify features into five groups: statistics features (S), cell features (C), layout features (L), predicate features (P) and score features (Sc).

Statistics Features

- Average number of cells in rows acr. The number of cells in every rows is counted before table normalization.
- Average number of cells in columns acc. The number of cells in every columns is counted before table normalization.
- ratio of acr/acc.
- Deviation of the number of cells in rows, and the deviation of the number of cells in columns.
- Average cell length, and deviation of cell length.
- Within-row length consistency, and Within-column length consistency [22].

Cell Features

- number of cells containing HTML tags (*th, img, b, a* respectively).
- Percentage of numeric cells.
- Percentage of alphabetical cells, date cells, string cells and empty cells.
- Within-row type consistency and within-column type consistency [22].

Layout Features

- row number of first row in un-normalized table.
- Ratio of cells containing HTML tag *th* in the first row to the all cells containing HTML tag *th*.
- Ratio of cells containing HTML tag *b* in the first row to the all cells containing HTML tag *b*.

- the number of columns with different cell type in the first row with other cells in the same column.

Predicate Features

- Weighted average position of predicate in row $wapr$. if $T(i,j) \in \cup_{e \in E} A(e)$, we call $T(i,j)$ is a predicate, which will form a set $S = \{T(i,j)|T(i,j) \in \cup_{e \in E} A(e)\}$. Then $wapr = \frac{1}{|S|} \cdot \sum_{T(i,j) \in S} i \cdot f_{attr}(T(i,j))$
- Weighted average position of predicate in column $wapc$. $wapc = \frac{1}{|S|} \cdot \sum_{T(i,j) \in S} j \cdot f_{attr}(T(i,j))$.
- Weighted deviation position of predicate in row, i.e. the deviation of $i \cdot f_{attr}(T(i,j))$ for all $T(i,j) \in S$
- Weighted deviation position of predicate in column, i.e. the deviation of $j \cdot f_{attr}(T(i,j))$ for all $T(i,j) \in S$

Score Features

- maximal and average row score for the first row, i.e. $\max_{1 \le i \le n}(\max_{c \in C_i} s_{row}(i,c))$, $\frac{1}{n}\sum_{i=1}^{n} \max_{c \in C_i} s_{row}(i,c)$.
- maximal and average column score for columns, i.e. $\max_{1 \le i \le n}(\max_{c \in C_i} s_{col}(i,c))$, $\frac{1}{n}\sum_{i=1}^{n} \max_{c \in C_i} s_{col}(i,c)$.
- maximal and average overall score for columns, i.e. $\max_{1 \le i \le n} score(i)$, $\frac{1}{n}\sum_{i=1}^{n} score(i)$.
- position of the subject column p_s, i.e. $\arg\max_{1 \le i \le n} score(i)$.
- position of the column with largest column score p_{lc}, i.e. $\arg\max_{1 \le i \le n}(\max_{c \in C_i} s_{col}(i,c))$.
- position of the object column with smallest column score p_{sc}, i.e. $\arg\min_{i \in objs_{col}}(\max_{c \in C_i} s_{col}(i,c))$.
- $|p_s - p_{lc}|$
- $p_s - p_{sc}$

With all these features, we train classifiers to classify *wikitables* into three categories *Genuine Table with Header*, *Genuine Table without Header* and *Non-genuine Table*.

4.3   RDF Extraction

In this section, we extract RDF from two type of tables *Genuine Table with Header* and *Genuine Table without Header*.

For *Genuine Table with Header*, table headers could be used as predicates in RDF triples. So with the detected subject column sub_{col} and object columns $objs_{col}$, we can easily extract RDF from tables: $\{\langle s_{ik}, p_j, o_{ij}\rangle | 1 \le i \le m, k = sub_{col}, j \in objs_{col}, p_j = header \ of \ the \ j^{th} \ column\}$.

For *Genuine Table without Header*, we mine the predicate between subject column and object columns for RDF triples. Given subject column and object columns, we extract all pairs on the same row i: $\{(e_{ij}, e_{ik})|j = sub_{col}, k \in objs_{col}, 1 \le i \le m\}$. Since *infobox* in articles have been parsed into triples in knowledge base, we can query the relation for each pair in the knowledge base. Obviously, some pairs would not be found in the knowledge base for the knowledge base is incomplete. But we can still select the proper predicate for subject column and each object column by major voting. Then, the triples could be extracted from the tables.

5   Experiments

We crawled 6,618,544 unique articles from *Hudong Baike* at Jan 10, 2015, and then extracted about 764,292 distinct *wikitables* from the articles according to the HTML tags. After ill-formed and small table ($< 2 \times 2$) filtering, we finally obtain 757,282 tables for our RDF extraction.

5.1   Column Scoring Evaluation

In order to evaluate the performance of column scoring, we randomly selected 76 tables, and labeled 76 subject columns (denoted as L_s), and 156 object columns (denoted by L_o). Our approach labeled the subject column and object columns in each tables by column's overall score, which are denoted by M_s and M_o respectively. Then, the precision, recall, and the F measure for subject column could be calculated by $precision = \frac{|L_s \cap M_s|}{|M_s|} = 94.44\%$, $recall = \frac{|L_s \cap M_s|}{|L_s|} = 94.44\%$, and $F_1 = \frac{2 \cdot precision \cdot recall}{precison + recall} = 94.44\%$. Similarly, the precision, recall, and the F measure for object column is $precision = 93.3\%$, $recall = 71.8\%$, and $F_1 = 81.2\%$. The recall for object column is relatively low, since many object column headers occur very infrequently as attributes of entities. But we still can extract sufficient triples from the tables.

We also compared the performance for subject column detection by using overall score, row score and column score, the results are shown in Table 2.

Table 2. The performance of subject column detection by using different scoring method

Method	Prec	Rec	F1
row score	88.41%	84.72%	86.51%
column score	88.73%	87.50%	88.11%
overall score	94.44%	94.44%	94.44%

Table 3. The F_1 of table classification with different classifiers

	S	S+C	S+C+L	S+C+L+P	ALL
NB	0.612	0.590	0.606	0.603	0.704
J48	0.702	0.713	0.720	0.737	0.790
LC	0.691	0.699	0.740	0.707	0.736
RF	0.702	0.762	0.756	0.786	**0.825**

Obviously, the overall score which combines row score and column score reaches the best performance. Column score can obtain a better performance than row score. This is because in *Hudong Baike*, only 11% articles have *infobox*, which makes the attribute quite sparse.

5.2   Table Classification Evaluation

In order to evaluate table classification, we randomly selected 60 entities in 12 different fields (e.g. Nature, Culture, Art, Economy, Science, etc.), and extracted 229 *wikitables* from the corresponding articles. In these tables, we found 86 of them are *Non-genuine Tables*, 93 of them are *Genuine Tables*, including 50 *Genuine Table with Header*. All these tables are used as training data, and then we also randomly picked up 300 tables in random fields as the test data, where each category has about 1/3 tables.

We trained four classifiers, including *Naive Bayes*(NB), *Decision Tree J48*(J48), *Logistic Classifier*(LC), *Random Forest*(RF), with different feature composition, and the results are shown in Table 3.

From the table, we can find that (1) all features are useful for table classification, since the classifiers can reach the best in F-measure when using all features. (2) Random Forest performs the best, so it is selected as our table classifier.

Finally, we classified all the 757,282 tables, and obtained 441,500 *Genuine Tables* and 315,782 *Non-genuine Tables*. The *Genuine Tables* consist of 71,981 *Genuine Table with Header* and 369,519 *Genuine Table without Header*.

5.3   RDF Extraction Evaluation

When *Genuine Tables* are obtained, RDF triples could be extracted from them. However, there are no similar works in extracting RDF triples from tables in Chinese encyclopedias. We compared our approach to the following methods with some modification to fit the Chinese text.

- **Table Classification.** [22] mainly classifies tables into *Genuine Table* and *Non-genuine Table*, which didn't distinguish the tables with or without header row. So we firstly classify the tables into these two categories as in [22], and then used header detector in [21] to locate headers in tables. However, [21] used Probase [25] for table understanding, which has little knowledge in Chinese. So we replace the Probase with our knowledge base. We denote this method as tc_c, and our table classification as tc_f.
- **RDF Mining.** [14] used DBPedia to mine RDF from Wikipedia' tables directly, and it didn't need to detect subject column and object columns in advance. However, DBPedia also has limited knowledge in Chinese, so it is not suitable for RDF mining from tables in Chinese encyclopedias. We tried to mine the predicate between two cells from different columns by using our knowledge base, and then extracted same features as in [14] for classifiers to predict whether the triples are correct or incorrect. We denote this method as tm_c, and our RDF mining with subject column and object column detection as tm_f.

In addition, we also built a raw knowledge base for comparison from the category system and *infobox* without any incorrect relation removal as in Section 3. This knowledge base is denoted by kb_c, and our knowledge base with refinement is denote by kb_f. Therefore, we can have different combinations for these methods. In Table 4, we show the total number of RDF triples extracted from tables (denoted as *tNo.* in the table) and the number of new triples (denoted as *nNo.*) which don't appear in the *infoboxs* with different methods. That is to say, theses new triples could be used to populate the knowledge base. In addition, we randomly selected 500 triples from the extracted triples for human evaluation, and calculated the accuracy of triples for each method.

Table 4. The performance of different combination of methods, where tc_c, tm_c and kb_c are the table classification, triple mining and knowledge base for comparison, while tc_f, tm_f and kb_f are our approaches for table classification (with ternary classification), triple mining (with subject column and object column detection) and knowledge base (with taxonomy and class-attribute refinement)

Method	tNo.	nNo.	Acc	Method	tNo.	nNo.	Acc
$tc_c + tm_f + kb_c$	3,907,243	1,499,608	67.8%	$tc_c + tm_c + kb_f$	3,395,423	1,314,791	74.3%
$tc_c + tm_f + kb_f$	3,002,028	1,215,170	80.0%	$tc_f + tm_f + kb_c$	3,617,509	1,332,534	76.4%
$tc_f + tm_c + kb_f$	3,056,287	1,140,470	81.8%	$tc_f + tm_f + kb_f$	2,787,027	1,053,407	**90.2%**

From the table, we find that although the number of triples extracted by our approach is the smallest, our approach reaches the best performance with the largest accuracy. Moreover, the improvement in table classification, triple mining and knowledge base construction are all helpful to promote the final performance.

6 Related Work

Our work is most related to knowledge base construction and web table understanding.

6.1 Knowledge Base Construction

Several knowledge bases have been built from Web sources, such as DBpedia [3], YAGO [18], Freebase [4], NELL [5], Probase [25], and Knowledge Vault [8].

DBpedia [3] mainly utilized the Wikipedia for automatic construction of large knowledge bases. YAGO [18] links WordNet [17] and Wikipedia to form a large and extendable knowledge base by using the taxonomy from WordNet and facts from Wikipedia. Probase [25] can automatically inference an open-domain, probabilistic taxonomy from the entire web with an iterative learning algorithm. NELL [5] aims to build a never-ending language learner to iteratively promote beliefs in knowledge base through a semi-supervised learning method. Knowledge Vault [8] created a much bigger knowledge base by fusing together multiple extraction sources with prior knowledge derived from an existing KB.

However, all these existing knowledge bases contain very limited knowledge in Chinese, so we have to create a Chinese knowledge base from scratch. Recently, several works about the construction of Chinese knowledge base are proposed, such as

Zhishi.me [16] and XLore [24], but they can not be used for our problem directly. Zhishi.me focused on the infobox information extraction and Chinese LOD construction. XLore can utilize a classification-based method to correctly semantify the wikis'category systems, but it doesn't mine the probabilities of the relations.

6.2   Web Table Understanding

Many works have been proposed for web table understanding, including table classification [6,9,10], table annotation [7,11,12,20,29], knowledge population from tables [13,14,21,26,28], and more advanced applications involving web tables, such as table search, fact search engine [27], search join [1,2] and table summarization [15].

Our work is mainly related to table annotation and knowledge population from tables. Table Miner [29] annotated Web tables by using an incremental, bootstrapping learning approach seeded by automatically selected partial content from tables. [20] leveraged a database of class labels and relationships which are automatically extracted from the Web to recover semantics of tables. [12] annotated table cells with entities, table columns with types, and relations that pairs of table columns simultaneously through a graphical model. InfoGather [26] tries to promote information gathering tasks by considering both indirectly and directly matching tables. Furthermore, InfoGather+ [28] was proposed with a probabilistic graphical model to discover the semantic labels of columns and semantic matches between columns over all web tables collectively.

Our work is similar to [21] and [14]. However, [21] used Probase [25] for table understanding, which has little knowledge in Chinese, and [14] used DBPedia to mine RDF from Wikipedia' tables directly, without detecting subject column and object columns.

7   Conclusion

In this paper, we proposed an approach to mine RDF triples from tables in Chinese encyclopedias for knowledge base population. Since there is no knowledge base like DBpedia and Probase to help table understanding, we constructed a Chinese knowledge base from scratch through taxonomy mining and class attribute mining at first. Then, we extracted RDF triples from tables with the following steps: column scoring, table classification and RDF extraction.

In future, we would like to evaluate the performance improvement with the iteration between knowledge population and table understanding. In addition, we would also like to introduce crowdsourcing technologies [11] to promote the table understanding.

Acknowledgments. This work is supported by the Fundamental Research Funds for the Central Universities(2014QNA5008), Chinese Knowledge Center of Engineering Science and Technology(CKCEST), and the National Natural Science Foundation of China (No.61103099).

References

1. Bhagavatula, C.S., Noraset, T., Downey, D.: Methods for exploring and mining tables on wikipedia. In: Proceedings of the ACM SIGKDD Workshop on Interactive Data Exploration and Analytics, pp. 18–26. ACM (2013)
2. Bizer, C.: Search joins with the web. In: ICDT, p. 3 (2014)
3. Bizer, C., Lehmann, J., Kobilarov, G., Auer, S., Becker, C., Cyganiak, R., Hellmann, S.: Dbpedia-a crystallization point for the web of data. Web Semantics: Science, Services and Agents on the World Wide Web 7(3), 154–165 (2009)
4. Bollacker, K., Evans, C., Paritosh, P., Sturge, T., Taylor, J.: Freebase: a collaboratively created graph database for structuring human knowledge. In: SIGMOD, pp. 1247–1250. ACM (2008)
5. Carlson, A., Betteridge, J., Kisiel, B., Settles, B., Hruschka Jr., E.R., Mitchell, T.M.: Toward an architecture for never-ending language learning. In: AAAI, vol. 5, p. 3 (2010)
6. Crestan, E., Pantel, P.: Web-scale table census and classification. In: WSDM, pp. 545–554. ACM (2011)
7. Deng, D., Jiang, Y., Li, G., Li, J., Yu, C.: Scalable column concept determination for web tables using large knowledge bases. Proceedings of the VLDB Endowment 6(13), 1606–1617 (2013)
8. Dong, X., Gabrilovich, E., Heitz, G., Horn, W., Lao, N., Murphy, K., Strohmann, T., Sun, S., Zhang, W.: Knowledge vault: a web-scale approach to probabilistic knowledge fusion. In: SIGKDD, pp. 601–610. ACM (2014)
9. Fang, J., Mitra, P., Tang, Z., Giles, C.L.: Table header detection and classification. In: AAAI (2012)
10. Lautert, L.R., Scheidt, M.M., Dorneles, C.F.: Web table taxonomy and formalization. ACM SIGMOD Record 42(3), 28–33 (2013)
11. Li, G.: A human-machine method for web table understanding. In: Wang, J., Xiong, H., Ishikawa, Y., Xu, J., Zhou, J. (eds.) WAIM 2013. LNCS, vol. 7923, pp. 179–189. Springer, Heidelberg (2013)
12. Limaye, G., Sarawagi, S., Chakrabarti, S.: Annotating and searching web tables using entities, types and relationships. Proceedings of the VLDB Endowment 3(1–2), 1338–1347 (2010)
13. Mulwad, V., Finin, T., Joshi, A.: Automatically generating government linked data from tables. In: Working Notes of AAAI Fall Symposium on Open Government Knowledge: AI Opportunities and Challenges, vol. 4 (2011)
14. Muñoz, E., Hogan, A., Mileo, A.: Using linked data to mine rdf from wikipedia's tables. In: WSDM, pp. 533–542. ACM (2014)
15. Nguyen, T.T., Nguyen, Q.V.H., Weidlich, M., Aberer, K.: Result selection and summarization for web table search. In: ICDE (2015)
16. Niu, X., Sun, X., Wang, H., Rong, S., Qi, G., Yu, Y.: Zhishi.me - weaving chinese linking open data. In: Aroyo, L., Welty, C., Alani, H., Taylor, J., Bernstein, A., Kagal, L., Noy, N., Blomqvist, E. (eds.) ISWC 2011, Part II. LNCS, vol. 7032, pp. 205–220. Springer, Heidelberg (2011)
17. Stark, M.M., Riesenfeld, R.F.: Wordnet: an electronic lexical database. In: Proceedings of 11th Eurographics Workshop on Rendering, vol. 37. MIT Press (1998)
18. Suchanek, F.M., Kasneci, G., Weikum, G.: Yago: a core of semantic knowledge. In: WWW, pp. 697–706. ACM (2007)
19. Tong, H., Faloutsos, C., Pan, J.Y.: Fast random walk with restart and its applications. In: ICDM, pp. 613–622. IEEE Computer Society (2006)

20. Venetis, P., Halevy, A., Madhavan, J., Paşca, M., Shen, W., Wu, F., Miao, G., Wu, C.: Recovering semantics of tables on the web. Proceedings of the VLDB Endowment **4**(9), 528–538 (2011)
21. Wang, J., Wang, H., Wang, Z., Zhu, K.Q.: Understanding tables on the web. In: Atzeni, P., Cheung, D., Ram, S. (eds.) ER 2012 Main Conference 2012. LNCS, vol. 7532, pp. 141–155. Springer, Heidelberg (2012)
22. Wang, Y., Hu, J.: A machine learning based approach for table detection on the web. In: WWW, pp. 242–250. ACM (2002)
23. Wang, Z., Li, J., Li, S., Li, M., Tang, J., Zhang, K., Zhang, K.: Cross-lingual knowledge validation based taxonomy derivation from heterogeneous online wikis. In: AAAI (2014)
24. Wang, Z., Li, J., Wang, Z., Li, S., Li, M., Zhang, D., Shi, Y., Liu, Y., Zhang, P., Tang, J.: Xlore: a large-scale english-chinese bilingual knowledge graph. In: International Semantic Web Conference (Posters & Demos), pp. 121–124 (2013)
25. Wu, W., Li, H., Wang, H., Zhu, K.Q.: Probase: a probabilistic taxonomy for text understanding. In: SIGMOD, pp. 481–492. ACM (2012)
26. Yakout, M., Ganjam, K., Chakrabarti, K., Chaudhuri, S.: Infogather: entity augmentation and attribute discovery by holistic matching with web tables. In: SIGMOD, pp. 97–108. ACM (2012)
27. Yin, X., Tan, W., Liu, C.: Facto: a fact lookup engine based on web tables. In: WWW, pp. 507–516. ACM (2011)
28. Zhang, M., Chakrabarti, K.: Infogather+: semantic matching and annotation of numeric and time-varying attributes in web tables. In: SIGMOD, pp. 145–156. ACM (2013)
29. Zhang, Z.: Start small, build complete: Effective and efficient semantic table interpretation using tableminer. Under Transparent Review: The Semantic Web Journal (2014)

Taxonomy Induction from Chinese Encyclopedias by Combinatorial Optimization

Weiming Lu[(✉)], Renjie Lou, Hao Dai, Zhenyu Zhang,
Shansong Yang, and Baogang Wei

College of Computer Science and Technology, Zhejiang University, Hangzhou, China
luwm@zju.edu.cn

Abstract. Taxonomy is an important component in knowledge bases, and it is an urgent, meaningful but challenging task for Chinese taxonomy construction. In this paper, we propose a taxonomy induction approach from a Chinese encyclopedia by using combinatorial optimizations. At first, **subclass-of** relations are derived by validating the relation between two categories. Then, integer programming optimizations are applied to find out **instance-of** relations from encyclopedia articles by considering the constrains among categories. The experimental results show that our approach can construct a practicable taxonomy from Chinese encyclopedias.

1 Introduction

Nowadays, knowledge bases such as DBPedia [1], YAGO [2], Freebase [3] and BabelNet [4] are becoming increasingly crucial in many fields of Artificial Intelligence. Taxonomy is the backbone of a knowledge base, which is a conceptual network including **subclass-of** relations between concepts and **instance-of** relations between concepts and instances.

DBpedia taxonomy was manually created from the most commonly used infobox templates in Wikipedia, which forms a shallow subsumption hierarchy. YAGO has a deep subsumption hierarchy, which takes the hyponymy relations of WordNet [5]. However, manual construction of taxonomies is extremely laborious, time consuming, and costly, which makes the scope and the scale of the taxonomies limited.

Recently, several researchers try to construct the taxonomies automatically. Pattern based methods are widely used to mine hypernym-hyponym relations from text corpora [6–9], but these methods often suffer from low recall and precision because of the noisy text and the low quality of patterns. Other works [10–12] mined the taxonomies from Wikipedia, which mainly learned the hypernym-hyponym relations from Wikipedia's category system. But the coverage is still limited, since they only used categories in the Wikipedia and ignored the numerous articles.

However, the knowledge bases in Chinese is quite limited currently. For example, DBpedia has become the central hub and reference point in the Web of data

J. Li et al. (Eds.): NLPCC 2015, LNAI 9362, pp. 299–312, 2015.
DOI: 10.1007/978-3-319-25207-0_25

in English, but the coverage of Chinese knowledge in DBpedia is restricted since Wikipedia has only 824,693 articles. Therefore, the construction of Chinese taxonomy is an urgent, meaningful but challenging task. Several works began to construct the Chinese taxonomies, where [13–16] were mainly based on Chinese encyclopedias (Hudong and Baidu Baike), while [17,18] identified the hypernym-hyponym relation from free text.

In this paper, we also try to address the Chinese taxonomy induction problem based on encyclopedias. In addition to mine the taxonomy from the category system of encyclopedias, we also mine the hypernym-hyponym relations from articles to form a more wide-coverage and fine-grained taxonomy. The core idea of our approach is that we leverage the taxonomy learned from category system to improve the coverage of the hypernyms for articles while guarantee the high precision.

The rest of the paper is organized as follows. Section 2 discusses the related works about taxonomy induction. Section 3 describes our approach in detail. Then, we evaluate our approach on real-world datasets in Section 4, and conclude the paper in Section 5.

2   Related Work

Previous works about taxonomy induction mainly focus on pattern-based methods and encyclopedia-based methods.

Hearst [6] is a pioneer of using manually constructed lexical patterns like "X such as Y" to harvest hyponym-hypernym relations. In order to address the limitation of small number of hand-crafted patterns, LASER [7] iteratively discovered new patterns through a sequential pattern mining framework by taking in a set of seed Hearst patterns. Probase [19] can take advantage of the existing knowledge they already learned to discover more syntactic patterns, and then to acquire more knowledge. Other patterns like word-class lattices [8,9] and dependency path [20] can also be learned to extract hypernym-hyponym relations.

Wikipedia is very useful in automatic taxonomy construction. WikiTaxonomy [10] was generated by automatically assigning is-a and not-is-a labels to the relations between categories in Wikipedia. However, it only used a set of lightweight heuristics for the assignment. Furthermore, WordNet was used in [11] to restructure the WikiTaxonomy. On the other hand, KOG [12] modeled subsumption detection task as a binary classification problem.

Machine-learning and purely distributional approaches also contributed to the task of hypernym discovery. [21] proposed an approach to distinguish hypernyms and co-hyponyms by a linear support vector machine with distributional features. [22] proposed a supervised approach with the selective distributional inclusion hypothesis for hypernymy detection. Probabilistic models were used to incorporate multiple evidences from hyponym and coordination cues for semantic taxonomy induction [23–25]. TAXIFY [26] was proposed to learn a taxonomy from a domain-specific corpus. It first uses Hearst patterns to collect initial set of *is-a* relations, and then improves the recall and precision by a clustering-based inference procedure and incorrect edges detection. [27] considered the hypernym extraction problem as a sequential classification task, which combined linguistic, definitional and graph-based information.

Recently, several works started to construct the Chinese taxonomies. Zhishi.me [13] focused on the infobox information extraction and Chinese LOD construction, but it did not construct the taxonomy between concepts. [14] built an ontology based on the category system and infobox templates in Hudong and Baidu Baike. However, they only used some simple heuristic methods to refine the category system, which would bring many wrong sub-concept relations to the ontology since sub-category relations between categories are not strict sub-concept relations. For instance, in the category system of Hudong Baike, 海洋(Ocean) contains sub-categories 海底隧道(channel tunnel) and 填海工程(reclamation), 计算机病毒(Computer Virus) has super-category 计算机安全(Computer Security), and 昆虫(insert) contains sub-category 昆虫学(insectology). They all would bring wrong sub-concept relations to the ontology. XLore [15] utilized a classification-based method to correctly semantify the wikis' category systems. [16] extracted candidate hypernyms from multiple sources, and applied a statistical ranking model to select correct hypernyms. [17,18] identified the hypernym-hyponym relation by using the word-embedding-based semantic projections between words and their hypernyms. Zheshi.schema [28] extracted semantic relations between categories from a large number of popular Chinese social Web sites. However, they all don't utilize the relatedness between hypernyms when inducing the hypernym-hyponym relations.

Multilingual information could also be used to reinforce the performance of taxonomy induction. For example, a cross-lingual knowledge validation based model was proposed in [29] to iteratively reinforce the performance of taxonomy derivation. MENTA [30] induced multilingual taxonomies from all editions of Wikipedia and WordNet.

The closely-related previous works are WiBi [31] and [16]. WiBi presented an approach to create an integrated taxonomy of Wikipedia pages and categories. In their work, the category taxonomy and page taxonomy could be enriched mutually, but they only used statistics information for taxonomy induction. [16] extracted candidate hypernyms from multiple sources, and applied a statistical ranking model to select correct hypernyms. However, they directly used the ranking models such as Support Vector Machine and Logistic Regression on features of candidate hypernyms without considering the relations among candidate hypernyms. Our approach mutually considers the linguistic features, structural features and the taxonomy have been learned into a unified learning model for taxonomy induction. In other words, our approach also leveraged the learned relations among candidate hypernyms to enhance the taxonomy induction.

3   The Proposed Approach

In this section, we formally defined our taxonomy induction problem at first, and then addressed this problem by a two step approach including **subclass-of** relation induction and **instance-of** relation induction.

3.1   Problem Formulation

In encyclopedia, each article can be considered as an instance, and then can be represented as a 5-tuple $a = \{L(a), A(a), C(a), P(a), T(a)\}$, where $L(a)$ is the title of article a, $A(a)$ is the set of linked articles of a, $C(a)$ is the catalog of a, $P(a)$ is the set of properties of the article's infobox, and $T(a)$ is the set of category tags of a. Figure 1 shows a sample article from Chinese encyclopedias.

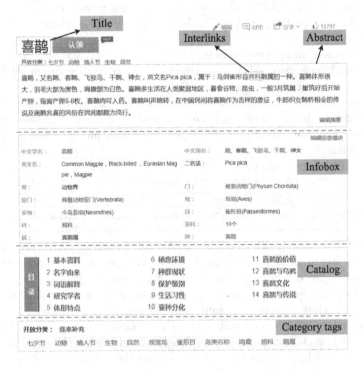

Fig. 1. Sample Article Page from Chinese Encyclopedias

Then, our taxonomy induction task for Chinese encyclopedias is illustrated in Figure 2.

From the figure, we observed that many category tags of an article are hypernyms of the article, and these tags may have **subclass-of** relation among them. Taking *magpie* as an example in Figure 2, *Songbird, Ornamental bird, Animal,*

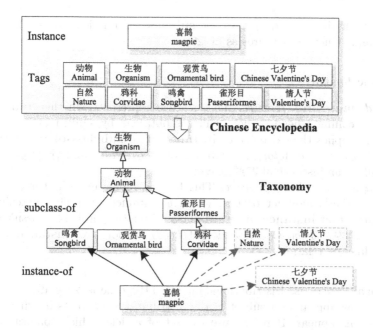

Fig. 2. The taxonomy induction task from Chinese Encyclopedia

etc. are hypernyms of *magpie*, and *Songbird* has **subclass-of** relation with *Animal*. However, there are also noisy tags, such as *Valentine's Day*, which only has related topic with *magpie*, so we should filter them out.

These category tags in the encyclopedia form a category system to provide navigational links to all articles in a hierarchical way, However, they do not form a real subsumption hierarchy, which includes *isa* and *notis* relations between categories. If the hypernym-hyponym relations between category tags can be mined, it could be very helpful for **instance-of** relation induction. For instance, if *Songbird* and *Ornamental bird* are selected as hypernyms of *magpie*, then *Animal* and *Organism* should also be the hypernyms of *magpie*, while *Valentine's Day* is not likely to be the hypernym for *magpie*, since *Animal* and *Organism* are the super classes of *Songbird* and *Ornamental bird*, and *Valentine's Day* and *Songbird* can not be the hypernyms of *magpie* simultaneously.

Therefore, our taxonomy induction problem are defined as follows. Given an encyclopedia with articles $\mathcal{A}$ and categories $\mathcal{C}$, we would like to induce the **subclass-of** relations between categories in $\mathcal{C}$, and then leverage the relatedness of categories to induce the **instance-of** relation between category tags and articles.

3.2 Subclass-of Relation Induction

The **subclass-of** relation induction problem is treated as a binary classification problem: given two category tags $c_i \in \mathcal{C}$ and $c_j \in \mathcal{C}$, we try to train a classifier to predict whether c_i and c_j have the **subclass-of** relation. we use linguistic features and structural features as in [29].

Linguistic Features.

1. **Head matching feature.** This feature represents whether two labels have a common head or not, which is obtained by suffix matching. Usually, it implies the existence of **instance-of** or **subclass-of** relations, e.g. 中国演员*(Chinese Actor)* is a subclass of 演员*(Actor)*, and 牡丹鹦鹉*(Peony parrot)* is an instance of 鹦鹉*(parrot)*.
2. **Modifier matching feature.** This feature represents whether one label is the other's modifier or not, which is obtained by prefix matching. It indicates that **instance-of** or **subclass-of** relation probably doesn't exist, e.g. 食品安全*(Food safety)* and 食品*(Food)* do not have **instance-of** and **subclass-of** relations.

Structural Features. Distributional Inclusion Hypothesis [22] states that more specific terms appear in a subset of the distributional contexts in which more general terms appear. Here, we use the set of articles which connect to the category as the distributional context of a category, and use Normalized Google Distance [32] to calculate the diversity between two categories. The structural features for category $c_i \in C$ and $c_j \in C$ are listed in Table 1. Then, a simple function $f(x) = \frac{1}{1+x}$ is used to normalize the structural features to [0,1].

After we obtain a six-dimensional features for two categories, we train a SVM classifier to predict the validity of **subclass-of** relation between categories.

Table 1. The structural features for categories

Feature	Calculation	Comment
Article Diversity	$d_a(c_i,c_j) = \frac{\max(|A(c_i)|,|A(c_j)|)-\log(|A(c_i)\cap A(c_j)|)}{\log(|A|)-\log(\min(|A(c_i)|,|A(c_j)|))}$	$A(c)$ is the set of articles in the category c, and $|A|$ is the total number of articles
Property Diversity	$d_p(c_i,c_j) = \frac{\max(|P(c_i)|,|P(c_j)|)-\log(|P(c_i)\cap P(c_j)|)}{\log(|P|)-\log(\min(|P(c_i)|,|P(c_j)|))}$	$P(c)$ is the set of properties of $A(c)$, and $|P|$ is the total number of properties.
Category Diversity	$d_t(c_i,c_j) = \frac{\max(|T(c_i)|,|T(c_j)|)-\log(|T(c_i)\cap T(c_j)|)}{\log(|T|)-\log(\min(|T(c_i)|,|T(c_j)|))}$	$T(c)$ is the set of category tags of $A(c)$, and $|T|$ is the total number of category tags.
Catalog Diversity	$d_c(c_i,c_j) = \frac{\max(|C(c_i)|,|C(c_j)|)-\log(|C(c_i)\cap C(c_j)|)}{\log(|C|)-\log(\min(|C(c_i)|,|C(c_j)|))}$	$C(c)$ is the set of words in catalogs of $A(c)$, and $|C|$ is the total number of words in catalogs.

3.3   Instance-of Relation Induction

In the **instance-of** relation induction task, we collect hypernym candidates for each instance at first, and then select the correct hypernyms with the help of the relatedness among hypernym candidates.

In addition to the category tags, we also collect the head words of the title of the article a as the hypernym candidates, denoted as $\mathcal{H}_a$. For instance, 鹦鹉 *(parrot)* is the head word of 牡丹鹦鹉 *(Peony parrot)*, so it is also collected as a hypernym candidate. Then, we train a classifier with pairs $\langle a, h \in \mathcal{H}_a \rangle$, with features similar to those for **subclass-of** relation classification by replacing c_i, c_j with a, h for the structural features. That is to say, besides the linguistic features, the structural features $d_a(a, h), d_p(a, h)$, $d_t(a, h)$ and $d_c(a, h)$ are calculated by Normalized Google Distance, and then a SVM classifier was also trained on the normalized features.

The probability output could be regarded as the coherence between the article a and the hypernym candidate h, which is denoted by $coh(a, h)$. Although $coh(a, h)$ could be used to predict the correct hypernyms, however, it ignores the relatedness among hypernym candidates. Here, we denote the results of the **subclass-of** relation induction by two sets: $Subset = \{\langle c_i, c_j \rangle | c_i \ subclass\text{-}of \ c_j\}$ and $Mutex = \{\langle c_i, c_j \rangle | c_i \ not\text{-}subclass\text{-}of \ c_j \ and \ c_j \ not\text{-}subclass\text{-}of \ c_i\}$, and the probability output of the classifier in section 3.2 is regarded as the coherence between two categories, denoted as $coh(c_i, c_j)$.

Then, we exploit the Integer Programming method to introduce the knowledge from category tags into the **instance-of** relation induction. Let $y_i \in \{0, 1\}$ be an indicator variable specifying whether $h_i \in \mathcal{H}_a$ is a hypernym of a, then the optimization problem can be formulated by a Mixed Integer Programming *(MIP)* optimization and Integer Quadratic Programming *(IQP)* optimization, where *MIP* utilizes the *Subset* and *Mutex* constrains, while *IQP* relaxes the *Mutex* constrain, since *Mutex* is too restrict. This is because we simply consider $Mutex = \mathcal{C} \times \mathcal{C} - Subset$ in our implementation.

1. **MIP** by utilizing the *Subset* and *Mutex* constraints.

$$\textbf{maximize} \quad \sum_{h_j \in \mathcal{H}_a} y_j \cdot coh(a, h_j) - \sum_{\langle h_j, h_k \rangle \in Subset} \zeta_{jk} - \sum_{\langle h_j, h_k \rangle \in Mubset} \xi_{jk}$$

$$\textbf{s.t.} \quad \forall h_j, h_k \in \mathcal{H}_a, y_j \in \{0, 1\}, \zeta_{jk}, \xi_{jk} \geq 0$$

$$y_k \geq y_j - \zeta_{jk}, \langle h_j, h_k \rangle \in Subset$$

$$y_j + y_k \leq 1 + \xi_{jk}, \langle h_j, h_k \rangle \in Mutex$$

Here, ζ_{jk} is the penalty for violation of *Subset* constraint. $y_k \geq y_j - \zeta_{jk}, \langle h_j, h_k \rangle \in Subset$ means if h_j is a subclass of h_k, and when $y_j = 1$, then y_k should be 1 for a smaller ζ_{jk}. That is to say, if h_j is selected as hypernym, then h_k who is the super class of h_j should also be selected as a hypernym.

ξ_{jk} is the penalty for violation of *Mutex* constraint. $y_j + y_k \leq 1 + \xi_{jk}, \langle h_j, h_k \rangle \in Mutex$ means if h_j and h_k are mutually exclusive to be super classes of an instance, then $y_j = y_k = 1$ would make a penalty $\xi_{jk} \geq 1$. That is to say, h_j and

h_k can not be selected as hypernyms simultaneously. If they are both selected, then the penalty $\xi_{jk} \geq 1$.

2. **IQP** by relaxing the *Mutex* constraint with the penalty for small $coh(h_k, h_j)$.

$$\textbf{maximize} \sum_{h_j \in \mathcal{H}_a} [\| \mathcal{H}_a \| \cdot y_j \cdot coh(a, h_j) +$$

$$\lambda \cdot \sum_{h_k \in \mathcal{H}_a} y_k \cdot (1 - y_j) \cdot (1 - coh(h_k, h_j))]$$

$$\textbf{s.t.} \quad \forall h_j \in \mathcal{H}_a, y_j \in \{0, 1\},$$

$$\forall \langle h_j, h_k \rangle \in Subset, y_k - y_j \geq 0$$

where $\sum_{h_k \in \mathcal{H}_a} y_k \cdot (1 - y_j) \cdot (1 - coh(h_k, h_j))$ indicates that when $coh(h_k, h_j)$ is small, then if h_k is selected as a hypernym, then h_j is encouraged not to be selected. This makes $\sum_{h_k \in \mathcal{H}_a} y_k \cdot (1 - y_j) \cdot (1 - coh(h_k, h_j)) = 1 - coh(h_k, h_j) > 0$. λ is the tradeoff parameter for precision and recall, and larger λ could get higher precision but lower recall. The experiments in Section 4.3 shows this phenomenon.

We solve the above optimization problem by using *IBM CPLEX* optimizer[1].

4   Experimental Evaluation

4.1   Experiment Settings

Data Set. *Hudong Baike*[2] and *Baidu Baike*[3] are the two largest collaboratively Chinese encyclopedias. In order to evaluate our approach, we crawled about 3.3M articles with 14963 categories from *Hudong Baike*, and 5.3M articles with 4639 categories from *Baidu Baike* in Jan 2015. We observed that *Hudong Baike* generally has more tags than *Baidu Baike*, but has more noise. For instance, 喜鹊*(magpie)* has tags of 鸦科*(Corvidae)*, 鸟类*(Bird)*, 雀形目*(Passeriformes)*, 民俗*(Folklore)* and 动物*(Animal* in *Baidu Baike*, while *Hudong Baike* has other more tags such as 鸣禽*(Songbird)*, 观赏鸟*(Ornamental Bird)*, 情人节*(Valentine's Day)* and 七夕节*(Chinese Valentine's Day)*. Obviously, *Hudong Baike* has more useful tags such as 鸣禽*(Songbird)* and 观赏鸟*(Ornamental Bird)*, but also introduces some noisy tags such as 情人节*(Valentine's Day)* and 七夕节*(Chinese Valentine's Day)* as the hypernyms of 喜鹊*(magpie)*. In statistics, *Hudong Baike* has about 6 tags with precision of 52.55% in average, while *Baidu Baike* has about 4 tags with precision of 56.21% in average.

From these data, we respectively formed two datasets $Hudong_D$ and $Baidu_D$ with human annotation as the ground truth. For each dataset, we randomly sampled 1400 pairs of *(category, sub-category)* and 1400 pairs of *(article, category tag)* for training classifiers, and then used the classifiers to validate all pairs in

---

[1] http://www-01.ibm.com/software/commerce/optimization/cplexoptimizer/

[2] http://www.baike.com

[3] http://www.baike.baidu.com

the *Hudong* and *Baidu Baike*. For each instance, we further used combinatorial optimizations to determine the **instance-of** relations. For testing, we randomly sampled 1000 articles with more than 3 tags as the testing data for **instance-of** relation induction, and sampled 500 categories with its direct sub-categories as the testing data for **subclass-of** relation induction.

Additionally,we also asked annotators to rate the category tags for each article according to the *specificity*. Larger *specificity* indicates the hypernym is more specific for the instance. For example, Table 2 shows the tags of *magpie* with the corresponding rates, which can be denoted by $\{h_1 : r_1, h_2 : r_2, ..., h_n : r_n\}$, where r_i is the rate for candidate h_i according to the specificity. In the table, we could find *Nature*, *Valentine's Day* and *Chinese Valentine's Day* are not the hypernyms of *magpie*, and *Songbird* and *Ornamental Bird* are more specific hypernyms than *Animal* and *Organism*.

Table 2. A user annotated example for **instance-of** relation

instance	category tags and its rates
magpie	Organism:1, Animal:2, Passeriformes:3, Songbird:4, Ornamental bird:4, Corvidae:4, Nature:0, Valentine's Day:0, Chinese Valentine's Day:0

Evaluation Metrics. We used precision, recall, F1 and accuracy to evaluate the **subclass-of** and **instance-of** relation induction. The pairs which are predicted to be the correct hypernym-hyponym relations by models are denoted by M_t, and the others are denoted by M_f. By comparing with the human labeled data sets H_t (the pairs with correct hypernym-hyponym relations) and H_f (the pairs with incorrect hypernym-hyponym relations), *precision, recall, F1* and *accuracy* could be defined as: $Prec = |M_t \cap H_t|/|M_t|$, $Rec = |M_t \cap H_t|/|H_t|$, $F_1 = 2 \cdot Prec \cdot Rec/(Prec + Rec)$, and $Acc = (|M_t \cap H_t| + |M_f \cap H_f|)/(|H_t| + |H_f|)$. Taking into the consideration of hypernym *specificity*, we also defined the weighted version of precision, recall, F1 and accuracy: $Prec^w = \sum_{x \in M_t \cap H_t}(x.r + 1)/\sum_{x \in M_t}(x.r + 1)$, $Rec^w = \sum_{x \in M_t \cap H_t}(x.r + 1)/\sum_{x \in H_t}(x.r + 1)$, $F_1^w = 2 \cdot Prec^w \cdot Rec^w/(Prec^w + Rec^w)$ and $Acc^w = (\sum_{x \in M_t \cap H_t}(x.r + 1) + \sum_{x \in M_f \cap H_f}(x.r + 1))/\sum_{x \in H_t \cup H_f}(x.r + 1)$.

In addition, we used Normalized Discounted Cumulative Gain(NDCG) [33] to evaluate our approach when we treated the hypernym prediction results as a ranked list.

Concretely speaking, given an instance a, its annotated list $\{h_1 : r_1, h_2 : r_2, ..., h_n : r_n | r_i \geq r_j, i < j\}$ can be used to calculate the IDCG (Ideal Discounted cumulative gain) $= r_1 + \sum_{i=2}^{n} \frac{r_i}{\log_2(i)}$. The comparative methods re-rank the list according to y_i and $coh(a, h_i)$. That is to say, the tags with $y_i = 1$ and larger $coh(a, h_i)$ are ranked in the front, like $\{h_{i_1} : r_{i_1}, h_{i_2} : r_{i_2}, ..., h_{i_n} : r_{i_n} | y_{i_1} = ... = y_{i_t} = 1, y_{i_{t+1}} = ... y_{i_n} = 0, coh(a, h_{i_m}) \geq coh(a, h_{i_n}), m < n\}$. Then, its DCG is calculated by $r_{i_1} + \sum_{j=2}^{n} \frac{r_{i_j}}{\log_2(j)}$. So $NDCG = \frac{DCG}{IDCG}$. For instance, if the tags of *magpie* in Table 2 is ranked as a list of *Songbird, Ornamental bird, Nature, Passeriformes, Corvidae, Organism, Animal, Valentine's Day, Chinese Valentine's Day*, then $IDCG = 4 + 4/\log_2(2) + 4/\log_2(3) + 3/\log_2(4) + 2/\log_2(5) +$

$1/\log_2(6) = 13.27$, $DCG = 4 + 4/\log_2(2) + 0/\log_2(3) + 3/\log_2(4) + 4/\log_2(5) + 1/\log_2(6) + 2/\log_2(7) = 12.32$, so $NDCG = 12.32/13.27 = 0.928$.

4.2 Performance Evaluation

In our experiments, we found that the precisions of **subclass-of** relation classification in *Hudong Baike* and *Baidu Baike* are 97.56% and 98.11% respectively, which are high enough to be the knowledge for **instance-of** relation induction.

Table 3 shows the average results of the **instance-of** relation induction, where we also evaluated the performance of hypermyn candidates only from article titles, and used *specificity* as weights for performance evaluation.

From the table, we observe that although *MIP* can achieve the highest precision, but the recall is very low. This is because *MIP* has very strict constraints, especially for the *Mutex* constrains. *IQP* achieves the best performance with F1 measurement. In addition, article's title includes high-quality hypernym candidates because of the high $Prec^h$, Rec^h, Acc^h and $F1^h$. Table 4 shows an specific example for the **instance-of** relations for *Scarlet Macaw*.

From the table, we find (1) *SVM* can not verify the relations between entity and high-level classes well, because of the great diversity of the high-level classes. But low-level classes have higher rate, so the NDCG of *SVM* is greater than *MIP* in Table 3. (2) *IQP* can select more low-level classes than *MIP*, since the derived **subclass-of** relations can not cover every valid relations, so the *Mutex* could be noisy, if we just take *Mutex* = {*pairs* $\notin$ *Subset*}. In this example, since the **subclass-of** relation between 鹦鹉 *(Parrot)* and 攀禽 *(Zygodactyl)* is not obtained, this makes them can not be simultaneously selected as hypernyms in *MIP*.

Table 3. The performance of **instance-of** relation induction, where $*^h$ means the measurements are only considering the hypernym candidates from the head of title, $*^w$ means it is the weighted version of the measurements

methods	Baidu Baike			Hudong Baike		
	SVM	MIP	IQP	SVM	MIP	IQP
Prec	72.85%	80.41%	68.70%	54.62%	59.62%	57.44%
Rec	68.16%	38.56%	92.84%	54.93%	24.10%	90.13%
Acc	67.79%	60.14%	72.20%	52.18%	51.38%	59.58%
F1	70.42%	52.12%	**78.97%**	54.78%	34.32%	**70.16%**
$Prec^h$	99.01%	100.00%	98.34%	83.08%	93.10%	80.22%
Rec^h	81.70%	54.47%	96.34%	50.00%	25.00%	67.60%
Acc^h	81.78%	56.59%	94.96%	53.57%	40.71%	62.14%
$F1^h$	89.53%	70.53%	**97.33%**	62.43%	39.42%	**73.37%**
$Prec^w$	87.78%	91.24%	85.46%	77.43%	79.76%	78.90%
Rec^w	68.87%	36.90%	93.86%	57.69%	23.71%	92.10%
Acc^w	68.52%	48.47%	92.90%	55.56%	38.13%	75.56%
$F1^w$	77.19%	52.55%	**89.46%**	66.12%	36.55%	**85.00%**
NDCG	92.12%	92.10%	**92.89%**	81.98%	80.19%	**82.12%**

Table 4. An example of *instance-of* relations, where the underline indicates the selected hypernyms for the instance

	instance-of relations for 五彩金刚鹦鹉 (Scarlet Macaw)
SVM	鹦形目(Psittaciformes), <u>鸟类(Birds)</u>, <u>动物(Animal)</u>, <u>攀禽</u>(Zygodactyl), <u>自然</u>(Nature), <u>鹦鹉</u>(Parrot), 生物(Organism)
MIP	鹦形目(Psittaciformes), <u>鸟类(Birds)</u>, <u>动物(Animal)</u>, <u>攀禽</u>(Zygodactyl), <u>自然</u>(Nature), <u>鹦鹉</u>(Parrot), 生物(Organism)
IQP	鹦形目(Psittaciformes), <u>鸟类(Birds)</u>, <u>动物(Animal)</u>, <u>攀禽</u>(Zygodactyl), <u>自然</u>(Nature), <u>鹦鹉</u>(Parrot), 生物(Organism)

4.3 Parameter Setting

Parameter λ in *IQP* influences the performance of **instance-of** relation induction. According to the objective function of *IQP*, larger λ could result in larger precision but lower recall. We solved the *IQP* with different λ, and the results are shown in Figure 3.

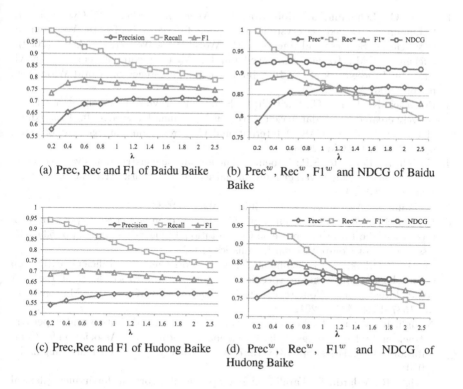

(a) Prec, Rec and F1 of Baidu Baike

(b) Prec^w, Rec^w, F1^w and NDCG of Baidu Baike

(c) Prec,Rec and F1 of Hudong Baike

(d) Prec^w, Rec^w, F1^w and NDCG of Hudong Baike

Fig. 3. Performance with different parameter λ for Baidu Baike and Hudong Baike

From the figure, we can see that recall decreases and precision increases when λ increases. Therefore, we selected $\lambda = 0.6$ for both *Hudong Baike* and *Baidu Baike* as it can reach the best $F1$.

5   Conclusion and Future work

In this paper, we propose a taxonomy induction approach from a Chinese encyclopedia by combinatorial optimizations. In future, we plan to take the semantic similarity among articles into account, since similar articles are most likely to take the similar categories as their hypernyms.

Acknowledgments. This work is supported by the Fundamental Research Funds for the Central Universities(2014QNA5008), Chinese Knowledge Center of Engineering Science and Technology(CKCEST), and the National Natural Science Foundation of China (No.61103099).

References

1. Bizer, C., Lehmann, J., Kobilarov, G., Auer, S., Becker, C., Cyganiak, R., Hellmann, S.: Dbpedia-a crystallization point for the web of data. Web Semantics: Science, Services and Agents on the World Wide Web **7**(3), 154–165 (2009)
2. Suchanek, F.M., Kasneci, G., Weikum, G.: Yago: a core of semantic knowledge. In: Proceedings of the 16th International Conference on World Wide Web, pp. 697–706. ACM (2007)
3. Bollacker, K., Evans, C., Paritosh, P., Sturge, T., Taylor, J.: Freebase: a collaboratively created graph database for structuring human knowledge. In: Proceedings of the 2008 ACM SIGMOD International Conference on Management of Data, pp. 1247–1250. ACM (2008)
4. Navigli, R., Ponzetto, S.P.: Babelnet: The automatic construction, evaluation and application of a wide-coverage multilingual semantic network. Artificial Intelligence **193**, 217–250 (2012)
5. Fellbaum, C.: Wordnet: An electronic database (1998)
6. Hearst, M.A.: Automatic acquisition of hyponyms from large text corpora. In: Proceedings of the 14th Conference on Computational Linguistics, vol. 2, pp. 539–545. Association for Computational Linguistics (1992)
7. Li, T., Chubak, P., Lakshmanan, L.V.S., Pottinger, R.: Efficient extraction of ontologies from domain specific text corpora. In: Proceedings of the 21st ACM International Conference on Information and Knowledge Management, pp. 1537–1541. ACM (2012)
8. Navigli, R., Velardi, P.: Learning word-class lattices for definition and hypernym extraction. In: Proceedings of the 48th Annual Meeting of the Association for Computational Linguistics, pp. 1318–1327. Association for Computational Linguistics (2010)
9. Navigli, R., Velardi, P., Faralli, S.: A graph-based algorithm for inducing lexical taxonomies from scratch. In: IJCAI, pp. 1872–1877 (2011)
10. Ponzetto, S.P., Strube, M.: Wikitaxonomy: a large scale knowledge resource. In: ECAI (2008)

11. Ponzetto, S.P., Navigli, R.: Large-scale taxonomy mapping for restructuring and integrating wikipedia. In: IJCAI (2009)
12. Wu, F., Weld, D.S.: Automatically refining the wikipedia infobox ontology. In: WWW. ACM (2008)
13. Niu, X., Sun, X., Wang, H., Rong, S., Qi, G., Yu, Y.: Zhishi.me - weaving chinese linking open data. In: Aroyo, L., Welty, C., Alani, H., Taylor, J., Bernstein, A., Kagal, L., Noy, N., Blomqvist, E. (eds.) ISWC 2011, Part II. LNCS, vol. 7032, pp. 205–220. Springer, Heidelberg (2011)
14. Wang, Z., Wang, Z., Li, J., Pan, J.Z.: Knowledge extraction from chinese wiki encyclopedias. Journal of Zhejiang University SCIENCE C 13(4), 268–280 (2012)
15. Wang, Z., Li, J., Wang, Z., Li, S., Li, M., Zhang, D., Shi, Y., Liu, Y., Zhang, P., Tang, J.: Xlore: a large-scale english-chinese bilingual knowledge graph. In: International Semantic Web Conference (Posters & Demos), vol. 1035, pp. 121–124 (2013)
16. Fu, R., Qin, B., Liu, T.: Exploiting multiple sources for open-domain hypernym discovery. In: EMNLP, pp. 1224–1234 (2013)
17. Fu, R., Guo, J., Qin, B., Che, W., Wang, H., Liu, T.: Learning semantic hierarchies via word embeddings. In: Proceedings of the 52th Annual Meeting of the Association for Computational Linguistics: Long Papers, vol. 1 (2014)
18. Ruiji, F., Guo, J., Qin, B., Che, W., Wang, H., Liu, T.: Learning semantic hierarchies: A continuous vector space approach. IEEE/ACM Transactions on Audio, Speech, and Language Processing 23(3), 461–471 (2015)
19. Wu, W., Li, H., Wang, H., Zhu, K.Q.: Probase: a probabilistic taxonomy for text understanding. In: SIGMOD. ACM (2012)
20. Snow, R., Jurafsky, D., Ng, A.Y.: Learning syntactic patterns for automatic hypernym discovery. Advances in Neural Information Processing Systems 17 (2004)
21. Weeds, J., Clarke, D., Reffin, J., Weir, D., Keller, B.: Learning to distinguish hypernyms and co-hyponyms. In: Proceedings of COLING, pp. 2249–2259 (2014)
22. Roller, S., Erk, K., Boleda, G.: Inclusive yet selective: supervised distributional hypernymy detection. In: COLING (2014)
23. Snow, R., Jurafsky, D., Ng, A.Y.: Semantic taxonomy induction from heterogenous evidence. In: Proceedings of the 21st International Conference on Computational Linguistics and the 44th Annual Meeting of the Association for Computational Linguistics, pp. 801–808. Association for Computational Linguistics (2006)
24. Zhang, F., Shi, S., Liu, J., Sun, S., Lin, C.-Y.: Nonlinear evidence fusion and propagation for hyponymy relation mining. In: Proceedings of the 49th Annual Meeting of the Association for Computational Linguistics: Human Language Technologies, vol. 1, pp. 1159–1168. Association for Computational Linguistics (2011)
25. Bansal, M., Burkett, D., de Melo, G., Klein, D.: Structured learning for taxonomy induction with belief propagation, pp. 1041–1051 (2014)
26. Alfarone, D., Davis, J.: Unsupervised learning of an is-a taxonomy from a limited domain-specific corpus. CW Reports (2014)
27. Espinosa-Anke, L., Ronzano, F., Saggion, H.: Hypernym extraction: combining machine-learning and dependency grammar. In: Gelbukh, A. (ed.) CICLing 2015. LNCS, vol. 9041, pp. 372–383. Springer, Heidelberg (2015)
28. Wang, H., Wu, T., Qi, G., Ruan, T.: On publishing chinese linked open schema. In: Mika, P., Tudorache, T., Bernstein, A., Welty, C., Knoblock, C., Vrandečić, D., Groth, P., Noy, N., Janowicz, K., Goble, C. (eds.) ISWC 2014, Part I. LNCS, vol. 8796, pp. 293–308. Springer, Heidelberg (2014)

29. Wang, Z., Li, J., Li, S., Li, M., Tang, J., Zhang, K., Zhang, K.: Cross-lingual knowledge validation based taxonomy derivation from heterogeneous online wikis. In: Twenty-Eighth AAAI Conference on Artificial Intelligence (2014)
30. de Melo, G., Weikum, G.: Menta: Inducing multilingual taxonomies from wikipedia. In: Proceedings of the 19th ACM International Conference on Information and Knowledge Management, pp. 1099–1108. ACM (2010)
31. Flati, T., Vannella, D., Pasini, T., Navigli, R.: Two is bigger (and better) than one: the wikipedia bitaxonomy project. In: ACL, pp. 945–955 (2014)
32. Cilibrasi, R.L., Vitanyi, P.M.B.: The google similarity distance. IEEE Transactions on Knowledge and Data Engineering 19(3), 370–383 (2007)
33. Järvelin, K., Kekäläinen, J.: Ir evaluation methods for retrieving highly relevant documents. In: Proceedings of the 23rd Annual International ACM SIGIR Conference on Research and Development in Information Retrieval, pp. 41–48. ACM (2000)

Recognition of Person Relation Indicated by Predicates

Zhongping Liang$^{(\boxtimes)}$, Caixia Yuan, Bing Leng, and Xiaojie Wang

School of Computer Science, Beijing University of Posts and Telecommunications,
Beijing, China
dreamfish_liang@163.com, {yuancx,xjwang}@bupt.edu.cn,
lb19900314@gmail.com

Abstract. This paper focuses on recognizing person relations indicated by predicates from large scale of free texts. In order to determine whether a sentence contains a potential relation between persons, we cast this problem to a classification task. Dynamic Convolution Neural Network (DCNN) is improved for this task. It uses frame convolution for making uses of more features efficiently. Experimental results on Chinese person relation recognition show that the proposed model is superior when compared to the original DCNN and several strong baseline models. We also explore employing large scale unlabeled data to achieve further improvements.

Keywords: Person relation indicated by predicate · Dynamic Convolution Neural Network (DCNN) · Frame convolution

1 Introduction

The goal of Open Relation Extraction (ORE) is automatically extracting relation triples from large scale of free texts without consulting a prespecified relation vocabulary (Banko et al. 2007) [1]. For instance, the triple <Tom, met, Jim> in the sentence "Tom met Jim" implies an open relation.

Relation between persons is an important subclass of entity relation (ER). One of the most frequent ways to express person relations in Chinese is indicated by predicate phrases. Predicate phrases have been used to indicate a variety of relation types. On the one hand, they can indicate static relationships between persons. For example, the predicate phrase "是...哥哥 (is the brother of)" in the sentence "Tom是Jim的哥哥(Tom is the brother of Jim)" expresses a kinship between Tom and Jim. On the other hand, predicate phrases in sentences may also describe dynamic relationships between persons. For example, the predicate phrase "见到(meet)" in the sentence "Tom见到Jim (Tom met Jim)" expresses that Tom meets with Jim at some time. However, not all predicate phrases do indicate "true" relations. For example, in the sentence "听说这本书是Jim的 (Tom heard that this book belonged to Jim)", the predicate "听说 (hear about)" does not indicate any relation between Tom and Jim.

© Springer International Publishing Switzerland 2015
J. Li et al. (Eds.): NLPCC 2015, LNAI 9362, pp. 313–324, 2015.
DOI: 10.1007/978-3-319-25207-0_26

Extracting person relations indicated by predicates is a restricted type of ORE. Since entity type is limited to person, and linguistic form of relation is limited to predicate phrase, it is therefore an easy job to extract a relation triple from a sentence with the form of <person1, verbal phrases, person2> if we confirm this sentence contains a person relation indicated by predicate. But judging whether a sentence contains a relation of this form is still very difficult because the types of relations are not limited. The focus here is therefore to recognize sentences containing this type of relation triples.

Several approaches have been explored to ORE of which the most relevant work to ours is ReVerb proposed by Fader et al. [5]. In their work, the longest verb phrases in sentences are firstly extracted by matching the POS sequence pattern of some predefined rules. Secondly, the verb phrases are filtered by some predefined constraints. Then entity pairs are extracted from both sides of the verb phrase to build relation triples of the form <entity1, verb phrase, entity2>. Finally, these triples are feed to logistic regression classifier with several simple manual features to judge whether a triple indicates a relation. The area under precision and recall curve of ReVerb is 0.47. However, unlike English, verb phrases in Chinese sentences are not always located between the two entities. Wang, et al. [6] investigated 671 predicate phrase based relation sentences and find that in 21 sentences (3.13%) relation mentions locate on the left side of the two entities, in 352 sentences (52.45%) relation mentions locate between the two entities and in 298 sentences (44.41%) relation mentions locate on the right side of the two entities. That is to say, mentions of predicate phrase based relations in Chinese sentences may locate on arbitrary side of the two entities.

This paper introduces a novel approach to recognize if there is a predicate phrase based relation in a Chinese sentence, no matter where the relation mention locates. The main inspiration of our work comes from recent advances in using Dynamic Convolution Neural Network (DCNN) to model sentences (Kalchbrenner et al. 2014) [8]. DCNN does not rely on a parse tree and handles input sentences of variable length. It induces a feature graph over the sentence capturing short and long range relevant structure features. It has been proved that DCNN outperforms other traditional approaches in both sentiment prediction and question classification. We improve DCNN in several ways and transplanted it into our task. Experimental results show that our model outperforms the original DCNN and some other baseline approaches.

2   Related Work

Previous close relation extraction aims at assigning each entity pair in a sentence a relation type. Zelenko et al. (2003) [9] proposed to extract relations by computing kernel functions between parse trees. Zhou et al. (2005) [10] explored various features in relation extraction, these features include word, entity type, mention level, overlap, phrase chunking, dependency tree, parse tree and gazetteers. Mintz et al. (2009) [11] proposed a distant supervised learning method. They first employed freebase to weakly annotate a large scale of training set, then trained a logistic regression classifier with lexical and syntactical features. The application of close relation extraction is limited because the relation types are predefined and hard to be extended.

For other open relation extraction methods, lexical pattern approaches such as Snowball (Agichtein and Gravano, 2000) [3] and KnowItAll (Etzioni et al. 2003) [4] start with a small number of seed instances or patterns, then expand instance set and pattern set with bootstrap learning. The initial instances are first used with a large corpus to extract a new set of patterns; these new patterns are then used to extract a new set of instances; these new instances are then used to extract more patterns; the two steps are alternated in an iterative fashion until no new patterns and instances are generated or the iteration time reaches a given threshold. The assumption under lexical pattern approaches is that relation mentions have common patterns in lexical level. However, arbitrary verbs in Chinese sentences are potential mentions of predicate phrase based relations, they don't share common patterns in lexical level. Thus bootstrap learning can't extract relation indicated by predicates. Parse Tree approaches such as WOE (Fei et al.) [7] get high performance using the shortest dependency path in a parse tree. Systems using parse tree features are not practical. On the one hand, parsers are usually trained with domain-limited Treebanks, they make more parsing errors when confronted with the diversity; on the other hand, time complexity of parsers is explosive, which keeps these systems from being applied to large scale of web text.

Recently, some explorations that using deep neural network to serve Natural Language Processing (NLP) tasks began to emerge. The Deep Neural Networks with Multi-task Learning (Collobert and Weston 2008, Collobert et al. 2011) [13,14] was proposed to solve the sequence labelling problems. Their network first projects multiple features of each word into their respective vectors. Then these projected feature vectors are fed to a convolution layer and a max over time layer, followed by a softmax output layer. Their network using large scale of unlabeled data outperforms other traditional approaches in Part-of-Speech Tagging, Chunking, Named Entity Recognition and Semantic Role Labeling. The Dynamic Convolutional Neural Network (DCNN) (Kalchbrenner et al. 2014) [8] firstly projects words to their vectors, secondly alternates wide convolution layers and dynamic k-max pooling layers to get a rich set of inner representations. Finally these representations are feed to a fold layer and a full connected layer, followed by output softmax layer. DCNN outperforms other traditional approaches in both sentiment prediction and question classification.

3 The Model

This section first describes the structure of our revised DCNN model[1], then lists the differences between our model and the original DCNN, and finally explains some details in training.

---

[1] Code available at https://github.com/dreamfish-liang/FDCNN/

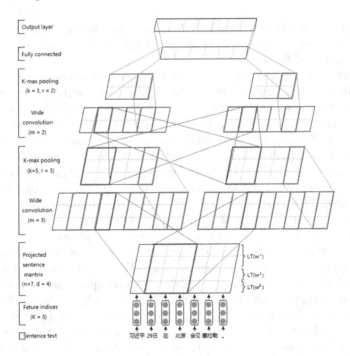

Fig. 1. The overall structure of improved DCNN

Fig.1 shows the overall network structure of our revised model for recognizing person relation indicated by predicates. It includes several forward layers. The network receives a sentence as input in the first layer, and the input sentence is projected to a matrix by lookup table operation in the second layer. This projected matrix, which includes multiple features of each word in the input sentence, is then fed to two sets of convolution and pooling layers (4 layers) to generate a rich set of inner representations. A full connection operation over the last pooling results is used to produce a fixed length sentence embedding for the variable length sentence in final hidden layer. And the network finally feeds the sentence embedding into a softmax classifier.

In Fig.1, the length of the exampled input sentence is 7. The number of different features extracted from each word is 3, and the dimension of each feature vector is respectively 2, 1 and 1 (totally 4). There are 2 convolution layers and 2 k-max pooling layers. Convolution is marked with red line, pooling is marked with green line and fully connection is marked with blue line. The widths of the convolution window at the two layers are respectively 3 and 2. The convolution output matrices have row numbers of 3 and 2. The dynamic k-max pooling layers have values k of 5 and 3.

The main differences between the above model and the original DCNN are:

1. The original DCNN only employs word as feature, while we import multiple features besides word.

2. Instead of one-dimensional convolution adopted by the original DCNN, a special two-dimensional convolution called frame convolution is used in our network.

3. The fold layer in the original DCNN is removed in our model, and is replaced by a smooth reduction produced by frame convolution.

Details of the improvements and some other descriptions are given in following parts.

3.1 Transforming Feature Indices into Vectors

The original DCNN employs only word feature, which limits its ability to capture diverse information from the input sentence. In our improved model, each word in the sentence can be represented with a couple of exact K different features. The idea that employing multiple features for each word comes from Collobert et al. (Collobert and Weston 2008, Collobert et al. 2011) [13,14]. These features can be word itself, stem, POS, indicator indicating if a word is in a given dictionary, or something else reasonable. The first layer of the network, called lookup table layer, projects the overall K different features of each word to their respective feature vector.

Formally, for the i-th feature w^i of each word w, there is a dictionary D^i and a Feature Embedding matrix $W^i \in \mathbb{R}^{|D^i| \times d^i}$ corresponding with it, where $|D^i|$ indicates the size of dictionary D^i, d^i is the dimension of the i-th feature's vector. We use w^i_j to indicate the i-th feature of the j-th word in a sentence. A lookup table operation which transforms a feature to a represented vector is described as follows:

$$LT(w^i_j) = W^{i^T} \cdot \phi(w^i_j) \tag{1}$$

where $\phi(w^i_j) \in \mathbb{R}^{d^i}$ is the binary one-hot representation of w^i_j indicating absence or presence of this feature. After lookup table layer, we obtain a frame vector representation x_j of word w_j by concatenating all result feature vectors one-by-one:

$$LT(w_j) = \left(LT(w^1_j)^T, \ldots, LT(w^K_j)^T \right)^T \tag{2}$$

where K indicates the number of different features.

Through a lookup table layer, an input sentence S is transform into a sentence matrix x:

$$x = (x_1, x_2, \ldots, x_n) \tag{3}$$

where $x_t = LT(w_t)$, $\forall t$ $x_t \in \mathbb{R}^d$, $d = \sum_{i=1}^{K} d^i$, and n is the length of the input sentence. The row size of matrix x which depends on the number of different features and the dimension of each feature vector is constant, while the column size of matrix x which depends on the length of input sentence is mutable.

3.2   Frame Convolution

Instead of one-dimensional convolution which applies convolution operation row by row in the original DCNN, we introduce a frame convolution here. Frame convolution, which is a particular two-dimensional convolution, is an operation between a weight matrix and a sequence of input column vectors, resulting in another sequence of output vectors. The idea of frame convolution mainly comes from Time-Delay Neural Network (TDNN, Waibel et al. 1989) [15]. In frame convolution, the value of output sequence vectors at time t denoted by $o(t)$ is:

$$o(t) = M\left(x_{t-\frac{m-1}{2}}{}^T, \ldots, x_{t+\frac{m-1}{2}}{}^T\right)^T + b \tag{4}$$

where $M \in \mathbb{R}^{r \times (m \times d)}$ is the weight matrix, $b \in \mathbb{R}^r$ is the bias term, m is the size of convolution window, and r which is a parameter can be set by user is the dimension of the output vector $o(t)$.

Convolution can be divided into two types, wide convolution and narrow convolution. Convolution between weight matrix of column size m and input sequence of length n results an output sequence of length $m + n - 1$ in wide convolution and $m - n + 1$ in narrow convolution. Kalchbrenner et al. [8] have explained that applying the weights M in a wide convolution has some advantages over applying them in a narrow one. Therefore we adopt a wide one in our experiments.

3.3   Dynamic K-max Pooling

To settle the problem of sentence of various length which makes the sentence inadequate to be feed to a traditional neural network, a pooling layer is necessary. A max over time operation takes out the max value of each row (Collobert and Weston 2008, Collobert and Weston 2011) [13,14], while a k-max pooling operation takes out the top k max values of each row maintaining their original order (Kalchbrenner et al. 2014) [8]. In frame convolution, each column of output matrix is acquired from combination of m continuous columns in the input layer. Therefore, each column of the output matrix can also be regard as a represented m-gram. It's plausible that treating each column of the output matrix as an atomic element when applying k-max pooling operation is better than breaking it down. This character performs especially obvious when words employ multiple features. Following this intuition, in our architecture, we apply k-max pooling column-wise. That is we select k columns with largest 2-norm value and keep their order. Also, the pooling parameter k is not fixed, but is dynamically selected in order to allow for a smooth extraction. A dynamic k-max pooling operation is a k-max pooling operation where k is a function of the length of the sentence and the depth of the network. We simply let k be a linear function as follows:

$$k_l = \max\left(k_{top}, n + \lfloor \frac{(k_{top} - n) \times l}{L} \rfloor\right) \tag{5}$$

where n is the length of input sentence, l is the number of the current convolutional layer to which the pooling is applied, L is the total number of convolutional layers in the network, and k_{top} is the fixed pooling parameter for the top-most convolutional layer.

3.4   One-Dimensional Convolution vs Frame Convolution

Fig. 2 illustrates the differences made by replacing one-dimensional convolution with frame convolution. One-dimensional convolution is shown in the left side, and frame convolution is shown in the right side. Both one-dimensional convolution and frame convolution employ the same two features. The first two rows in both input matrices are projected from the first feature, and the third row is projected from the second feature. Convolution is marked with red line, pooling is marked with green line.

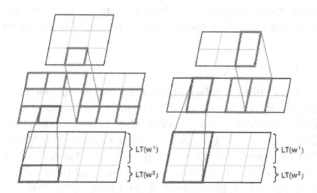

Fig. 2. The comparison of one-dimensional convolution and frame convolution

The one-dimensional convolution is an operation between a vector of weights and a vector of inputs, and is calculated between weight matrix and input matrix row by row. Convolutions of different features are computed independently. Contrast to one-dimensional convolution, in frame convolution every column of input matrix is regarded as an entirety so that units of two features is computed together during every convolution operation. In this way, different features is combined by the operation of convolution, which provides the possibility of modeling correlations and constraints between different features.

After convolution, k-max pooling operation is again acted on the input matrix row by row in one-dimensional convolution structure. In this way, the order of different values may be disrupted when several rows are taken into account although order in each row can be maintained easily. Consequently, values in the same column of result may indicate features of different words, which may disturb structure information of features. However, reviewing the frame convolution, a column is regard as a whole so that not only the relative order between rows but also that in the same row can be kept at the same time. It will help when modeling sentence.

3.5 Training

In our implement, hyperbolic tangent *tanh* is selected as non-linear function between every convolution layer and pooling layer. The network is trained by minimizing a loss function over the training data using stochastic gradient descent by backpropagation. And the objective function includes an $L2$ regularization term over the parameters.

4 Experiments

In this section, we first describe the data set on which we conduct experiments. Then we describe the three experiments respectively. We compare our model with some other models in first experiment, investigate the efficient of making use of multiple features of our model in second experiment, and explore employing large scale unlabeled data to achieve further improvement in third experiment.

4.1 Data Set

Since there was no public available corpus on extracting person relation indicated by predicate structure, we built a corpus for it. A large number of news documents from SohuNews were crawled. Raw texts are then segmented into sentences. Each sentence was segmented by a Chinese-Word-Segment (CWS) tool and a Part-Of-Speech (POS) tagger provided by FudanNLP (Xipeng et al. 2013)[16]. We randomly selected some sentences, and manually labeled if the sentence contains exactly a predicate indicated relation between two persons.

We finally get totally 10000 samples, where 3107 positive samples (including a predicate indicated relation between two persons) and 6893 negative samples. In the following experiments, we separate all the 10000 samples into two parts, 9000 for training and 1000 for testing. The ratios of positive samples and negative samples in both parts are same.

4.2 Experiment-1

In our first experiment, we propose to make a comparison between traditional classifiers and DCNNs to prove that DCNNs are effective enough to model sentences. For traditional classifier, we choose Maximum Entropy (ME), Naïve Bayes (NB), and Support Vector Machine (SVM) as control experiments. For DCNNs, we implement both one-dimensional convolution schema and frame-convolution schema. The most primary feature of a sentence is word itself, so we simply employ bag-of-word (BOW) feature for all above classifiers. Note for DCNNs, we replace all person names by a special identifier <PER> in all sentences of training and testing set.

At the meantime, we compare DCNN with traditional relation classification approach. Zhou et al. (2005) [10] employed word, entity type, mention level,

overlap, phrase chunking, dependency tree, parse tree and gazetteers for relation classification task. Entities in our problem are known and limited to persons, so features such as entity type, mention level and overlap are helpless. We
here employ word and parse tree feature. The sentences are parsed with Stanford Parser (Klein et al. 2003) [18]. And we use SVMLight with Tree Kernels
(Joachims et al. 1999) [19] (Moschitti et al. 2004, 2006) [20,21].

For both 1D-DCNN and F-DCNN, the dimension of word vector is set to 40,
the number of convolution layers is 2, the number of convolution filters each layer
is respectively 5 and 7, the size of convolution windows each layer is respectively
5 and 5, the number of hidden unit of fully connected layer is 10. For F-DCNN,
the row number of convolution result matrices is respectively 20 and 10. The
result is shown in Table 1.

Table 1. Result of Experiment-1

Approach	Pre. (%)	Rec. (%)	F1 (%)
NB	64.19	49.84	56.11
ME	57.46	57.82	57.64
SVM	73.33	38.65	50.62
SVMLight	75.86	49.20	59.68
1D-DCNN	62.50	55.91	59.02
F-DCNN	64.53	58.14	61.17

In Table 1, approaches are list in the first column, and precision, recall and
F1-score are described in other columns respectively.

We see that the DCNNs significantly outperforms NB, ME and SVM in F1-
score with the same bag-of-word feature. This is because that besides words,
DCNN can get information about combination of words and grammars and
their orders. Further on, F-DCNN outperforms traditional Tree Kernel approach
(SVMLight). This may be caused by two reasons. Firstly, parsers are usually
trained with domain-limited Treebanks, they make more parsing errors when
confronted with web texts, and parse errors directly influence the classification
results; secondly, when a tree is large, it's hard for a kernel function to capture
the most key structure from the parse result. In addition, parsing is extremely
time consuming, and improper to deal with large scale of web text.

F-DCNN also performs better than 1D-DCNN. Since every positive sample
in our data set expresses a person relation indicated by predicates. Therefore,
the common characteristic of all positive samples is that there is a dependency
path connecting the two persons and the predicate phrase. A higher F1-score
implies that frame convolution may have a greater ability to capture the structure information of sentences than a one-dimensional one.

4.3 Experiment-2

In this experiment, we explore more useful features to improve performance of our model. At the same time, we also intent to compare frame convolution schema network with one-dimensional convolution schema network on the performance of making use of multiple features. Since predicate in a sentence is always a verb phrase, POS feature might be a type of useful feature. The two persons are participants of relations, so making the two person names high light and labeling each word with relative offset from the two persons may also helpful. So we try these features in this experiment. The dimension of vectors for word, POS, person name high light and word's offsets from the first and second person names are respectively 40, 10, 5, 5 and 5 (totally 65). Other parameters are the same with what in the first experiment. The experiment results are shown in Table 2.

Table 2. Result of Experiment-2

Approach	Pre. (%)	Rec. (%)	F1 (%)
1D-DCNN-W	62.50	55.91	59.02
1D-DCNN-WP	64.88	61.98	63.39
1D-DCNN-WPP	59.33	62.93	61.08
1D-DCNN-WPPO	67.38	60.06	63.51
F-DCNN-W	64.53	58.14	61.17
F-DCNN-WP	68.51	59.10	63.46
F-DCNN-WPP	65.56	63.25	64.39
F-DCNN-WPPO	72.62	63.57	67.80

In Table 2, W denotes word; WP denotes word together with POS; WPP denotes word POS and person name high light (The value of this feature is 'yes' if a word is a person name, and 'no' otherwise.); WPPO denotes word, POS, person name high light and word's offsets from each of the two persons.

F-DCNN performs better than 1D-DCNN at any case. F-DCNN significantly out-performs 1D-DCNN when employing multiple features. It is due to two main reasons. One is that one-dimensional convolution doesn't blend different features of the same word together since the convolution operation between input sentence matrix and weight matrix is calculated row by row. Another reason is that pooling in 1D-DCNN is calculated row by row, and this breaks the relative positions of different features in high layer of the network. Missing order information amount different features may result in a higher error rate.

4.4 Experiment-3

The scale of our data set is totally 10000 samples, with a word vocabulary of size around 30000. We calculate out that more than 20000 (78.62%) words appear no more than 3 times and more than half (57.19%) appear only once. This may lead to a drift of overfitting. Therefore, in our last experiment, we propose to explore employing large scale unlabeled data to improved performance. Unlabeled data again comes from SohuNews Web documents with a total size of 5GB and is handled with FudanNLP by CWS and POS. We replace all person names with a particular identifier <PER>, and train a word embedding of dimension 40 using Google open source word2vec (Mikolov et al. 2013) [17]. The parameters of DCNN are the same as what is in Experiment-2. Besides, the word vectors are initialized with what we obtain by word2vec, and fixed while training.

Table 3. Result of Experiment-3

Approach	Pre. (%)	Rec. (%)	F1 (%)
F-DCNN-WPPO	72.62	63.57	67.80
F-DCNN-WPPO+UN	73.47	65.49	69.25

In Table 3, F-DCNN-WPPO is the same as what is in table 2, and UN denotes using unlabeled data.

We see that, large scale of unlabeled data improves the performance in both precision and recall. After trained with large scale of unlabeled data by word2vec, words which are similar in syntax and semantic are close in their embedding space. This information is indeed helpful for recognizing person relation indicated by predicate.

5 Conclusion

This paper investigated an improved DCNN model to detect whether a sentence contains a relation. The present method extended 1D convolution to 2D. Improved model achieves better performance than original model. We also find that large number of unlabeled data is helpful. Although the model employs multiple features of each word, it fails in employing global features of a sentence. It might be a future work.

Acknowledgments. This work was partially supported by National Natural Science Foundation of China (No.61273365, No.61202248), discipline building plan in 111 base (No.B08004) and Engineering Research Center of Information Networks, Ministry of Education.

References

1. Banko, M., Cafarella, M.J., Soderland, S., Broadhead, M., Etzioni, O.: Open Information Extraction from the Web. In: IJCAI, pp. 2670–2676 (2007)
2. Weston, J., Bordes, A., et al.: Connecting language and knowledge bases with embedding models for relation extraction. arXiv preprint arXiv:1307.7973 (2013)

3. Agichtein, E., Gravano, L.: Snowball: Extracting relations from large plain-text collections. In: Proceedings of the Fifth ACM Conference on Digital Libraries. ACM (2000)
4. Etzioni, O., Cafarella, M., Downey, D., et al.: Unsupervised named-entity extraction from the web: An experimental study. Artificial Intelligence **165**(1), 91–134 (2005)
5. Fader, A., Soderland, S., Etzioni, O.: Identifying relations for open information extraction. In: Proceedings of the Conference on Empirical Methods in Natural Language Processing, pp. 1535–1545. Association for Computational Linguistics (2011)
6. Wang, M., Li, L., Huang, F.: Semi-supervised Chinese Open Entity Relation Extraction. In: Proceedings of 2014 IEEE 3rd International Conference on Cloud Computing and Intelligence Systems, TP391.1 (2014)
7. Wu, F., Weld, D.S.: Open information extraction using Wikipedia. In: ACL 2010 Proceedings of the 48th Annual Meeting of the Association for Computational Linguistics, pp. 118–127 (2010)
8. Kalchbrenner, N., Grefenstette, E., Blunsom, P.: A convolutional neural network for modelling sentences. arXiv preprint arXiv:1404.2188 (2014)
9. Zelenko, D., Aone, C.: Richardella A: Kernel methods for relation extraction. The Journal of Machine Learning Research **3**, 1083–1106 (2003)
10. GuoDong, Z., Jian, S., Jie, Z., et al.: Exploring various knowledge in relation extraction. In: Proceedings of the 43rd Annual Meeting on Association for Computational Linguistics, pp. 427–434. Association for Computational Linguistics (2005)
11. Mintz, M., Bills, S., et al.: Distant supervision for relation extraction without labeled data. In: Proceedings of the Joint Conference of the 47th Annual Meeting of the ACL and the 4th International Joint Conference on Natural Language Processing of the AFNLP, pp. 1003–1011. Association for Computational Linguistics (2009)
12. Bengio, Y., Ducharme, R., Vincent, P., et al.: A neural probabilistic language model. The Journal of Machine Learning Research **3**, 1137–1155 (2003)
13. Collobert, R., Weston, J.: A unified architecture for natural language processing: Deep neural networks with multitask learning. In: Proceedings of the 25th International Conference on Machine Learning, pp. 160–167. ACM (2008)
14. Collobert, R., Weston, J., Bottou, L., et al.: Natural language processing (almost) from scratch. The Journal of Machine Learning Research **12**, 2493–2537 (2011)
15. Waibel, A., Hanazawa, T., Hinton, G., Shikano, K., Lang, K.: Phoneme recognition using time-delay neural networks. IEEE Transactions on Acoustics, Speech, and Signal Processing **37**, 328–339 (1989)
16. Qiu, X., Zhang, Q., Huang, X.: FudanNLP: A Toolkit for Chinese Natural Language Processing. In: Proceedings of Annual Meeting of the Association for Computational Linguistics (ACL) (2013)
17. Mikolov T., Chen K., Corrado G., et al.: Efficient estimation of word representations in vector space. arXiv preprint arXiv:1301.3781 (2013)
18. Klein, D., Manning, C.D.: Accurate Unlexicalized Parsing. In: Proceedings of the 41st Meeting of the Association for Computational Linguistics, pp. 423–430 (2003)
19. Joachims, T.: Making large scale SVM learning practical. Universit Dortmund (1999)
20. Moschitti, A.: Making Tree Kernels Practical for Natural Language Learning. In: EACL, vol. 113(120), p. 24 (2006)
21. Moschitti A.: A study on convolution kernels for shallow semantic parsing. In: Proceedings of the 42nd Annual Meeting on Association for Computational Linguistics, p. 335. Association for Computational Linguistics (2004)

Target Detection and Knowledge Learning for Domain Restricted Question Answering

Mengdi Zhang$^{(\boxtimes)}$, Tao Huang, Yixin Cao, and Lei Hou

Department of Computer Science and Technology,
Tsinghua University, Beijing 100084, China
mdzhangmd@gmail.com, {huangtao,caoyinxin,houl10}@mails.tsinghua.edu.cn

Abstract. Frequent Asked Questions(FAQ) answering in restricted domain has attracted increasing attentions in various areas. FAQ is a task to automated response user's typical questions within specific domain. Most researches use NLP parser to analyze user's intention and employ ontology to enrich the domain knowledge. However, syntax analysis performs poorly on the short and informal FAQ questions, and external ontology knowledge bases in specific domains are usually unavailable and expensive to manually construct. In our research, we propose a semi-automatic domain-restricted FAQ answering framework SDFA, without relying on any external resources. SDFA detects the targets of questions to assist both the fast domain knowledge learning and the answer retrieval. The proposed framework has been successfully applied in real project on bank domain. Extensive experiments on two large datasets demonstrate the effectiveness and efficiency of the approaches.

Keywords: Frequent Asked Questions · Domain restricted · Domain knowledge learning · Target-word · Question answering

1 Introduction

With the blossom of web-based services, some companies and organizations post online Frequent Asked Questions(FAQ) pages(a list of typical questions and resolved answers pairs), to provide consumers with timely information to common concerns. However, the proliferation of questions tends to overwhelm the user in skimming. Therefore, building a domain-restricted automated FAQ answering framework is in urgent need.

FAQ, a list of typical questions and experts' answers, offers users an online resource to respond to common concerns. The inclusion of FAQ pages enhances consumers' experience and increases potential purchases [14,16,15]. Take the the scenario in a bank FAQ page for example: a novice customer inquires *"Is it free to use mobile bank?"*("我想开通手机银行, 花钱吗?"), then the FAQ page match a common question called *"Does the 95959 mobile-bank charge?"*("95959 手机银行是否收费?") and returns the responding answer to the user. Obviously, the more timely and precise the FAQ answering service responds, the more possible the user choose this company's the products.

© Springer International Publishing Switzerland 2015
J. Li et al. (Eds.): NLPCC 2015, LNAI 9362, pp. 325–336, 2015.
DOI: 10.1007/978-3-319-25207-0_27

However, it is not an easy task. Questions asked by users are usually short, informal and expressed in different words to refer to the same concept. Most researches use syntax and ontology information to analyze questions. However, those questions are usually too short and informal to be well parsed, and ontology resources are usually unavailable and costly to construct. As a subclass problem of Question Answering, FAQ answering in restricted domain often exhibits different challenges compared to traditional question answering in open domain:

Question Understanding. To understand the questions, most researches like [2,25] use syntactic parsing via some language-specific parsers. However, the performance of the whole system is limited by the precision of the parser. First, the parser relies on a specific language, which implies a difficulty to extend to other languages. Second, expressed in spoken language in most cases, the questions from users couldn't be parsed well via existing parser tools. Third, those questions are quite short to contain complete syntactic elements. Previous studies show that users' questions are usually limited in length [5,7]. In our corpora, the average question lengths are about 20 and 10 Chinese characters respectively (as showing in Fig. 2). Thus, question analysis methods that heavily rely on syntax may not be appropriate in here.

Vocabulary Gap. Due to diverse background knowledge, users don't share the same vocabulary with each other as well as the experts. For example, the user's question *"Since my friend is abroad, can I call the bank to send money to him?"*("朋友在国外, 我打电话到银行可以给他汇钱吗?") actually refers to the formal question in FAQ *"Does the telephone banking's transfer support foreign currency?"*("电话银行对外转账是否支持外币?"). Considerable methods focus on employing an ontology to represent the questions and answers [27,23,2]. Through domain ontology, semantic distance is calculated between words of user's question and FAQ. However, the domain-specific ontology may be unavailable, and is difficult to construct and reuse. Therefore, how to bridge the gap without external resources remains a challenge.

Despite of the above challenges, once a practical FAQ answering framework is successfully formed, its impact for commerce is tremendous. Based on these observations, we propose the semi-automated domain-restricted FAQ answering framework(SDFA) that addresses all above challenges: SDFA collects users' questions from the system log to expand question in FAQ; proposes the target-word concept to understand question; fast learn a lightweight knowledge structure from the FAQ source itself to further bridge the vocabulary gap; all the gradually obtained information and knowledge are carefully designed as evidence to the final answer retrieval. The contributions of our work can be summarized as follows:

- We propose a novel semi-automatic framework SDFA to tackle the restricted domain FAQ answering issue.
- We design a semi-automatic pipeline to learn domain knowledge by clustering the FAQ, instead of from extra resources.

– Our framework has been successfully applied in real project on bank domain. Extensive experiments on two large datasets demonstrate the effectiveness and efficiency of the approaches.

2 Problem Definition

In this section, we first formalize the FAQ answering problem, and then demonstrate our proposed techniques.

2.1 Problem Definition

We present required definitions and formulate the problem of FAQ answering in restricted domain. Without loss of generality, we assume the domain FAQ corpus S has been well collected as a input, along with another input, user's question q. Our goal is to return the best matched answer in FAQ for a user's query.

Definition 1 *QA Pair*. *QA Pair* $p_i =< Q_i, A_i >$ *is a pair of a typical question Q_i and its responding answer A_i. FAQ corpus S is a set of QA Pair, i.e., $S = \{p_i | i = 1, 2.., n\}$.*

Both the query and question in FAQ are one interrogative sentence in a list of words. From observation, some words are informative enough to stand for the whole question, while some words are less important, even noising. To detect question's intention, we distinguish those words by define a concept called target-word:

Definition 2 *Target-word*. *We define the target-word w^t as a word which can stand for the main meanings of the question, i.e., user's intention. There are usually more than one target-word in a question. We represent a question $Q_i = \{w_1, w_2, ..w_m\}$ in a ranked list of target-words $Q_i^t = \{w_1^t, w_2^t, .., w_k^t\}$, where $k \leq m$.*

Cases of target-words are showing in early example: *"mobile bank"* ("手机银行") and *"free"*("收费") are actually the intentions of the user and should be treated as two target-words of the query. With a further analysis on the relations between words, two kinds of target-word can be summarized: one is actually service's name, and the other is the service's property. We define them as the domain knowledge.

Definition 3 *Domain Knowledge*. *We define a domain knowledge structure embedded in the questions: service category and its properties, as $< C, P_1, P_2, ..P_c >$. A question can be categorized by this two-layer labels. Domain knowledge $K = \{< C_i, P_{i1}, P_{i2}, ..P_{ic} > | i = 1, 2, ...d\}$ is defined for FAQ corpus S, where d means the number of services included in S.*

The domain knowledge in early example of user's question in bank domain is the service name *"mobile-bank"* ("手机银行"), and its properties *"charge"* ("收费"). Ultimately, the goal of our task is to find a QA Pair whose question is similar to the query, and return the responding answer in that QA Pair to the user. In terms of similarity, it means they are both questions about the same category's same properties, and share similar question's ranked target-word list. Based on these definitions, we define the task of utilizing domain knowledge in FAQ to answer user's query as follow:

Problem 1 *Data-driven FAQ Answering. Given a FAQ corpus $S = \{< Q_i, A_i > | i = 1, 2.., n\}$ and a user's query q, the goal is to firstly detect the target-word and learn the domain knowledge K from S, and finally find a list of QA Pair $p \in S$ for q, ranked by a function $Score(q, p)$ which measures the similarity between p and q based on the target-words and domain knowledge obtained previously.*

3   Framework

To tackle the problem described above, we propose a semi-automatic FAQ answering framework SDFA. Fig.1 shows the proposed framework which consists of a series of offline preparation on FAQ and online operations on user's query. Besides a common preprocess module, the offline mainly consists of target-word detection and domain knowledge learning, and the online mainly consists of the question analyzing and answer retrieval, which utilize the offline outputs like target-word model and domain knowledge. Additionally, after the final answer returning to user, if the user is satisfied with the final answer, then his/her query will be added to the FAQ corpus, as extension of the corresponding question.

The preprocessing are completed by traditional NLP techniques. Thus target-word detection, domain knowledge learning and answer retrieval are three major modules, and we'll present the details in the following section.

3.1   Target-Word Detection

In this section we model the target-word to detect user's intention. What kind of word should be recognized as target-word? Naturally, we should firstly exclude all obvious non-informative words by text preprocessing, including the removal of polite words, 1 length Chinese words and some personal pronouns. Ruling out these must-not-be word, we get the candidates of target-words.

Given a word, according to the structural and linguistic characteristics of questions, both lexical and semantic features are used to pick up the the target-word, including (1)the word itself, (2)word location in the question, (3)word length, (4)term frequency in the retrieval corpus, (5)POS tag of the word itself, (6)POS tag of the word ahead and (7)POS tag of the word after.

We use **Logistic Regression** [6] to assess the probability of each word to be a target-word in the question. In logistic function, the probability of a word w to be target-word is defined as follow:

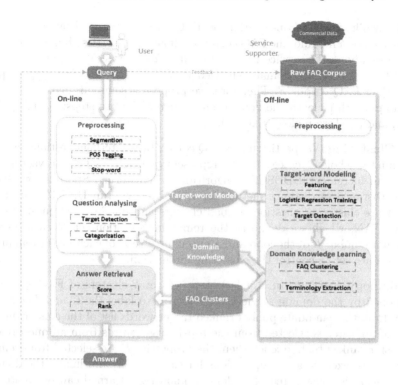

Fig. 1. An overall architecture for SDFA.

$$P(Y = 1|X = x_w) = \frac{1}{1 + e^{-(\alpha x_w + \beta)}} \qquad (1)$$

where x_w is the feature vector describing attributes of word w; α are weights of the attributes and β is a bias, both of which are learned by maximizing the objective function O:

$$O(\alpha, \beta) = \prod_{w \in W} P(Y = 1|X = x_w)^{y_w} P(Y = 0|X = x_w)^{1-y_w} \qquad (2)$$

where y_w is the label of w in the training set.

Through the target identification module, we rank the words in question according to their target-word probability, highlighting the informative points of the question.

3.2 Domain Knowledge Learning

Domain Knowledge is crucial to Restricted Domain Question Answering(RDQA). Recent researches use ontology in domain to calculate semantic distance between words when matching query to FAQ. However, external ontology knowledge bases in specific domains are usually unavailable. Additionally, manually building a domain knowledge base is laboriously expensive and highly subjective.

Meanwhile, through observation on FAQ's target-words obtained from the target-detection module, the FAQ corpus itself actually carries abundant domain knowledge about the services: service's name and service's properties. As we define before, they are domain knowledge $K = \{< C_i, P_{i1}, P_{i2}, ..P_{ic} > | i = 1, 2, ...d\}$ in FAQ corpus S. Therefore, we propose a semi-automatic and data-driven approach to learn the domain knowledge K by two steps: (1) FAQ clustering, and (2) Terminology Extraction.

FAQ Clustering. We partition the FAQ corpus by clustering questions in FAQ, to obtain the whole C_i categories. We represent questions as TF-IDF vectors and employ an incremental DBSCAN algorithm for clustering. Specially, we select Cosine Distance as distance metrics for DBSCAN; we run DBSCAN in a few cycles to re-clustering questions in bad clusters and un-clustered questions left by DBSCAN, which we define as the remaining question set of each loop; a maximum number L of the loop and a minimum size R of each loop's remaining questions are set to control the clustering circulation.

Terminology Extraction. By last step, we get k clusters of FAQ questions partitioned by the domain services. This step is to label the C and P for each cluster from the candidate pool of ranked target-words list of the cluster: firstly, the C of a cluster is selected from the top-1 target-words from all questions in the cluster ranked by frequency then, the properties P are selected from remain target-words; execute above operations for each cluster, and finally the domain knowledge K is constructed. The domain knowledge learned can be treated as a two-layer terminologies structure of services scattered in FAQ, as showing in Table 1. Each entry in the domain knowledge responds to a specific service, consisting of a category name C and its properties P_i.

Table 1. A snippet domain knowledge of banking

Category	Properties
phone-bank	query open-account register close-account
电话银行	查询 开户 注册 销户
text-bank	password binding query
短信银行	密码 绑定 查询
e-pay	query close-account register remittance
e支付	查询 注销 注册 汇款
noble-metal	sale price buying specification
贵金属	销售 零售价 购买 规格
fund	investment custody subscribing purchase
基金	定投 托管 认购 申购

Indexed by the learned domain knowledge, each question in FAQ is off-line labeled with a set of C and P IDs if it contains the corresponding terms in K, an equally categorization process. The input query is on-line categorized in the same way.

3.3   Retrieval Model

Given a question, the main goal of the retrieval model is to find the most relevant QA Pair in FAQ and output the answer as the final answer. Based on the target and category information obtained previously, the probability of relevance between the query and each candidate QA Pair is calculated through a target-word based BM25 algorithm. Thus, we can rank the QA Pairs in the probability of relevance, and then strategically return matched answer according to the query's categories. For example, if a query contains only one category label, we will return the top1 answer to the user; if a query contains multiple category labels, we will return the top3 or top5 answers depending on the circumstances.

The retrieval model includes two steps: (1) categorize user's query based on domain knowledge, and (2) find the related QA Pair by the advanced BM25 algorithm.

Candidate Documents. Clearly, the relevant documents should have the same categories and properties. For example, the question *"How to log off personal mobile bank?"*(" 如何注销个人手机银行?") will be classified to *"Mobile Bank"* concept with *"Log off"* action, which should be retrieved under FAQs whose concept and action label are the same. Therefore, only a subset of the whole collection that shares the common category and property with the question is worth retrieving, which we called candidate documents.

Target-word Based BM25. BM25 [17] is a probabilistic relevance framework and has been widely used in text-retrieval area. It considers each term of the question as an independent unit, and the final probability of relevance between the document and the question is proved to be proportional to the weighted sum of all the terms. The weight of term is computed in traditional BM25 as follows.

$$w_i(tf) = w_i(idf) * \frac{tf}{tf + K} \tag{3}$$

In Eq.3, $w_i(idf) = log\frac{N-n_i+0.5}{n_i+0.5}$ is a close approximation to the classical *idf*, where N is the size of the whole collection of the documents and n_i is the number of documents containing term t_i. For K in Eq.3,

$$K = k_1 * (1 - b + b * \frac{dl}{avdl}) \tag{4}$$

dl is the document length and $avdl$ is the average length of all the documents. Note that b is used to control the document length normalization, and k_1 is to smooth the special term frequency. Particularly, for high K, increments in tf contribute significantly to the relevance, whereas for low K, the additional contribution of another occurrence of the word decreases rapidly.

As the term weight, TF-IDF provides a statistical perspective; however, target words indicate the semantic importance. To model this feature, we define the final probability of relevance as follows:

$$P(rel|q,p) \propto \sum_{q,tf} \lambda_i * w_i(tf) \tag{5}$$

where

$$\lambda_i = \begin{cases} \eta & tf_i > 0, i \in q \\ -\gamma & tf_i > 0, i \notin q \end{cases} \qquad (6)$$

indicates the degree of being the target of the question. To be specific, if the term is a target word of the question, it will be rewarded; otherwise, when the term that only occurs in the QA pair is also a target word, it will be punished. To differentiate the importance of standard question and extended question, we use $\widehat{tf} = \tau * tf$ instead of the normal term frequency, where $\tau = 2, 1, 2$ denotes standard question, extended question and standard answer respectively.

4 Experiments

4.1 Data Preparation

In the experiment, we take bank as the restricted domain. Since there is no benchmark dataset for bank domain, we construct two datasets from two different banks. Each dataset contains both the standard FAQ and the extended questions in spoken language, collected from the corresponding bank's consult log. In this way, each standard question in FAQ is corresponding to a couple of extended questions (after the system is running, the user's query can be added to the extended question and automatically accumulate). Both the datasets come from the needs of real projects in life. The total statistics of two datasets are showing in Table 2 and the distributions of question's length in these two FAQ set are showing in Fig.2.

Table 2. Statistics on Datasets.

	#QA Pairs	#Extended Questions	#Test Set	#Target-word Train Set
Bank1	48,495	127,026	4,336	2,272
Bank2	2,399	42,404	5,536	500

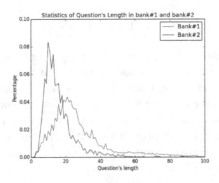

Fig. 2. The Distribution of Question's Length.

4.2 Evaluation Measures

Given a test question, our proposed methods return a list of QA Pair ranked in descending order of probability of relevance. Generally, we extract the answer from the top QA Pair as the expected answer. As such, we are typically interested in the *Precision@*1, which measures the percentage of the results whose top-1 answer is correct. Similarly, the *Precision@*5 measures the percentage of the results where the correct answer showing among top-5. Additionally, for those not hitting the top, we calculate the Mean Reciprocal Rank(MRR), the multiplicative inverse of the rank of the first correct answer: $MRR = \frac{1}{|Q|} \sum_{i=1}^{|Q|} \frac{1}{rank_i}$. Consider rank position $rank_i$, of the first relevant QA pair in the candidate rank list, then the Reciprocal Rank score is $\frac{1}{rank_i}$ and MRR is the mean RR across multiple test questions. The more close the MRR score to 1, the more possible the correct answer in the result is near to the top.

4.3 Experiments

To evaluate the effectiveness of each single module and the whole framework, we gradually designed a series of experiments and judged the performance based on the metrics as described in previous section.

In the target detection module, we use Logistic Regression to train the target-word model from a manually labeled question set and then detect the target-word of FAQ set. The second step is to learn domain knowledge and categorize each question. we run DBSCAN clustering on the FAQ set and then manually label each cluster with domain terminology guarded by the target-word detected from the first step. The DBSCAN clustering requires two parameters ε and $minPt$, which stands for physical distance and the minimum number of points required to form a dense region respectively. In our experiment, ε is set between 0.40 and 0.45, and $minPt$ is set as 100 for bank1 and 30 for bank2. The maximum clustering loop L is set as 5 for bank1 and 3 for bank2, and the minimum remains R is both set as 10%.

For the FAQ retrieval step, we execute experiments on five strategies:

(1)BM25. In the traditional BM25 model, we empirically initialize the model's parameters in Eq.4: the default parameter values we set are k1=1.2 and b=1 as recommended in [17].

(2)$BM25^t$. In the target-word based BM25 model, we set the K1 and b as same in (1). We use the target-word score as the term's weight of BM25 score in Eq.5 and we set the reword η between 1.2-1.5 in Eq.6.

(3)$BM25^t$+Class. We first learn domain knowledge from the FAQ corpus by clustering the QA Pair, obtaining the categories and their properties. Then use these learned two-layer domain knowledge to categorize each QA Pair's type in the FAQ corpus and the input question's type and finally execute the same retrieval process as (2), which means we map the question to certain document space then retrieve the best matched document in this small data space.

(4)$BM25^t$+Class+Punish. After executing the same process of strategy (3), we get a list of candidate QA Pairs, ranking in the positive relevance with the query. Before returning the tops to the user, we rerank this list by punishing the negative difference as described in Eq.6. We set the punish $-\gamma$ as -1.

(5)Cosine. As a baseline, we represent the query and each question of FAQ in TF-IDF vector and calculate cosine similarity of vectors as the ranking basis.

4.4   Results and Analysis

The overall results are showing in Table 3. As we can see, our method, the target-word based BM25 utilizing the domain knowledge and considering the term punishment, performs best on two datasets. Each method shows a consistent performance on all three evaluation measures. For simplicity, we only take the $Precision@1$ here for detailed analysis. The Cosine method performs steadily but limited: its $Precision@1$ is about 40% on two datasets. Traditional BM25 performs better than Cosine. After considering the target-word, the $BM25^t$ has averagely improved by 2.5 percent. Interestingly, $BM25^t$ with categorization utilizing the domain knowledge is a little bit lower than the $BM25^t$ itself. However, the search time, though not showing here for the limited room, has reduced 50%. It indicates that domain knowledge can map the query to a much small search space to improve the efficiency, with a slight risk of filtering some correct answer simultaneously. Finally, by punishing the negative difference between the query and candidate QA Pairs, the rank of the correct answer can be further improved.

Our proposed strategy, the $BM25^t$+Class+Punish, performs the best on the two different bank datasets, which proves the practical and robustness of our framework.

Table 3. Overall results

Method	Bank1			Bank2		
	Precision@1	*Precision@5*	*MRR*	*Precision@1*	*Precision@5*	*MRR*
Cosine	41.3%	64.5%	55.7%	45.4%	68.1%	57.1%
BM25	61.1%	79.4%	68.2%	62.8%	84.3%	70.3%
$BM25^t$	63.6%	81.7%	70.0%	64.2%	87.0%	73.9%
$BM25^t$ +Class	63.5%	81.3%	69.8%	64.1%	86.7%	73.6%
$BM25^t$+Class+Punish	66.6%	84.1%	73.9%	65.3%	88.2%	74.6%

5   Related Work

There are several lines of researches that are related to our work, and we present some of the related literatures as follows:

Frameworks and architecture of domain restricted QA system [13,12,22,3] have been proposed, mostly based on text, yet rarely concerned with question answer pairs format data in FAQ [4]. Instead of extracting answers from free text, the QA systems on FAQ focus on retrieving the most relevant QA Pair in respect to the user's question. In FAQ systems, the main categories include NLP-based [23,26,24], statistical-based [9,10] and template-based methods [18,19,20]. Researches on question similarity calculation include cosine similarity on TF-IDF vectors, BM25 and etc. [21,1,8]. Our work focuses upon bank service FAQ answering system and combines NLP and statistic methods.

There is one closely related work [11], which proposes a cluster-based retrieval system Fract. Fract clusters the query logs into predefined FAQ categories and extract weight scores of potentially occurring words from the clusters by using LSA techniques. During retrieval time, Fract extracts important terms from query by parsing and expand these terms with the potentially occurring words to help in ranking relevant FAQs. Differently, we propose a new concept called Target-word to detect user's intention and cluster the FAQ corpus, instead of the query logs, to learn the domain knowledge. Both these information and knowledge are carefully designed as evidence combined into an adjusted target-word based BM25 score function to retrieve final answer.

6 Conclusion

In our research, we propose a semi-automatic domain-restricted FAQ answering framework SDFA, without relying on any external resources. SDFA detects the targets of questions to assist both the fast domain knowledge learning and the answer retrieval. The proposed framework has been successfully applied in real project on bank domain. Extensive experiments on two large datasets demonstrate the effectiveness and efficiency of the approaches.

Acknowledgments. The work is supported by 973 Program (No. 2014CB340504), NSFC-ANR (No. 61261130588), Tsinghua University Initiative Scientific Research Program (No. 20131089256), Science and Technology Support Program (No. 2014BAK04B00), and THU-NUS NExT Co-Lab.

References

1. Achananuparp, P., Hu, X., Zhou, X., Zhang, X.: Utilizing sentence similarity and question type similarity to response to similar questions in knowledge-sharing community. In: Proceedings of QAWeb 2008 Workshop, vol. 214 (2008)
2. Basili, R., Hansen, D.H., Paggio, P., Pazienza, M.T., Zanzotto, F.M.: Ontological resources and question answering. In: Workshop on Pragmatics of Question Answering, held jointly with NAACL 2004. Citeseer (2004)
3. Benamara, F.: Cooperative question answering in restricted domains: the webcoop experiment. In: Proceedings of the Workshop Question Answering in Restricted Domains, within ACL (2004)
4. Burke, R.D., Hammond, K.J., Kulyukin, V., Lytinen, S.L., Tomuro, N., Schoenberg, S.: Question answering from frequently asked question files: Experiences with the faq finder system. AI Magazine **18**(2), 57 (1997)
5. Chau, M., Fang, X., Liu Sheng, O.R.: Analysis of the query logs of a web site search engine. JASIST **56**(13), 1363–1376 (2005)
6. Cox, D.R., Snell, E.J.: Analysis of binary data, vol. 32. CRC Press (1989)
7. Croft, W.B., Cook, R., Wilder, D.: Providing government information on the internet: Experiences with thomas. In: DL (1995)
8. Guo, X.Y., Zhang, B.F., Fang, A.G., Zhou, C.F.: Research on question similarity algorithm for intelligent question answering system and its implementation. Computer Applications **2**, 449–452 (2005)

9. Jijkoun, V., de Rijke, M.: Retrieving answers from frequently asked questions pages on the web. In: Proceedings of the 14th ACM International Conference on Information and Knowledge Management, pp. 76–83. ACM (2005)

10. Kim, H., Seo, J.: High-performance faq retrieval using an automatic clustering method of query logs. Information Processing & Management **42**(3), 650–661 (2006)

11. Kim, H., Seo, J.: Cluster-based faq retrieval using latent term weights. IEEE Intelligent Systems **2**, 58–65 (2008)

12. Ma, Z., Sun, A., Yuan, Q., Cong, G.: A tri-role topic model for domain-specific question answering (2015)

13. Mollá, D., Vicedo, J.L.: Question answering in restricted domains: An overview. Computational Linguistics **33**(1), 41–61 (2007)

14. Mustaffa, S., Beaumont, N.: The effect of electronic commerce on small australian enterprises. Technovation **24**(2), 85–95 (2004)

15. Page, C., Lepkowska-White, E.: Web equity: a framework for building consumer value in online companies. Journal of Consumer Marketing **19**(3), 231–248 (2002)

16. Pei, Y., Xue, W., Li, D., Chang, J., Su, Y.: Research on the relationship between empowerment and sales performance in e-commerce service industry enterprises. In: LISS 2014, pp. 959–967. Springer (2015)

17. Robertson, S.E., Walker, S.: Some simple effective approximations to the 2-poisson model for probabilistic weighted retrieval. In: Proceedings of the 17th Annual International ACM SIGIR, pp. 232–241. Springer (1994)

18. Sneiders, E.: Automated faq answering: continued experience with shallow language understanding. In: Papers from the 1999 AAAI Fall Symposium on Question Answering Systems, pp. 97–107 (1999)

19. Sneiders, E.: Automated question answering using question templates that cover the conceptual model of the database. In: Andersson, B., Bergholtz, M., Johannesson, P. (eds.) NLDB 2002. LNCS, vol. 2553, pp. 235–239. Springer, Heidelberg (2002)

20. Sneiders, E.: Automated faq answering with question-specific knowledge representation for web self-service. In: 2nd Conference on Human System Interactions, HSI 2009, pp. 298–305. IEEE (2009)

21. Song, W., Feng, M., Gu, N., Wenyin, L.: Question similarity calculation for faq answering. In: Third International Conference on Semantics, Knowledge and Grid, pp. 298–301. IEEE (2007)

22. Vila, K., Ferrández, A.: Model-driven restricted-domain adaptation of question answering systems for business intelligence. In: Proceedings of the 2nd International Workshop on Business Intelligence and the WEB, pp. 36–43. ACM (2011)

23. Wang, F., Teng, G., Ren, L., Ma, J.: Research on mechanism of agricultural faq retrieval based on ontology. In: Ninth ACIS International Conference on Software Engineering, Artificial Intelligence, Networking, and Parallel/Distributed Computing, SNPD 2008, pp. 955–958. IEEE (2008)

24. Winiwarter, W.: Adaptive natural language interfaces to faq knowledge bases. Data & Knowledge Engineering **35**(2), 181–199 (2000)

25. Xu, K., Zhang, S., Feng, Y., Zhao, D.: Answering natural language questions via phrasal semantic parsing. In: Zong, C., Nie, J.-Y., Zhao, D., Feng, Y. (eds.) NLPCC 2014. CCIS, vol. 496, pp. 333–344. Springer, Heidelberg (2014)

26. Yang, S.Y., Chuang, F.C., Ho, C.S.: Ontology-supported faq processing and ranking techniques. Journal of Intelligent Information Systems **28**(3), 233–251 (2007)

27. Zajac, R.: Towards ontological question answering. In: Proceedings of the Workshop on Open-Domain Question Answering, vol. 12, pp. 1–7. ACL (2001)

Short Papers

An Improved Algorithm of Logical Structure Reconstruction for Re-flowable Document Understanding

Lin Zhao, Ning Li[✉], Xin Peng, and Qi Liang

Department of Computer, Beijing Information Science
and Technology University, Beijing, China
zhaolin_0124@126.com, ningli.ok@163.com

Abstract. The basic idea of re-flowable document understanding and automatic typesetting is to generate logical documents by judging the hierarchical relationship of physical units and logical tags based on the identification of logical paragraph tags in re-flowable document. In order to overcome the shortages of conventional logical structure reconstruction methods, a novel logical structure reconstruction method of re-flowable document based on directed graph is proposed in this paper. This method extracts the logical structure from the template document and then utilizes directed graph's single-source shortest path algorithm to filter out redundant logical tags, thus solving the problem of logical structure reconstruction of a document. Experimental results show that the algorithm can effectively improve the accuracy of logical structure recognition.

Keywords: Logical structure reconstruction · Document understanding · Logical tags

1 Introduction

With the popularity of electronic documents, re-flowable document has been used more and more extensively. Previous format documents and the most widely used re-flowable documents at present have to deal with logical structure reconstruction of the document [1-2]. For format document (Figure 1), the region and positional relationship of diagrams, graphics, tables and text information are first cut apart and judged automatically through layout analysis, and then the physical geometry structure will be mapped to logical structure through layout understanding. But re-flowable document does not contain high-level logical structure. The high-level structure needs to be deduced from the low-level elements (Figure 2).

So if the information structure of the re-flowable document can be accurately identified, it will have a great significance for document understanding, especially for the applications such as document retrieval and format check of document. Chinese and international researchers have carried out studies on of logical structure reconstruction of documents, including the method based on rules (i.e. the methods are proposed by LeBourgeois [3], Rosenfeld [4], Hu [5]), and the method based on statistics (i.e. Brugger [6] and Palmero [7]), the method based on FSA(i.e. Song Haosu [8]).

© Springer International Publishing Switzerland 2015
J. Li et al. (Eds.): NLPCC 2015, LNAI 9362, pp. 339–346, 2015.
DOI: 10.1007/978-3-319-25207-0_28

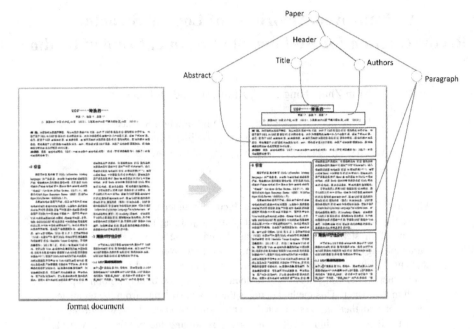

Fig. 1. Format document understanding

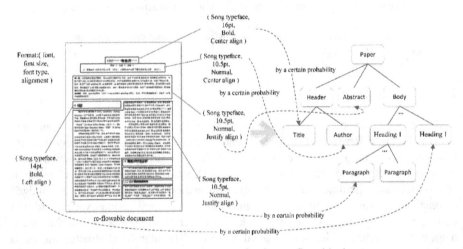

Fig. 2. Logical structure reconstruction in a re-flowable document

Rule-based reconstruction method is an iterative algorithm, which tries to find out an optimal global solution based on text features. Statistics-based reconstruction method requires a lot of training sets and is more dependent on the training text corpus. FSA-based reconstruction method, may include some states containing errors due to possible typesetting errors, so they can result in an infinite state if added to the FSA. Based on the above situation, a directed graph-based logical structure reconstruction method for documents on the basis of FSA logical structure reconstruction method is presented in this paper.

2   The Emergence of Logical Structure Reconstruction Method Based on Directed Graph

2.1   Logical Tag Structure Extraction from a Template Document

Logical tags, such as "CAT", "CAB", etc., represent the logical roles of paragraphs in a document. Logical structure reconstruction of a document needs a reconstruction rule which is determined by extracting logical tag structure from template document.

Extracting logical tag structure is to determine the sequence of logical document tags [9], i.e. the sequence of document roles, which provides a reconstruction basis for logical structure reconstruction of a document. To take thesis as an example, assuming the thesis with a maximum of Heading 3, then the sequence of logical structure tags extracted from the template document is shown in Figure 3.

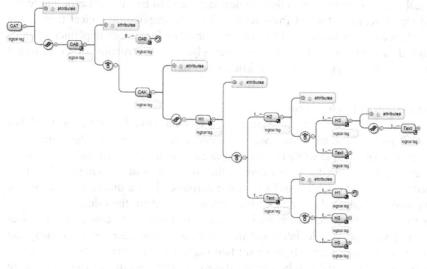

Fig. 3. Hierarchical structure of logical document tags

The meanings of the elements in the Figure 3 are as follows. CAT(Chinese Abstract Title), CAB(Chinese Abstract Body), CAK(Chinese Abstract Keywords), H1 (Heading 1), H2(Heading 2), H3(Heading 3), Text(Normal(body)). The attributes in the Figure 3 are shown in Figure 4.

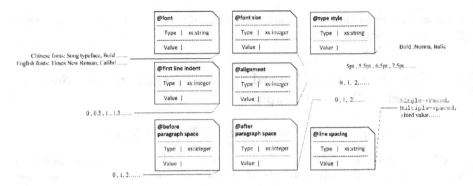

Fig. 4. The attributes in the Figure 3

2.2   Construction Method of Directed Graph of a Document

The logical tags of a paragraph shall be first identified when constructing nodes of the directed graph [10]. Since some logical structures have similar format, misjudgment of logical tags may occur. Therefore, logical tags need to be filtered according to the logical tag structure extracted from section 2.1. The impossible logical tags of the paragraph need to be filtered out to make preparations for further implementation of the logical structure reconstruction. Accordingly, the construction of the directed graph nodes can be divided into the following two steps:

Determination of Logical Tags

By section 2.1, every paragraph in a document corresponds to a format vector which has eight components. Then the to-be-investigated document dictionary in which the values of the components of the to-be-investigated document are recorded and the template document dictionary in which the values of the components of the template document are recorded are then established. Every value corresponds to a different branch which is composed of similar values and the paragraph which contains the value.

For the paragraph which needs logical tag identification, the score of each logical tag in the paragraph is calculated and the tag with highest score is most likely the logical tag of the paragraph. If there are two highest scores, they should be seen as candidate logical tags and will be further filtered in latter selection process. The score of a logical tag of a paragraph is the sum of the scores of its logical components of this paragraph format vector. The score of a logical tag equals the degree of similarity of the branch times the number of occurrences of the logical tag in the branch times the weight of the component of the logical tag. The weight value of the component is evaluated by the frequency of occurrence of each component's value in the document format, the higher the frequency of occurrence, the greater the corresponding weight. The basic algorithm of logical tag identification is shown as follows:

```
1:  for p in P do
2:     for l in L do
3:        for c in C do
```

```
4:        for d  in dic  do
             w       c
5:        if F (p) = d  && s(d ,w) = max{s(d ,d 's branches)}
              c       w      w         w  w
6:          then score (l) = s(w,d )*e(l,d 's index)*w(l,c)
                       c          w      w
```

7: $score(l) = \sum_{c \in C} score_c(l)$

```
8: if score(l ) = max(score(l)) then
            i
9:    return l
              i
```

where s (d_w, w) is the degree of similarity membership degree of the vocabulary d_w of the to-be-investigated document and the vocabulary w of the template document. e(l, d_w's index) is the number of occurrences of the logical tag l in the vocabulary d_w. w(l, c) is the weight of the logical tag l on the component c, $score_c(l)$ is the score of the logical tag l on the component c, score(l) is the total score of the logical tag l on all the components.

Filtering of Logical Tags
Logical tag filtering has adopted the same approach. Logical tags are filtered by combining the logical tags of the previous paragraph with the logical tags of the template. By default, the logical tag of the first paragraph shall be retained and the subsequent paragraphs will be examined one by one. The logical tags left by the previous paragraph shall be checked if it is the expected tag of the current paragraph according to the sequence of logical tags in the template. If it is, then the logical label shall be retained. If not, then the tag shall be filtered out. If no result is the expected tag of the current paragraph, then all the judgment results of the previous paragraph shall be filter out, and the logical tags of the current paragraph shall be kept. However, the logical tag is not unique. After logical tag filtering, some paragraphs may have only one tag, other paragraphs may have multiple logical tags due to multiple logical tag judgment results. For this situation, probability of occurrence of a paragraph shall be added, it will be detailed introduction in section 2.3.

The nodes of the directed graph were obtained from previous paragraphs. The edges of the directed graph are used to connect each node so as to show the relationship of the nodes. The logical tag left by each paragraph is the corresponding tag of the paragraph in the graph, and each node of the previous paragraph points to every node of the next paragraph.

2.3   Reconstruction Method of the Logical Structure of a Document

The single-source shortest-path algorithm describes a way to find out the shortest path from a certain source point s $\in$ V to the rest vertices in V in a known directed-weighted graph G = (V, E). In the graph G, the weighting function w is defined, w:E $\rightarrow$ R is the mapping from the edge to real weight value. We adopt Bellman-Ford algorithm [11] here. The shortest path from u to v is defined as:

$$\delta(u,v) = \begin{cases} \min\{w(p):u \rightarrow v\} & \text{a pathway from u to v} \\ \infty & \text{else} \end{cases} \tag{1}$$

At the beginning, the path length of the source point u is 0. Meanwhile, the path lengths of all other vertices are assigned as ∞, which indicates that the paths going to all these vertices (except u and v) are unknown. After the continuous updating of values, the shortest sequence which is the shortest path from u to v will be obtained eventually. The basic idea is as follows:

```
1: for i ← 1 to |V[G]| - 1 do
2:    for EDGE (u, v) in E[G] do
3:       if d[v] > d[u] * 1 / w(u, v)
4:          then d[v] ← d[u] * 1 / w(u, v)
5:          [v] ← u
6: for EDGE (u, v) in E[G]    do
7:    if d[v] > d[u] + w(u, v)
8:       then return FALSE
9: return TRUE
```

The weight of the path is the sum of all weight values of its edges in this algorithm. In order to be more suitable for the logical structure reconstruction of a document, we redefine the path length. The path length is the reciprocal of the probability of existence of the path, the smaller the length is, the bigger the probability of existence of the path is, the bigger the possibility of the corresponding logical structure is. In this algorithm, the calculation of the weight on the graph edge is as follows: if the starting point of the edge is l_i and the ending point of the edge is l_j, then the weight w (l_i, l_j) of the edge is equal to the reciprocal of the probability of occurrence P (l_i, l_j) of corresponding logical tag of the starting point and ending point. Then the number of occurrences of various logical tags after each tag is calculated. C (l_i, l_j) is the number of occurrences of l_j after l_i, C(l_i) is the total number of occurrences of all logical tags after li, then the probability of occurrence of l_j after l_i is P (l_i, l_j) = C (l_i, l_j) / C (l_i). Then the single-source shortest path algorithm can thus be obtained. Since there may be multiple shortest paths, there may exist more than one corresponding logical document structure.

3   Experiment and Related Analysis

The corpus used by the experimental set is built by our university. We extract 500 documents from the corpus, which are the theses of undergraduate students including but not limited to "CAT", "CAK", "H1", "H2", "H3", "Text" and then carry out artificial identification and labeling on their paragraphs human. And then we select 200 documents randomly to set up learning sample set and the rest 300 documents are used to establish the test sample set. For the learning sample set, the logical structure reconstruction method based on directed graph is applied to do statistics. Then this method is applied to the test sample set which is reconstructed using the logical structure reconstruction of a document to collect results. The evaluation criterion of logical structure reconstruction of a document is the proportion of the paragraphs whose logical tags are determined by the logical structure reconstruction method in all the paragraphs. Precision, Recall and Balance F value are used to evaluate the evaluation criterion of the logical paragraph tags. It is defined as follows:

$$P = \frac{|\{relevant\ paragraphs\} \cap \{retrieved\ paragraphs\}|}{|\{retrieved\ paragraphs\}|} \tag{2}$$

$$R = \frac{|\{relevant\ paragraphs\} \cap \{retrieved\ paragraphs\}|}{|\{relevant\ paragraphs\}|} \tag{3}$$

$$F = \frac{2 \times P \times R}{P + R} \tag{4}$$

For the identification method of logical tags here we use the method mentioned in section 2.3 and logical tags identification method based on VSM. The identification results are shown in Table 1.

Table 1. Identification results of logical tags of a paragraph before and after the improvement

Paragraph role	Before the improvement			After the improvement			F change
	P	R	F	P	R	F	
CAT	97.62%	86.52%	91.73%	96.54%	86.12%	91.03%	-0.70%
CAK	98.33%	80.64%	88.61%	98.43%	82.36%	89.68%	+1.07%
H1	95.41%	78.42%	86.08%	96.65%	79.86%	87.45%	+1.37%
H2	94.39%	79.43%	86.26%	95.01%	79.23%	86.40%	+0.14%
H3	94.47%	76.28%	84.40%	95.40%	78.18%	85.93%	+0.53%
Text	98.54%	80.47%	88.59%	96.23%	76.89%	85.48%	-3.11%

As can be seen from the table 1, the precision and recall rate of the improved logical paragraph tag identification are high, indicating that it can accurately find the most similar logical paragraph formatting tags and can also find out a logical tag's paragraph as many as possible. We apply T-Test to P (before the improvement and after the improvement) in Table1, and the results are shown in Table 2.

Table 2. Paired Samples Test

Pair1	Paired Differences							
	Mean	Std. Deviation	Std. Error Mean	95% Confidence Interval of the Difference		t	df	Siq. (2-tailed)
				Lower	Upper			
before - after	8.33E-04	1.36E-02	5.55E-03	-1.34E-02	1.51E-02	0.15	5	0.887

4   Conclusions

The algorithm of logical structure reconstruction has great significance for document understanding. However, because of many factors are considered in the proposed algorithm, so there is a certain limitation in the study. Among them, it is a lack of consideration for complex structure in the logical structure reconstruction algorithm. In addition, there is no direct connection between the probability of solution and the probability of shortest path. Therefore, we still make every endeavor to resolve these problems further. In the next research work, it is an important research perspective for how to apply the recognized document information structure to document format checking.

Acknowledgement. This paper is supported by the Project of Construction of Innovative Teams and Teacher Career Development for Universities and Colleges Under Beijing Municipality (No.IDHT20130519), and the general program of science and technology development project of Beijing Municipal Education Commission (No.KM201511232013).

References

1. Mao, S., Rosenfeld, A., Kanungo, T.: Document structure analysis algorithms: a literature survey. In: Electronic Imaging 2003, International Society for Optics and Photonics, pp. 197–207 (2003)
2. Namboodiri, A.M., Jain, A.K.: Document structure and layout analysis. In: Digital Document Processing, pp. 29–48. Springer, London (2007)
3. Wu, Z., Mitra, P., Giles, C.L.: Table of contents recognition and extraction for heterogeneous book documents. In: Document Analysis and Recognition 12th International Conference, 2, pp. 1205–1209 (2013)
4. Sonka, M., Hlavac, V., Boyle, R.: Image processing, analysis, and machine vision. Cengage Learning (2014)
5. Hu, T.: New Methods for Robust and Efficient Recognition of the Logical Structures in Documents. IIUFUniversité de Fribourg, Switzerland (1994)
6. Satkhozhina, A., et al.: Non-manhattan layout extraction algorithm. In: Proceedings of SPIE-IS&T Electronic Imaging, 86640A (2013)
7. Belaïd, A., D'Andecy, V.P., Hamza, H., Belaïd, Y.: Administrative document analysis and structure. In: Biba, M., Xhafa, F. (eds.) Learning Structure and Schemas from Documents. SCI, vol. 375, pp. 51–71. Springer, Heidelberg (2011)
8. Song, H., Li, L., Zhang, W.: Application of VSM model to document structure identification. Journal of Beijing Information Science and Technology University (Natural Science Edition) 6, 66–69 (2011)
9. Jin, C.: Determine Algorithm of logical order in document layout based on directed graph. Microcomputer Information 12, 292–293 (2008)
10. Peng X., Li, N.: Improved VSM algorithm for judging paragraph logic label. Journal of Beijing Information Science and Technology University (Natural Science Edition), 19–24 (2014)
11. Nepomniaschaya, A.S.: An associative version of the bellman-ford algorithm for finding the shortest paths in directed graphs. In: Malyshkin, V.E. (ed.) PaCT 2001. LNCS, vol. 2127, pp. 285–292. Springer, Heidelberg (2001)

Mongolian Inflection Suffix Processing in NLP: A Case Study

Xiangdong Su, Guanglai Gao$^{(\boxtimes)}$, Yupeng Jiang, Jing Wu, and Feilong Bao

College of Computer Science, Inner Mongolia University,
Hohhot 010021, People's Republic of China
csggl@imu.edu.cn

Abstract. Inflection suffix is an important morphological characteristic of Mongolian words, since the suffixes express abundant syntactic and semantic meanings. In order to provide an informative introduction of it, this paper implements a case study on it. Through three Mongolian NLP tasks, we disclose the following information: (1) views of inflection suffix in NLP tasks, (2) Inflection suffix processing ways, (3) Inflection suffix effects on system performance and (4) some suffix related conclusion.

Keywords: Mongolian · Inflection suffix · NLP · Case study

1 Introduction

Mongolian is an agglutinative language which normally ranks as a member of the Altaic language family, a family whose principal members are Turkish, Mongolian and Manchu (with Korean and Japanese listed as possible relations). There are about seven million Mongol speakers in the world, including two million in Mongolia, more than twice that number in Inner Mongolia and other parts of China, and another half million or so in the Buryat and Kalmyk Republic and elsewhere in Russia. The Mongols have written their language in several different scripts, the oldest and most durable of which, called the classical Mongol script, was introduced almost 800 years ago under Genghis Khan. It originated with the Sogdo-Uighur alphabet and has been revised several times. It derives the traditional Mongolian, which is used in Inner Mongolian Autonomous Region of China now. Their differences mainly concentrate on the following aspects [1]. Firstly, the glyphs of the same character between them are completely different even in the same position of one word. Secondly, the glyphs of the traditional Mongolian characters have diacritics, such as dots, while the corresponding glyphs of the classical Mongolian characters do not. The Mongolian used in Mongolia is called Cyrillic Mongolian, which was introduced in the 1940's [2]. It has the same pronunciation as traditional Mongolian. But its written form differs from traditional Mongolian. It uses the letters from the Russian alphabet. This paper introduces Mongolian according tradition Mongolian used in Inner Mongolian Autonomous Region of China. Without otherwise specified, Mongolian means traditional Mongolian in this paper.

© Springer International Publishing Switzerland 2015
J. Li et al. (Eds.): NLPCC 2015, LNAI 9362, pp. 347–352, 2015.
DOI: 10.1007/978-3-319-25207-0_29

The main features of Mongolian are a system of vowel harmony, agglutination and the SOV word order. Mongolian does not distinguish gender, has no definite article, and has only a very limited plural system. As other Altaic languages, there is hardly any difference between nouns and adjectives. There are 30 Mongolian characters. Each character has as many as three different glyphs (visual forms) depending on whether the character appears in an initial, medial, or final position of a word [3]. In some cases, some characters have the same glyph. In appearance, some glyphs are a part of other glyphs. The characters in words are connected along the baselines. Mongolian documents conform to a top-down writing style. The column order is from left to right.

Mongolian words are formed by attaching suffixes to roots. The suffix falls into two groups: derivational suffix and inflection suffix. Derivational suffix is also called the word-building suffix. They are added to the root and give the original words new meanings. The root adding one or more derivation suffixes is called a stem. Inflection suffix is also called word-changing suffix. They are added to the stems and give the original words grammatical meanings. These suffixes serve to integrate a word into sentence. Since the role difference between derivation suffix and inflection suffix, many NLP tasks pay more attention to the inflection suffix processing. The motivation of this paper elaborates how inflection suffix works in Mongolian NLP through a case study. The following aspects will be discussed: (1) views of inflection suffix in NLP tasks, (2) inflection suffix processing ways, (3) case effects on system performance and (4) some suffix related conclusion. The case study is taken out on Mongolian translation, Mongolian syntactic analysis, and Mongolian information retrieval. The experiments reveal some important points. We also refer some related works.

The remainder of this paper is organized as follows. Section 2 describes inflection suffix. Section 3 describes inflection suffix in Mongolian Syntactic Analysis. Section 4 describes inflection suffix in Mongolian information retrieval. Section 5 describes inflection suffix in Mongolian machine translation. Section 6 draws the conclusion.

2   Inflection Suffix

Inflectional suffix can be divided into plural suffixes, case suffixes, reflexive suffixes, voice suffixes, aspect suffixes and mood suffixes. Among this 6 group suffixes above, plural suffixes, case suffixes and reflexive suffixes only for nominal class; Voice suffixes, aspect suffixes and mood suffixes only for verb class. The nominal class includes noun, adjective, numeral, time-place word and pronoun. Figure 1 shows the structure of nominal word. The case suffixes express the relationship between two words and phrases, and the reflexive suffixes express the possessive relationship between two parts of sentence.

---

root + plural suffixes + case suffixes + reflexive suffixes

Fig. 1. The structure of nominal word

Case suffixes, reflexive suffixes and plural suffixes connected to the stem through a Narrow Non-Break Space (U+202F, Latin.:"-") [4]. For instance, in a word " ᡐᡆᠢᠮᠣᠨᠭᠭᠤᡐᡆᠸ/ " (transliteration: "vlaganbagatvr-vn", means: "Ulaanbaatar's"), the " ᡐᠣᠸ/ " (-vn) is case suffixes. Case suffixes in Mongolian divided into eight types. These are Nominative Case, Accusative Case, Genitive Case, Dative-Locative Case, Ablative Case, Comitative Case, Instrumental Case and Directive Case. For case grammar, one stem is allowed to append one or two case(s) to from a new word. Table 1 demonstrates some examples of case suffixes.

Table 1. Examples of case suffixes

Stem	Case	Type	English Translation
ᡐᡆᠢᠮᠣᠨᠭᠭᠤᡐ (vlaganbagatvr)	ᠣᠸ/(-vn)	Genitive	Ulaanbaatar's
	ᠠᠪᠡ/(-tv)	Dative-Locative	in Ulaanbaatar
	ᡐᠠᠸ/(-aqa)	Ablative	from Ulaanbatar
ᠣᠪᠠᠸ/ ᠙ᠠᠪᠡ (wbama deguu)	ᠠᠷᠡ ᠪᠠᠸ/(-tai-ban)	Comitative	Obama with his brother
ᡐᠣᠵᠣᠸ/ᠣᠸ/ (svrvn_a)	ᠷᠠ(-yi)	Accusative	let Surna
	ᠪᠠᠸ/(-ber)	Instrumental	through Surna

The reflexive suffix helps to integrate nominal word with nominal syntagma, nominal word with verb syntagma or verbal word with verbal syntagma in a sentence coupling them to the subject.

3 Inflection Suffix in Mongolian Syntactic Analysis

Case is one kind of inflection suffix. Case inflection is the phenomenon that adding a case to a noun, adjective, or pronoun that expresses the syntactic relation of the word to other words in the sentence. Take the sentence " ᡐᠣᠵᠣᠣ ᠣᠸ/ ᠊ᠣᠢᠮᠣᠣᠢᠣᠸ/ ᠷᠠ ᠊ᠣᠢᠮᠣᠣᠢᠣ ᠙ᠣᠸᠣᠸ/ ᠈ ᠮᠠᠣᠸᠣᠸ/ ᠊ᠣᠵᠣᠣᠢ ᠈ ᠪᠠᠸ ᠙ᠣᠵᠣᠣᠢᠣ ᠈᠊ " for example. Here, the word " ᠷᠠ" in the sentence is called accusative case (a constituent which appears after " ᠊ᠣᠵᠣᠣᠢ " and indicates that " ᠊ᠣᠵᠣᠣᠢ " is an object); the word " ᠷᠠ" in is called nominative case (a constituent which appears after " ᠊ᠣᠢᠮᠣᠣᠢᠣᠸ/ " and indicates that " ᠊ᠣᠢᠮᠣᠣᠢᠣᠸ/ " is a subject). In Mongolian grammar, the attached stem and the case are considered as a whole word. However, treating them individually bring significant benefit to syntactic analysis.

In order to utilize case information, X. Su et al. in [5] treated the stem and the case suffix in a word as two parts in syntactic analysis and annotated them individually in dependency treebank development. The annotation types of the cases take the case function into consideration. The treebank includes eight kinds of analytical dependencies which relate to the case units.

Meanwhile, X. Su et al. in [6] carried out an experiment on the Mongolian dependency treebank, which contains 400 sentences (13028 annotated words) from Inner Mongolian daily. There are 4460 distinct words and 1548 distinct stems. Labeled attachment score and dependency accuracy achieve 85.0% and 82.6% individually.

4   Inflection Suffix in Mongolian Information Retrieval

There are a few works focusing on the Mongolian information. Part of them investigated the suffix effect in this NLP task. G. Gao in [7] removed all the inflectional suffixes (includes "case") in the words before indexing since they assumed that stemming the Mongolian can not only effectively improve search efficiency, but also reduce index storage space. They constructed a stem and suffix dictionary and used it to remove the inflection suffixes. The dictionary includes 299 suffixes. Since Mongolian words include multiple suffixes, Y. Jin in [8] took three different strategies to processing the corpus before indexing, including (1) no suffix processing, (2) removing the inflection suffixes (including "case"), and (3) removing all the suffixes (derivation suffix and inflection suffix). They compared the retrieval performances when different suffix processing strategies were used. J. Yue in [9] calculated the occurrence of inflection suffix in Mongolian corpus, and used this information to construct the suffix set.

For the two words which have the same stem and different cases, they should be treated as relevance in information retrieval. Suppose that we do not remove the cases before indexing, it is happened that, when we use one of them as the query word, the retrieval result will not include the other one. This is unexpected. Therefore, removing the case in preprocessing section can solve this problem.

In the study, we carry out an experiment to test the effect of inflection suffix in Mongolian information. 27345 Mongolian documents are collected from the Inner Mongolian Newswire. We process the corpus with three methods: (1) no suffix processing, (2) removing inflection suffixes and (3) removing all suffixes. Then, we construct three indexes using the resultant corpus after preprocessing. The index number is 125796 before suffix processing. And the number of index terms decreased to 74001 when inflection suffix processing is used. It is suggested that Mongolian suffix processing can effectively reduce the number of term. And the retrieval efficiency improved in some extent.

The retrieval performance is listed in Table 2. It is clear that inflectional suffix process achieves the best performance at different recall level. This implies that the inflection suffixes should be removed in Mongolian information retrieval. However, derivation suffixes should not be removed, since they give the stems new meanings.

5   Inflection Suffix in Mongolian Translation

As noted, among the many kinds of suffixes, inflectional suffix changes the formation of the word but does not change the meaning of the word. In the process of machine translation, removing the control marks (U202f) is necessary. For an example: a noun will append a case suffix when it acts as a subject, but append another case suffix when it acts as an object. We have to remove the control marks to get the stem of the word to reveal the word semantics, and to reduce the extent of data sparseness. For the template-based machine translation, we need to remove the control marks and the suffixes follow by them. That is because we have to match the Mongolian words in

Mongolian-Chinese dictionary, only if we get off the suffixes of plural and subordinative relationship, the words can be matched correctly.

Table 2. The effectiveness of inflection processing

Recall	No Suffix Processing	Removing Inflection Suffix	Removing All Suffixes
0.0	1	1	1
0.1	0.8101	0.8155	0.8155
0.2	0.7791	0.7992	0.8012
0.3	0.7523	0.7712	0.7712
0.4	0.7114	0.7231	0.7210
0.5	0.6543	0.7012	0.6980
0.6	0.5367	0.5958	0.5731
0.7	0.4367	0.5120	0.4872
0.8	0.3214	0.3964	0.3823
0.9	0.1928	0.2428	0.2516
1.0	0.1002	0.1484	0.1329
Avg	0.5723	0.6096	0.6031

For statistical machine translation, there are two ways to process the U202f control mark. One is to remove the control marks and the suffixes following them directly; the other one is to replace the control mark with space, then the word and the suffix will be taken as two separate words. Both of the way can reveal the real meaning of the words, and alleviate data sparse to get better performance of alignment and translation. The second one will take more abundant information to interference the alignment and translation. We conduct a translation experiment with statistical machine translation mode to evaluate these two ways. The test set includes 1000 bilingual sentence pairs.

Table 3 lists the evaluation result. It is clear that treating the inflection suffix as individual unit makes the system perform best. It also proves that inflection suffix is quite useful in Mongolian translation. Comparing with no inflection suffix processing, removing the inflection suffix slight improves the system. It reduces the sparsity of data. But removing the derivation suffixes changes the meanings of the words.

Table 3. The performance comparison of inflection processing in Mongolian-Chinese translation

Method	BLEU
no inflection suffix processing	21.88
removing the inflection suffix	21.96
treating inflection suffix as individual unit	23.49

6   Conclusion

This paper investigates the Mongolian inflection suffix in NLP tasks. The key findings are as follows: Firstly, inflection suffixes express abundant syntactic meanings. Especially, case suffixes play important roles in syntactic analysis. Secondly, there are three common ways to process the inflection suffix in NLP tasks, including treating them as parts of words, treating them as individual syntactic units, and removing them in the preprocessing section of NLP task. Thirdly, removing the inflection suffix brings a slight improvement in Mongolian information retrieval. Using the inflection suffixes as individual units improves the performance of translation. Furthermore, the inflection suffix inflection can be integrating into more Mongolian NLP tasks. Inflection suffix deserves more investigation to make full of its syntactic functions.

Reference

1. Wei, H., Gao, G.: A Keyword Retrieval System for Historical Mongolian Document Images. International Journal on Document Analysis and Recognition (IJDAR) **17**, 33–45 (2014)
2. Tserenpil, D., Kullmann, R.: Mongolian Grammar. Admon Co. Ltd., Mongolia (2008)
3. Qinggeertai: Traditional Mongolian grammar. Inner Mongolian Press, Huhhot, China (1992)
4. Quejingzabu: Mongolian Coding. Inner Mongolia University Press (2000)
5. Su, X., Gao, G., Yan, X.: Development of traditional mongolian dependency treebank. In: Sun, M., Zhang, M., Lin, D., Wang, H. (eds.) CCL and NLP-NABD 2013. LNCS, vol. 8202, pp. 247–256. Springer, Heidelberg (2013)
6. Su, X., Gao, G., Yan, X.: Dependency parsing for traditional mongolian. In: Proceedings of the 2013 International Conference on Asian Language Processing, pp. 181–184. IEEE Computer Society, Washington, DC (2013)
7. Gao, G., Jin, W., Long, F., Hou, H.: A first investigation on mongolian information retrieval. In: Proceedings of the 2nd International Workshop on Evaluating Information Access (EVIA) (2008)
8. Jin, Y.: Research of Mongolian Information Retrieval Model Based on Markvo Random Field. Inner Mongolia University (2011)
9. Yue, J.: Study on the Methods in the Selection of Retrieval Unit in Mongolian Information Retrieval System. Inner Mongolia University (2011)

Resolving Coordinate Structures for Chinese Constituent Parsing

Yichu Zhou, Shujian Huang[✉], Xinyu Dai, and Jiajun Chen

State Key Laboratory for Novel Software Technology, Nanjing University,
Nanjing, China
zhouyc@nlp.nju.edu.cn, {huangsj,daixinyu,chenjj}@nju.edu.cn

Abstract. Coordinate structures are linguistic structures consisting of
two or more conjuncts, which usually compose into larger constituent
as a whole unit. However, the boundary of each conjunct is difficult to
identify, which makes it difficult to parse the whole coordinate and larger
structures. In labeled data, such as the Penn Chinese Tree Bank (CTB),
coordinate structures are not labeled explicitly, which makes solving the
problem more complicated. In this paper, we treat resolving coordinate
structures as an independent sub-problem of parsing. We first define
coordinate structures explicitly and design rules to extract the coordinate structures from labeled CTB data. Then a specifically designed
grammar is proposed for automatic parsing of coordinate structures. We
propose two groups of new features to better model coordinate structures in a shift-reduce parsing framework. Our approach can achieve a
15% improvement in F-1 score on resolving coordinate structures.

Keywords: Coordinate structure · Grammar · Shift-reduce · Phrase
similarity

1 Introduction

Over the past decades, the Chinese constituent parsing task has been rapidly
improved. However, there are still several structures that can not be parsed
correctly. One of the most difficult structures is the coordinate structure. Kummerfeld et al.[4] showed that the coordinate structures cause 10% of the total
errors in Chinese parsing. So, resolving the coordinate structures is critical for
improving the performance of Chinese parsing.

In linguistics, a coordinate structure is a complex, frequently occurring type of
syntactic structure which links together two or more elements, known as conjuncts
or conjoins. Identifying these conjuncts may need high order information from
other conjuncts, which may highly increase the complexity of a parsing system.

In this paper, we resolve coordinate structures by separating this task from
the overall parsing process to make it an independent sub-task. To investigate
the problem independently, there are some problems to be solved. First of all, the
commonly used syntactic human labeled data (Penn Chinese Tree Bank[1] (CTB))

---

[1] http://www.cis.upenn.edu/~chinese/

© Springer International Publishing Switzerland 2015
J. Li et al. (Eds.): NLPCC 2015, LNAI 9362, pp. 353–361, 2015.
DOI: 10.1007/978-3-319-25207-0_30

does not contain explicit label information for coordinate structures. So there is no explicitly labeled data for training. Secondly, coordinate structures have various numbers of conjuncts and may be nested which makes searching and modeling not trivial.

We analyze possible cases of coordinate structures from both the tree bank data and the CTB label guidance and derive three extraction rules to manually convert a CTB style constituent tree to another tree style which can describe coordinate structures (Section 3). We propose to adapt a context free grammar from Hara [3] to describe multi-conjuncts or nested coordinate structures. Then, we propose two groups of features to model the validity of a single conjunct and similarity between conjuncts, respectively. The proposed grammar and features could be easily integrated into a standard shift-reduce parser to perform efficient search (Section 4).

Experiments are conducted on CTB to verify our proposed solutions (Section 5). The results show that our proposed grammar and features could improve the F1 score of coordinate structures by 15%.

2   Related Work

There have been several researches about coordinate structures in English and other languages. Popel et al. [9] discussed the different representations of coordinate structures in different formats and different languages (not including Chinese).

Hara et al. [3] used a grammar to construct a coordination tree and used edit graphs to evaluate the similarity among the possible conjuncts. They did not evaluate the validity of each single conjunct. They used a simple chart parsing algorithm to generate the possible coordination trees, which searches a significant large number of wrong coordinations.

Ogren [8] introduced language model as the main feature into the detection task. Although language model probabilities could give a implicit and rough indication of structural similarity among conjuncts, it is not sufficient to determine a valid coordinate structure.

Maier and Kübler [5] focused on classifying the punctuations as whether it is the separator of a coordinate structure or not. But no experiments are presented about the identification of coordinate structures in the paper.

3   Coordinate Structure in Chinese

CTB does not label coordinate structures in an explicit way, which gives no explicit target for the identification task. Our first step is to extract correct coordinate structures from labeled trees.

According to the definition of coordinate structures from the bracketing guidlines of CTB [10], the coordinate structures are divided into 3 different levels: **Word Level**, **Phrase Level** and **Clause Level**. In these three levels of coordinate structure, the **Clause Level** is much more complicated than the other

Table 1. Notions of symbols

Symbols	Explanation
conjunct	The conjunct of a coordinate structure.
CC	All possible conjunction words, as listed in Xue et al. [10]
ETC	The Chinese word for ETC (" 等").
PU	Punctuations act as the separator of conjuncts, e.g. ", "

two levels and requires sentence level information to resolve. So at present, we do not consider the clause level in this paper.

For the other two levels, we design 3 rules to extract different coordinate structures from the CTB trees. For simplicity, we first define some notions, which are used in the this section (Table 1). The extracting rules are listed as follows:

- **Rule 1**: Extract subtree structures of the following form:
  conjunct {PU conjunct} PU conjunct[ETC]
- **Rule 2**: Extract subtree structures of the following form:
  conjunct {PU conjunct} [PU] CC conjunct [ETC]
- **Rule 3**: Extract subtree structures whose children are leaf nodes with the same POS labels, or node with the POS labels CC or PU.

In these rules, content in [•] can only appear zero or once; content in {•} can appear zero or more times. Rule 2 handles a special case in Chinese which is showed in Figure 1. Examples of other rules are also showed in Figure 1.

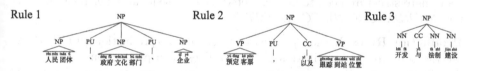

Fig. 1. Examples of 3 different rules

4   Learning to Resolve Coordinate Structures

In this section, we present our methods to resolve coordinate structures. We separate the identification of coordinate structures from the parsing process to be an independent sub-problem. So complex and higher order features could be used in the process.

The general framework is a shift-reduce parsing framework with a perceptron learner [1]. We propose a grammar specifically designed for Chinese coordinate structures (Section 4.1). We propose features to evaluate the validity of a conjunct and the similarity between conjuncts (Section 4.2).

4.1   Grammar of Chinese Coordinate Structures

Hara [3] proposed a grammar for English coordinate structures. We modify it to adjust the specialty of Chinese. We call a parse tree of this grammar a *coordination tree*.

Our grammar, which can cover both nested and flat cases of coordinate structures, is composed of non-terminals (Table 2(a)) and productions (Table 2(b)).

Table 2. Grammar tables

(a) Non-terminals

S	Start symbol
COORD	Complete coordination
COORDX	Partially-built coordination
N	Non-coordination
CJT	Conjunct
CC	conjunction words like " 和"
W	Any word
ETC	only for " 等"
SEP	Connector of conjuncts other than CC

(b) Productions

$$S \longrightarrow COORD$$
$$S \longrightarrow N$$
$$N \longrightarrow W$$
$$N \longrightarrow COORD\ N$$
$$N \longrightarrow COORD$$
$$N \longrightarrow W\ N$$
$$CJT \longrightarrow N$$
$$COORD \longrightarrow COORDX$$
$$COORD \longrightarrow COORDX\ ETC$$
$$CC \longrightarrow SEP\ CC$$
$$COORDX \longrightarrow CJT\ CC/SEP\ CJT$$
$$COORDX \longrightarrow CJT\ CC/SEP\ COORDX$$

Non-terminals. In Table 2(a). *CC* represents coordination words. *COORD* represents a complete coordinate structure and *COORDX* represents a partially coordinate structure need to be completed. *N* represents all inner nodes except for *COORDX*, *COORD*, *CC* and *CJT*.

Production Rules. In these production rules, the two productions of *COORDX* are the core productions in this grammar. These two productions are used to describe both nested and flat cases of coordinate structures. An example of a *coordination tree* is illustrated in the Figure 2.

4.2   New Features

In this section, we focus on the new features that can model coordinate structures. We split these new features into two different groups: **structural/semantic similarity** and **conjunct validity** . According to our experiment results, only similarity related features between two spans is not strong enough to decide if these two spans should be conjuncted. An important source of errors is the wrong identification of a single conjunct. So, we use the **conjunct validity** to evaluate if the given span is a valid span. A valid span means this span can constitute a syntax node in the CTB tree.

There are two different information sources for the new features. First, inspired by the recent success of the distributed word representation in many NLP tasks, we use word embeddings to describe the semantic similarity and

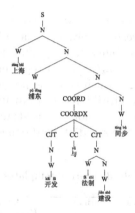

Fig. 2. Example of coordination tree

structural similarity. Another information source is statistical results from CTB data, we use this information to evaluate the validity of conjunct.

We define some notions as follows (w_i represent the i-th word):

- S is a sentence in the form: $w_1 w_2 \cdots w_i \cdots w_j \; CC \; w_k \cdots w_t \cdots w_n$
- $w_i \cdots w_j$ is the first conjunct of S
- $w_k \cdots w_t$ is the second conjunct of S

Similarity. As usual word embeddings, we represent each word as d-dimensional vector $e_i \in R^d$ and use the cosine value of two words to describe the similarity between the two words. For two given spans, we calculate the **semantic similarities** in two perceptive: (i) the average word similarity of the two spans based on different alignments which can be left-most and right-most; (ii) word similarity between the span and context of the two spans. Context of a span is the adjacent words of the span. As an example, the left-most alignment feature is calculated as follows:

$$sim\_align(e_{i,j}, e_{t,k}) = \underset{\substack{i \le q \le j \\ k+q \le t}}{\text{average}}(\cos(e_q, e_{k+q})) \tag{1}$$

where cos is the cosine function to calculate the cosine value of two vectors; *average* is the average function to calculate mean value.

For **structural similarity**, the features are almost the same, except that POS tags are used to replace the words.

Conjunct Validity. Another aspect of information that describes coordinate structures is the validity of a single conjunct. We use statistical probability tables calculated from the CTB tree bank to evaluate this **conjunct validity**. The base idea is calculating the conditional probability from the spans extracted from CTB trees. We use different probability tables which condition on words from the previous, current or next span, respectively. For example, $P_{left\_coherence}(w_{i+1}|w_i)$ represents the conditional probability of w_{i+1} conditioned on w_i which is in the same span of w_{i+1}, while $P_{left\_split}(w_i|w_{i-1})$ represents the conditional probability of w_i conditioned on w_{i-1} which is in the previous spans of w_i.

Complete list of **semantic similarity** and **conjunct validity** features are listed in Table 3. For simplicity, we do not list the **structural similarity** features which use POS tag embeddings instead of word embeddings for calculation.

Table 3. Feature templates

	semantic similarity	conjunct validity
left CJT	$\text{sim\_align}(e_{i,j}, e_{k,k+j-i+1})$; $\cos(e_i, e_k); \cos(e_i, e_{i-1})$; $\cos(e_{i-1}, e_i) - \cos(e_{i-1}, e_k)$;	$P_{left\_coherence}(w_{i+1}|w_i)$; $P_{left\_split}(w_i|w_{i-1})$; $P_{left\_boundary}(w_i)$;
right CJT	$\text{sim\_align}(e_{i,t-k+1}, e_{k,t})$; $\text{sim\_align}(e_{j-(k-t+1),j}, e_{k,t})$; $\cos(e_j, e_t); \cos(e_t, e_{t+1})$; $\cos(e_j, e_{t+1}) - \cos(e_t, e_{t+1})$;	$P_{right\_coherence}(w_k|w_{k-1})$; $P_{right\_split}(w_{k+1}|w_k)$; $P_{right\_boundary}(w_k)$;
COORD	$\cos(averge(e_{i,j}), averge(e_{t,k}))$; $bool(w_{i,j} == w_{t,k})$;	$P_{cond}(w_{i,j}|w_{i-1}, w_{t+1})$; $P_{cond}(w_{k,t}|w_{i-1}, w_{t+1})$;

5   Experiments

We conduct our experiments on Penn Chinese Tree Bank (CTB 5.1) [2] data sets and adapt the same training-test split as described in Zhang and Clark[11]. The embeddings of words and POS tags are trained on data set composed of Chinese gigaword[2] and CTB data using word2vec tools [6]. An in-house implemented Shift-Reduce parser is used as the baseline parser in out experiments.

5.1   Extraction of Coordinate Structures

We apply the three rules (Section 3) to all CTB trees to get well defined coordinate structures, which serve as the learning targets. Among all 18,776 sentences, there are 5,830 sentences which has at least one coordinate structure. The total number of coordinate structures is 8,255, suggesting that a large portion of sentences have multiple coordinate structures. These multiple structures in one sentence may increase the difficulty of resolving coordinate structures. The numbers of coordinate structure with different syntactic tags are showed in the Table 4. As we can see, most of the coordinate structures are under the tag *NP*, *VP* and *QP*. Considering the difficulty of resolving tag *NP* and *VP*[4], resolving the coordinate structures under *NP* and *VP* is much more difficult.

Table 4. Syntax tags distribution of coordinate structures

VCD	ADJP	UCP	ADVP	VP	CLP	PP	DNP	QP	LCP	NP	IP	CP
2	50	5	21	1054	2	13	2	216	10	6880	0	0

Table 5. Experiments results

(a) The different results of sentences with and without coordinate structure

	all test	with coordinate structure	without coordinate structure
Recall	0.7880	0.7849	0.7919
Precision	0.8437	0.8321	0.8585
F-measure	0.8149	0.8078	0.8239

(b) Coordination tree result

	Recall	Precision	F-measure
baseline	0.6339	0.6418	0.6378
new features	0.6717	0.6730	0.6723
basic features	0.6683	0.6708	0.6696
basic + new featrues	0.7775	0.7836	0.7805

5.2 Effects on the Parsing Process

There are no previous experimental results that demonstrate the influence of Chinese coordinate structures[3]. After we get the well defined coordinate structures, we conduct two experiments to check out the real influence of Chinese coordinate structures on the parsing process. Firstly, we separated the CTB test data into two parts, each sentence of the first part has at least one coordinate structure while the sentences of the second part have no coordinate structures at all. Then, we parse and score the two parts separately using the same training data set and the results are showed in the Table 5(a). As we can see, the sentences without coordinate structures get 1.6% higher score in F-measure than the sentences with coordinate structures, which proves coordinate structures have great effect on the overall parsing task.

Secondly, we apply the same extracting rules(Section3) to the output result of our baseline parser to check out how many coordinate structures can be parsed correctly by traditional parsers. This parser achieved 63% recall and 64% precision. This result tells us that the state-of-art parser can only parse about 63% coordinate structures correctly. We use this result as our baseline results in the following experiments.

5.3 Resolving Coordinate Structures

Scoring methods of *coordination tree* is similar to the traditional scoring methods in parsing. But in this *coordination tree*, we only score the spans of the node *COORD* and do not count other spans which is not related to the coordinate structures. We trained *coordinate tree* with a perceptron learner on different

---

[2] https://catalog.ldc.upenn.edu/LDC2005T01

[3] Ng and Curran[7] has showed the influence of coordinate structures on dependency parsing.

features set. *Base features* means the traditional feature template described in Zhang and Clark[12]. *New features* means the features we discussed in this paper. We also conduct an experiment on the combination of these two feature sets.

As showed in the Table 5(b), using *base features* and *new features* separately can only achieve a little improvement. While when we combine these features, we can achieve 15% improvement in F-measure. This indicates that some coordinate structures can simply be remembered (*base features*) by the model while other coordinate structures need more information (*new features*) to resolve.

6 Conclusion and Future Work

In this paper, we discuss the problem of coordinate structures in Chinese constituent parsing. We separate the problem of identifying the coordinate structure from parsing task. We present how to extract coordinate structures from CTB style trees according to their definitions. Then we presented a framework to solve the identification problem, which includes a specifically designed grammar and newly designed features. Our new features focusing on evaluating coordinate structures include two different groups: **similarity between conjuncts** and **conjunct validity** .

Experiment results show these new features have advantages on modeling coordinate structures. With these features and the grammar, we achieved 15% improvements on detecting coordinate structures.

Acknowledgments. This work is supported by the National Natural Science Foundation of China (Grant No. 61300158, 61170181), the Jiangsu Provincial Research Foundation for Basic Research (Grant No. BK20130580).

References

1. Collins, M.: Discriminative training methods for hidden markov models: theory and experiments with perceptron algorithms. In: Proceedings of the ACL 2002 Conference on Empirical Methods in Natural Language Processing, vol. 10, pp. 1–8. Association for Computational Linguistics (2002)
2. Graff, D., Chen, K.: Chinese gigaword. LDC Catalog No.: LDC2003T09, ISBN 1, 58563–58230 (2005)
3. Hara, K., Shimbo, M., Okuma, H., Matsumoto, Y.: Coordinate structure analysis with global structural constraints and alignment-based local features. In: Proceedings of the Joint Conference of the 47th Annual Meeting of the ACL and the 4th International Joint Conference on Natural Language Processing of the AFNLP, vol. 2, pp. 967–975. Association for Computational Linguistics (2009)
4. Kummerfeld, J.K., Tse, D., Curran, J.R., Klein, D.: An empirical examination of challenges in chinese parsing. In: ACL (2), pp. 98–103 (2013)
5. Maier, W., Kübler, S.: Are all commas equal? detecting coordination in the penn treebank. In: The Twelfth Workshop on Treebanks and Linguistic Theories (TLT 2012), p. 121 (2013)
6. Mikolov, T., Chen, K., Corrado, G., Dean, J.: Efficient estimation of word representations in vector space (2013). arXiv preprint arXiv:1301.3781

7. Ng, D., Curran, J.R.: Identifying cascading errors using constraints in dependency parsing
8. Ogren, P.V.: Improving syntactic coordination resolution using language modeling. In: Proceedings of the NAACL HLT 2010 Student Research Workshop, pp. 1–6. Association for Computational Linguistics (2010)
9. Popel, M., Marecek, D., Stepánek, J., Zeman, D., Zabokrtskỳ, Z.: Coordination structures in dependency treebanks. In: ACL (1), pp. 517–527 (2013)
10. Xue, N., Xia, F., Huang, S., Kroch, A.: The bracketing guidelines for the penn chinese treebank (3.0) (2000)
11. Zhang, Y., Clark, S.: A tale of two parsers: investigating and combining graph-based and transition-based dependency parsing using beam-search. In: Proceedings of the Conference on Empirical Methods in Natural Language Processing, pp. 562–571. Association for Computational Linguistics (2008)
12. Zhang, Y., Clark, S.: Transition-based parsing of the chinese treebank using a global discriminative model. In: Proceedings of the 11th International Conference on Parsing Technologies, pp. 162–171. Association for Computational Linguistics (2009)

P-Trie Tree: A Novel Tree Structure
for Storing Polysemantic Data

Xin Zhou[✉]

Department of Economic Engineering, Kyushu University, Fukuoka 812-8581, Japan
thebestzx@163.com

Abstract. Trie tree, is an ordered tree data structure that is used to store a dynamic set or associative array where the keys are usually strings. It makes the search and update of words more efficient and is widely used in the construction of English dictionary for the storage of English vocabulary. Within the application of big data, efficiency determines the availability and usability of a system. In this paper, I introduce p-trie tree-a novel trie tree structure which can be used for polysemantic data which are not limited to English strings. I apply p-trie to the storage of Japanese vocabulary and evaluate the performance through experiments.

Keywords: P-trie · Trie tree · Polysemy · Time stamp · Binary search tree

1 Introduction

1.1 Favorite Words and High Frequency Words

Everyone has favorite words which are often spoken out or written in the letters or papers. For some people, the favorite word is "sorry", because they are gentle and always be modest for other people. And for some other people, one of their favorite words is "amazing", which is used in "Having an amazing day" or "That is amazing". For most people, favorite words are never really given much thought to, they use these words unconsciously, only when they are asked such questions about "what are your favorite words? ", they come to recall some words. Even when they tell their favorite words, they don't realize how much they use them.

Favorite words is one manifestation of high frequency words. The statistic of high frequency words is not only used for exploring people's psychological status, but also used for natural language processing, language study and text mining, etc.In text mining, high frequency words are helpful to obtain the substance of the text.

Not only English, any kind of language has favorite words. For example, in Japanese, we call it "口癖"(the pronunciation is "kuchiguse").

1.2 Kuchiguse

"Kuchiguse" can help to know people's psychological trend and character. For example, when two people are talking, if one people always say "たしかに"("tashikani"),

© Springer International Publishing Switzerland 2015
J. Li et al. (Eds.): NLPCC 2015, LNAI 9362, pp. 362–371, 2015.
DOI: 10.1007/978-3-319-25207-0_31

it implies that although the people seem to be understanding another people's sayings, but actually he is not interested in the topic, he wants to finish the talking as soon as possible. If one people always say "忙しい"("isogashii"), but he is not really so busy, we can infer that the people wants other people to recognize his existence and sometimes he is a little lonely.

1.3　NLP and Trie Tree

Natural language processing (NLP) is an interdiscipline related with computer science, artificial intelligence, and computational linguistics, its task is to enable computers to derive meaning from natural language input. For Japanese, some important steps of NLP, such as semantics-based words segmentation, can be solved by the Part-of-Speech Analyzer Mecab[1].

In NLP, a vital problem is how to store words. For English, we can use trie tree to store words. A trie tree has a number of advantages over a binary search tree (BST for short). It can also be used to replace a hash table, over which it has several advantages as follows:

1. For looking up data in a trie tree, the time complexity is $O(m)$ (where m is the length of a search string).
2. There are no collisions of different keys in a trie tree.
3. There is no need to provide a hash function.
4. A trie tree can provide an alphabetical ordering of the information of nodes.

Trie tree does have some drawbacks as well. The main drawback is the more require ment of space than a hash table.

1.4　Polysemy

"Polysemy" means one word has multiple meanings. In this paper, we give it more implications which are as follows.

1. One pronunciation has different words. For example, "選考", "先行", "専攻" and "せんこう" have the same pronunciation of "senkou".
2. One data item has different manifestations. This situation is often used in the field of cryptology. For example, a string "10" may have some concealed information which are comprised of three strings:"10", "IT" and "do". If we don't know the concealed information, we can't get to know the actual representation of the original data.
3. One word has different manifestations of other words without regard for pronunciation. For example, "make" may have one actual meaning selected from three words: "make", "more", "more", "made".

I propose a deformed trie tree named p-trie tree(p-trie for short). "P" means "polysemy". P-trie can store any kind of data, not limited to words. It can efficiently help us to search out high frequency words and the actual representations of original data.

In this paper, we apply p-trie to the storage of Japanese vocabulary.

The existing trie tree is unavailable for Japanese because of the difference of language structures. In English, any word is composed of 26 characters, every character has the same length of one digit. In Japanese, words are composed of "kanji" , "hiragana" or "katagana" , "kanji"'s with equal length may have different lengths of hiragana which represents the pronunciation, same words may have different forms, sometimes words are written by "kanji", sometimes by "hiragana".

The paper is organized as follows. Section 2 gives the related works. Section 3 explain the structure of p-trie. Section 4 discusses the application of p-trie. Section 5 is the conclusion.

2   Related Works

Mecab[1] is a very popular Part-of-Speech Analyzer in the field of Japanese language processing. It has a number of useful functions. We often use the functions of Lexical analysis, N-Best solutions output, word by word output, pronunciation ("hiragana" or "katakana") output, etc. Sosuke Amano et al.[4] do frequency statistics of Japanese food names from the foodlog. Because for one food, maybe there are several different names . They can use small numbers of words to describe most of the Japanese foods. Through building various kinds of libraries for frequently appearing words, Yuki Akiyama et al.[5] improve the accuracy of the comparison and identification of Japanese words. Tomoya Noro et al.[6] use a graph-based method to rank Japanese words, this is the first step toward building a Japanese defining vocabulary. These are some latest publications in the field of Japanese Language Processing, within these papers, the storage mode of Japanese words is not discussed as a key point.

All the time, there are a lot of research papers concerning on tree structure. Based on the idea of partitioning T into a set of linked small tries, each of which can be maintained efficiently, Jesper Jansson et al.[2] propose a new technique for maintaining a dynamic trie T of size at most 2 w nodes under the unit-cost RAM model with a fixed word size w. Kevin Leckey et al.[3] consider a more realistic model where words are generated by a Markov source. By a novel use of the contraction method combined with moment transfer techniques they prove a central limit theorem for the complexity of radix sort and for the external path length in a trie tree.

3   P-Trie Tree

3.1   Definition

P-trie tree is a novel trie-tree. It is an ordered tree data structure that is used to store a dynamic set or associative array where the keys can be of any type. The position in the tree defines the key with which it is associated. Any node not only links to its child nodes, but also links to a data structure called the information of this node("Info" for short). All the descendants of a nod have a common prefix of the string associated with that node, and the root is associated with the empty string.

Compared with trie-tree, p-trie has some differences.

1. The keys in the nodes can be of any type. According to different conditions, the number of types of keys and the values for the corresponding keys may be different.
2. Any node also links to a data structure called "Info". According to different conditions, "Info" may be vector, binary search tree or others.
3. "Info" can contain uncorrelated data which are of different types or different meanings. The data in "Info" are not limited to have relation with the keys in the nodes.

3.2   Comparison with Map (Red-Black Tree)

Map is one kind of red-black tree, it can also store a dynamic set or associative array. Compared with Map, p-trie has some advantages.

1. Map has the limitation of capacity. P-trie has a much larger capacity.
2. Map has a mechanism of automatically sorting of the data. But for other different language's vocabulary, it is very different for us to compare which is smaller between two words. For example, in a Map, "務める"("つとめる") is smaller than "収集"("しゅうしゅう"). But according to the order of Japanese phonetic alphabet, "収集" should be in front of "務める". In p-trie, we store words by the order of phonetic alphabet, it is easy to compare two words.

3.3   Application in Japanese

In p-trie, every node terminates some words comprised of the Japanese characters in the depth-first search route from the root's child node to this node. Each node contains a subtree which stores the different appeared words of the same pronunciation In the subtree, nodes are arranged by the sequence of time stamps which denotes the first time of appearance of this written form, the node with time stamp 0 is the first child, node with time stamp 1 is the second, and so on. Each node's subtree stores some information ("info") of words terminated at this node, "Info" contains four main parts: the string of the word in the written form of hiragana, kanji or katakana; the time stamp; the frequency and the vector of numbers of sentences in which the word appears.

Based on the p-trie tree structure, we can do Japanese paragraph analysis. The fundamry ental of paragraph analysis is sentence analysis, and the fundamental of sentence analysis is word analysis.

Case 1. Given 13 words, assume that words appear chronologically and each word is in separated different sentences. We can insert them to construct a p-trie tree.

The words are:　せんこう　先行　選考　専攻　学問
　　　　　　　　　生協　せんこう　成果　生活　セイカ
　　　　　　　　　先行　成果　せんこう

For a Japanese word, there are three kinds of written forms: hiragana, katakana and kanji. For one hiragana, maybe there are several different kanjis with different meanings. Although kanjis are different, they also have the same hiragana. In some Japanese advertisements, hiragana can also be written in the form of katakana, the meaning is same, just the written forms are changed.

To classify different words which have the same hiragana(then same pronunciation), we give every word a time stamp which is ordered by the time of appearance starting from 0. In p-trie, we use hiragana characters to express words, katakana is deemed to be a different form of the same hiragana. For one hiragana, maybe there is only one word, then the time stamp is 0, maybe there are more than one written form, then time stamp is the time order number of the first appearance of this form.

The information of all the words in case 1 are shown in Table 1.

Table 1. Information of words

Japanese word	Time stamp	frequency	Sentence No.
せんこう	0	1	1
先行	1	1	2
選考	2	1	3
専攻	3	1	4
学問	0	1	5
生協	0	1	6
せんこう	0	2	7
成果	0	1	8
生活	0	1	9
セイカ	1	1	10
先行	1	2	11
成果	0	2	12
せんこう	0	3	13

Based on the definition of p-trie, we can construct a p-trie as shown in Fig. 1.

As shown in the p-trie, root node is an empty node which is denoted as "R", from the left to right, the Japanese characters of child nodes are arranged by the order in Japanese phonetic alphabet which is shown in Fig. 3. Every pane represents "Info". In a pane, the first line is the word, the second, third and fourth line separately represents the time stamp, frequency and the vector of sentence numbers.

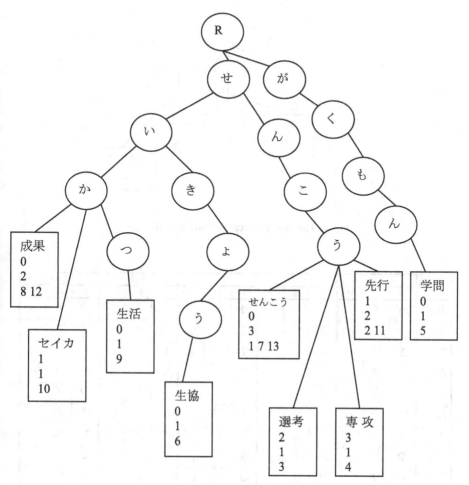

Fig. 1. The p-trie tree of case 1

The data structure of "Info" is decided by the actual applications. If we want to search by the content of characters, we shall construct a p-trie in Fig.1, if we want to search by the frequency, the "Info" shall be constructed as a binary search tree in which the time complexity of search a data item by the key (in this context is frequency) is O(logN) where N is the number of nodes in the BST. Fig.2 shows the BST of "Info" for the hiragana "せんこう".

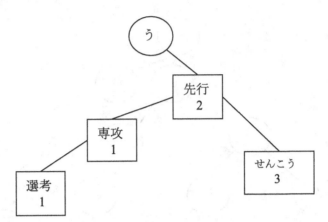

Fig. 2. The binary search tree structure of "Info"

段 行	あ段			い段			う段			え段			お段		
	平	片	罗马	平	片	罗马	平	片	罗马	平	片	罗马	平	片	罗马
あ行	あ	ア	a	い	イ	i	う	ウ	u	え	エ	e	お	オ	o
か行	か	カ	ka	き	キ	ki	く	ク	ku	け	ケ	ke	こ	コ	ko
さ行	さ	サ	sa	し	シ	si	す	ス	su	せ	セ	se	そ	ソ	so
た行	た	タ	ta	ち	チ	ti	つ	ツ	tu	て	テ	te	と	ト	to
な行	な	ナ	na	に	ニ	ni	ぬ	ヌ	nu	ね	ネ	ne	の	ノ	no
は行	は	ハ	ha	ひ	ヒ	hi	ふ	フ	hu	へ	ヘ	he	ほ	ホ	ho
ま行	ま	マ	ma	み	ミ	mi	む	ム	mu	め	メ	me	も	モ	mo
や行	や	ヤ	ya	い	イ	i	ゆ	ユ	yu	え	エ	e	よ	ヨ	yo
ら行	ら	ラ	ra	り	リ	ri	る	ル	ru	れ	レ	re	ろ	ロ	ro
わ行	わ	ワ	wa	い	イ	i	う	ウ	u	え	エ	e	を	ヲ	wo
か行	が	ガ	ga	ぎ	ギ	gi	ぐ	グ	gu	げ	ゲ	ge	ご	ゴ	go
さ行	ざ	ザ	za	じ	ジ	zi	ず	ズ	zu	ぜ	ゼ	ze	ぞ	ゾ	zo
た行	だ	ダ	da	ぢ	ヂ	di	づ	ヅ	du	で	デ	de	ど	ド	do
は行	ば	バ	ba	び	ビ	bi	ぶ	ブ	bu	べ	ベ	be	ぼ	ボ	bo
は行	ぱ	パ	pa	ぴ	ピ	pi	ぷ	プ	pu	ぺ	ペ	pe	ぽ	ポ	po
													ん	ン	n

Fig. 3. Japanese phonetic alphabet

In Fig.3, we can see the order is numbered from left to right and from top to bottom. For hiragana, there are 71 distinct characters, the first one is "あ" and the last one is "ん". There are four additional short syllables "っ","ゃ","ゅ","ょ" and one long syllable "ー" which are not shown in phonetic alphabet. Katakana is deemed to be a special written form of the same hiragana. Therefore, there are totally 76 characters.

In fig. 1, we can see for the same root "せ","い" is in the left of "ん";"セイ カ" is another form of "せいか".

3.4   Other Applications

As discussed in section "Introduction", p-trie can be used in the field of cryptology.

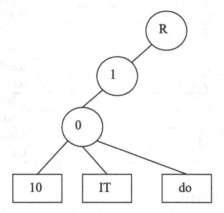

Fig. 4. A part of p-trie which stores actual representations

Fig.4 gives a part of a p-trie in which the written string "10" has three types of actual representations. For example, given a two-tuples ("10", 1) in which "1" is the index of the actual value in the "Info", if we have this p-trie, we can easily get the string "IT", otherwise, we cannot know the actual value of the original string.

4   Experiments

We build a file called "words.txt" in which each line is consisted of several Japanese words with the same pronunciation and the last one is the hiragana. In "words.txt", there are 110 words. We randomly generate a text file named "in.txt" which contains 10000000 lines, each line is a word, all of the words are from "words.txt". We construct a p-trie by the words in "in.txt" and do operations of insert and search in the p-trie.

We use C++ to do the experiments, the development environment is Microsoft Visual C++ 2010. All experiments were run on a computer with an Intel COREi5 2.7 GHz CPU, 4GB RAM, running Windows 7 OS.

4.1   Results

In the p-trie, we insert strings and search strings. If the string is existing, we output the time stamp and frequency. Sometimes, the string is not in the p-trie, but it's a prefix of a word which is in the p-trie, "かく" is an example for this case.

In table 2, we list the operations and the corresponding running time. In Init(), we build the map of Japanese phonetic alphabet and get the hiragana for every word which is in "words.txt". In ReadIn(), we read in the data in "in.txt" whose size is 68176 KB.

Table 2. Experimental results

Operation	Running time (unit: seconds)
Init()	0.005000
ReadIn();	546.100000
insert "ありがとう"	0.000000
search "ありがとう"	0.000000
search "かく"	0.000000
search "かく s"	0.000000
search "かくしんかく"	0.000000
search "大学"	0.000000
search "因る"	0.000000
search "合う"	0.000000
search "昇る"	0.000000

4.2   Performance Analysis

In the p-trie, every node has at most 76 child nodes which are arranged by the order of Japanese phonetic alphabet. Each child node has a subtree in which the nodes are arranged by the order of time stamp.

Define the length of characters of the hiragana of a word as the length of the word.

The time complexity of insert, search and update is $O(m)$ where m is the length of the word. The space complexity is $O(76^n)$ where n is the maximum length of word in the p-trie. P-trie tree uses huge space to achieve a faster running time.

5   Conclusion

In this paper, we discussed a novel data structure named p-trie tree. The most remarkable advantage of p-trie tree is it can effectively store polysemantic data. The time performance of most operations on p-trie tree is very good. The requirement of space is huge but nowadays hard drive is cheap and easy to get.

P-trie tree is not only available for the storage of vocabulary of languages, but also available for other applications with the same or similar data structure, for example, the application of information security. For different applications, the structure of p-trie could have a little change, such as the number of child nodes, the information of child nodes, etc. New applications of p-trie tree could be future challenges.

References

1. http://mecab.googlecode.com/svn/trunk/mecab/doc/index.html
2. Jansson, J., Sadakane, K., Sung, W.-K.: Linked Dynamic Tries with Applications to LZ-Compression in Sublinear Time and Space. Algorithmica **71**(4), 969–988 (2015)
3. Leckey, K., Neininger, R., Szpankowski, W.: A Limit Theorem for Radix Sort and Tries with Markovian Input (2015). CoRR abs/1505.07321
4. Amano, S., Ogawa, M., Aizawa, K.: Frequency statistics of words used in Japanese food records of FoodLog. In: UbiComp Adjunct 2014, pp.547–552 (2014)
5. Akiyama, Y., Shibasaki, R.: A method for identifying japanese shop and company names by spatiotemporal cleaning of eccentrically located frequently appearing words. In: Adv. Artificial Intellegence (ADVAI 2012), pp. 562604:1–562604:18 (2012)
6. Noro, T., Tokuda, T.: Ranking words for building a japanese defining vocabulary. In: IJCNLP 2008, pp. 679–684 (2008)

Research on the Extraction of Wikipedia-Based Chinese-Khmer Named Entity Equivalents

Qing Xia[1,2], Xin Yan[1,2(✉)], Zhengtao Yu[1,2], and Shengxiang Gao[1,2]

[1] School of Information Engineering and Automation,
Kunming University of Science and Technology, Kunming 650051, China
kg_yanxin@sina.com
[2] The Intelligent Information Processing Key Laboratory,
Kunming University of Science and Technology, Kunming 650051, China

Abstract. Named entity equivalent has been playing a significant role in the processing of cross-language information. However limited by the corpora resource, few in-depth studies have been made on the extraction of the bilingual Chinese-Khmer named entity equivalents. On account of this, this paper proposes a Wikipedia-based approach, utilizes the internal web links in Wikipedia and computes the feature similarity to extract the bilingual Chinese-Khmer named entity equivalents. The experimental result shows that good effect has been achieved when the entity equivalents are acquired through the internal web links in Wikipedia with F value up to 90.67%. Also it shows that the result is quite favorable when the bilingual Chinese-Khmer named entity equivalents are acquired through the computation of feature similarity, turning out that the method proposed in this paper is able to give better effect.

Keywords: Named entity equivalents · Chinese-Khmer bilingual · Wikipedia · Transliteration model · Translation model

1 Introduction

Name entity equivalents have been widely applied in the processing of natural language. In a MTS, the targeted processing of named entities has played an important role in the improvement of the overall translation quality, therefore it's of great application value. Currently most of the researches on the acquisition of named entity equivalents are focused on the adoption of the parallel corpora to mine the named entity equivalents with high accuracy rate having been achieved[1]. In Literature[2], it has proposed such a method to extract the relevant named entity equivalents from the source language corpus. Meng[3] has extracted the named entity equivalents through transliteration, while Huang[4] proposes a multi-feature-based minimum cost approach to extract automatically the named entity equivalents. However since the resource of parallel corpora is extremely limited, it will cost a lot. Compared with the parallel corpora, the comparable corpus can be obtained more conveniently with more abundant contents. Cao[5] proposes an approach to mine the Chinese-English translation equivalence from Chinese web pages. However the effect is not so favorable through such an approach

© Springer International Publishing Switzerland 2015
J. Li et al. (Eds.): NLPCC 2015, LNAI 9362, pp. 372–379, 2015.
DOI: 10.1007/978-3-319-25207-0_32

with few entity equivalents having been acquired. Although this method has brought some positive effect when comparable corpus is utilized to acquire the bilingual named entity equivalents[6], it never takes the characteristics of named entity itself into account. However since the Chinese- Khmer bilingual corpus resources are extremely limited on internet, currently no researches have ever been made on the extraction of Chinese- Khmer named entity equivalences.

On Wikipedia, all of the entries are provided with links to the other languages, as the correlation between different languages and the same entity is implied in the organizational structure[7]. Firstly, utilize the characteristic of an entry, which is correlated to different languages on Wikipedia to acquire the correlation between the entry in source language and that in target language, whose entry is the same with that in the source language. Then decode the character according to the correlation with the target language to acquire the Chinese- Khmer named entity equivalences. In order to compensate for the dependency of this internal web link-based extraction method on the link information and to obtain more Chinese- Khmer named entity equivalences, this paper utilizes the characteristic of Wikipedia that different language descriptions on the same entity are comparable corpora to extract the Chinese- Khmer named entity equivalences by computing the feature similarity in transliteration and in translation with the combination of the characteristic that personal name and place name is quite similar in pronunciation and that the organization name is featured with translation characteristics.

2   Analysis on the Wikipedia Page Structure

The analysis on the page structure reveals that the entry information in source language is always covered in the web title of the page that contains this entry. Since all of the entries on Wikipedia have been provided with links to the other languages, the link analysis shows that the hyperlink of an entry to the other language always contains the information of the other language for this entry. In this way, both of the Chinese information and the Khmer information of an entry might be found on the same Wikipedia page. Then through the further analysis on the Chinese content and the Khmer content of an entry on the same Wikipedia page, it reveals that the Chinese content and the Khmer content are featured with the following characteristics such as entity co-occurrence and similar description content etc.

3   Extraction of Chinese- Khmer Named Entity Equivalences Based on the Internal Links on Wikipedia

The Chinese entries with links to Khmer can be obtained on Wikipedia pages when Chinese is the source language and Khmer is the target language. Take the Chinese entry of "China" as an instance. The web title for this entry on the Chinese web page is "China- Wikipedia, the free encyclopedia" with the header information having been stored between the html tags of <title> and </title> in the page source. Actually the header information for the Entry "China" is "<title> China- Wikipedia, the free encyclopedia</title>" in the page source. However on Wikipedia, the page source of each entry only consists of an html tag <title></title>. Then the extraction of the information between "<title>" and "- Wikipedia, the free encyclopedia</title>" also indicates the entry information in Chinese.

Since Wikipedia provides different language links to the same entry, then the link of the entries containing Khmer link information will be started with "km" in a webpage source file. For example, the Khmer link to the entry of "China" is indicated as "href=//km.wikipedia.org/wiki/%E1%9E%85%E1%9E%B7%E1%9E%93" in the webpage source file, where the link contains a UTF-8 encoded character string besides the domain name address that is linked to. Followed with the decoded character string, it's the relevant entry in Khmer corresponding to the Chinese entry.

A bilingual Chinese- Khmer entry will be obtained after html tag filtering is conducted on the extracted bilingual Chinese-Khmer information. However such a bilingual Chinese- Khmer word pair always contains some entries that are not named entities. Since numerous researches have been made on the Chinese named entity recognition by now, this paper then adopts the Chinese named entity recognition method to extract the Chinese-Khmer named entity equivalence from the bilingual Chinese-Khmer word pairs. The process for the extraction of Chinese-Khmer named entity equivalences based on links is shown in Figure 1 below.

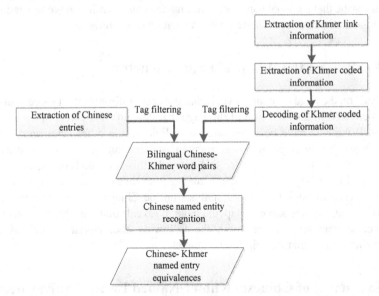

Fig. 1. Process for the extraction of Chinese-Khmer named entity trans-lingual equivalence based on links

4　Extraction of Chinese- Khmer Named Entity Equivalence Based on Feature Similarity

On Wikipedia, the link-based extraction of named entity equivalence is subject to the link information of this entry in the other languages. However since numerous named entities on Wikipedia are not expressed by entries, there won't be any link available in the other languages corresponding to them. However they can still be found in the introduction of the entries. Through the analysis on the introduction of an entry in different languages, it reveals that the introduction of the same entry in Chinese and in Khmer is almost the same. Therefore in this paper, such entries are considered as the comparable corpus. Firstly, this paper identifies the named entities in Chinese and in Khmer through the named entity recognition method. Then according to the different types of the named entities, this paper separately utilizes the characteristics of transliteration and translation to compute the similarity in the candidate named entity equivalence for the purpose to obtain finally the Chinese-Khmer named entity equivalence with the specific steps provided as below:

(1) Acquire the textual description of an entry in Chinese on Wikipedia.
(2) Acquire the corresponding textual description of an entry in Khmer according to the hyperlink to this entry in Khmer.
(3) Filter the Chinese and Khmer texts of this entry.
(4) Conduct named entity recognition on the Chinese text.
(5) Conduct named entity recognition on the Khmer text.
(6) Compute the similarity in the candidate named entity equivalences.
(7) Acquire the bilingual Chinese-Khmer named entity equivalences.

4.1   Characteristics of Transliteration

Due to the linguistic characteristics of Khmer, there's a high degree of similarity between the production of the personal name and place name in Khmer and the production of them in Chinese. In view of this, this paper would like to acquire the personal name and place name equivalences through the computation on the phonetic similarity in the candidate named entity equivalences. In allusion to the characteristics of transliteration[8-9], this paper utilizes a transliteration probability dictionary obtained through the training of the statistical machine translation model to generate a transliteration probability dictionary.

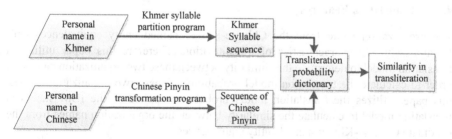

Fig. 2. Flow chart for the calculation of similarity in transliteration

As shown in Figure 2, the generated transliteration probability dictionary can be used directly to compute the similarity in the personal name and place name from the Chinese-Khmer named entity equivalences. Three steps are provided as below according to this computing method:

(1) Decompose the candidate personal name or place name equivalence in Khmer into syllable sequences.

(2) Transform the personal name or place name in Chinese into Pinyin sequences.

(3) Use the transliteration probability dictionary to assess the similarity of the candidate Chinese-Khmer personal name or place name equivalence in transliteration.

Regarding the process to calculate the similarity in the candidate Chinese-Khmer personal name or place name equivalence[10], this paper takes the personal name equivalence as an instance. Assume that there's a candidate Chinese-Khmer personal name equivalence, which contains Cp, the personal name entity in Chinese and Kp, the personal name entity in Khmer. Firstly, transform Cp into a Pinyin sequence, $ne_c =$ $\{ c_1, c_2, ... c_n \}$. For example, "WangJunJie" can be transformed into "wang/jun/jie". As to Kp, the personal name entity in Khmer, it can be decomposed into kcc syllable sequences, $ne_k = \{k_1, k_2, ... k_n\}$. For example, " វ៉ាងជុនជៀ (WangJunJie)" can be transformed into "វ៉ា/ជុន/ជៀ.Apply the generated nec, the Chinese Pinyin sequence and nek, the Khmer syllable sequence that has been generated to the transliteration probability dictionary so as to calculate the similarity between the candidate Chinese-Khmer name equivalences, which are Cp and Kp in transliteration with the computational formula (1) provided as below:

$$Score(ne_c, ne_k) = \frac{\sum_{i=1}^{n} \sum_{j=1}^{m} \left(P(c_i \mid k_j) + P(k_j \mid c_i) \right)}{m+n} \tag{1}$$

Where ne_c is the Pinyin sequence of Cp, the personal name entity in Chinese and n indicates the length. c_i is the ith syllable in the Sequence ne_c. ne_k is the syllable sequence of Kp, the personal name entity in Khmer and m is the length. Meanwhile k_j is the jth syllable in Sequence ne_k, $P(c_i|k_j)$ is the probability that c_i can be translated into k_j, while $P(k_j|c_i)$ is the probability that k_j can be translated into c_i.

4.2   Translation Features

The organization name from the Chinese-Khmer named entity equivalences isn't featured with the characteristics of transliteration. Therefore this paper utilizes a translation model to determine the similarity between these two organization names. In order to calculate the translation model probability of these two organization names, this paper utilizes the translation model probability applied in the IBM statistical translation model to calculate the similarity between the organization names from the candidate Chinese-Khmer named entity equivalences.

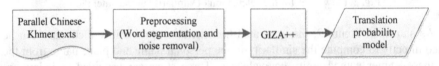

Fig. 3. The process for the generation of a translation probability model

Figure 3 shows the flow chart for the generation of a translation probability model with the parallel Chinese-Khmer alignment corpora acting as the input data and the translation probability model acting as the output data.

The steps to train a translation probability model are provided as below:

(1) Pre-process the parallel Chinese-Khmer aligned texts to remove the noise and interference.

(2) Align the words with the help of GIZA++.

(3) Calculate the translation probability.

As shown in Formula (2), use the generated translation probability dictionary to calculate the similarity of the organization names from the candidate Chinese-Khmer named entity equivalences in the translation features:

$$Score_{trans}(ne_c, ne_k) = \frac{2 \times \sum_{i=1}^{n} \sum_{j=1}^{m} \Pr(k_i \mid c_j)}{m+n} \tag{2}$$

Where ne_c represents the named entity in Chinese, n is the number of the words in ne_c, the named entity in Chinese, ne_k represents the named entity in Khmer, m is the number of the words in ne_k, the named entity in Khmer, k_i is the ith word in ne_k and c_j refers to the jth word in ne_c.

5   Experiment and Assessment

This paper chooses totally 5000 Chinese entries from Wikipedia including such named entities as personal name, place name and organization name. Then tag manually the named entities from these 5000 Chinese entries, the corresponding Khmer and the entity types, choosing randomly 500 entries, which are also the named entities with Chinese content and with Khmer content as the comparable corpus, where both of the Chinese entities and Khmer entities are tagged.

Table 1. Number of Chinese entries in Wikipedia

Entries	Personal name	Place name	Organization name	Non-entity
5000	1448	1836	1672	44

Table 2. Number of entities in the comparable corpus

Chinese corpus (500)			Khmer corpus (500)		
Personal name	Place	Organization name	Personal name	Place name	Organization name
361	539	452	212	487	284

Table 3. Number of Chinese-Khmer named entity equivalences extracted based on links

Assessment	Total entities	Personal name	Place name	Organization name
Input	4956	1448	1836	1672
Output	4238	1184	1658	1396
Recall rate	85.5%	81.7%	90%	83.4%
Rate of accuracy	96.5%	89.7%	99.5%	98.9%
F value	90.67%	85.5%	94.5%	90.5%

In the experiment, where the extraction of the Chinese-Khmer named entity equivalences is made based on the links ,as shown in the analytical table 3, the number of the extracted Chinese-Khmer named entity equivalences is 4238, which is less than 4956, the number of the inputted Chinese entities. Through the analysis on the extracted Chinese-Khmer named entity equivalences, it reveals that no equivalences in Khmer have been extracted to correspond to 671 Chinese entities. It's because that no Khmer links have been available on Wikipedia to these 671 Chinese entities. Moreover although 47 pairs of Chinese-Khmer named entity equivalences have been extracted, they fail to be covered in the effective set of named entity equivalence since all of them have been recognized as non-entities during the named entity recognition, including 28 pairs of organization names, indicating that the recognition on the organization name is yet to be improved. Regarding the rate of accuracy, high accuracy is achieved in the place name and organization name compared with the personal name, whose accuracy is quite low. The reason is that family name is not covered in the Chinese name, while it has been attached to the Khmer name after the extraction.

In the experiment, where the Chinese-Khmer named entity equivalence is extracted based on the feature similarity, this paper has extracted separately the personal name, the place name and the organization name based on the characteristics of transliteration and the translation features.

Table 4. Performance for the extraction based on feature similarity

Category	Characteristics	Recall rate	Rate of accuracy	F value
Personal name	Transliteration	65%	81%	75%
	Translation	87%	45%	59%
Place name	Transliteration	50%	71%	59%
	Translation	79%	75%	77%
Organization name	Transliteration	7%	4%	5%
	Translation	42%	24%	31%

As shown in the analytical table 4, the result is not satisfactory when the extraction of organization name equivalence is made based on the characteristics of transliteration and the translation features. The main reason is that most of the organization names are involved with both of the transliteration and translation. Therefore the pure utilization of the characteristics of transliteration or the translation features won't bring about good effect. Moreover some of the organization names are shown in the abbreviated form, which will also bring some interference to the extraction of equivalences.

6   Conclusions

Since the link-based extraction of Chinese-Khmer named entity equivalence is made based on the multi-language characteristic of Wikipedia and the structural features of its web page.Also since the feature similarity-based extraction of Chinese-Khmer named entity equivalence is able to adopt different features based on the different types of named entities, good effect will be achieved in the extraction of personal name equivalence and the place name equivalence. In this way, it will makes up for the deficiency in the link-based approach that depends too much on the link information. However as to the extraction of the organization name, the extraction performance will be improved significantly if both of the characteristics of transliteration and the translation features can be integrated in the extraction.

References

1. Ru, K., Xu, J., Zhang, Y., Wu, P.: A method to construct chinese-japanese named entity translation equivalents using monolingual corpora. In: Zhou, G., Li, J., Zhao, D., Feng, Y. (eds.) NLPCC 2013. CCIS, vol. 400, pp. 164–175. Springer, Heidelberg (2013)
2. Chen, H.X., Yin, C.Y., Chen, J.J.: An Approach to Extract Nam ed Entity Translingual Equivalence. Journal of Chinese information 22(4), 55–60 (2008)
3. Meng, H., Lo, W.K., Chen, B., et al.: Generating phonetic cognates to handle named entities in English-Chinese cross- language spoken doeument retrieval. In: Proceedings of the Automatic Speech Recognition and Understanding Workshop, Trento, pp. 311–314 (2001)
4. Huang, F., Vogel, S., Waibel, A.: Automatic extraction of named entity translingual equivalence based on multi-feature cost minimization. In: Proceedings of Association of Computational linguistics, Sapporo, pp. 9–16 (2003)
5. Cao, G.H., Gao, J.F., Nie, J.Y.: A system to mine large-scale bilingual dictionaries from monolingual web pages. In: Proceedings of MT Summit XI, Copenhagen, Denmark, pp. 57–64 (2007)
6. Lee, L., Aw, A., Zhang, M., et al.: Em-based hybrid model for bilingual terminology extraction from comparable corpora. In: Proceedings of the 23rd International Conference on Computational Linguistics: Posters, pp. 639–646. Association for Computational Linguistics (2010)
7. Yu, K., Tsujii, J.: Bilingual dictionary extraction from wikipedia. In: Machine Translation Summit XII, Ottawa, Canada (2009)
8. Kim, J., Hwang, S., Jiang, L., et al.: Entity Translation Mining from Comparable Corpora: Combining Graph Mapping with Corpus Latent Features (2012)
9. Udupa, R., Saravanan, K., Kumaran, A., et al.: Mint: a method for effective and scalable mining of named entity transliterations from large comparable corpora. In: Proceedings of the 12th Conference of the European Chapter of the Association for Computational Linguistics, pp. 799–807. Association for Computational Linguistics (2009)
10. Li, L., Wang, P., Huang, D., et al.: Minning English-Chinese Named Entity Pairs from Comparable Corpora. ACM Transactions on Asian Language Information Processing 10(4) (2011)

Bilingual Lexicon Extraction with Temporal Distributed Word Representation from Comparable Corpora

Chunyue Zhang and Tiejun Zhao[(⊠)]

School of Computer Science and Technology,
Harbin Institute of Technology, Harbin, China
cyzhang@mtlab.hit.edu.cn, tjzhao@hit.edu.cn

Abstract. Distributed word representation has been found to be highly effective to extract a bilingual lexicon from comparable corpora by a simple linear transformation. However, *polysemous words* often vary their meanings at different time points in the corresponding corpora. A single word representation which is learned from the whole corpora can't express the temporal change of the word meaning very well. This paper proposes a simple solution which exploits the temporal distributed word representation for *polysemous words*. The experimental results confirm that the proposed solution can offer better performance on the English-to-Chinese bilingual lexicon extraction task.

1 Introduction

Over the years, the automatic extraction of bilingual lexicons (BLE) from comparable corpora, where documents are not direct translations but share a topic or domain, has attracted many researchers. In this field, how to represent a word is an essential problem. In recent years, **Distributed Word Representation** [1,15], which is often called **word embedding**, has been extensively studied. Word embedding projects discrete words to a dense low-dimensional and continuous vector space where co-occurred words are located close to each other. Often the word embedding is learned from a big text corpora. In [11], inspired by the linear relation in the bilingual scenario, a linear transform is learned to project semantically identical words from a language to another with word embedding.

However, many words are *polysemous*. When occurring in the time sequential corpora, they often vary their meaning at different time points. For example, the word *apple* in Chinese was nearly the name for some fruit in the corpora twenty years ago, but recently it's more possible to refer a technology company. A single word embedding which is learned from the whole corpora can't express the change of the word meaning with the time very well.

In fact, comparable corpora often are automatically collected from some specific multilingual information source such as Wikipedia[1] and Xinhua News

---

[1] https://www.wikipedia.org

© Springer International Publishing Switzerland 2015
J. Li et al. (Eds.): NLPCC 2015, LNAI 9362, pp. 380–387, 2015.
DOI: 10.1007/978-3-319-25207-0_33

Agency[2], whose corpora are contents updated constantly [5] at different time stamps. And the different word embedding can be trained from the corpora at the different time stamp. Furthermore, different transform matrices will be learned from the word embedding which is obtained from comparable corpora with the different time stamp. So here are two natural questions to motivate this paper:

- How does the word embedding with the different time stamp affect the quantity of learned bilingual lexicon?
- How can BLE get better performance by exploiting the word embedding with the different time stamp?

In this paper, we propose a solution by exploiting the **Temporal Distributed Word Representation** to learn a more accurate translation matrix. Specifically, in this work:

- firstly we divide the sub-corpora set into the corresponding windows according to the different time stamp,
- after fixing a common vocabulary in every corpora window, then we learn the different word embedding from the corresponding corpora,
- then we concatenate these different word embedding into a single new word embedding,
- finally, we learn a new linear transform matrix from the new word embedding.

2 Background: Linear Translation Transformation

The bilingual lexicon extraction provided in [11] learns a linear transform from the source language to the target language by the linear regression. During the training period, suppose we are given a set of bilingual word pairs and their associated word embeddings $\{x_i, z_i\}_{i=1}^n$, and $x_i \in R^{d_1}$ is the word embedding of word i in the source language, $z_i \in R^{d_2}$ is the word embedding of its translation. The objective function is as follows:

$$\hat{\mathbf{W}} = \underset{W \in R^{d_2 \times d_1}}{\operatorname{argmin}} \sum_{i=1}^n \|W x_i - z_i\|^2 \tag{1}$$

During the prediction period, given a new source word embedding x, the standard way to retrieve its translation word in the target language is to return the nearest neighbour (in terms of cosine similarity measure) of mapped $z = \hat{W}x$ from the set of word embedding of the target language.

---

[2] http://www.news.cn/english/

3 Temporal Distributed Word Representation

In [10], the authors proposed a skip-gram model to learn word embedding in which aims at predicting the context words with the word in the central position. Formally, the training process maximizes the following likelihood function with a word sequence $w_1, w_2, \ldots, w_N$:

$$\frac{1}{N} \sum_{i=1}^{N} \sum_{-C \leq j \leq C} logP(w_{i+j}|w_i) \qquad (2)$$

Obviously, the word embedding learned depends on the training word sequence, i.e. the corpora. In the bilingual scenario, this corpora is the source (target) side of the comparable corpora. And often comparable corpora collected from the Internet is often labeled with the time stamp. So one can train the different word embedding for the same word with the corresponding corpora at the different time stamp. And every word embedding trained from the different corpora can represent the word meaning in a specific time slot. So exploiting these different word embeddings can represent the multiple meanings for a polysemous word better. In this paper we therefore propose the **Temporal Distributed Word Representation** which concatenates these different word embedding into a single one.

Mathematically, suppose we are given a list of sub-corpora ordered by their time stamp $C = \{C_1, C_2, \ldots, C_T\}$. For a word we can learn T different word embedding theoretically. However, we found the quality of word embedding is very poor when the size of training corpora is small. So we propose three kinds of temporal distributed word representations which can be trained from large scale corpora:

- **Sliding-Window Temporal Distributed Word Representation (STWR)**
  After empirically setting the predefined size of the corpora window M and the sliding step k, we first divide the T sub-corpora into $N = \lceil \frac{T-M}{k} \rceil$ windows. Note that $T - M$ does not have to be divisible by k and the last window can have a smaller size than M. Then we can train the skip-gram model on the window corpora set $\{SC_1, \ldots, SC_i, \ldots, SC_N\}$, where $SC_i = \{C_{1+k(i-1)}, \ldots, C_{M+k(i-1)}\}$. For a word w, we can get a word embedding list $\{sw_1, \ldots, sw_i, \ldots, sw_N\}$. Finally, we concatenate the word embedding in the list in order, i.e.

  $$\textbf{STWR}(w) = \oplus_{i=1}^{N} sw_i \qquad (3)$$

  where $\oplus$ means vector concatenation operator.
- **Accumulated Temporal Distributed Word Representation (ATWR)**
  In order to express the meaning of a word in a global time slot, we can accumulate the small sub-corpora into a larger one according to the time stamp.

As predefining M and k in **SWTR**, we can get the accumulated corpora set $\{AC_1,\ldots,AC_i,\ldots,AC_N\}$, where $AC_i = \{C_1,\ldots,C_{M+(i-1)k}\}$ and $N = \lceil\frac{T-M}{k}\rceil$. For a word w, we can get a word embedding list $\{aw_1,\ldots,aw_i,\ldots,aw_N\}$. Finally, we concatenate the word embedding in the list in order, i.e.

$$\mathbf{ATWR}(w) = \oplus_{i=1}^{N}aw_i \tag{4}$$

- **Ensemble Temporal Distributed Word Representation(ETWR)**
From the definition above, **STWR** can represent the meaning of a word in a local time slot, and **ATWR** can represent a global meaning conversely. So it's natural to ensemble this two representations. We define the **Ensemble Temporal Word Representation** as follows:

$$\mathbf{ETWR}(w) = \oplus_{i=1}^{N}sw_i \oplus aw_N \tag{5}$$

where sw_i holds the word embedding trained from the slide-window corpora, and aw_N holds the word embedding learned from the whole corpora AC_N.

4 Experiment and Results

4.1 Experimental Settings

In this paper, we carry on the bilingual lexicon extraction task in the English-to-Chinese direction. For the comparable corpora, we use the English Gigaword Corpus (LDC2009T13) and the Chinese Gigaword Corpus (LDC2009T27). In order to align the two comparable corpora better, we select the part of the two corpus published by Xinhua News Agency which contains news articles from January 1995 to December 2008. So we have 14 corpus at the different year in every language.

For **STWR** and **ATWR**, we set window size $M = 10$ and sliding step $k = 2$, then get three corpora $\{SC_1,SC_2,SC_3\}$ for every language and also get three corpora $\{AC_1,AC_2,AC_3\}$ for every language where AC_1 is identical to SC_1. Details of every corpora are reported in Table 1. We intersect the vocabulary sets of $\{SC_1,SC_2,SC_3\}$ as the common vocabulary. After that, we get an English common vocabulary consisting of 92335 English words and a Chinese common vocabulary consisting of 143621 Chinese words.

From every language, we use the same setup to train the skip-model. We use the word2vec toolkit[3] to learn a 200-dimensional word embedding. We just consider the words occurred at least 10 times, and set a context windows of 3 words to either side of the target. Other hyper-parameters follows the default software setup.

To obtain a bilingual training lexicon between English-to-Chinese, we use an in-house dictionary which consists of 55668 English words and 137420 Chinese words. Firstly, we filter the dictionary with the intersection set of vocabulary of

---

[3] https://code.google.com/p/word2vec

Table 1. The sizes of the monolingual training dataset for **STWR** and **ATWR**

		Training Tokens	Vocabulary Size
EN	SC_1/AC_1	222M/222M	106K/106K
	SC_2/AC_2	237M/274M	112K/122K
	SC_3/AC_3	245M/326M	116K/136K
CH	SC_1/AC_1	223M/223M	163K/163K
	SC_2/AC_2	238M/275M	168K/181K
	SC_3/AC_3	266M/346M	177K/205K

$\{SC_1, SC_2, SC_3\}$. From the filtered dictionary, we randomly select 2500 different English words and their translation as test set, and the left dictionary as the training set. Details of the train set and test set are listed in Table 2.

Table 2. The statistics for the train set and test set

	Train Set	Test Set
Entries	64692	10576
Words	14413	2500
Avg	4.48	4.23

4.2 Results

We choose the approach in [11] as our baseline where a single embedding is learned from $AC3$. The performance is measured by accuracy of translation retrieval list of the test set at Top-k, where k is set $\{1, 5, 10\}$. Here accuracy means if there is one candidate in the Top-k list occurs in the reference list, the translation will be right. Results of **STWR**, **ATWR** and **ETWR** are given in Table 3. In all tables in this paper, the $AC3$ row represents the baseline approach.

From Table 3, a continuous performance improvement from $\{AC_1, AC_2, AC_3\}$ can be seen. It shows that using a larger corpus can learn better word embedding. And **ATWR** achieves significant improvements over the baseline $AC3$. **ATWR** increases the accuracy at Top-5 from 0.225 to 0.255. From Table 3, the performance achieved by **STWR** is also improved significantly over the baseline $AC3$ and is comparable with **ATWR**. As can be seen, exploiting the local word embedding at the different time stamp is effective for extracting bilingual lexicon. Finally, from Table 3, the **ETWR** performs best. This representation achieves an improvement of near 15% over the baseline $AC3$ at Top-5.

Furthermore, we also use the metric **unnormalized precision** which can measure the times of the correct translation occurring in the Top-k translation list for a word like used in information retrieval. Obviously, for a word with multiple translations in the test set, higher precision of an approach means better performance for *polysemous words*. In order to compare all the representations in this paper fairly, we choose the intersection set of words correctly predicted at the Top-10 with the representation learned with $AC3$,**ATWR**,**STWR** and

Table 3. The performance of **ATWR,STWR** and **ETWR**

Representation	ACC@1	ACC@5	ACC@10
AC1	0.096	0.204	0.251
AC2	0.110	0.216	0.280
AC3	0.126	0.225	0.292
SC1	0.096	0.204	0.251
SC2	0.103	0.209	0.269
SC3	0.108	0.221	0.274
ATWR	**0.131**	**0.255**	**0.314**
STWR	**0.132**	**0.251**	**0.307**
ETWR	**0.136**	**0.260**	**0.316**

Table 4. The average precision of L234 and L5 at Top-10

Representation	Prec@L234	Prec@L5
AC3	1.217	1.436
ATWR	1.233	1.513
STWR	1.272	1.532
ETWR	**1.281**	**1.544**

ETWR as the evaluation set. Then we choose two subsets $\{L234, L5\}$ from this evaluation set according to the number of translation for a word, where $L234$ means the word has 2, 3 or 4 translations and $L5$ means the word has at least 5 translations. In this setting, $L234$ has 253 words and $L5$ has 261 words. From the Table 4, we can see the performance of all the temporal word representations proposed in this paper can outperform the baseline $AC3$, and the **ETWR** gains the most significant improvement.

5   Related Works

In the BLE task from comparable corpora, most of the previous methods are based on the distributional hypothesis that a word and its translation tend to appear in similar contexts across languages [2–4,13]. Based on this assumption, generally an unsupervised standard approach [9] using **Co-occurrence Word Representation** calculates the context similarity and then extract word translation pairs with high similarity.

Another interesting word representation is topic word representation in [16,17]. They train a cross-language topic model on the document-aligned comparable corpora. It attempts to abrogate the need of seed lexicon. However, the bilingual topic representation must be learned from aligned documents.

Recently some supervised approaches have been tried to solve this task. An linear classifier in [7] and a Random Forest classifier [8] are used to automatically decide if two words in source language and target language are translated each other. In [11], a linear transform is learned to project semantically identical words from one language to another. In this approach, the word is represented with a continues and dense vector i.e. word embedding. It is surprising that this approach achieved a high accuracy on a bilingual word translation than the standard approach.

All the above methods just learn the translation for a word with a single representation. Its use is problematic when a word has several translations. Discovering multiple senses embedding per word type is the focus of [6,12]. Compared with these context-based approaches, our method is based on the observed fact on the corpora that the word meaning often varies at the different time slot. We exploit the time information to learn multiple word representations.

Temporal information is firstly used in [14]. And a similar approach in [7] uses the frequency distribution on the corpora at the different time stamp which is estimated and as a feature of a classifier. Compared with these methods, our approach exploits temporal distributed word representation which is more robust and continuous.

6 Conclusions

We presented a simple but effective method that exploiting the temporal distributed word representation to learn the linear transform matrix. Three temporal distributed word representations are used for this purpose. This method can learn multiple translations for polysemous words better. Experiments conducted on an English-Chinese comparable corpora indicate that the three temporal word representations all improve the baseline significantly and **ETWR** performs best. By measuring the average unnormalized precision in the Top-10 list, it's better shown that the temporal distributed word representation is effective for the translation of polysemous words.

Acknowledgments. This work is supported by the project of National Natural Science Foundation of China (61173073, 61272384) and International Science and Technology Cooperation Program of China (2014DFA11350).

References

1. Bengio, Y., Courville, A., Vincent, P.: Representation learning: A review and new perspectives. IEEE Transactions on Pattern Analysis and Machine Intelligence **35**(8), 1798–1828 (2013)
2. Chiao, Y.C., Zweigenbaum, P.: Looking for candidate translational equivalents in specialized, comparable corpora. In: Proceedings of the 19th International Conference on Computational Linguistics, vol. 2, pp. 1–5. Association for Computational Linguistics (2002)
3. Emmanuel, M., Hazem, A.: Looking at unbalanced specialized comparable corpora for bilingual lexicon extraction. In: Proceedings of the 52nd Annual Meeting of the Association for Computational Linguistics (ACL), pp. 1284–1293 (2014)
4. Fung, P., Yee, L.Y.: An ir approach for translating new words from nonparallel, comparable texts. In: Proceedings of the 17th International Conference on Computational Linguistics, vol. 1, pp. 414–420. Association for Computational Linguistics (1998)

 5. Huang, D., Zhao, L., Li, L., Yu, H.: Mining large-scale comparable corpora from Chinese-English news collections. In: Proceedings of the 23rd International Conference on Computational Linguistics: Posters, pp. 472–480. Association for Computational Linguistics (2010)
 6. Huang, E.H., Socher, R., Manning, C.D., Ng, A.Y.: Improving word representations via global context and multiple word prototypes. In: Proceedings of the 50th Annual Meeting of the Association for Computational Linguistics: Long Papers, vol. 1, pp. 873–882. Association for Computational Linguistics (2012)
 7. Irvine, A., Callison-Burch, C.: Supervised bilingual lexicon induction with multiple monolingual signals. In: HLT-NAACL, pp. 518–523. Citeseer (2013)
 8. Kontonatsios, G., Korkontzelos, I., Tsujii, J., Ananiadou, S.: Using a random forest classifier to compile bilingual dictionaries of technical terms from comparable corpora. In: Proceedings of the 14th Conference of the European Chapter of the Association for Computational Linguistics: Short Papers, vol. 2, pp. 111–116 (2014)
 9. Laroche, A., Langlais, P.: Revisiting context-based projection methods for term-translation spotting in comparable corpora. In: Proceedings of the 23rd International Conference on Computational Linguistics, pp. 617–625. Association for Computational Linguistics (2010)
10. Mikolov, T., Chen, K., Corrado, G., Dean, J.: Efficient estimation of word representations in vector space (2013). arXiv preprint arXiv:1301.3781
11. Mikolov, T., Le, Q.V., Sutskever, I.: Exploiting similarities among languages for machine translation (2013). arXiv preprint arXiv:1309.4168
12. Neelakantan, A., Shankar, J., Passos, A., McCallum, A.: Efficient non-parametric estimation of multiple embeddings per word in vector space (2015). arXiv preprint arXiv:1504.06654
13. Rapp, R.: Automatic identification of word translations from unrelated english and german corpora. In: Proceedings of the 37th Annual Meeting of the Association for Computational Linguistics on Computational Linguistics, pp. 519–526. Association for Computational Linguistics (1999)
14. Schafer, C., Yarowsky, D.: Inducing translation lexicons via diverse similarity measures and bridge languages. In: Proceedings of the 6th Conference on Natural Language Learning, vol. 20, pp. 1–7. Association for Computational Linguistics (2002)
15. Turian, J., Ratinov, L., Bengio, Y.: Word representations: a simple and general method for semi-supervised learning. In: Proceedings of the 48th Annual Meeting of the Association for Computational Linguistics, pp. 384–394. Association for Computational Linguistics (2010)
16. Vulić, I., De Smet, W., Moens, M.F.: Identifying word translations from comparable corpora using latent topic models. In: Proceedings of the 49th Annual Meeting of the Association for Computational Linguistics: Human Language Technologies: short papers. vol. 2, pp. 479–484. Association for Computational Linguistics (2011)
17. Vulić, I., Moens, M.F.: Detecting highly confident word translations from comparable corpora without any prior knowledge. In: Proceedings of the 13th Conference of the European Chapter of the Association for Computational Linguistics, pp. 449–459. Association for Computational Linguistics (2012)

Bilingually-Constrained Recursive Neural Networks with Syntactic Constraints for Hierarchical Translation Model

Wei Chen[(✉)] and Bo Xu

IDMTech, Institute of Automation, Chinese Academy of Sciences, Beijing, China
{wei.chen.media,xubo}@ia.ac.cn

Abstract. Hierarchical phrase-based translation models have advanced statistical machine translation (SMT). Because such models can improve leveraging of syntactic information, two types of methods (leveraging source parsing and leveraging shallow parsing) are applied to introduce syntactic constraints into translation models. In this paper, we propose a bilingually-constrained recursive neural network (BC-RNN) model to combine the merits of these two types of methods. First we perform supervised learning on a manually parsed corpus using the standard recursive neural network (RNN) model. Then we employ unsupervised bilingually-constrained tuning to improve the accuracy of the standard RNN model. Leveraging the BC-RNN model, we introduce both source parsing and shallow parsing information into a hierarchical phrase-based translation model. The evaluation demonstrates that our proposed method outperforms other state-of-the-art statistical machine translation methods for National Institute of Standards and Technology 2008 (NIST 2008) Chinese-English machine translation testing data.

1 Introduction

Hierarchical phrase-based models [1] have advanced statistical machine translation (SMT) by employing hierarchical rules. Formally, a hierarchical phrase-based model is a synchronous context-free grammar that is learned from a bitext without any syntactic information. Thus, such models can be considered to be a shift in the formal machinery of syntax-based translation systems without any linguistic commitment, which enables their convenient and extensive application.

Numerous studies have leveraged syntactic information in SMT systems. Some of these studies have introduced linguistic syntax via source parsing to direct word reordering. For example, [2] used dependency tree to add syntactic cohesion. [3] proposed to parse and to translate jointly by taking tree-based translation as parsing. [4] propose a nonparametric Bayesian method for inducing Part-of-Speech (POS) tags in dependency trees to improve the performance of statistical machine translation. Such methods are performed within the unit of tree nodes and efficiently address some mistakes such as word reordering in SMT. However, they cause data sparseness and are vulnerable to parsing errors because

© Springer International Publishing Switzerland 2015
J. Li et al. (Eds.): NLPCC 2015, LNAI 9362, pp. 388–395, 2015.
DOI: 10.1007/978-3-319-25207-0_34

of their strict constraints on the parse tree. Other studies have employed shallow parsing (also chunking) to justify the selection of translation rules. [5] presented a chunk-to-string translation model where the decoder generates a translation by first translating the words in each chunk, then reordering the translation of chunks. [6] present a hierarchical chunk-to-string translation model, which can be seen as a compromise between the hierarchical phrase-based model and the tree-to-string model. The constraints on the syntactic information that are applied in these studies are significantly weaker. However, these methods tend to suffer from the conflict between the different definitions of phrases in SMT and traditional chunking methods: phrases in SMT are grammar-free, whereas traditional chunking methods require phrases to be intact in terms of grammar.

In this paper, we present a bilingually-constrained recursive neural network (BC-RNN) model to combine the merits of the two types of studies. First, we propose a standard recursive neural network (RNN) model to perform supervised learning to determine how to parse phrases and how to represent phrases in a continuous vector space of features [7] for source- and target- languages, respectively. A simple softmax layer is employed in this model to predict syntactic categories (also chunk labels). Second, we propose a bilingually-constrained learning model to fine-tune the parameters of the standard RNN to improve the accuracy of the representation and chunk labels of the phrases. Finally, by leveraging the BC-RNN model, we extract information about parsing and chunking from the source sentence and efficiently add extra syntactic features to state-of-the-art hierarchical phrase-based translation systems.

Using the 2008 National Institute of Standards and Technology (NIST) Chinese-English MT translation test set, the results of the experiments demonstrate that our model can significantly improve the performance of hierarchical phrase-based translation models and outperform other state-of-the-art SMT methods that leverage syntactic information.

2 Bilingually-Constrained RNN

In this section, we describe the structure of the BC-RNN model. We also define the objective function and the inferences of the parameters of the BC-RNN model.

2.1 The BC-RNN Model

Assume that we are given the phrase $w_1 w_2 ... w_m$; it is projected onto a list of vectors $(x_1, x_2, ..., x_m)$ using word vector representation. The standard RNN learns the parsing tree and the distributed representation of the phrase by recursively combining two child vectors in a bottom-up manner. Given the distributed representation p of the phrase $w_1 w_2 ... w_m$, it is convenient to add a simple softmax layer to predict chunk labels, such as NP and VP. The details of structure prediction and Category Classifier using standard RNN can be got from [7] and we don't introduce it in this paper.

Given two standard RNN models, we propose a bilingually-constrained optimization to fine-tune the parameters of both standard RNN models. The structure of the BC-RNN model is illustrated in Figure 1.

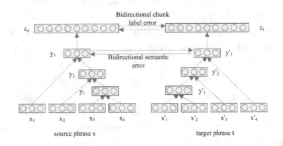

Fig. 1. Structure of bilingually-constrained recursive neural network

2.2   The Objective Function

To fine-tune the standard RNN models for the source and target languages, for a bilingual phrase pair (s, t), two types of errors are involved:

(1) Semantic error: this is quantified in terms of the semantic distance of the distributed representation p_s and p_t of the bilingual phrase pair (s, t) [9].

Because word embeddings for two languages are learned separately and located in different vector spaces, a transformation must be performed to calculate the semantic distance. Thus, the semantic distance is bidirectional: there is both the distance between p_t and the transformation of p_s, and the distance between p_s and the transformation of p_t. Consequently, the total semantic error becomes

$$E_{sem}(s, t; \theta) = E_{sem}(s|t, \theta) + E_{sem}(t|s, \theta) \tag{1}$$

where θ denotes the parameters of the BC-RNN model and we calculate $E_{sem}(s|t, \theta)$ using the Euclidean distance:

$$E_{sem}(s|t, \theta) = \frac{1}{2}||p_t - f(W_{en}^{ch} p_s + b_{en}^{ch})||^2 \tag{2}$$

$E_{sem}(t|s, \theta)$ can be calculated in exactly the same manner.

(2) Chunk label error: this is quantified by the difference between the predicted chunk labels of the distributed representations p_s and p_t.

After applying the simple softmax layer of each standard RNN models, the output vector representations c_s and c_t denote the probability distribution of the chunk labels. Consequently, similar to semantic error, the total chunk label error becomes bidirectional as follows:

$$E_{chunk}(s, t; \theta) = E_{chunk}(s|t, \theta) + E_{chunk}(t|s, \theta) \tag{3}$$

where θ denotes the parameters of the BC-RNN model and $E_{chunk}(s|t,\theta)$ is calculate as follows:

$$E_{chunk}(s|t,\theta) = \frac{1}{2}||c_t - f(W^{ch}_{chunk}p_s + b^{ch}_{chunk})||^2 \tag{4}$$

Thus, for a bilingual phrase pair (s,t), the joint error is

$$E(s,t;\theta) = \alpha E_{sem}(s,t;\theta) + (1-\alpha)E_{chunk}(s,t;\theta) \tag{5}$$

The hyper-parameter α weights the semantic and chunk label errors. The final BC-RNN objective function over the phrase pairs training set (S,T) becomes:

$$J_{BC-RNN} = \sum_{(s,t)\in(S,T)} E(s,t;\theta) + \frac{\lambda}{2}||\theta||^2 \tag{6}$$

2.3 Parameter Inference

The parameter θ can be divided into the source-side parameter θ_s and the target-side parameter θ_t [9]. We apply the stochastic gradient descent (SGD) algorithm to optimize each parameter. Word vector representations θ_L are initialized with the DNN toolkit Word2Vec [8] using large-scale monolingual data, and other parameters are randomly initialized. The details of optimization of the parameters can be got from [9].

3   A Hierarchical Phrase-Based Translation Model that Leverages Syntactic Information

In this section, we leverage two types of syntactic information in the hierarchical phrase-based translation model.

3.1   Feature1: The Score of the Parse Tree

First, we calculate the score of this tree by applying the fine-tuned parameter θ_s^* in the BC-RNN model.

Given the fine-tuned parameters W_s^{r*} and b_s^{r*}, the distributed representation of each nonterminal in this tree is calculated as

$$p = f(W_s^{r*}[c_1 : c_2] + b_s^{r*}) \tag{7}$$

where the concatenation of two children $[c_1 : c_2]$ is provided by word vector representation. Then, we calculate the parsing score similarly to [7] as follows:

$$s = W_{score}p \tag{8}$$

Let $T(y_i)$ denote the set of spans coming from all nonterminal nodes of this parse tree. The total parsing score of this tree is calculated as the sum of the scores of each span:

$$s_{parse}(f, e, a) = \sum_{d \in T(y_i)} s_d(c_1, c_2) \tag{9}$$

where "f" represents the target sentence and "e" represents the source sentence and "a" represents the word alignment. The more details of parsing score can be got from [7] and we don't introduce it in this paper.

3.2 Feature2: Hierarchical Rules with Syntactic Categories

We now discuss how to extract chunk-based hierarchical rules, as Basic phrases are defined using the same heuristic as in previous systems [11][12]. Chunk-based hierarchical phrases are extracted as follows:

1. If $< f_{j_1}^{j_2}, e_{i_1}^{i_2} >$ is a basic phrase with a chunk label c (source-side only), then a rule is extracted:

$$X \rightarrow < f_{j_1}^{j_2}, e_{i_1}^{i_2} > \tag{10}$$

2. Assume that $X \rightarrow < \alpha, \beta >$ is a rule with $\alpha = \alpha_1 f_{j_1}^{j_2} \alpha_2$ and $\beta = \beta_1 e_{i_2}^{i_1} \beta_2$. Let $C(X)$ denotes the set of chunk labels of the nonterminals in this rule, then we update the rule as:

$$X \rightarrow < \alpha_1 X_1 \alpha_2, \beta_1 X_1 \beta_2 > \qquad X_1 \in C(X) \cup c \tag{11}$$

The chunk labels are tagged using the fine-tuned parameter θ_s^* in the BC-RNN model. We evaluate the distribution of these rules in the same manner as [13].

Leveraging chunk-based hierarchical rules, we introduce information about chunk labels into the hierarchical translation model. In the translating decoding process, we select a penalization of incorrect chunk labels as our loss function and add a penalization term to each incorrect decision. Let T_X denote the set of applied hierarchical rules in the decoding process. The penalization can be derived as

$$s_{chunk}(f, e, a) = \sum_{X \in T_X} \sum_{n \in N(X)} 1\{c(n) \notin C(X)\} \tag{12}$$

where $N(X)$ represents the set of the nonterminals of the hierarchical rule X and $c(n)$ represents the chunk labels (source-side only) of the subphrase that covers the nonterminal n and is given by BC-RNN model.

The Hierarchical Translation Model with Syntactic Information. Finally, we introduce both the features into the standard hierarchical translation model [1]. The formula can be derived as follows:

$$s_{trans}(f, e, a) = \sum_i \lambda_i log(s_i(f, e, a))$$
$$+ \lambda_{parse} log(s_{parse}(f, e, a)) + \lambda_{chunk} log(s_{chunk}(f, e, a)) \tag{13}$$

where $s_i(f, e, a)$ represent the traditional features used in standard hierarchical translation model which is as same as [1]. The weights λ_i and λ_{parse} and λ_{chunk} are learned via minimum error-rate training [14] using the development dataset.

4 Experiments

4.1 Data Preparation and Tools

We got training data for source-side standard RNN model from the standard Chinese Treebank (CTB) 6.0, which has 780k words with several categories. The English Treebank corpus (ETB), which contains 2,881k tokens is used to train the target-side standard RNN model.

We used the NIST training set for Chinese-English translation tasks excluding the Hong Kong Law and Hong Kong Hansard as the training data, which contains 470K sentence pairs. For the training data set, we first performed word alignment in both directions using GIZA++ toolkit [10] then refined the alignments using "final-and". We trained a 5-gram language model with modified Kneser-Ney smoothing on The English Gigaword corpus, section AFP which contains 611,506,174 words. we employ an out-of-the-box toolkit Moses (v3.0) framework and minimum error rate training [14] to train and tune the feature weights of SMT systems. We used our in-house Chinese-English data set as the development set and used the 2008 NIST Chinese-English MT test set (1859 sentences) as the test set. Our evaluation metric is BLEU-4.

We employed the Stanford Chinese word segmentation tools to segment the Chinese sentences in the training and testing process.

4.2 Machine Translation Performance

First, we evaluate the performance of our method and compare it with other state-of-the-art methods, including a phrase-based machine translation model [11], a standard hierarchical phrase-based machine translation model [1], a tree-to-string machine translation model that leverages source parsing [4], and a chunk-to-string machine translation model that leverages shallow parsing [6]. The comparison of the performance is shown in Table.1.

Table 1. Translation performance of different methods on NIST 2008

Methods	NIST2008 %
phrase-based	23.25
hierarchical phrase-based	23.94
Tree-to-string	24.1
chunk-to-string	24.8
Feature1 only	25.56
Feature2 only	25.35
Feature1+Feature2	26.02

Table 2. Translation performance using different chunkers on NIST 2008

Methods	NIST2008 %
Feature2 with Left corner PCFG	23.5
Feature2 with Standford parser	24.0
Feature2 with BC-RNN	25.35

The results indicate that using either the score for parsing tree or the chunk-based penalization function can effectively improve the performance of the standard hierarchical translation model. When we integrate both features, the model outperforms the other translation model and can significantly improve the performance of machine translation.

Moreover, for additional analysis, we use the traditional chunker in the Feature2 instead of our BC-RNN model and compare the translation performance with our method in Table.2. The method "Left corner PCFG" is obtained from [15]. The Stanford parser is an out-of-the-box parsing system [16] with the latest version.

The results show that our BC-RNN model can integrate syntactic information into hierarchical translation models more effectively and accurately than traditional chunkers.

5   Conclusion

In this paper, we propose a bilingually-constrained RNN model to introduce high-quality syntactic information into the standard hierarchical translation model. We combine the merits of the two types of studies and propose a bilingually-constrained tuning to improve the quality of syntactic information. The evaluation demonstrate that our method outperforms other state-of-the-art SMT systems.

References

1. Chiang, D.: Hierarchical phrase-based translation. Computational Linguistics **33**(2), 201–228 (2007)
2. Cherry, C.: Cohesive phrase-based decoding for statistical machine translation. In: ACL 2008, pp. 72–80 (2008)
3. Liu, Y., Liu, Q.: Joint parsing and translation. In: ACL 2010, pp. 707–715 (2010)
4. Tamura, A., Watanabe, T., Sumita, E., et al.: Part-of-speech induction in dependency trees for statistical machine translation. In: ACL (1) 2013 (2013)
5. Watanabe, T., Sumita, E., Okuno, H.G.: Chunk-based statistical translation. In: ACL 2003, pp. 303–310 (2003)
6. Feng, Y., Zhang, D., Li, M., et al.: Hierarchical chunk-to-string translation. In: Association for computational linguistics 2012, pp. 950–958 (2012)
7. Socher, R., Manning, C.D., Ng, A.Y.: Learning continuous phrase representations and syntactic parsing with recursive neural networks. In: NIPS-2010 Deep Learning and Unsupervised Feature Learning Workshop 2010, pp. 1–9 (2010)

8. Mikolov, T., et al.: Distributed representations of words and phrases and their compositionality. In: Advances in Neural Information Processing Systems 2013 (2013)
9. Zhang, J., Liu, S., Li, M., Zhou, M., Zong, C.: Bilingually-constrained phrase embeddings for machine translation. In: EMNLP 2014 (2014)
10. Och, F.J., Ney, H.: Improved statistical alignment models. In: Proceedings of ACL, pp. 440–447 (2000)
11. Koehn, P., Och, F.J., Marcu, D.: Statistical phrase-based translation. In: Proceedings of the 2003 NAACL (2003)
12. Chiang, D.: A hierarchical phrase-based model for statistical machine translation. In: ACL 2005 (2005)
13. Liu, Y., Liu, Q., Lin, S.: Tree-to-string alignment template for statistical machine translation. In: ACL 2006 (2006)
14. Och, F.J.: Minimum error rate training in statistical machine translation. In: ACL 2003 (2003)
15. Manning, C.D., Carpenter, B.: Probabilistic parsing using left corner language models. In: 5th ACL International Workshop (1997)
16. Klein, D., Manning, C.D.: Accurate unlexicalized parsing. In: ACL, pp. 423–430 (2003)

Document-Level Machine Translation Evaluation Metrics Enhanced with Simplified Lexical Chain

Zhengxian Gong[✉] and Guodong Zhou

School of Computer Science and Technology,
Soochow University, Suzhou 215006, China
{zhxgong,gdzhou}@suda.edu.cn,
http://nlp.suda.edu.cn

Abstract. Document-level Machine Translation (MT) has been drawing more and more attention due to its potential of resolving sentence-level ambiguities and inconsistencies with the benefit of wide-range context. However, the lack of simple yet effective evaluation metrics largely impedes the development of such document-level MT systems. This paper proposes to improve traditional MT evaluation metrics by simplified lexical chain, modeling document-level phenomena from the perspectives of text cohesion. Experiments show the effectiveness of such method on evaluating document-level translation quality and its potential of integrating with traditional MT evaluation metrics to achieve higher correlation with human judgments.

Keywords: MT · Evaluation metrics · Lexical chain · Document-level MT

1 Introduction

According to modern trends in linguistics, a text rather than individual words or fragments of sentences is preferred as the unit of communication [1]. Therefore, more researchers propose to build document-level Statistical Machine Translation (SMT) systems in recent years [8,11,17–19]. However, most of their experimental results show moderate or even minimal improvement despite of a great deal of efforts. Some authors doubted the failure of current MT evaluation metrics to objectively reflect the changes at document level.

Automatic evaluation metrics have close relation to MT and current SMT especially depends on them since SMT usually requires automatic metrics to tune parameters. Some automatic MT metrics, such as BLEU [15] and METEOR [2], give evaluation by measuring the amount of segment overlap between MT output and reference at sentence level. And document/system-level evaluation scores can be obtained by accumulating (not simply averaging) sentence-level scores. Obviously, such accumulation loses some document-level information, such as the difference of sentences. Thus it is unlikely that a document-level SMT system can obtain significant improvements guided by such inappropriate evaluation metrics.

© Springer International Publishing Switzerland 2015
J. Li et al. (Eds.): NLPCC 2015, LNAI 9362, pp. 396–403, 2015.
DOI: 10.1007/978-3-319-25207-0_35

A document is made up of sentences, but there exist separate principles of text-construction, beyond the rules for making sentences. Document-level metrics should have the capability of identifying text-level rather than sentence-level change. In recent years, automatic document-level MT metrics have been drawing more and more attention. Gimenez et al. [7] propose a MT evaluation metric based on Discourse Representation Theory [12], which uses co-reference and discourse relations to assess the quality of MT output. However, their metric fails to achieve a higher correlation with human assessments than sentence-level metrics. In order to measure cohesion for translated text, Wong and Kit propose to use lexical cohesion devices and obtain positive experimental results [20] . To better estimate translation adequacy, Rubino et al. propose to do quality estimation for machine translation (without references) by bilingual topic models and show some promising results [16].

A text is a "communicative occurrence which meets seven standards of textuality: cohesion, coherence, intentionality, acceptability, informativity, situationality and intertextuality" [4]. According to this definition, cohesion and coherence are important standards of textuality. Coherence interprets meaning connectedness in the underlying text while cohesion can be formulated quite explicitly on the basis of grammatical and lexical properties. This paper proposes simple yet effective cohesion score to measure text cohesion via lexical chain. Our experimental results show the number of matching lexical chain between reference and MT output has close relevance to the translation quality of full text.

2   Related Work

As the most famous evaluation metric, BLEU is based on n-gram matching. Alternatively, METEOR is based on unigram alignment of references and system translations. METEOR is explicitly designed to improve the correlation with human judgments at sentence level [2]. Document-level BLEU or METEOR score can be generated by aggregating sentences in a document rather than simply averaging scores at sentence level.

Wong and Kit [20] propose to build document-level MT metrics by using lexical cohesion devices. Lexical cohesion devices refer to content words (stopwords are removed) that reiterate once or more times in a document. In their study, the higher ratio of such content words in machine translated text means stronger lexical cohesion.

Text cohesion refers to top-level characteristics of text while document-level BLEU/METEOR score shows the degree to which the detailed information in the original text is conveyed in MT output. Therefore, it is natural to build document-level metrics by extending traditional MT evaluation metrics with document-level feature scores. Thus, Wong and Kit [20] built document-level metrics by extending traditional metrics with lexical cohesion scores as follows:

$$H = \alpha \times CS_{doc} + (1 - \alpha) \times G_{mdoc} . \qquad (1)$$

where CS_{doc} means lexical cohesion score and G_{mdoc} refers to document-level BLEU/METEOR score, and α is a weight controlling their proportion.

The success of [20] is due to such observation: most SMT systems tend to build less lexical cohesion than human translators. However such observation seems to be controversial. Carpuat and Simard [6] show: MT output tend to have more incorrect repetition than human translation when MT systems especially trained on smaller corpora. Thus, metrics in [20] cannot distinguish such "false" cohesion devices.

3   Evaluation Data

Table 1 shows the evaluation data for this study, including Multiple-Translation Chinese Part 2 (LDC2003T17, MTC2 for short) and Multiple-Translation Chinese Part 4 (LDC2006T04, MTC4 for short). The MTC2 consists of 878 source sentences, translated by 4 human translators (references) as well as 3 MT systems. The MTC4 consists of 919 source sentences, translated by 4 human translators (references) as well as 6 MT systems.

Table 1. Evaluation data

Data Set	MTC2	MTC4
Source language	Chinese	Chinese
Target language	English	English
Systems	3	6
♮ Documents	100	100
♮ Sentences	878	919
♮ References	4	4
♮ Genre	Newswire	Newswire

Besides, each machine translated sentence on the MTC4 and MTC2 was evaluated by 2 to 3 human judges for their adequacy and fluency on a 5-point scale. To avoid the bias in the distributions of different judges' assessments in the evaluation data, we normalize these scores following Blatz et al. [5].

Due to the lack of document-level human assessments on the two evaluation data sets, document-level human assessments are averaged over sentence scores, weighted by sentence length. This method is also adopted by famous Metrics-MaTr (the NIST Metrics for Machine Translation Challenge) and approximated in [7] and [20].

4   Text Cohesion Representing by Simplified Lexical Chain

The major problem in [20] is only to measure the cohesion of MT output and completely ignore the one of references. In another word, we think the cohesion score between MT output and references should be consistent. In this study, we use simple yet effective lexical chain to measure lexical cohesion at document level. The basic idea is to compute the number of matching lexical chains respectively in reference and MT output.

4.1 Simplified Lexical Chain

Traditional lexical chain is the sequence of semantically related words [14]. During the process of chaining, special thesaurus, such as WordNet and HowNet, is often used to help recognize synonyms and hyper/hypo-nyms. For the general-purpose, this study only focuses on reiterating words including stem-matched words. Another difference is that our lexical chain mainly used to record the difference of position for each content word. To distinguish our lexical chain from traditional lexical chain, we call it as simplified lexical chain.

The establishing procedure of simplified lexical chain is simple. For each document d:

(1) extracting all unique content words occurring at the first sentence, then constructing an array for each word. Here such array corresponds to one lexical chain. And the element in the chain records its location(sentence identity). All these chains are stored into a hash table, ht, shown in Fig.1.
(2) for each successive sentence, if content word w or stem(w) has existed in ht, then inserting its location information into the corresponding chain, else constructing a new chain for w and inserting this chain into ht.
(3) removing chains from ht which only contain one word.

So each content word which appears more than one time at different sentences will have a lexical chain. For example, Fig.1 shows a lexical chain $LC1$ for the word "die" (perhaps with different morphology) which occurs at the 1st, 2nd and 3rd sentence. There are several lexical chains in one document, thus a hash table ht is utilized to organize all these chains. For clarity, ht is called as lexical-chain index. In this hash table, keys are content words and values refer to lexical chains.

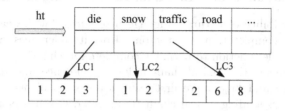

Fig. 1. The structure of the lexical-chain index for one document

4.2   The Characteristics of Lexical-Chain Index

We constructed lexical-chain index for each document on our evaluation data, including 4 human translations (references) and all MT output on different corpus in advance. After that, we carefully studied these chains and achieved the following observations.

The First Observation: lexical chains extracted from reference are more consistent than the ones from MT. Table 2 lists 5 translation versions for one source document. The first 4 columns (E01-E04) correspond to references and the last column refers to one MT output. One word with a list of position represents a lexical chain. The upper part of Table 2 shows some matching lexical chains, which appear in all references. Due to the flexibility of expression, there exist a few un-matching chains shown on the lower part of Table 2. On the whole, the matching number of lexical chain in references greatly exceeds the one in MT output. Moreover, the chains in references are very consistent both in frequency and location while the chains extracted from MT output have large difference.[1]

The Second Observation: it is not always right that lexical cohesion of human translation exceeds the one of MT output.

On the whole evaluation data, we count the total number of lexical chains extracted from human and MT output respectively. The average number of chains extracted from human translation (2111) is greater than the one of MT output (1999) on MTC4, that means lexical cohesion devices existed in human translation more than the one in MT output, which is consistent to the observation described in [20]. But the number of lexical chain extracted from each MT system on MTC2 (2380) exceeds the one from human translations (2030). So the assumption in [20] is not right on MTC2.

Based on above observations, we know there exist some differences in lexical-chain index even all of them extracted from equivalent references due to the complexity and flexibility of linguistics. But they are more stable and consistent among human translations than those in MT output.

4.3   Text Cohesion Scores Based on the Matching of Lexical Chain

Due to high flexibility of natural language utterances, few lexical chains from MT output can completely match the ones from its references. So we need to design a reasonable function to permit incomplete matching.

For a document, the lexical-chain index in reference and in MT output denoted as ht_{ref} and ht_{mt} . We first extract the word of one chain in ht_{mt}, and find its corresponding lexical chain ID in ht_{ref}. Given a pair of matching lexical chain of ht_{ref} and ht_{mt} is LC_r and LC_t. LC_r contains m elements and

---

[1] According to the LDC manual, the ranking for the manual translations is $E01 > E02 > E03 > E04$, so the matching lexical chain in E04 has slight difference in position to other references.

Table 2. An Example of lexical-chain indexes extracted from different translation versions for the same source document

References				Machine
E01	E02	E03	E04	Translation
Government:4,5	Government:4,5	Government:4,5	Government:4,5	-
Ople:2,4	Ople:2,4	Ople:2,4	Ople:2,4	-
Ambassador:2,4	Ambassador:2,4	Ambassador:2,4	Ambassador:2,4	Ambassador:1,2
Australian:2,6,7	Australian:2,6,7	Australian:2,6,7	Australian:2,6,7	-
Attack:1,8	Attack:1,8	Attack:1,8	Attack:1,8	Attack:1,8
Terrorist:1,6,8	Terrorist:1,6,8	Terrorist:1,6,8	Terrorist:1,6,8	Terrorist:1,6,8
Threat :1,5,6,8	Threat:1,5,6,8	Threat:1,5,6,8	Threat:1,5,6	Threat:1,6,8
Manila:0-2,6	Manila:0-2,6	Manila:0-2,6	Manila:0,1,6	-
Embassy:0-3,5-7	Embassy:0-3,5-7	Embassy:0-3,5-7	Embassy:0-3,5-7	Embassy:0,1
Reopen:0,1,3	Reopen:0,1,3	Reopen:0,1,3	Reopen:0,1,3	Reopen:0,1
Australium:0,1,3	Australium:0,1,3	Australium:0,1,3	Australium:0,1,3,6	Australium:0-3,6-7
Philippine:1,2,4	Philippine:1,2,4,5	Philippine:1,2,4,5	Philippine:2,4,5	Philippine:1,4
Week:1,3	Week:1,3	Week:1,3	Week:1,3	Week:1,3
Shut:1,8	Close:1,8	Shut:1,3,5,8	Close:1,3,5,8	Close:1,3,5,8
-	European:3,7	European:3,7	Eu:3,7	Open:0,1,5,6
-	Union:3,7	Union:3,7	-	Canada:3,6
-	-	So-call:1,5	So-call:1,5	China:1,4
-	Express:4,8	Officer:6,7	Mission:7,8	Chinese:2,5
		Due:1,8		Collaboration:1,3,5

LC_t contains n elements, but only $m'(m' <= m)$ element both occur in LC_r and LC_t , then the cohesion score of LC_t can be calculated by the following formula:

$$CS_i = m'/m . \qquad (2)$$

CS_i only refers to one pair of matching chain. If one chain cannot be found in its reference, the chain is invalid ("false"). Suppose ht_{mt} contains K lexical chains, we punish such "false" cohesion by averaging K. Given the number of matching chain is L, the final cohesion score assigned to ht_{mt} is calculated as follows:

$$Doc_{cohesion} = (\sum_{i=1}^{L} CS_i)/K . \qquad (3)$$

For example, one lexical-chain in column of "E01" in Table 2 is "Ambassador:2,4" while its matching chain in MT shown in Table 2 is "Ambassador:1,2", so the CS value of this chain is 1/2(only 1 item is matching). We use this policy to calculate CS value and finally to obtain cohesion score for the whole document. We choose the best $doc_{cohesion}$ against 4 references.

5   Experiments

Following the formula 1, we build document-level metrics by combining document-level BLEU or METEOR with text cohesion score calculated by the formula 3. Especially, the gradient ascending algorithm described in [13] is utilized to automatically tune the weight α.

Table 3. The Correlation of different metrics with human assessments at document level

DataSet	MTC2		MTC4	
Metrics	Pearson	Kendall	Pearson	Kendall
BLEU	0.0994	0.0449	0.5862	0.4256
METEOR	0.3069	0.2037	0.7390	0.5180
$Doc_{cohesion}$	0.1284	0.0609	0.6891	0.4601
HBLEU	0.1240	0.0698	0.6551	0.4800
HMETEOR	0.3107	0.2103	0.7467	0.5244

The results of our proposed metrics are shown in Table 3. To our surprise, the solely use of lexical cohesion already outperforms document-level BLEU, but it still subordinates to METEOR. The hybrid BLEU (HBLEU) scores rise from 42.56% to 48.00% on Kendall score on MTC4 and with a similar increase on MTC2. Furthermore, differing with the results in [20], our hybrid METEOR (HMETEOR) scores also obtain a moderate rise (0.64%- 0.67%) both on MTC4 and MTC2.

6   Conclusion

This paper describes how to modeling text cohesion for MT output based on simplified lexical chains. We successfully build reasonable document-level evaluation metrics by extending traditional MT evaluation metrics with text cohesion score based on lexical chains.

Since important words will be repeated in one text, lexical chains can not only model text cohesion but also highlight key words. So our proposed metrics can obtain very significant improvements for BLEU and also give moderate improvements for METEOR.

In the future work, we will explore how to estimate more document-level features, such as co-reference matching, and hope our study can bring more inspirations to document-level SMT.

Acknowledgments. This research is supported by the National Natural Science Foundation of China under grant No.61305088 and No.61401295.

References

1. Al-Amri, K.H.: Text-linguistics for Students of Translation. King Saud University (2007)
2. Banerjee, S., Lavie, A.: METEOR: an automatic metric for MT evaluation with improved correlation with human judgments. In: Proceedings of the ACL Workshop on Intrinsic and Extrinsic Evaluation Measures for Machine Translation and/or Summarization, pp. 65–72 (2005)
3. Barzilay, R., Lapata M.: Modeling local coherence: an entity-based approach. In: Proceedings of ACL, pp. 141–148 (2008)
4. Beaugrande, R.D., Dressler, W.U.: Introduction to Text Linguistics. Longman, London (1981)
5. Blatz, J., Fitzgerald, E., Foster, G., Gandrabur, S., Goutte, C., Kulesza, A., Sanchis, A., Ueffing, N.: Confidence Estimation for Machine Translation. Technical report, Natural Language Engineering Workshop Final Report (2003)
6. Carpuat, M., Simard, M.: The trouble with SMT consistency. In: Proceedings of the 7th Workshop on Statistical Machine Translation, pp. 442–449 (2012)
7. Gimenez, J., Marquez, L., Comelles, E., Castellon, I., Arranz, V.: Document-level automatic MT evaluation based on discourse representations. In: Proceedings of WMT and MetricsMATR, pp. 333–338 (2010)
8. Gong, Z.X., Zhang, M., Zhou, D.: Cache-based document-level statistical machine translation. In: Proceedings of EMNLP, pp. 909–919 (2011)
9. Guzman, F., Joty, S., M'arquez, L.: Using discourse structure improves machine translation evaluation. In: Proceedings of ACL, pp. 687–698 (2014)
10. Halliday, M.A.K., Hasan, R.: Cohesion in English. Longman, London (1976)
11. Hardmeier, C., Nivre, J., Tiedemann, J.: Document-wide decoding for phrase-based statistical machine translation. In: Proceedings of EMNLP, pp. 1179–1190 (2012)
12. Kamp, H., Reyle, U.: From Discourse to Logic. Introduction to Model Theoretic Semantics of Natural Language, Formal Logic and Discourse Representation Theory. Kluwer Academic Publishers, Dordrecht (1993)
13. Liu, D., Gildea, D.: Source-language features and maximum correlation training for machine translation evaluation. In: Proceedings of NAACL, pp. 41–48 (2007)
14. Morris, J., Hirst, G.: Lexical Cohesion Computed by Thesauri Relations as an Indicator of the Structure of Text. Computational Linguistics 17(1), 21–48 (1991)
15. Papineni, K., Roukos, S., Ward, T., Zhu, W.J.: BLEU: a method for au-tomatic evaluation of machine translation. In: Proceedings of ACL, pp. 311–318 (2002)
16. Rubino, R., Jos'e, G.C.S., Foster, J., Specia, L.: Topic models for translation quality estimation for gisting purposes. In: Proceedings of the XIV Machine Translation Summit, pp. 295–302 (2013)
17. Tiedemann, J.: Context adaptation in statistical machine translation using models with exponentially decaying cache. In: Proceedings of the 2010 Workshop on Domain Adaptation for Natural Language Processing, pp. 8–15 (2010)
18. Xiao, T., Zhu, J.B., Yao, S.J., Zhang, H.: Document-level consistency verification in machine translation. In: Proceedings of MT Summit XIII, pp. 131–138 (2011)
19. Xiong, D.Y., Ding, Y., Zhang, M., Tan, C.L.: Lexical chain based cohesion models for document-level statistical machine translation. In: Proceedings of EMNLP, Seattle, Washington, USA, pp. 1563–1573 (2013)
20. Wong, B.T.M., Kit, C.: Extending machine translation evaluation metrics with lexical cohesion to document level. In: Proceedings of EMNLP, pp. 1060–1068 (2012)

Cross-Lingual Tense Tagging
Based on Markov Tree Tagging Model

Yijiang Chen[✉], Tingting Zhu, Chang Su, and Xiaodong Shi

School of Information Science and Engineering, Xiamen University, Xiamen 361005, China
cyj@xmu.edu.cn

Abstract. In this paper, we transform the issue of Chinese-English tense conversion into the issue of tagging a Chinese tense tree. And then we propose Markov Tree Tagging Model to tag nodes of the untagged tense tree with English tenses. Experimental results show that the method is much better than linear-based CRF tagging for the issue.

Keywords: Chinese-English tense conversion · Markov tree tagging model · Tense tree · Untagged tense tree · Tagged tense tree

1 Related Work

Chinese-English machine translation systems need to deal with the issue of tense, otherwise it will affect the accuracy and fluency of translation results [1]. Gong etc. [2] used 2005 MT NIST data to find that tense translation has big effect.[1]

People have proposed several solutions to tense conversion. The common strategy is the use of manual rules [3] [4]. For example, if a temporal phrase is "昨天 (yesterday)", the sentence is tagged as "past" [4]. Ma [5] uses rules to acquire Chinese tenses and aspects, and then uses the predefined mapping relations to map them to English tenses. Methods based on manual rules are usually too rough and arbitrary.

Another strategy is the use of statistics methods [2] [5] [6].Yang Ye [5] uses the linear conditional random field (CRF) classifier to tag the verb sequence with English tenses. The advantage of this method is that it can cover complicated phenomenon of language, and consider tagging all verbs from the whole. Gong etc. [2] use N-grams of intra tense corpus and inter tense corpus to guess tenses. Gong etc. [6] use SVM method for tense classification. The method is combined with a statistical MT system.

In all these methods, interdependence of tenses between main clauses and subordinate clauses is not fully expressed. A compound sentence is composed of some clauses, and there are constraints of tenses between main clauses and subordinate clauses.

In this paper, we propose a method for Chinese-English tense conversion based on tense trees. We also propose a statistical machine learning method called Markov tree tagging model (MTTM). Firstly, we construct an untagged tense tree according to the parse tree of a Chinese sentence; secondly, we use MTTM to tag all nodes of the untagged tense tree with English tenses.

© Springer International Publishing Switzerland 2015
J. Li et al. (Eds.): NLPCC 2015, LNAI 9362, pp. 404–412, 2015.
DOI: 10.1007/978-3-319-25207-0_36

2   Tense Trees

The existing tense processing methods did not take full advantage of hierarchical information in parse trees. For this, we propose the concept of tense tree.

An untagged tense tree (UTT) is a reduced and flattened tree with syntactic and semantic information related to tense, which is converted from a parse tree.

A tagged tense tree (TTT) is an untagged tree plus tags, in which "IP, CP and verb" nodes are tagged with one of 16 English tenses and other nodes are tagged with "NONE". IP or CP is the label of a clause in a parse tree.

Untagged tense tree and tagged tree are collectively called tense tree. For example, Fig. 1 is bilingual word alignment between a Chinese sentence and its English translation. Fig. 2 is its untagged tense tree, and Fig. 3 is its tagged tense tree.

We believe that a clause (labeled with IP or CP) can have a tense, which is usually the tense of the main verb of the clause. We define the main verb of a clause (labeled with IP or CP) as the first verb which is directly covered by the label IP (or CP) of the clause. In Fig. 3, the verb "看见 (saw)" is the main verb of CP (but not IP).

Then we transform the issue of tense conversion into that of tagging tense tree.

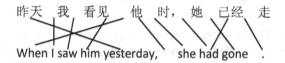

Fig. 1. Bilingual word alignment

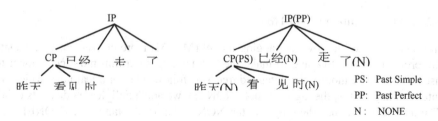

Fig. 2. Untagged tense tree   **Fig. 3.** Tagged tense tree

We construct the untagged tense tree during the traversal of the parse tree of a Chinese sentence. Attributes of nodes in a tense tree are as follows. (1) Attributes of a verb node include the verb itself, POS, whether being a modal verb, and its phase.(2) Attributes of a IP or CP node include the label(IP or CP), the label of the parent in the parse tree, and the type of the temporal phrase node in the clause. (3) Attributes of a temporal phrase node include the tense of the temporal phrase, point or period type of the temporal phrase, and whether occurring time earlier than reference time.(4) Attributes of a temporal adverb node include the adverb itself and its type [7]. (5)Attributes of a temporal auxiliary word node include the word itself, such as "了,着,过".

3   Markov Tree Tagging Model

3.1   Introduction to Tree Tagging Models

For the need to solve hierarchical issues, people try to tag nodes of a tree. For example, Fig. 4(a) is an untagged tree T, the tag $x_i \in H_i$. Fig. 4(b) is a tagged tree T[X].

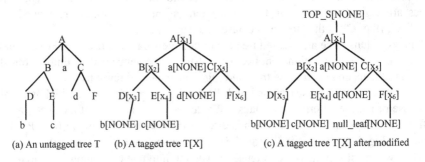

(a) An untagged tree T   (b) A tagged tree T[X]   (c) A tagged tree T[X] after modified

Fig. 4. A sample of tree tagging

Inspired by Head-Driven parsing method proposed by Collins [8], we propose a machine learning method called Markov tree tagging model (MTTM). The model has global consideration for best tagging. However, the algorithm of MTTM is very different from Head-Driven parsing method.

3.2   The Definition of MTTM

In order to make Markov tree tagging model (MTTM) more convenient to calculate the probability of a tree and to make the tree tagging algorithm more convenient to use recursion, we modify the untagged tree T as follows:(1) For a leaf node which we intend to tag (that is, the tag set is not {NONE}), we add a null_leaf node as its child. The null_leaf node has the only one tag NONE (that is, the tag set is {NONE}). (2) Add a root node TOP_S for the tree T. The tagging set of TOP_S is {NONE}. We can prove that the probability of the tree T is not changed. Shown as the Fig.4, we get Fig.4(c) after modifying Fig.4(b).

Tree T[X] can be generated by context free grammars. In the following statement, uppercase letters A, B, ..., Z represent internal nodes (i.e. non-terminals) of the untagged tree, and lowercase letters a,b,...,w_1,w_2,...represent leaf nodes (i.e. terminals). In addition, we use the Greek alphabet α_1, α_2, ... to represent any node (non-terminal or terminal). In Fig.4(b), a production of the non-terminal A[x_1] is A[x_1] → B[x_2]a[NONE]C[x_5], a[NONE] is a terminal, and TOP_S[NONE] is a non-terminal.

MTTM is defined as follows: Assume there is an untagged tree T, in which the tag of the node i is $x_i \in H_i$, and tags of all internal nodes (not including leaf nodes because their tagging set is {NONE}) constitute the tagging vector X=< $x_1, x_2, ... , x_n$ > $\in H_1 \times H_2 \times ... \times H_n$. Then the best tag X^* of the tree T is

$$X^* = \underset{X \in H_1 \times H_2 \times \ldots \times H_n}{argmax} P(X/T) = \underset{X \in H_1 \times H_2 \times \ldots \times H_n}{argmax} P(T[X]/T)$$
$$= \underset{X \in H_1 \times H_2 \times \ldots \times H_n}{argmax} P(T[X]) = \underset{X \in H_1 \times H_2 \times \ldots \times H_n}{argmax} P(T[X]/TOP\_S[NONE])$$

In the following, we discuss how to decompose and calculate $P(T[X]/$ $TOP\_S[NONE])$ and how to use dynamic programming algorithm to calculate $\underset{X}{max} P(T[X]/TOP\_S[NONE])$ and $\underset{X}{argmax} P(T[X]/TOP\_S[NONE])$ effectively.

3.3   Left-Most Derivation of the Tagged Tree T[X]

A tagged tree T[X] corresponds to a left-most derivation $r_1, r_2, \ldots, r_n$. We set the following rules: (1) The number i is the number of an internal node which corresponds to the left side of the production r_i. Node i is written as $Y_i[x_i]$. So we give each internal node a unique number. For example, TOP_S[NONE] is the node $Y_1[x_1]$. (2) A sub-tree of the tree T[X] whose root is $Y_i[x_i]$ is called $T_i[x_i]$. The leftmost derivation of the sub-tree $T_i[x_i]$ is $r_i, r_{i+1}, \ldots, r_j$.

Assuming $Y_i[x_i]$ is known, the sub-tree $T_i[x_i]$ is not unique because its internal node $Y_{i+1}[x_{i+1}]$, $Y_{i+2}[x_{i+2}]$, $\cdots$, $Y_j[x_j]$ will vary with different values of $x_{i+1}, \ldots, x_j$.

Obviously, a tagged tree T[X] = $T_1[NONE]$, its root is $Y_1[x_1]$ = TOP_S [NONE], and its leftmost derivation is $r_1, r_2, \ldots, r_n$, that is, i =1, j=n. n is the number of internal nodes of T[X].

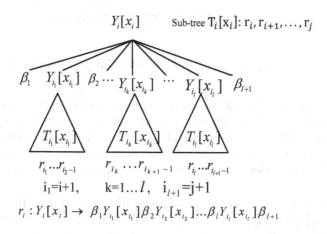

$$i_1 = i+1, \qquad k=1 \ldots l, \qquad i_{l+1} = j+1$$
$$r_i : Y_i[x_i] \to \beta_1 Y_{i_1}[x_{i_1}] \beta_2 Y_{i_2}[x_{i_2}] \ldots \beta_l Y_{i_l}[x_{i_l}] \beta_{l+1}$$

Fig. 5. Leftmost derivation of $T_i[x_i]$

Assume the production $r_i: Y_i[x_i] \to \beta_1 Y_{i_1}[x_{i_1}] \beta_2 Y_{i_2}[x_{i_2}] \ldots \beta_l Y_{i_l}[x_{i_l}] \beta_{l+1}$, β_q is a string which is composed of one or more terminals, q=1 … l +1, (β_q can be null), $Y_{i_k}[x_{i_k}]$ is the kth non-terminal of the right hand side of the production r_i, k=1… l, and the number of its corresponding internal node in tree T[X] is i_k. l is the number of the non-terminals of the right hand side of the production r_i. When l =0, production r_i has no non-terminal. Because the production sequence $r_i, r_{i+1}, \ldots, r_j$ is the left-most derivation of the tree $T_i[x_i]$ which begins with the non-terminal $Y_i[x_i]$, r_{i+1} is the direct derivation of the non-terminal $Y_{i_1}[x_{i_1}]$, that is $r_{i+1}: Y_{i_1}[x_{i_1}] \to \lambda$, and $i_1 = i+1$. Let $i_{l+1} = j + 1$, then $r_{i_k} \ldots r_{i_k-1}$ is left-most derivation of the sub-tree $T_{i_k}[x_{i_k}]$ which begins with the non-terminal $Y_{i_k}[x_{i_k}]$. Details are shown as Fig.5.

3.4   Using Dynamic Programming to Find the Best Tagged Tree

In the following, we discuss how to use dynamic programming to find the best tagged tree. We use dynamic programming twice, represent with σ and δ respectively, and use the method of accumulator.

$\sigma(Y_i[x_i])$ = the maximum probability of the sub-tree $T_i[x_i]$ when $x_i \in H_i$ is known That is, $\sigma(Y_i[x_i]) = \max\limits_{x_{i+1} \in H_{i+1}, x_j \in H_j} P(T_i[x_i]/Y_i[x_i])$)

Assume $\hat{T}_i[x_i]$ is the sub-tree which maximizes the probability $P(T_i[x_i]/Y_i[x_i])$. That is, $\hat{T}_i[x_i] = \operatorname*{argmax}\limits_{x_{i+1} \in H_{i+1}, x_j \in H_j} P(T_i[x_i]/Y_i[x_i])$.

Specially, when $i = 1, Y_1 = TOP\_S$, $x_1 \in H_1 = \{NONE\}, T[X] = T_1[NONE]$, then probability of the best tagged tree $T[X^*]$ is

$$\max\limits_{X \in H_1 \times H_2 \times \ldots \times H_n} P(T[X]/TOP\_S[NONE])$$
$$= \max\limits_{x_2 \in H_2, \ldots, x_m \in H_n} P(T_1[NONE]/TOP\_S[NONE]) = \sigma(Y_1[x_1]) \qquad (1)$$

And, $$T[X^*] = \operatorname*{argmax}\limits_{X \in H_1 \times H_2 \times \ldots \times H_n} P(T[X]/TOP\_S[NONE])$$

$$= \operatorname*{argmax}\limits_{x_2 \in H_2, \ldots, x_n \in H_n} P(T_1[NONE]/TOP\_S[NONE]) = \hat{T}_1[x_1] \qquad (2)$$

That is to say, $\hat{T}_1[x_1]$ is the best tagging tree, and its probability is $\sigma(Y_1[x_1])$. We use a recursive method to post-order traverse the untagged tree T, and calculate $\sigma(Y_1[x_1])$ and $\hat{T}_1[x_1]$, $\sigma(Y_2[x_2])$ and $\hat{T}_2[x_2]$,..., $\sigma(Y_n[x_n])$ and $\hat{T}_n[x_n]$. Note that, $\sigma(Y_1[x_1])$ and $\hat{T}_1[x_1]$ are the last ones to calculate, because $Y_1[x_1] = TOP\_S[NONE]$ is the root of tree T, which is the last node to visit during the post-order traversal.

In the following, we discuss how to calculate $\sigma(Y_i[x_i])$ and $\hat{T}_i[x_i]$. Because

$P(T_i[x_i]/Y_i[x_i]) = P(r_i, r_{i+1} \ldots, r_j/Y_i[x_i]) = P(r_i/Y_i[x_i]) * P(r_{i+1} \ldots, r_j/r_i, Y_i[x_i])$

$\approx P(r_i/Y_i[x_i]) * \prod_{k=1}^{l} P(r_{i_k} \ldots r_{i_k-1}/r_i, Y_i[x_i])$ // $\approx$ is independence assumption

$\approx P(r_i/Y_i[x_i]) * \prod_{k=1}^{l} P(r_{i_k}, \ldots, r_{i_k+1}/Y_{i_k}[x_{i_k}])$

$= P(r_i/Y_i[x_i]) * \prod_{k=1}^{l} P(T_{i_k}[x_{i_k}]/Y_{i_k}[x_{i_k}])$

Then $\sigma(Y_i[x_i]) = \max\limits_{x_{i+1} \in H_{i+1}, x_j \in H_j} P(T_i[x_i]/Y_i[x_i])$

$$\approx \max_{x_{i+1} \in H_{i+1}, x_j \in H_j} P(r_i/Y_i[x_i]) * \prod_{k=1}^{l} P(T_{i_k}[x_{i_k}]/Y_{i_k}[x_{i_k}])$$

$$= \max_{x_{i_k} \in H_{i_k}, k=1...l} \{P(r_i/Y_i[x_i])$$

$$* \prod_{k=1}^{l} \max_{x_{i_k+1} \in H_{i_k+1}, ..., x_{i_{k+1}-1} \in H_{i_{k+1}-1}} \{P(T_{i_k}[x_{i_k}]/Y_{i_k}[x_{i_k}])\}$$

$$= \max_{x_{i_k} \in H_{i_k}, k=1...l} \{P(r_i/Y_i[x_i]) * \prod_{k=1}^{l} \sigma(Y_{i_k}[x_{i_k}])\} \tag{3}$$

We decompose r_i using Markov chain to calculate $P(r_i/Y_i[x_i])$ in (3) as follows.

Assume r_i is: $Y_i[x_i] \to \alpha_1^i[x_1^i]\alpha_2^i[x_2^i] ... \alpha_m^i[x_m^i]$, $\alpha_q^i[x_q^i]$ can be either the internal node (i.e., non-terminal) or the leaf node (i.e. terminal), q=1,2,...,m. m is the number of non-terminals in the right hand side of r_i . We add two terminals STOP in the right hand side of the rule. That is,

$$\alpha_0^i[x_0^i] = STOP[NONE], \alpha_{m+1}^i[x_{m+1}^i] = STOP[[NONE],$$
$$r_i : Y_i[x_i] \to \alpha_0^i[x_0^i]\alpha_1^i[x_1^i]\alpha_2^i[x_2^i] ... \alpha_m^i[x_m^i]\alpha_{m+1}^i[x_{m+1}^i].$$

This modification will not change the probability of the production and the tree.

Then $\sigma(Y_i[x_i]) = \max_{x_{i_k} \in H_{i_k}, k=1...l} \{P(r_i/Y_i[x_i]) * \prod_{k=1}^{l} \sigma(Y_{i_k}[x_{i_k}])\}$

$$= \max_{x_{i_k} \in H_{i_k}, k=1...l} \{P(\alpha_0^i[x_0^i]\alpha_1^i[x_1^i]\alpha_2^i[x_2^i] ... \alpha_m^i[x_m^i]\alpha_{m+1}^i[x_{m+1}^i]/Y_i[x_i]) \prod_{k=1}^{l} \sigma(Y_{i_k}[x_{i_k}])\}$$

$$\approx \max_{x_{i_k} \in H_{i_k}, k=1...l} \{\{P(\alpha_0^i[x_0^i]/Y_i[x_i]) \prod_{q=1}^{m+1} P(\alpha_q^i[x_q^i]/Y_i[x_i], \alpha_{q-1}^i[x_{q-1}^i])\} \prod_{k=1}^{l} \sigma(Y_{i_k}[x_{i_k}])\}$$

$$= \max_{x_{i_k} \in H_{i_k}, k=1...l} \{\{\prod_{q=1}^{m+1} P(\alpha_q^i[x_q^i]/Y_i[x_i], \alpha_{q-1}^i[x_{q-1}^i])\} * \prod_{k=1}^{l} \sigma(Y_{i_k}[x_{i_k}])\} \tag{4}$$

Let $\rho(\alpha_q^i[x_q^i]) = \begin{cases} 1 & when\ \alpha_q^i[x_q^i]\ is\ a\ leaf\ node \\ \sigma(Y_{i_k}[x_{i_k}]) & when\ \alpha_q^i[x_q^i] = Y_{i_k}[x_{i_k}] \end{cases}$ $q = 1,2,...,m+1$

Then formula (4) is as follows

$$\sigma(Y_i[x_i]) = \max_{x_{i_k} \in H_{i_k}, k=1...l} \prod_{q=1}^{m+1} \{P(\alpha_q^i[x_q^i]/Y_i[x_i], \alpha_{q-1}^i[x_{q-1}^i])\rho(\alpha_q^i[x_q^i])\}$$

$$= \max_{x_q^i \in H_q^i, q=1,2,...,m+1} \prod_{q=1}^{m+1} \{P(\alpha_q^i[x_q^i]/Y_i[x_i], \alpha_{q-1}^i[x_{q-1}^i])\rho(\alpha_q^i[x_q^i])\} \tag{5}$$

In the following, we use dynamic programming again to calculate the value of formula (5). This step is similar to the Viterbi algorithm of HMM, because it is to get the best tagging $< x_1^i, x_2^i, \ldots x_m^i, x_{m+1}^i >$ of the sequence $< \alpha_1^i, \alpha_2^i, \ldots \alpha_m^i, \alpha_{m+1}^i >$.

(1) Initialization

$$\delta_0 \left(Y_i[x_i], x_0^i \right) = 1, x_0^i \epsilon \{NONE\}$$

(2) Derivation

For $q = 1,2, \ldots, m+1, x_q^i \epsilon H_q^i$,

$$\delta_i \left(Y_i[x_i], x_q^i \right)$$
$$= \underset{x_{q-1}^i \epsilon H_{q-1}^i}{max} \delta_{q-1} \left(Y_i[x_i], x_{q-1}^i \right) * P\left(\alpha_q^i[x_q^i] / Y_i[x_i], \alpha_{q-1}^i[x_{q-1}^i] \right) * \rho\left(\alpha_q^i[x_q^i] \right)$$

Storage the backtracking path

$$\psi_q \left(Y_i[x_i], x_q^i \right)$$
$$= \underset{x_{q-1}^i \epsilon H_{q-1}^i}{argmax} \delta_{q-1} \left(Y_i[x_i], x_{q-1}^i \right) * P\left(\alpha_q^i[x_q^i] / Y_i[x_i], \alpha_{q-1}^i[x_{q-1}^i] \right) * \rho\left(\alpha_q^i[x_q^i] \right)$$

(3) Assume the most likely sequence of the backtracking is

$\hat{x}_0^i, \hat{x}_1^i, \ldots, \hat{x}_m^i, \hat{x}_{m+1}^i$, then,

$$\hat{x}_{m+1}^i = \underset{x_{m+1}^i \epsilon H_{m+1}^i}{argmax} \delta_{m+1} \left(Y_i[x_i], x_{m+1}^i \right) \text{ (Actually, } x_{m+1}^i \epsilon H_{m+1}^i = \{NONE\})$$
$$\hat{x}_{q-1}^i = \psi_q \left(Y_i[x_i], x_q^i \right), q = m+1, m, \ldots, 1$$

Now, the maximum probability of the sub-tree $T_i[x_i]$ whose root is $Y_i[x_i]$ is

$$\sigma(Y_i[x_i]) = \underset{x_{m+1}^i \epsilon H_{m+1}^i}{max} \delta_{m+1} \left(Y_i[x_i], x_{m+1}^i \right)$$

Then $\sigma(Y_i[x_i]) = \delta_{m+1} \left(Y_i[x_i], x_{m+1}^i = NONE \right)$ because $x_{m+1}^i \epsilon H_{m+1}^i = \{NONE\}$
So, under the circumstance that the value of x_i is known, $\psi(Y_i[x_i]) =< \hat{x}_0^i, \hat{x}_1^i, \ldots, \hat{x}_m^i, \hat{x}_{m+1}^i >$ is the best tag of r_i in the sub-tree $\hat{T}_i[x_i]$ whose root is $Y_i[x_i]$.

Assume that the number of internal nodes of the untagged tree is N, the maximum size of the tagging sets is H=max($|H_i|$), and the maximum number of child nodes of an internal node (including STOP node) is M, then the time complexity is O(N*M*H³).

4   Use Markov Tree Tagging Model to Tag Tense Trees

Training:

(1) Construct the untagged tense tree according to the parse tree of a sentence.
(2) Manually tag tenses of IP, CP and verb nodes of the untagged tense tree T, and construct the tagged tense tree T[X].
(3) Calculate probability parameters, which will be used to calculate the probability of a tagged tense tree in decoding.

Decoding:

(1) Construct the untagged tense tree according to the grammar tree of a sentence.
(2) Use MTTM to get the tense tree T[X*] as the final result.

$$X^* = \underset{\substack{\text{the English tense of IP,CP,verb node} \\ \text{et aggi ng set}}}{\operatorname{argmax}} P(T[X]/T) = \underset{\substack{\text{he English tense of IP,CP,verb node} \\ \text{et aggi ng set}}}{\operatorname{argmax}} P(T[X])$$

The tagging set of IP, CP and verb nodes is {16 tenses}, and that of other nodes is {NONE}. Attribute values of node i in T can be regarded as Y_i in MTTM.

5   Experimental Results and Discussion

5.1   Experimental Data

We manually tag *Chinese TreeBank 6.0* with 16 English tenses according to its English translation *English Chinese Translation Treebank v 1.0*. The chtb_0110.fid ~ chtb_0139.fid is training set, and chtb_0140.fid ~ chtb_0144.fid is the test set.

We use accuracy [5] to measure the whole performance, and use precision, recall and F_measure to measure the performance of each tense.

5.2   Experimental Results

We use MTTM to tag untagged tense trees of the test set with 16 English tenses, convert the experimental results into the 3 tenses (past, present, future), and compare them with the experimental results of Ye [5], because they only contain 3 tenses.

Table 1 is the comparison of the verb tagging results between our method and the method of Ye [5]. The accuracy of our method is better than that of Ye. The precision and recall of present tense are very low in Ye [5], and our method has a substantial increase. Because future tense appears rarely in the tense corpus and the test set, the F-measure of future tense of Yang Ye and our method are very low.

Table 1. The experimental comparison of tagging verbs between our method and Yang Ye [5]

	Accuracy of Ye[5]=58.21%			Accuracy of our method = 66.0%		
	Precision	Recall	F-measure	Precision	Recall	F-measure
past	67.57%	79.55%	72.10%	55.1%	81.4%	65.7%
present	42.50%	27.48%	32.07%	80.5%	61.9%	70.01%
future	29.66%	25.56%	21.56%	38.5%	22.7%	28.6%

6   Conclusions

The experimental results show that the verb accuracy of tense tagging based on tense tree improves 7.79% compared to that of Ye which is based on sequence. This comparison shows that hierarchical interdependence is important to tense tagging, while methods based on sequence are difficult to express hierarchical interdependence. Though linear conditional random field is a very good machine learning method, the correct expression of hierarchical interdependence is more important.

Acknowledgments. Supported by the National Natural Science Foundation of China (No. 61075058).

References

1. Liu, Q., Yu, S.: Discussion on the difficulties of Chinese-English machine translation. In: Huang, C.L. (ed.) International Conference on Chinese Information Processing, pp. 507–514. Tsinghua University Press, Beijing (1998)
2. Gong, Z., Zhang, M., Tan, C., Zhou, G.: N-gram-based tense models for statistical machine translation. In: Proceedings of EMNLP, East Stroudsburg, PA, USA, pp. 276–285. Association for Computational Linguistics (ACL) (2012)
3. Chen, J., Dai, X., Chen, J., Wang, Q.: Processing of Tense and Aspect in Chinese-English Machine Translation. Application Research of Computers **21**(3) (2004)
4. Hongmei, M.: Research on the Representation and Application of Chinese Context in Chinese-English Machine Translation [Ph.D. Thesis], Changsha, Hunan: National University of Defense Technology (2002)
5. Ye, Y., Zhang, Z.: Tense tagging for verbs in cross-lingual context: a case study. In: Dale, R., Wong, K.-F., Su, J., Kwong, O.Y. (eds.) IJCNLP 2005. LNCS (LNAI), vol. 3651, pp. 885–895. Springer, Heidelberg (2005)
6. Gong, Z., Zhang, M., Tan, C., Zhou, G.: Classifier-based tense model for SMT. In: Proceedings of the 24th International Conference on Computational Linguistics, PP. 411–420 (2012)
7. Lu, J., Ma, Z.: Comments on Function Word of Modern Chinese, pp. 106–141. Peking University Press, Beijing (1999)
8. Collins, M.: Head-Driven Statistical Models for Natural Language Parsing. Computational Linguistics (2003)

Building a Large-Scale Cross-Lingual Knowledge Base from Heterogeneous Online Wikis

Mingyang Li [(✉)], Yao Shi, Zhigang Wang, and Yongbin Liu

Department of Computer Science and Technology,
Tsinghua University, Beijing 100084, China
{meeya.yx,syontheway,wangzigo,yongbinliu03}@gmail.com

Abstract. Cross-Lingual Knowledge Bases are very important for global knowledge sharing. However, there are few Chinese-English knowledge bases due to the following reasons: 1) the scarcity of Chinese knowledge in existing cross-lingual knowledge bases; 2) the limited number of cross-lingual links; 3) the incorrect relationships in semantic taxonomy. In this paper, a large-scale Cross-Lingual Knowledge Base(named XLORE) is built to address the above problems. Particularly, XLORE integrates four online wikis including English Wikipedia, Chinese Wikipedia, Baidu Baike and Hudong Baike to balance the knowledge volume in different languages, employs a link-discovery method to augment the cross-lingual links, and introduces a pruning approach to refine taxonomy. Totally, XLORE harvests 663,740 classes, 56,449 properties, and 10,856,042 instances, among of which, 507,042 entities are cross-lingually linked. At last, we provide an online cross-lingual knowledge base system supporting two ways to access established XLORE, namely a search engine and a SPARQL endpoint.

Keywords: Knowledge base · Cross-lingual linking · Taxonomy pruning

1 Introduction

As the Web is evolving to a highly globalized information space, knowledge sharing across different languages attracts increasing attentions. Multi-lingual knowledge bases have significant applications such as information retrieval, machine translation and deep question answering. DBpedia, by extracting structured information from Wikipedia[1], is a multi-lingual knowledge base covering many domains and becomes the nucleus of Linked Open Data[2]. YAGO, MENTA and BabelNet are other famous large multi-lingual knowledge bases.

However, most non-English knowledge is pretty scarce. The knowledge distribution across different languages is highly unbalanced in Wikipedia-based knowledge bases. For instance, DBpedia contains 4.58 million English instances but no Simplified Chinese dataset published. On the other hand, the Chinese

---

[1] http://www.wikipedia.org
[2] http://linkeddata.org

© Springer International Publishing Switzerland 2015
J. Li et al. (Eds.): NLPCC 2015, LNAI 9362, pp. 413–420, 2015.
DOI: 10.1007/978-3-319-25207-0_37

Hudong Baike[3] and Baidu Baike[4], both containing more than 11 million articles, are even larger than the English Wikipedia. If a knowledge base could be established based on both English Wikipedia and Chinese Hudong Baike, more Chinese-English knowledge can be generated.

We try to build a large-scale cross-lingual knowledge base generated from four heterogeneous online wikis, i.e. English Wikipedia, Chinese Wikipedia, Hudong Baike and Baidu Baike. This non-trivial task poses the challenges as follows: 1) Limited English-Chinese cross-lingual links within Wikipedia (i.e. 4.6%). How could we find enough Chinese-English owl:sameAs relations? 2) Noisy subsumption relations in the category systems, e.g. "Wikipedia-books-on-people", which is actually *subClassOf* "Books", is mistaken as sub-category of "People". How could we detect those incorrect semantic relations?

To tackle these issues, we propose a unified framework to build a Chinese-English knowledge base from four heterogeneous online wikis in three steps: 1) extract wiki dataset 2) extend cross-lingual link set 3) prune taxonomy. The generated KB, named **XLORE**[5], contains 663,740 classes, 56,449 properties and 10,856,042 instances. Specifically, we make the following contributions: (1) We extend cross-lingual link set by employing a cross-lingual knowledge linking discovery approach for class and instance, and by analyzing templates in Wikipedia for property. (2) We prune the original taxonomy, which is extracted from wiki category system, to retrieve more precise *subClassOf* and *instanceOf* relations. (3) An online-system supporting keyword search and SPARQL endpoint is provided for public access to our knowledge base.

2 Preliminaries

Online Wikis. Nowadays, Wikipedia is the largest data store of human knowledge. It has hold over 35 million articles in 288 languages by 2015. Baidu Baike and Hudong Baike are the most content-rich among the large-scale monolingual Chinese wikis currently. Hudong Baike contains more than 12 million articles until 2015 while Baidu Baike maintains over 11 million articles.

Wikis usually provide two important elements with potential semantic information, category system and articles. Here, we define an encyclopedia wiki as: $W = < C, A >$, where C denotes categories, A denotes articles. A category system represents the relations between categories as a tree by *subCategoryOf*.

Wiki Pages. Articles from the wiki sources are similar in structure. An article a can be defined as follow: $a = < Ti(a), Ab(a), Li(a), Ib(a), C(a), U(a) >$ where $Ti(a), Ab(a), Li(a), Ib(a), C(a), U(a)$ denote title, abstract, links, infobox, category tags, url of article a.

Notably, infoboxes in articles are generated based on certain templates recommended by Wikipedia. An infobox template collects attributes describing

---

[3] http://www.baike.com
[4] http://baike.baidu.com
[5] http://xlore.org

similar entities, e.g. infobox of film 冰雪奇缘 (Frozen) is normalized by *Template:Infobox film*, and we denote the infobox template used in article a as $T(a)$. However, attribute labels in templates are usually different from those displayed on the webpage. Thus, an infobox-attribute is defined as $p =< tl, dl, v >$, where tl and dl denote the label in template and web page, and v is the attribute value.

Cross-lingual Links. In Wikipedia, cross-lingual links help readers switch to preferred languages. If an entity containing both Chinese article a_z and English article a_e, then a_z and a_e are *cross-linked*. Infobox templates may also be cross-linked. For a pair $cl \in CL$, $cl =< L_z, L_e >$, where L_z and L_e denote the entity's cross-lingual links in Chinese and English.

Knowledge Base. A knowledge base is a formal specification of a group of entities. Our knowledge base is described as a 4-tuple: $KB =< C, P, I, H^C >$, where C, P, I are the sets of classes, properties, and instances respectively, and H^C represents the class hierarchy. Semantic relations include *subClassOf, instanceOf, relatedClassOf, relatedTopicOf*.

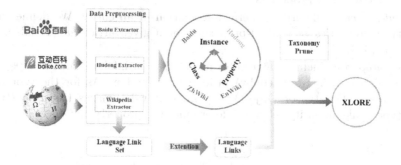

Fig. 1. Procedure of Building Our Cross-Lingual Knowledge Base XLORE

A Cross-Lingual Knowledge Base(CLKB) is a database conforming to a cross-lingual ontology, and is often integrated from various sources based on cross-lingual links. Thus CLKB is defined as: $CLKB =< KB_z, KB_e >$, where KB_z and KB_e denote the knowledge bases in Chinese and English.

Cross-Lingual Knowledge Base Building is to build a CLKB assembling knowledge from several English and Chinese wiki sources. Specifically, first build monolingual knowledge base from each W_i, then enrich the existing cross-lingual links, further refine the taxonomy, and finally integrate datasets in different languages based on the cross-lingual links as shown in Fig. 1.

3   Semantic Data Extraction

Semantic data extraction aims to achieve a structured dataset from the input wikis. Specifically, we extract classes from category system, instances according to articles, and properties based on infoboxes.

Class Extraction. A class is defined as a type of similar instances. For example, the class of instance 冰雪奇缘 (Frozen) is 电影 (Film). In general, a class is semantic related with other classes by $subClassOf$ relation. Such relations comprise a class hierarchy H^C which presents the backbone of an ontology. In a wiki, a category groups several articles and also has $subCategoryOf$ relation with others. Therefore we can extract classes based on existing category system.

To gain a more precise H^C in a wiki, we discard categories matching the following conditions: 1) There are auxiliary categories in Wikipedia, which help arrange specific articles or category pages. For example, *Lists of artists* or *Food templates.* 2) Some categories are related to only one article. According to the definition of class, such categories are less representative types, therefore it's unwise to retain them as classes.

Property Extraction. A property is defined as an attribute of an entity. We divide properties into two types: Object Property, whose value is an individual, such as 导演 (directed by); Datatype Property, whose value is a literal text, such as 出生日期 (birth date). Considering both content and infobox of an article, we extract two kinds of properties:

General-properties describe general information of an entity. We define three Datatype properties as general-properties for a given article a: (1) label; (2) abstract; (3) URL.

Infobox-properties includes attributes acquired from infobox, such as 上映时间 (release date), 导演 (directed by) in a movie's infobox. As for the type of a property (i.e. Datatye or Object), a plain text value marks the property as Datatype while an entity reference determines the property as Object.

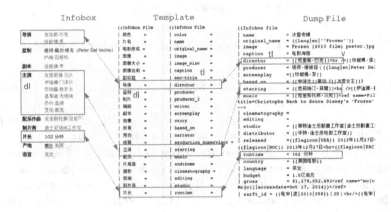

Fig. 2. Comparison of display label and template label in *Frozen* infobox

In Wikipedia, attribute labels displayed in webpage infoboxes are inconsistent with those in the dump file. Fig. 2 gives a case-control of display labels and template labels in 冰雪奇缘 (Frozen)'s infobox. The attribute label 主演 displayed on webpage is different from the label *starring* extracted from dump file. However, readers generally consider the display label as an attribute label. Therefore, we employ the infobox template to bridge this gap: replace the template label in dump file by its matched display label. Then we sweep special characters in labels such as hyphen "-" or dot "•" in Wikipedia and ":" or "*" in Baidu. Furthermore, properties correspond to only one instance are discarded.

Instance Extraction. An article describes a unique entity in the world. Therefore we can extract an article as an instance. We harvest four types of information during this stage. (1) General-properties of instance, including title as label property value, first paragraph as abstract property value and HTTP URL as URL property value; (2) Infobox-properties which are acquired via extracting from the infobox in the article; (3) *articleOf* relation with categories listed at the bottom of article page. For example, 冰雪奇缘 (Frozen) is an article of category 美国电影作品 (American films); (4) Reference relation with other instances according to links in the content, such as 冰雪女王 (The Snow Queen).

4 Cross-lingual Integration

We construct a CLKB with obtained structured data in the following three steps:

Cross-lingual Linking is to match the same entity (i.e., class, property and instance) in two languages. For class and instance, we utilize the linkage factor graph model in [6] to extend the original 227 thousand cross-lingual Chinese-English links in Wikipedia, and achieve 215 thousand links between English Wikipedia and Baidu Baike. For property, because Infobox-properties have no obvious cross-lingual links, we achieve the links using infobox templates:

1. Given two cross-linked templates, T_e and T_z, find the display labels mapping to the same template label. That is, if a template label tl_e in T_e is equal to a template label tl_z in T_z, $< dl_e, dl_z >$ are cross-lingual property labels;
2. Given two cross-linked articles, a_e and a_z, and their infobox templates, $T_e(a_e)$ and $T_z(a_z)$, for p_e in $T_e(a_e)$ and p_z in $T_z(a_z)$, if $p_e.tl$ is equal to $p_z.tl$, $< p_e.dl, p_z.dl >$ are cross-lingual property labels;
3. Given two cross-linked articles and their Infobox-properties P_e and P_z, for $p_e \in P_e$ and $p_z \in P_z$, $p_e.dl$ and $p_z.dl$ are cross-lingual when: (1)for datatype properties, $sim(p_e.v, p_z.v) > threshold$, where $sim(a, b)$ is a similarity function. (2)for object properties, $p_e.v$ and $p_z.v$ refer to the same entity.

Wikidata Integration. In order to integrate all wikis, we unify classes, instances or properties describing the same thing from four sources, and distribute a unique identifiers. For instance, we merge instances by the following steps: (1) Merge all instances extracted from wikis by title. (2) If a Chinese instance has a cross-linked English instance, that is, to an L_z, if there is $< L_z, L_e >$ in CL, make them as one instance. (3) Identify all instances, including both monolingual and cross-lingual, by IDs. The processes of unifying class and property are the same as instance.

Taxonomy Prune. There is inevitably noise in the taxonomy since we combine multi-source information without verification. Therefore, we introduce the method from [8] to detect the correct *subClassOf* and *instanceOf* relations from *subCategoryOf* and *articleOf*. The ideal result after pruning is a tree, whose edges, nodes, and leaves respectively denote semantic relations, classes and instances. However, since getting rid of incorrect entity relations without consideration of integrity, a forest result is inevitable. To retain integrity of semantic relation, we define two types of new relations: *relatedClassOf* for cut instance-class relations and *relatedTopicOf* for pruned class-class relations.

5 Result

Here we show statistic results of our CLKB, XLORE, and introduce the developed system based on XLORE dataset.

Table 1. Statistics of Elementary Extraction Result

	Enwiki	Zhwiki	Hudong	Baidu
#Class	982,432	159,705	31,802	1300
#Instance	4,304,113	662,650	5,590,751	5,622,404
#Property	43,976	18,842	1187	139,634

Table 2. Statistics of XLORE

	Classes		Instances		Properties	
English	639,020	96.26%	3,879,121	38.79%	15,380	27.24%
Chinese	88,615	13.35%	7,409,519	68.25%	51,618	91.44%
Cross-lingual	63,895	9.63%	432,598	3.98%	10,549	18.69%
Total	663,740	-	10,856,042	-	56,449	-

Knowledge Base Overview. We collect the resources from four online wikis, English and Chinese Wikipedia dump files in May, 2014, Hudong html pages until May, 2014, and Baidu html pages until September, 2014. Each of the wikis has three types of information, which can be utilized for constructing our knowledge base, namely, category system, specific articles, and attributes of articles. Table 1 shows the results we get after elementary extraction on 4 different wiki sources.

After fusing the heterogeneous sources, we harvest a cross-lingual knowledge base with 663,740 classes, 56,449 properties, and 10,856,042 instances respectively. With different methods of extraction and language link discovery, these three kinds of entries show different results in languages. We give a breakdown of both Chinese knowledge and English knowledge in Table 2.

Web Access to XLORE. We organize XLORE in Openlink Virtuoso, and provide a platform, which locates on http://xlore.org, to present an intuitive visualization in the forms of instance, class and property. URIs http://xlore. org/type/id (type could be *class, instance, property*) are created to identify each

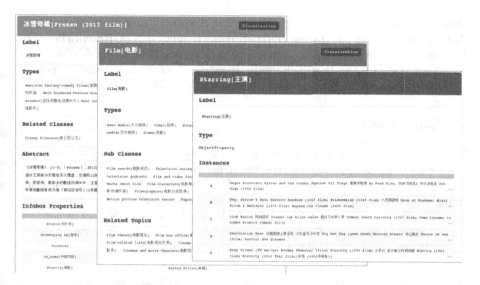

Fig. 3. Sample Pages of Instance, Class and Property

entry . Fig. 3 shows sample pages of the integrated data. Language could be switched, which is convenient for both English speaking and Chinese speaking users. Besides these user-friendly pages, we provide two ways to access our knowledge base. For general users, they can send a query by inputing related text into searchbox to get probable related entries. To present practicable result, a fuzzy-query strategy is employed over all entries. We as well provide SPARQL interface for professional users to query our knowledge base. Users can choose the language tags of their desired results by **"filter(langMatches(?label),"en"))"** or **"filter(langMatches (?label),"zh"))"**.

6   Related Work

In this section, we introduce some related knowledge bases.

Chinese Knowledge Bases. Zhishi.me[5] is the first published Chinese large-scale Linking Open Data, which acquires structural information from Chinese Wikipedia, Baidu Baike and Hudong Baike. Similarly, Wang et.al learns an ontology based on category system and properties from Hudong Baike [7]. XLORE is an extension of CKB in multi-language. Utilizing the rich content of multiple online-wikis, XLORE gathers abundant valuable semantic information.

Cross-lingual Knowledge Bases. DBpedia [4] is one of the most widely-used [1,3] cross-lingual knowledge base in the world. It extracts various kinds of structured information from Wikipedia and employs the multi-lingual characteristic of Wikipedia to generate 97 language versions of content. Universal WordNet(UWN) [2] is a large multi-lingual lexical knowledge base built from

WordNet and Wikipedia through sophisticated knowledge extraction, link prediction, information integration, and taxonomy induction. XLORE enriches non-English things by employing other language-version wikis, which eliminates disadvantages of using Wikipedia only. It extracts more classes and properties automatically and validates precise semantic relations by a pruning approach.

7   Conclusion

This paper presents an approach of building a Chinese-English CLKB from multiple wiki sources. We extract structured information and unify data format. Then a cross-lingual link set is generated and expanded to help combine the bilingual sources. To refine our dataset, we also conduct pruning work on taxonomy. Finally, we acquire a CLKB containing 663,740 classes, 56,449 properties, and 10,856,042 instances. Currently, an online-system supporting keyword search and SPARQL query is provided to access the knowledge base.

Acknowledgement. The work is supported by 973 Program (No. 2014CB340504), NSFC-ANR (No. 61261130588), NSFC (No.61402220), Tsinghua University Initiative Scientific Research Program (No. 20131089256), Science and Technology Support Program (No. 2014BAK04B00), and THU-NUS NExT Co-Lab.

References

1. Fernández-Tobías, I., Cantador, I., Kaminskas, M., Ricci, F.: A generic semantic-based framework for cross-domain recommendation. In: Proceedings of the 2nd International Workshop on Information Heterogeneity and Fusion in Recommender Systems, pp. 25–32. ACM (2011)
2. de Melo, G., Weikum, G.: Uwn: a large multilingual lexical knowledge base. In: Proceedings of the ACL 2012 System Demonstrations, pp. 151–156. Association for Computational Linguistics (2012)
3. Mendes, P.N., Daiber, J., Jakob, M., Bizer, C.: Evaluating dbpedia spotlight for the tac-kbp entity linking task. In: Proceedings of the TACKBP 2011 Workshop, vol. 116, pp. 118–120 (2011)
4. Mendes, P.N., Jakob, M., Bizer, C.: Dbpedia: a multilingual cross-domain knowledge base. In: LREC, pp. 1813–1817 (2012)
5. Niu, X., Sun, X., Wang, H., Rong, S., Qi, G., Yu, Y.: Zhishi.me - weaving chinese linking open data. In: Aroyo, L., et al. (eds.) ISWC 2011, Part II. LNCS, vol. 7032, pp. 205–220. Springer, Heidelberg (2011)
6. Wang, Z., Li, J., Wang, Z., Tang, J.: Cross-lingual knowledge linking across wiki knowledge bases. In: Proceedings of the 21st International Conference on World Wide Web, pp. 459–468. ACM (2012)
7. Wang, Z., Wang, Z., Li, J., Pan, J.Z.: Building a large scale knowledge base from chinese wiki encyclopedia. In: Pan, J.Z., et al. (eds.) JIST 2011. LNCS, vol. 7185, pp. 80–95. Springer, Heidelberg (2012)
8. Wang, Z., Li, J., Li, S., Li, M., Tang, J., Zhang, K., Zhang, K.: Cross-lingual knowledge validation based taxonomy derivation from heterogeneous online wikis. In: Twenty-Eighth AAAI Conference on Artificial Intelligence (2014)

Refining Kazakh Word Alignment Using Simulation Modeling Methods for Statistical Machine Translation

Amandyk Kartbayev[✉]

Laboratory of Intelligent Information Systems,
Al-Farabi Kazakh National University, Almaty, Kazakhstan
a.kartbayev@gmail.com

Abstract. Word alignment play an important role in the training of statistical machine translation systems. We present a technique to refine word alignments at phrase level after the collection of sentences from the Kazakh-English parallel corpora. The estimation technique extracts the phrase pairs from the word alignment and then incorporates them into the translation system for further steps. Although it is a pretty important step in training procedure, an word alignment process often has practical concerns with agglutinative languages. We consider an approach, which is a step towards an improved statistical translation model that incorporates morphological information and has better translation performance. Our goal is to present a statistical model of the morphology dependent procedure, which was evaluated over the Kazakh-English language pair and has obtained an improved BLEU score over state-of-the-art models.

Keywords: Word alignment · Optimization · Kazakh morphology · Word segmentation · Machine translation

1 Introduction

In this paper, we present the work done for improving a baseline statistical machine translation (SMT) system from an agglutinative Kazakh language to English. In this pair of translation, English word correspond to Kazakh suffixes that can fit more than one of the suffixes to the word. For instance, using the Kazakh lemma el - 'state' we can generate 'eldin' - 'of the state', 'elge' - 'to the state' and so on. The Kazakh language, which is the majority language in the Republic of Kazakhstan, has poor open resources and there are a small available parallel corpora unlike to other languages that more widely used as English or French. The parallel corpora for this work is 70k sentences for Kazakh and English, much bigger one than this corpus we had before.

In previous work[1], we described an approach to word alignment intended to address these problems. A research more relevant to that work was done by Bisazza and Federico[2]. The main goal of this research, different from the previous works, that is to make proposals that increase the expected benefit from

© Springer International Publishing Switzerland 2015
J. Li et al. (Eds.): NLPCC 2015, LNAI 9362, pp. 421–427, 2015.
DOI: 10.1007/978-3-319-25207-0_38

a word alignment estimation, rather than learning a morphology of Kazakh text for a processing activities. The given model is estimated from two principles: a morphological processing technique that gives the correct word segmentation, and a alignment model that determines the correct segment alignment in another case. Simulated results shows that this model has a potential to decrease a sparse words level, and to reach an overall consistency, which was observed during an evaluation.

The idea of using word alignment as a problem of determining correspondence at the word level was introduced by Brown et al.[3] is becoming as one of crucial components of all statistical machine translation approaches. Word alignment refinement can't be seen as a form of the relationship between word alignment quality and translation quality, what is well explained[4]. The common approaches of word alignment training are IBM Models and hidden Markov model (HMM)[5], which practically use expectation-maximization (EM) algorithm[6]. Compared to the word-based alignment models - in which a pair made of two words, each one from a different text that is certainly common to occur, but the words often are not a correct translation of each one - a phrase alignment model focuses on acquiring translations of phrases.

Phrase-based SMT systems usually train a phrase translation table, which may be produced after processing of word alignment and their probabilities for phrases. Phrase-based model has set of advantages over word-based alignment. At first, it naturally integrates context of the phrases and provide possibility to use these contexts in the translation. Eventually, a phrase is a consecutive sequence of words and the model allows the translation of unseen phrases unsupervised way. This makes the model generally applicable to similar language pairs we have learned.

We use Morfessor tool[7] to out grammatical features of word and can find the benefit of using morphological analysis in machine translation. We also explored rule-based morphological analyzer[8], which consist in deep language expertise and a exhaustive process in system development. For a comprehensive survey of the rule-based morphological analyze we refer a reader to the research by Altenbek[9] and Kairakbay[10].

The paper is structured as follows: Section 2 discusses the proposed model and describes the different segmentation techniques we study. And Section 3 presents our evaluation results.

2   Description of Our Method

Simulation modeling is too costly and time-consuming in this area of science as it exists. That we describe a new method that has to be more optimal and faster in some applications. In this approach, system starts a simple generic model and then incrementally replace its parts with more special pieces from a systematically organized processing components. Eventually, the system changes its subpart, and then automatically creates a new model or shows the step where further manual changes is necessary. For instance, we used the Helsinki Finite-State Toolkit (HFST)[11] to treat rule-based analyze and could to conduct a study of its benefits for morpheme based alignment. Also we use the GIZA++[12] tool, which intersects two word alignments and get an union of the alignments, finally it produces a nearly symmetric result. Our study is based on the set of experiments, which have the goal of most properly extraction a phrase table from the word alignment. That actually leads to higher BLEU scores[13] and rises overall translation quality by reduction the level of sparse phrases.

We suppose a phrase pair is denoted by (F, E) and with an alignment A, if any words f_j in F have a correspondence in a, with the words e_i in E. Formal definition can be described as follows: $\forall e_i \in E : (e_i, f_j) \in a \Rightarrow f_j \in F$ and $\forall f_j \in F : (e_i, f_j) \in a \Rightarrow e_i \in E$, clearly, there are $\exists e_i \in E, f_j \in F : (e_i, f_j) \in A$.

Generally, the phrase-based models are generative models that translate sequences of words in f_j into sequences of words in e_j, in difference from the word-based models that translate single words in isolation.

$$P(e_j \mid f_j) = \sum_{j=1}^{J} P(e_j, a_j \mid f_j) \tag{1}$$

Improving translation performance directly would require training the system and decoding each segmentation hypothesis, which is computationally impracticable. That we made various kind of conditional assumptions using a generative model and decomposed the posterior probability. In this notation e_j and f_i point out the two parts of a parallel corpus and a_j marked as the alignment hypothesized for f_i. If $a \mid e \sim ToUniform(a; I + 1)$, then

$$P\left(e_j^J, a_j^J \mid f_i^I\right) = \frac{f_i}{(I+1)^J} \prod_{j=1}^{J} p\left(e_j \mid f_{a_j}\right) \tag{2}$$

We extend the alignment modeling process of Brown et al. at the following way. We assume the alignment of the target sentence e to the source sentence f is a. Let c be the tag(from Penn Treebank) of f for segmented morphemes. This tag is an information about the word and represents lexeme after a segmentation process. This assumption is used to link the multiple tag sequences as hidden processes, that a tagger generates a context sequence c_j for a word sequence $f_j(3)$.

$$P\left(e_1^I, a_1^I \mid f_1^J\right) = P\left(e_1^I, a_1^I \mid c_1^J, f_1^J\right) \tag{3}$$

Then we can show Model 1 as(4):

$$P\left(e_i^I, a_i^I \mid f_j^J, c_j^J\right) = \frac{1}{(J+1)^I} \prod_{i=1}^{I} p\left(e_i \mid f_{a_i}, c_{a_i}\right) \tag{4}$$

We conduct an extensive experiment with using the descriptions mentioned above to construct dynamic models of alignment object in a SMT pipeline. The experiment included developing a comprehensive morpheme analysis that incorporates morphemes and text chunks to the phrase table. We create simulation models by moving actual linguistic objects onto the bijective space constraints, how it could be used to form a symmetric matrix mxn. A Kazakh-English surjective space consists of two "word" vectors with a certain number of more elementary parts $\{e_i, f_j\}$, and a set of linkages $\{x_{ij}\}$ between these vector elements. We require that all individual elements able to change their links, so every element may be linked to other one or more. Also we use the term "matrix", when we emphasize packing the probabilities $p_{i,j}(x_{ij}|e_i, f_j)$ to the building block, and the term "link constraints" emphasize the components of some grammar, a simulation model.

For describing how we organize vector of words, it is useful to see some detailed schemes of morphological segmentation. Our morphological segmentation has run Morfessor tool and HFST to each entry of the corpus. Accordingly, we take surface forms of the words and generate their all possible lexical forms. The schemes presented below are different combinations of outputs determining the removal of affixes from the analyzed words. We mainly focused on detection of a few techniques for the segmentation of word forms. In order to find an effective rule set we tested several segmentation schemes named S[1..5], some of which have described in the following table.

Table 1. The segmentation schemes

Id	Schema	Examples	Translation
S1	stem	el	state
S2	stem+case	el + ge	state + dative
S3	stem+num+case	el + der + den	state + num + ablativ
S4	stem+poss+	el + in	state + poss2sing
S5	stem+poss+case	el + i +ne	state + poss3sing + dative

While GIZA++ tool produces a competitive alignment between words, the Kazakh sentences must be segmented as we already have in the first step. Therefore our method looks like an word sequence labeling problem, the contexts can be presented as POS tags for the word pairs. Because Kazakh derivational suffixes cannot occur freely, only in conjunction with a word stem, so each input word was reduced to its lemma and POS tagged word parts.

Since we have applied morphological segmentation, we will use word translation probabilities as random variables. This enables IBM Model 1 integration with an associated prior distribution. Dirichlet prior Θ is placed on word translation probabilities are basically a parameters t_{ij} of a linear relation function. In the mathematical notation, integration is represented as a rectangle, a stylized covariant matrix. A relation is represented as a symmetric row, like a pipe. A policy controlling a process is represented as the sub-policies are represented as labels connected by curved arrows. Every relation comes from one element and goes into another. An advantage of a matrix is that it simplifies the creation of new models by allowing one to build up a more specialized submatrixes by replacing elements in a smaller one.

Many of the constraints we have added through our previous experiments with this method are the complex side of its benefits. For instance, the fact that relations accumulate a context that in the some stages of using our process has an access to a hierarchy of context free grammar.

In some modeling cases, the real benefit of modeling comes from the numerical result of a simulation after deeper analyses of the finite state loops, which generate patterns of rules. Usually, analysis proceeds as slowly as creating the model. If there were no technology to allow finite state analysis, that an advantage of the modeling would be lost. Although we have not enough combined this feature with the other components of the system, our further approaches will provide a scalable flexibility in model analysis as well. The study of the effect of the model is pretty difficult because a morpheme ambiguity influences to the overall result so much.

3   Evaluation

We evaluate the SMT with the phrase-based Moses[14] system on the Kazakh-English parallel corpus of approximately 60K sentences, which have a maximum of 100 morphemes. Our corpora consists of the legal documents from http:// adilet.zan.kz, a content of http://akorda.kz, and Multilingual Bible texts, and the target-side language models were trained on the MultiUN[15] corpora. We conduct all experiments on a single PC, which runs the 64-bit version of Ubuntu 14.10 server edition on a 4Core Intel i7 processor with 32 GB of RAM in total. All experiment files were processed on a locally mounted hard disk.

The model is implemented like a middle tier component, that processes the input alignment files in a single pass. Current implementation reuses the code from https://github.com/akartbayev/clir that conducts the extraction of phrase pairs and filters out low frequency items. After the processing all valid phrases will be stored in the phrase table and be passed further.

Therefore, we expect the accuracy of the alignment will be measured using precision, recall, and F-measure, we present equations given in the below; here, A represents the reference alignment; T, the output alignment; A and T intersection, the correct alignments.

$$pr = \frac{|A \cap T|}{|T|}, re = \frac{|A \cap T|}{|A|}, F - measure = \frac{2 \times pr \times re}{pr + re} \tag{5}$$

The system parameters were optimized with the minimum error rate training (MERT) algorithm [16], and we trained 5-gram language models with the IRSTLM toolkit[17] and then were converted to binary form using KenLM for a faster execution[18].

Table 2 shows metric scores, which were computed using the MultEval[19]: BLEU, TER[20] and METEOR[21]; the survey shows that translation quality measured by BLEU metrics is not strictly related with lower AER. The final values show that the model can work consistently to give a greater improvement, despite the independent assumptions.

Table 2. Metric scores for all systems

System	Precision	Recall	F-score	AER	BLEU	METEOR	TER
Baseline	57.18	28.35	38.32	36.22	30.47	47.01	49.88
Morfessor	71.12	28.31	42.49	20.19	31.90	47.34	49.37
Rule-based	89.62	29.64	45.58	09.17	33.89	49.22	48.04

4   Conclusions

In this work, we address a morpheme alignment problems concerned the Kazakh language. We compared our approach against a baseline of the Moses translation pipeline and have found it is able to obtain translation quality better than the baseline method by substantial level.

The system results can be transferred to other fields of application, where exists an alignment problem in natural language processing and the incorporation of word segments is useful. Subjects of future research include improvements in the phrase selection method and a context disambiguation. A special experiment with different learning methods may change the interpretation of the results. The improved model works at the same speed as the previous one, and gives an increase of about 3 BLEU in translation quality. This is a modest improvement, but we feel the potential of simulation modeling for this application, and we plan to conduct more sophisticated approaches in the future.

References

1. Bekbulatov, E., Kartbayev, A.: A study of certain morphological structures of Kazakh and their impact on the machine translation quality. In: IEEE 8th International Conference on Application of Information and Communication Technologies, Astana, pp. 1–5 (2014)
2. Bisazza, A., Federico, M.: Morphological pre-processing for Turkish to English statistical machine translation. In: International Workshop on Spoken Language Translation 2009, Tokyo, pp. 129–135 (2009)
3. Brown, P.F., DellaPietra, V.J., DellaPietra, S.A., Mercer, R.L.: The mathematics of statistical machine translation: Parameter estimation. Computational Linguistics **19**, 263–311 (1993). MIT Press Cambridge, MA
4. Moore, R.: Improving IBM word alignment model 1. In: 42nd Annual Meeting on Association for Computational Linguistics, Barcelona, pp. 518–525 (2004)
5. Vogel, S., Ney, H., Tillmann, C.: HMM-based word alignment in statistical translation. In: 16th International Conference on Computational Linguistics, Copenhagen, pp. 836–841 (1996)
6. Dempster, A., Laird, N., Rubin, D.: Maximum likelihood from incomplete data via the EM algorithm. Journal of the Royal Statistical Society. Series B **39**, 1–38 (1977). Wiley-Blackwell, UK
7. Creutz, M., Lagus, K.: Unsupervised models for morpheme segmentation and morphology learning. ACM Transactions on Speech and Language Processing **4**, article 3. Association for Computing Machinery, New York (2007)

8. Beesley, K.R., Karttunen, L.: Finite State Morphology. CSLI Publications, Palo Alto (2003)
9. Altenbek, G., Xiao-long, W.: Kazakh segmentation system of inflectional affixes. In: CIPS-SIGHAN Joint Conference on Chinese Language Processing, Beijing, pp. 183–190 (2010)
10. Kairakbay, B.: A nominal paradigm of the kazakh language. In: 11th International Conference on Finite State Methods and Natural Language Processing, St.Andrews, pp. 108–112 (2013)
11. Lindén, K., Axelson, E., Hardwick, S., Pirinen, T.A., Silfverberg, M.: HFST–framework for compiling and applying morphologies. In: Mahlow, Cerstin, Piotrowski, Michael (eds.) SFCM 2011. CCIS, vol. 100, pp. 67–85. Springer, Heidelberg (2011)
12. Och, F.J., Ney, H.: A Systematic Comparison of Various Statistical Alignment Models. Computational Linguistics **29**, 19–51 (2003). MIT Press, Cambridge, MA
13. Papineni, K., Roukos, S., Ward, T., Zhu, W.: BLEU: a method for automatic evaluation of machine translation. In: 40th Annual Meeting of the Association for Computational Linguistics, Philadephia, pp. 311–318 (2002)
14. Koehn, P., Hoang, H., Birch, A., Callison-Burch, C., Federico, M., Bertoldi, N., Cowan, B., Shen, W., Moran, C., Zens, R., Dyer, C., Bojar, O., Constantin, A., Herbst, E.: Moses: open source toolkit for statistical machine translation. In: 45th Annual Meeting of the Association for Computational Linguistics, Prague, pp. 177–180 (2007)
15. Tapias, D., Rosner, M., Piperidis, S., Odjik, J., Mariani, J., Maegaard, B., Choukri, K.h., Calzolari, N.: MultiUN: a multilingual corpus from united nation documents. In: Seventh conference on International Language Resources and Evaluation, La Valletta, pp. 868–872 (2010)
16. Och, F.J.: Minimum error rate training in statistical machine translation. In: 41st Annual Meeting of the Association for Computational Linguistics, Sapporo, pp. 160–167 (2003)
17. Federico, M., Bertoldi, N., Cettolo, M.: IRSTLM: an open source toolkit for handling large scale language models. In: Interspeech 2008, Brisbane, pp. 1618–1621 (2008)
18. Heafield, K.: Kenlm: faster and smaller language model queries. In: Sixth Workshop on Statistical Machine Translation, Edinburgh, pp. 187–197 (2011)
19. Clark, J.H., Dyer, C., Lavie, A., Smith, N.A.: Better hypothesis testing for statistical machine translation: controlling for optimizer instability. In: 49th Annual Meeting of the Association for Computational Linguistics, Portland, pp. 176–181 (2011)
20. Snover, M., Dorr, B., Schwartz, R., Micciulla, L., Makhoul, J.: A study of translation edit rate with targeted human annotation. In: Association for Machine Translation in the Americas, Cambridge, pp. 223–231 (2006)
21. Denkowski, M., Lavie, A.: Meteor 1.3: automatic metric for reliable optimization and evaluation of machine translation systems. In: Workshop on Statistical Machine Translation EMNLP 2011, Edinburgh, pp. 85–91 (2011)

A Local Method for Canonical Correlation Analysis

Tengju Ye[✉], Zhipeng Xie, and Ang Li

School of Computer Science, Fudan University, Shanghai, China
{yet13,xiezp,angli12}@fudan.edu.cn

Abstract. Canonical Correlation Analysis (CCA) is a standard statistical technique for finding linear projections of two arbitrary vectors that are maximally correlated. In complex situations, the linearity of CCA is not applicable. In this paper, we propose a novel local method for CCA to handle the non-linear situations. We aim to find a series of local linear projections instead of a single globe one. We evaluate the performance of our method and CCA on two real-world datasets. Our experiments show that local method outperforms original CCA in several realistic cross-modal multimedia retrieval tasks.

Keywords: Local linearity · Multivariate analysis · Cross-modal multimedia retrieval

1 Introduction

Canonical correlation analysis (CCA) is a statistical method of correlating linear relationships between two parts of multidimensional variables [11,17]. CCA can be regarded as the problem of finding two basis directions onto which the correlation between the projections of two variables is maximized. CCA is broadly used in unsupervised analysis since it does not require labeled data. The applications are therefore cross various areas, including natural language processing [9,18], neuronal data analysis [6], computer vision [20] and cross modal multimedia retrieval [8,15]. However, because of its linearity, when strong nonlinear relation occurs, CCA is often not applicable. Several methods were thus proposed to find nonlinear projections. However, most of the state-of-the-art non-linear improvements still aim to maximize the correlation of two variables in a single uniform projection.

In this paper, we propose a local linear model for CCA. Unlike the methods which aim to find uniform projections, we consider to construct several local projections, each of them maximizes correlation in a particular region of the dataset. The local projection relaxes the global linearity objective of CCA. In order to construct local CCA projection, we make use of the techniques from non-parametric kernel smoothing. The final correlation between two variables is smoothed combination of local correlation. Our results show that the method based on local linear projection outperforms the standard CCA in various real world information retrieval tasks.

2 Related Work

Rasiwasia [15] used canonical correlation analysis to solve the cross-modal multimedia retrieval problem. Instead of classical text-based information retrieving, Rasiwasia

© Springer International Publishing Switzerland 2015
J. Li et al. (Eds.): NLPCC 2015, LNAI 9362, pp. 428–435, 2015.
DOI: 10.1007/978-3-319-25207-0_39

made use of the rich multiple modalities information. Take text-image cross retrieving for example, they used Latent Dirichlet Allocation (LDA) method to process text corpus into a group of vectors and use SIFT method to extract images' features into another group of vectors. By applying canonical correlation analysis on these two kinds of vectors, it is able to retrieve related images from a text query or retrieve related texts from a given image.

Improvement of canonical correlation analysis for handling non-linear projections has been researched intensively. Kernel Canonical Correlation Analysis [4] maximizes correlation in higher dimension space with help of kernel trick. Neuronal networks are also introduced for solving the linearity drawback of CCA [2,19].

Another part of related work is the local models. Local principal component analysis by Kambhatla[13] proposed local models for PCA. Kambhatla's work partition the train data into disjoint regions and within each of which they construct linear models by PCA. Local Linear Embedding(LLE)[16] provides local linear model for dimensionality reduction. Local methods is also common in non-parametric analysis. Lee, Joonseok, et al.[14] proposed a local low rank model for matrix completion.

2.1   Background

Canonical Correlation Analysis(CCA)
Consider a pair of training vectors (we call they are in different views), $(X, Y) \in \mathbb{R}^{m \times r_1} \times \mathbb{R}^{m \times r_2}$ with corresponding covariance pair $(\Sigma_{11}, \Sigma_{22})$. Let Σ_{12} denote the cross-covariance of(X, Y). CCA aims to find a pair of directions (a, b) (called canonical components) on which the vectors' projection is maximally correlated, i.e.

$$(a, b) = \arg\max_{a,b} \frac{\text{cov}(a^T X, b^T Y)}{\sigma_{a^T X} \sigma_{b^T Y}} = \arg\max_{a,b} \frac{a^T \Sigma_{12} b}{\sqrt{a^T \Sigma_{11} a} \sqrt{b^T \Sigma_{22} b}} \tag{1}$$

Because the choice of re-scaling is arbitrary, the optimization of Eq. (1) is equivalent to maximizing the numerator subject to

$$a^T \Sigma_{11} a = 1, b^T \Sigma_{22} b = 1 \tag{2}$$

The optimization thus is transformed into

$$(a, b) = \text{argmax}_{a^T \Sigma_{11} a=1, b^T \Sigma_{22} b=1} \, a^T \Sigma_{12} b \tag{3}$$

For simplicity, we denote $u = a^T X, v = b^T Y$. Then by constructing the corresponding Lagrangian we get,

$$\mathcal{L} = \sum_{i=1}^{N} a^T (x_i - \bar{u})(y_i - \bar{v}) b - \frac{\lambda}{2}(a^T \Sigma_{11} a) - \frac{\theta}{2}(b^T \Sigma_{22} b) \tag{4}$$

Taking derivatives to a, b, we obtain the following equations,

$$\frac{\partial \mathcal{L}}{\partial a} = \sum_{i=1}^{N}(x_i - \bar{u})^T (y_i - \bar{v}) b - \lambda \Sigma_{11} a = 0 \tag{5}$$

$$\frac{\partial \mathcal{L}}{\partial b} = \sum_{i=1}^{N}(y_i - \bar{v})^T(x_i - \bar{u})a - \lambda \Sigma_{22}b = 0 \tag{6}$$

By using a^T times the Eq.(5) and using b^T times Eq.(6) and accompanied the constraint Eq.(2) we obtain,

$$\lambda = \theta = a^T \Sigma_{12}^w b \tag{7}$$

Obviously, the maximal λ is in fact the maximal correlation. By simplifying Eq.(7) and assuming Σ_{11} and Σ_{22} is invertible we obtain,

$$\Sigma_{11}^{-1}\Sigma_{12}b = \lambda a \tag{8}$$

$$\Sigma_{21}^{-1}\Sigma_{22}b = \lambda b \tag{9}$$

We then transform the Eq. (8) and Eq.(9) into matrix format,

$$\begin{pmatrix} \Sigma_{11}^{-1} & 0 \\ 0 & \Sigma_{22}^{-1} \end{pmatrix}\begin{pmatrix} 0 & \Sigma_{12} \\ \Sigma_{21} & 0 \end{pmatrix}\begin{pmatrix} a \\ b \end{pmatrix} = \lambda \begin{pmatrix} a \\ b \end{pmatrix} \tag{10}$$

Let B denotes $\begin{pmatrix} \Sigma_{11} & 0 \\ 0 & \Sigma_{22} \end{pmatrix}$, A denotes $\begin{pmatrix} 0 & \Sigma_{12} \\ \Sigma_{21} & 0 \end{pmatrix}$ and r denotes $\begin{pmatrix} a \\ b \end{pmatrix}$, finally, we transform Eq. (9) into the following form,

$$B^{-1}Ar = \lambda r \tag{11}$$

The problem left in Eq. (11) is a generalized eigenvalue problem.

Local Approach and Kernel Methods

In non-parametric statistics, kernel methods can be used to specify the local neighborhood by assigning weights to the points around a given point [10]. Let $K_h(p_0, p)$ denote a kernel function, h is the bandwidth parameter h > 0, p_0 is a query point. A large value of h implies that $K_h(p_0,)$ spread widely while a small value means it spread narrowly. The kernel function will calculate weight for each point within the neighborhood. Points within the local neighborhood will be assigned weights by the kernel function. The following is three popular kernels [14], uniform kernel, triangular kernel and Epanechnikov kernel.

$$K_h(p_0, p) := \mathbf{1}[d(p_0, p) < h]$$

$$K_h(p_0, p) := (1 - h^{-1}d(p_0, p))\mathbf{1}[d(p_0, p) < h]$$

$$K_h(p_0, p) := (1 - d(p_0, p)^2)\mathbf{1}[d(p_0, p) < h]$$

After the neighborhood is determined, result is calculated by Nadaraya–Watson [10] kernel weighted average.

$$\hat{f}(p_0) = \sum_i^N \frac{K_h(p_0, p_i)p_i}{\sum_j^N K_h(p_0, p_j)} \tag{12}$$

3   Local Linear Model

Most of the real-world data do not contain strong linearity globally. In order to handle non-linear situation. Our local model focuses on maximizing the correlation in a particular local region. Therefore, we aim to find a strategy to construct such a series of local projections and calculate distance between query vectors to the candidate vectors based on these local projections.

- **Weighted CCA:** In order to make use of the weights in the local region, we change the standard CCA to a new weighted form. Similarly ,we start with defining the pair of two projected vectors by $u = a^T X, v = b^T Y$ with the corresponding weighted averages:

$$\bar{u}^w = \frac{\sum_i w_i X_i}{\sum_i w_i}, \bar{v}^w = \frac{\sum_i w_i Y_i}{\sum_i w_i} \tag{13}$$

  For simplicity, we normalized the weights, let $\sum_i w_i = 1$. The weighted variance and the weighted covariance are,

$$var^w(u) = \sum_i^N w_i(a^T X - a^T \bar{u}^w)^2 = a^T \sum_i^N w_i(X_i - \bar{u}^w)^2 a = a^T \Sigma_{11}^w a \tag{14}$$

$$var^w(v) = b^T \Sigma_{22}^w b \tag{15}$$

$$cov^w(u,v) = \frac{1}{\sum w} a^T \sum_{i=1}^N w_i(X_i - \bar{u}^w)(Y_i - \bar{v}^w)b = a^T \Sigma_{12}^w b \tag{16}$$

  We now aim to maximize the weighted correlation in each region,

$$(a,b) = \text{argmax}_{a,b} \frac{a^T \Sigma_{12}^w b}{\sqrt{a^T \Sigma_{11}^w a}\sqrt{b^T \Sigma_{22}^w b}} \tag{17}$$

  The later derivation is similar as the standard CCA, we maximize the numerator, by constraining the weighted variances into unit ones.

$$var^w(u) = 1, var^w(v) = 1 \tag{18}$$

  The remaining steps are as same as standard CCA that we mentioned in Section 2.1 and we find then left a generalized eigenvalue problem again.

- **Local Method by Anchor Pairs:** In order to construct local projections, we firstly select q anchor pairs. Each anchor pair has its own region which is specified by the kernel function. Pairs in each region are also assigned weights by kernel function according to the distance to the anchors. Moreover, for each region, we apply the aforementioned weighted CCA to obtain a local projection. Fig.1.illustrates this idea.

- **Combining Strategy:** By local method with anchor pairs, suppose we have already selected q anchor pairs which are denoted by $(e_1, f_1) \dots (e_q, f_q), e \in \mathbb{R}^{m \times r_1}, f \in \mathbb{R}^{m \times r_2}$, we now obtain q local canonical components$< a_1, b_1 > \dots < a_q, b_q >$. Evaluation state of CCA is often the problem of retrieving top k nearest-data in different view. We denote $Q_1 \in \mathbb{R}^{r_1}$ as a query vector from view one.

The problem of calculating correlation from Q_1 to the vectors in $\mathbb{R}^{r_2}$ from q local projections is known as non-parametric regression. We propose using the aforementioned Nadaraya-Waston regression. Let $D(Q_1)$ denotes the final distance between the query vectors Q_1 to all other candidate vectors in $\mathbb{R}^{r_2}$. Let $\widehat{D}^i(Q_1)$ denotes the distance from Q_1 to all other candidate vectors which are calculated in the i_{th} local projection. Therefore, with Nadaraya-Waston regression, we obtain the final distance by Eq. (19).

$$D(Q_1) = \sum_i^q \frac{K_h(e_i, Q_1)}{\sum_j^q K_h(e_j, Q_1)} \widehat{D}^i(Q_1) \tag{19}$$

- **Local Canonical Correlation Analysis(LCCA):** We denote the i_{th} pair of data in the training set using (X_i, Y_i) in which X_i contains r_1 dimensions and Y_i contains r_2 dimensions. Let h_1 represent the kernel width for view X and h_2 represent the kernel width for view Y. An anchor pair is denoted by (e, f). K_{h_1} is a vector that stores the values from the kernel function with h_1 kernel width and K_{h_2} has the similarly meaning. We use a vector $w \in \mathbb{R}$ to store weights for training pairs. Distance function is denoted as d (). $< a_t, b_t >$ is the canonical components maximize the correlation in the local region around the t_{th} anchor pair. Algorithm 2-1 describes the training state of LCCA. Obviously, during the training state, the q local region are independent. Therefore, the q iteration can be computed parallel which great increases the algorithm speed. In the predicating state , we calculate the distances between the query vector with candidate vectors in each local spaces thus we get q distance vectors. With Eq. (19). , we combine the local distance vectors into the final distance vectors. Then we select the top k nearest candidate as the results for the query.

Algorithm 2-1 LCCA training state

Input: $X \in \mathbb{R}^{n \times r_1}, Y \in \mathbb{R}^{n \times r_2}, h_1, h_2, q$

for all $t = 1 \dots q$ in parallel do

$\quad (e_t, f_t) :=$ randomly selected traiing pair

$\quad$ for $i = 1 \rightarrow n$ do

$\qquad K_{h_1}^{e_t}[i] := (1 - d(e_t, X_i)^2)\mathbf{1}_{d(e_t, X_i) < h_1}$

$\qquad K_{h_2}^{f_t}[i] := (1 - d(f_t, Y_i)^2)\mathbf{1}_{d(f_t, Y_i) < h_2}$

$\qquad w_i = K_{h_1}^{e_t}[i]\, K_{h_2}^{f_t}[i]$

$\quad$ end for

$\quad (a_t, b_t) = \underset{a_t, b_t}{\text{argmax}} \dfrac{a_t^T \Sigma_{12}^w b_t}{\sqrt{a_t^T \Sigma_{11}^w a_t}\sqrt{b_t^T \Sigma_{22}^w b_t}}$

end for

Output: $< a_t, b_t >, t = 1 \dots q$

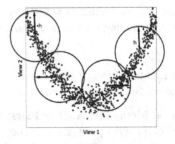

Fig. 1. Local Method by Anchor Pairs

4   Experiment

In this section, we perform experiments on two real world datasets to illustrate the performance of our algorithm. The task we set is cross modal information retrieval task. Cross-modal information retrieval can match queries from one modality to database entries from another modality. Each dataset contains training set and testing set, either set consists of paired vectors which are categorized into several classes. Based on LCCA and CCA, we measure the distance between a query of one modality to candidate result of the other modality. The performances are measured with mean average precision (MAP) which is widely used in information retrieval task. Since the Norm Correlation distance metric achieve the best performance in the experiments of Rasiwasiaet al. [15], we use this metric as our distance function in our later experiments.The kernel we used is the Epanechnikov kernel which is mentioned above. In each of the following dataset, we also compare the influence of different region sizes for LCCA.

4.1   Datasets

- "Wikipedia" is a dataset assembled from Wikipedia's "featured articles" by Rasiwasia et al. [15]. It contains 2866 documents which are random split into a training set with 2173 documents and a test set with 600 documents. The documents are categorized into 10 categories. Each text is represented as a topic histogram over 10 topics by LDA topic model, while each image is represented by a SIFT codebook of 128 codewords.
- "Chinese Web Portal" is collected by ourselves from several popular web portal sites in China. It contains 7033 web documents of paired texts and images, which belong to 11 categories. These documents are randomly divided into two parts: 70% for training set, and 30% for testing set. We use a popular Chinese segment tool IKAnalyzer[1] to separate the documents into "bags of words" and use LDA to extract 30 latent features from each text. Images are also represented as SIFT histograms by a SIFT codebook of 128 codewords.

4.2   Results and Analysis

Fig. 2. graphs the MAP performances achieved by LCCA and CCA. We set LCCA with two different local region size for each experiment. In "Wikipedia" dataset, we set the local region size to 2000 and 2100 for both image query evaluation and text query evaluation. The main tendency of LCCA grows with the increasing number of anchor points. Either in image queries or in text queries, LCCA always outperforms the CCA when there are more than ten anchors. Similar results appear in the "Chinese Web Portal" datasets, the main tendency of MAP scores increases with the increasing number of anchors.

---

[1] From http://code. google. com/p/ik-analyzer

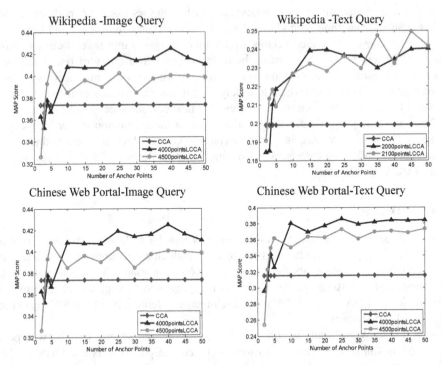

Fig. 2. MAP scores against the number of anchor points in different datasets.

5   Conclusions

We presented a novel local approach for canonical correlation analysis. Our proposed algorithm is called Local Canonical Correlation Analysis (LCCA) which can be easily implemented for parallel computing. The performance is evaluated in two different real world datasets. Our experiments indicate that LCCA outperforms the standard CCA in the cross modal information retrieval task. We also analyze LCCA's performance in terms of its locality (required training points in each local region) and number of required anchor points. Since the basic idea of LCCA is construct local regions and apply weighted CCA in the local region, our future work is plan to investigate the performance by applying other existing non-linear CCA(e.g. kernel CCA) in the local region.

Acknowledgements. This work is supported by National High-tech R&D Program of China (863 Program) (No. SS2015AA011809), Science and Technology Commission of Shanghai Municipality (No. 14511106802), and National Natural Science Foundation of China (No. 61170007). We are grateful to the anonymous reviewers for their valuable comments.

References

1. Akaho, S.: A kernel method for canonical correlation analysis (2006). arXiv preprint cs/0609071
2. Andrew, G., Arora, R., Bilmes, J., Livescu, K.: Deep canonical correlation analysis. In: Proceedings of the 30th International Conference on Machine Learning, pp. 1247–1255 (2013)
3. Asoh, H., Takechi, O.: An approximation of nonlinear canonical correlation analysis by multilayer perceptrons. In: ICANN 1994, pp. 713–716. Springer (1994)
4. Bach, F.R., Jordan, M.I.: Kernel independent component analysis. The Journal of Machine Learning Research 3, 1–48 (2003)
5. Barnard, K., Duygulu, P., Forsyth, D., De Freitas, N., Blei, D.M., Jordan, M.I.: Matching words and pictures. The Journal of Machine Learning Research 3, 1107–1135 (2003)
6. Bießmann, F., Meinecke, F.C., Gretton, A., Rauch, A., Rainer, G., Logothetis, N.K., Müller, K.R.: Temporal kernel CCA and its application in multimodal neuronal data analysis. Machine Learning 79(1–2), 5–27 (2010)
7. Bishop, C.M.: Pattern recognition and machine learning. Springer (2006)
8. Costa Pereira, J., Coviello, E., Doyle, G., Rasiwasia, N., Lanckriet, G.R., Levy, R., Vasconcelos, N.: On the role of correlation and abstraction in cross-modal multimedia retrieval. IEEE Transactions on Pattern Analysis and Machine Intelligence 36(3), 521–535 (2014)
9. Dhillon, P., Foster, D.P., Ungar, L.H.: Multi-view learning of word embeddings via CCA. In: Advances in Neural Information Processing Systems, pp. 199–207 (2011)
10. Friedman, J., Hastie, T., Tibshirani, R.: The elements of statistical learning. Springer series in statistics, vol. 1. Springer, Berlin (2001)
11. Hardoon, D.R., Szedmak, S., Shawe-Taylor, J.: Canonical correlation analysis: An overview with application to learning methods. Neural Computation 16(12), 2639–2664 (2004)
12. Hsieh, W.W.: Nonlinear canonical correlation analysis by neural networks. Neural Networks 13(10), 1095–1105 (2000)
13. Kambhatla, N., Leen, T.K.: Dimension reduction by local principal component analysis. Neural Computation 9(7), 1493–1516 (1997)
14. Lee, J., Kim, S., Lebanon, G., Singer, Y.: Local low-rank matrix approximation. In: Proceedings of the 30th International Conference on Machine Learning, pp. 82–90 (2013)
15. Rasiwasia, N., Costa Pereira, J., Coviello, E., Doyle, G., Lanckriet, G.R., Levy, R., Vasconcelos, N.: A new approach to cross-modal multimedia retrieval. In: Proceedings of the International Conference on Multimedia, pp. 251–260. ACM (2010)
16. Roweis, S.T., Saul, L.K.: Nonlinear dimensionality reduction by locally linear embedding. Science 290(5500), 2323–2326 (2000)
17. Thompson, B.: Canonical correlation analysis. Encyclopedia of Statistics in Behavioral Science (2005)
18. Vinokourov, A., Cristianini, N., Shawe-Taylor, J.S.: Inferring a semantic representation of text via cross-language correlation analysis. In: Advances in Neural Information Processing Systems, pp. 1473–1480 (2002)
19. Wand, M.P., Jones, M.C.: Kernel smoothing. CRC Press (1994)
20. Zheng, N., Loizou, G., Jiang, X., Lan, X., Li, X.: Computer vision and pattern recognition (2007)

Learning to Rank Microblog Posts
for Real-Time Ad-Hoc Search

Jing Li[1,2](✉), Zhongyu Wei[3], Hao Wei[1], Kangfei Zhao[1], Junwen Chen[4],
and Kam-Fai Wong[1,2]

[1] The Chinese University of Hong Kong, Shatin, N.T., Hong Kong
{lijing,hwei,kfzhao,kfwong}@se.cuhk.edu.hk
[2] MoE Key Laboratory of High Confidence Software Technologies, Beijing, China
[3] The University of Texas at Dallas, Richardson, TX, USA
zywei@hlt.utdallas.edu
[4] Tencent, Nanshan District, Shenzhen, China
zywei@hlt.utdallas.edu

Abstract. Microblogging websites have emerged to the center of information production and diffusion, on which people can get useful information from other users' microblog posts. In the era of Big Data, we are overwhelmed by the large amount of microblog posts. To make good use of these informative data, an effective search tool is required specialized for microblog posts. However, it is not trivial to do microblog search due to the following reasons: 1) microblog posts are noisy and time-sensitive rendering general information retrieval models ineffective. 2) Conventional IR models are not designed to consider microblog-specific features. In this paper, we propose to utilize learning to rank model for microblog search. We combine content-based, microblog-specific and temporal features into learning to rank models, which are found to model microblog posts effectively. To study the performance of learning to rank models, we evaluate our models using tweet data set provided by TERC 2011 and TREC 2012 microblogs track with the comparison of three state-of-the-art information retrieval baselines, vector space model, language model, BM25 model. Extensive experimental studies demonstrate the effectiveness of learning to rank models and the usefulness to integrate microblog-specific and temporal information for microblog search task.

Keywords: Microblogging analysis · Online social network · Information retrieval · Microblog search · Experimental study

1 Introduction

With the arrival of Web 2.0, microblogging websites, e.g. Twitter and Sina Weibo, now have become new and valuable source of information. As more

This work is partially supported by General Research Fund of Hong Kong (417112), RGC Direct Grant (417613), and Huawei Noah's Ark Lab, Hong Kong. We would like to thank Junjie Hu, Prof. Michael R. Lyu and anonymous reviewers for the useful comments. This work was done when Zhongyu Wei and Junwen Chen were at The Chinese University of Hong Kong.

© Springer International Publishing Switzerland 2015
J. Li et al. (Eds.): NLPCC 2015, LNAI 9362, pp. 436–443, 2015.
DOI: 10.1007/978-3-319-25207-0_40

and more people join in these networks, the microblog messages they posted at different time cover a wide range of topics. Microblog posts are valuable due to the following reasons. 1) Information on microblogging websites rapidly updates, people can obtain latest information of many time-sensitive events, such as breaking news about missing flight MH370 or the latest information about iPhone 6. 2) There are rich of some small but useful tips in microblog posts, such as how to ace an interview in Google, and how to find delicious food in Hong Kong. These tips spread widely and quickly over social networks, which may be ignored by conventional social medias like webpages and newspapers. 3) Microblog posts are short and mostly in daily conversational style, thus are easy to read. People can get what they want with a quick scan.

Due to these overwhelming data, designing search tools to distinguish interesting and relevant microblog posts is crucial. To make good use of these information, we study the task of real-time and ad-hoc search microblog search. The goal is to retrieve "interesting" and "new" microblogging messages. A message is "interesting" means that it is relevant and covers content what users want. Also, the retrieved posts should be new, because people post millions of messages every day and out-of-date information values nothing. In particular, the users' information needs is represented by a query at a specific time, since they wish to see most recent and relevant information.

An intuitive approach is to directly apply conventional information retrieval models for microblog search. However, the characteristics of microblog posts challenge the current information retrieval models. Firstly, microblog posts are short and noisy. There are full of informal texts and unedited contents in microblog messages. Secondly, microblog posts are time-sensitive. Therefore, different from conventional search engines, microblog search engine should answer a query by providing a list of not only interesting but also newer relevant microblog posts. Thirdly, current information retrieval models are not designed for microblog specific features, such as emoticons, hashtags, urls, etc. These features are very useful in measuring relevance and would help to improve the performance of microblog retrieval model.

In this paper, we extend learning to rank framework to rank and retrieve microblog messages. Learning to rank is a combination of machine learning techniques and traditional ranking model. It makes prediction based on the difference of features. Specifically, we implement four state-of-the-art Learning to Rank Models, i.e. MART [4], RankBoost [3], Coordinate Ascent [10] and Lambda MART [15]. Also, we combine content-based, microblog-specific and temporal features into the learning to rank models.

To study the effectiveness of learning to rank models and proposed features, we evaluate the performance of our model using tweet set provided by TREC2011 and TREC 2012 microblog track. TREC 2011 data is for training and TREC 2012 is for test. In the experiment, we compare our learning to rank models with three baselines, i.e. cosine similarity, language model and BM25, all of which are state of the art conventional IR models. The experimental results confirm the effectiveness of learning to rank models for this task and demonstrates the usefulness of microblog-specific temporal features in microblog retrieval. In particular, though we conduct experiment on Twitter dataset, it is trivial to utilize models and features proposed in this paper on other microblogging services, e.g. Sina Weibo and Tencent Weibo.

2   Related Work

Microblog retrieval has drawn tramendous attentions in recent years. Therefore, TREC introduced a track for ad-hoc microblog retrieval in 2011 [7,12,13]. Large tweet collections and annotations for various queries were released. Different approaches were investigated for microblog retrieval to overcome the special nature of microblog messages, e.g. short, noisy and time-sensitive characters of microblog posts. One of the main challenges in microblog retrieval is term mismatch due to short queries and short relevant messages, which renders conventional IR models ineffective. Researches tackled the term mismatch problem in microblog posts either by text clustering or query expansion. Another line of research in improving retrieval performance in microblog messages focused on using microblog-specific features to improve the performance of microblog retrieval [2,5,9,11].

Duan et al. [2] used learning to rank models for tweet search. They studied some microblog-specific features, namely embedded URLs, mentions, hashtags, retweet behaviors, etc. They constructed a corpus for evaluation containing 20 self-selected queries and relevant tweets were crawled from Twitter Search[1]. Experiment results indicated the effectiveness of several features, namely, mention, URL and length. As the best system in TREC microblog 2011, Metzler et al. [9] also utilized learning to rank model incorporating 8 basic features.

Although researchers have proved the effectiveness of microblog message re-ranking for improving search accuracy, afore-mentioned works have the following problems: 1) the training sets used in some study were too small to produce a stable ranker [2]; 2) Features in some works were not rich enough[5,9,11]; 3) All of these works simply employed some single ranking model which were also query-insensitive.

Different from the previous, our work focused on improving the microblog search result by using the state-of-art learning to rank models instead of the conventional information retrieval approaches. Having large dataset from Twitter, we can extract various features and learn a stable ranker based on our multi-view features, which greatly improve the performance of our system.

3   Learning to Rank Model

Learning to Rank models, combination of machine learning techniques and conventional ranking model, attract increasing interests in the area of information retrieval [6]. In this paper, we extend a learning to rank framework for our

---

[1] https://twitter.com/twittersearch

real-time ad-hoc microblog search task. As a supervised learning method, learning to rank model can be formulated as a machine learning problem. A training instance, is a feature-label pair $< x, y >$, where x denotes a feature vector and y is the label corresponding to the ranking for microblog post t given query q. The essential issue in this step is how to extract feature vector x for $< t, q >$ pair.

Suppose that $F(\mathbf{x})$ is a function mapping a list of feature vectors $\mathbf{x}$ into a list of scores, and the loss function $L(F(\mathbf{x}, \mathbf{y})$ evaluates the difference between prediction result $F(\mathbf{x})$ and golden truth $\mathbf{y}$.The training process is to optimize $L(F(\mathbf{x}, \mathbf{y})$ and find optimal solution $F^*(\mathbf{x})$ with training data $< x_1, y_1 >, < x_2, y_2 >, ..., < x_n, y_n >$. So, after training, for any given query q, learning to rank models can automatically rank a corpus of microblog posts $T = \{t_1, t_2, ..., t_m\}$ according to the predictions of $F(\mathbf{x})$ and our search engine returns the top-k microblog posts.

Previous works propose various learning to rank models with different definition of prediction functions $F(\mathbf{x})$ and loss functions $L(F(\mathbf{x}, \mathbf{y}))$, and proves them effective for many kinds of tasks. In this work, we utilize four state-of-the-art learning to rank models, i.e. MART [4], RankBoost [3], Coordinate Ascent [10] and Lambda MART [15].

Table 1 lists all the features we integrate to learning to rank models.

Table 1. Features Description

Feature Category	Feature Name	Feature Description
	CosSim	CosSim refers to the cosine similarity between t and q
Content-based	LM	Language Model denotes the uniform language model related to t and q
	BM25	BM25 represents the BM25 score of a microblog and a query
	#hashtags	#hashtags denotes the count of hashtags in t
	#match-hashtags	This is the number of words in q that also appear in t's hashtags.
Microblog-Specific	#words	This feature represents the number of words in t excluding stop words,
	Length	Length denotes #characters in t, including punctuations, emoticons, etc.
	#urls	#urls means the count of urls in t.
	#emoticons	#emoticons refers to the number of emoticons in t.
Temporal	gap	gap is the difference between q's query time and t's posting time.

4 Experiments

We tested our learning to rank model on the dataset TREC microblog track, and evaluated the results from two evaluation metrics : precision@30 and mean average precision (MAP). Through comparison with traditional information retrieval models Vector Space Model, BM25 and Language Model, we demonstrated our learning to rank approach remarkably can outperform all these baseline models.

4.1 Data Collection and Set-up

Our dataset was provided by TREC 2011 and 2012 microblog track[2]. It consists of two parts: the microblog posts corpus for search, and the queries and golden sets of year 2011 and 2012 for evaluation. The raw microblog post corpus

---

[2] http://trec.nist.gov/data/microblog.html

was crawled from the famous world-wide microblogging platform, Twitter. The queries and golden sets are available in TREC [3].

Table 2 summarizes the statistics of queries and golden set. There are 50 and 60 queries in 2011 and 2012, respectively. Every query contains one or multiple query keywords followed by a query time for ad-hoc search. For each query, the golden set annotates relevant or non-relevant information about 1,000 tweets in the corpus. Our learning to rank approaches used 2011 queries for learning while both our approaches and the baseline approaches were tested on the 2012 queries.

Table 2. The statistics of queries and golden set

	QS 2011	QS 2012
# of queries	50	60
# of annotated tweets	40,855	73,073
# of highly relevant tweets	558	2,572
# of all relevant tweets	2,864	6,286

The microblog track corpus contains tweets from Jan. 23rd to Feb. 8th, 2011, with 15,598,190 valid tweets. When a query is issued, only the relevant tweet posted before the query time will be returned as the result. According to the study of Wang et al.[14], considering the microblog posts after query time can in fact decrease the performance of microblog search system, because of microblog posts' time sensitive features.

To ensure the quality of microblog post retrieval from noisy and unstructured tweets, we preprocessed the raw tweet corpus as well as the queries via the following steps: 1) non-English tweets filtering; 2) tokenization and lemmatization; 3) case conversion (all letters were converted into lowercase); 4) Hashtag words duplication; 5) meaningless characters removing.

4.2 Experimental Results

We tested learning to rank models, with three state-of-the-art conventional IR models: Vector Space Model, Language Model and BM25, whose detailed information can be found in Manning et al. [8]. And following previous works relevant to BM25 models, we set $k_1 = 1.2$ and $b = 0.75$. The implementation of learning to rank models is based on RankLib [1] toolkit. The evaluation metrics were precision@30 and mean average precision (MAP for short), which are popular benchmarks in IR evaluation. Table 3 and Table 4 show the performance of base-line models and learning to rank models, respectively.

Table 3 indicates that conventional IR models performed badly and were not suitable for tweet retrieval. Among all the baselines, BM25 performed the best, because it can theoretically integrate the advantages of Vector Space Model and Language Model.

---

[3] https://github.com/lintool/twitter-tools/wiki/Tweets2011-Collection

By comparing Table 4 to Table 3, we observed that learning to rank models remarkably outperformed all the baseline models. Specifically, the four learning to rank models achieved at least 32% improvement on precision@30 and 24.75% improvement on mean average precision. This experiment proves the effectiveness of learning to rank models on microblog search task.

Table 3. Performance of Conventional IR Models

	Vector Space Model	Language Model	BM25 Model
Precision@30	10.11%	3.84%	**21.92%**
MAP	6.64%	3.64%	**13.74%**

Table 4. Performance of Learning to Rank Models

	MART Model	Rank Boost	Coordinate Ascent	Lambda Mart
Precision@30	40.00%	33.00%	**40.56%**	35.65%
MAP	**32.03%**	25.75%	31.52%	28.33%

4.3 Feature Study

To illustrate the impact of content-based features, microblog specific features and temporal features, we conducted an analysis about different groups of features combined with learning to rank models. Figure 1 and Figure 2 illustrate the performance of learning to rank models combined with different combination of features, i.e. only content-based feature, only microblog specific feature, only temporal feature, content+microblog feature, content+temporal feature, microblog+temporal feature, and the combination of all three kinds of features, evaluated by precision@30 and MAP, respectively.

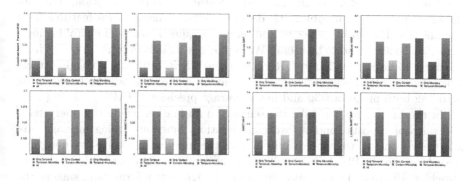

Fig. 1. Performance on Precision@30   **Fig. 2.** Performance on MAP

From the experimental result, we can learn the following results:

- Different features have different influence on learning to rank models. Models are more sensitive to content-based features than that to temporal and microblog-specific features. Coordinate Ascent model and MART model had good performance with content-based feature. With only content based features and microblog specific features, the models can perform nearly as good as the models with all features. While with only temporal and microblog feature, models perform badly. However, the model combining all features had the best performance. So all features we proposed were effective and useful to learning to rank models.
- Different models have different sensitivity to different features. Among all models, Coordinate Ascent models and MART model are more sensitive to temporal and content-based features, while MART model and Lambda MART model are more sensitive to microblog-specific features.

In the practice, different queries may have different requirements on the features. For example, users may want to search microblog posts for some news about a topic. So the system should focus more on the temporal features. Coordinate ascent model should be a good choice. In some other case, users may want to get information from the online discussions. So the system is supposed to return microblog posts having relevant hashtags where Lambda MART model may be a good choice.

5   Conclusion and Future Work

In this paper, we address the importance of microblog search and point out the challenges of handling microblogging data. We find that conventional information retrieval models are not applicable for microblog search because: 1) Microblogging messages are short and noisy; 2) Conventional IR models cannot take full advantage of the microblog features into consideration. Through detailed analysis of microblogging data, we extract many useful features from content-based, microblog-specific and temporal aspects to model microblog messages, which are found to be useful. We deploy our models to retrieve tweets from the tweet dataset provided by TREC 2011 and TREC 2012 microblogging track. The experimental results show that our models outperform all the baselines by doubling both precision@30 and mean average precision (MAP).

In the future, we will take the semantic information inside the query term and microblog posts content into consideration. Since microblog messages are short and concise, we should fully exploit the inherent meanings by designing microblog-specific PLSA (Probability Latent Semantic Analysis) or LDA (Latent Dirichlet Allocation) techniques. In addition, structure information, such as the social relationship network, can be used to improve the retrieval performance.

References

1. Dang, V.: Ranklib (2013)
2. Duan, Y., Jiang, L., Qin, T., Zhou, M., Shum, H.Y.: An empirical study on learning to rank of tweets. In: Proceedings of the 23rd International Conference on Computational Linguistics, pp. 295–303. Association for Computational Linguistics (2010)
3. Freund, Y., Iyer, R., Schapire, R.E., Singer, Y.: An efficient boosting algorithm for combining preferences. The Journal of Machine Learning Research 4, 933–969 (2003)
4. Friedman, J.H.: Greedy function approximation: a gradient boosting machine. Annals of Statistics, 1189–1232 (2001)
5. Han, Z., Li, X., Yang, M., Qi, H., Li, S., Zhao, T.: Hit at trec 2012 microblog track. In: Proceedings of the 21st Text REtrieval Conference (TREC) (2012)
6. Hang, L.: A short introduction to learning to rank. IEICE Transactions on Information and Systems 94(10), 1854–1862 (2011)
7. Lin, L., Efron, M.: Overview of the trec-2013 microblog track. In: Proceedings of the 23rd Text REtrieval Conference (TREC) (2013)
8. Manning, C.D., Raghavan, P., Schütze, H.: Introduction to Information Retrieval, vol. 1. Cambridge university press, Cambridge (2008)
9. Metzler, D., Cai, C.: USC/ISI at trec 2011: microblog track. In: TREC (2011)
10. Metzler, D., Croft, W.B.: Linear feature-based models for information retrieval. Information Retrieval 10(3), 257–274 (2007)
11. Obukhovskaya, Z., Pervyshev, K., Styskin, A., Serdyukov, P.: Yandex at trec 2011 microblog track. In: Proceedings of the 20th Text REtrieval Conference (TREC) (2011)
12. Ounis, I., Macdonald, C., Lin, J., Soboroff, I.: Overview of the trec-2011 microblog track. In: Proceedings of the 20th Text REtrieval Conference (TREC) (2011)
13. Soboroff, I., Ounis, I., Lin, J., Soboroff, I.: Overview of the trec-2012 microblog track. In: Proceedings of the 21st Text REtrieval Conference (TREC) (2012)
14. Wang, Y., Lin, J.: The impact of future term statistics in real-time tweet search. In: de Rijke, M., Kenter, T., de Vries, A.P., Zhai, C.X., de Jong, F., Radinsky, K., Hofmann, K. (eds.) ECIR 2014. LNCS, vol. 8416, pp. 567–572. Springer, Heidelberg (2014)
15. Wu, Q., Burges, C.J., Svore, K.M., Gao, J.: Adapting boosting for information retrieval measures. Information Retrieval 13(3), 254–270 (2010)

Fuzzy-Rough Set Based Multi-labeled Emotion Intensity Analysis for Sentence, Paragraph and Document

Chu Wang[1], Shi Feng[1,2(✉)], Daling Wang[1,2], and Yifei Zhang[1,2]

[1] School of Information Science and Engineering, Northeastern University, Boston, USA
wangchu@research.neu.edu.cn,
{fengshi,wangdaling,zhangyifei}@ise.neu.edu.cn
[2] Key Laboratory of Medical Image Computing, Northeastern University,
Ministry of Education, Shenyang 110819, People's Republic of China

Abstract. Most existing sentiment analysis methods focus on single-label classification, which means only a exclusive sentiment orientation (negative, positive or neutral) or an emotion state (joy, hate, love, sorrow, anxiety, surprise, anger, or expect) is considered for the given text. However, multiple emotions with different intensity may be coexisting in one document, one paragraph or even in one sentence. In this paper, we propose a fuzzy-rough set based approach to detect the multi-labeled emotions and calculate their corresponding intensities in social media text. Using the proposed fuzzy-rough set method, we can simultaneously model multi emotions and their intensities with sentiment words for a sentence, a paragraph, or a document. Experiments on a well-known blog emotion corpus show that our proposed multi-labeled emotion intensity analysis algorithm outperforms baseline methods by a large margin.

Keywords: Multi-labeled emotion analysis · Emotion intensity · Fuzzy-rough set

1   Introduction

Nowadays, a lot of papers have been published for analyzing sentiments in social media documents, and the previous work concerned with following different facets.

Sentiment Orientation. Most previous researches focused on sentiment orientation classification, i.e. classifying the subjective text into two-orientation (*positive* and *negative*), or three-orientation (*positive*, *neutral*, and *negative*).

Fine-Grained Emotion. Different above two or three categories sentiment, many current researches considered more sentiments such as *joy*, *hate*, *love*, *sorrow*, *anxiety*, *surprise*, *anger*, *expect* [5] called as fine-grained emotion.

Project supported by National Natural Science Foundation of China (61370074, 61402091), the Fundamental Research Funds for the Central Universities of China under Grant N140404012.

J. Li et al. (Eds.): NLPCC 2015, LNAI 9362, pp. 444–452, 2015.
DOI: 10.1007/978-3-319-25207-0_41

Previous work mainly focus on single-labeled sentiment analysis for either sentiment orientation or fine-grained emotion. In fact, multiple emotions may be coexisting in just one sentence, paragraph, or document. For the multi-labeled emotion classification, we propose an improved fuzzy-rough set based approach to detect the multi-labeled emotions and calculate their corresponding intensities in social media text. Experiment results using a well annotated blog emotion dataset show that our proposed algorithm significantly outperforms other baselines by a large margin.

2   Related Work

Multi-labeled emotion analysis can be regarded as multi-label learning problem. Zhang, et al. [8] did a comprehensive review for the multi-label learning algorithms. The typical problem transformation algorithm is called "Binary Relevance", which is proposed by Boutell, et al. [1]. Elisseeff, et al. [3] solved the multi-label data problem by improving the kernel SVM.

Although there are a lot of papers for multi-label learning and sentiment analysis, little work has been done for detecting the multi-labeled emotions, especially the emotion intensities in one social media post.

In this paper, we regard the problem as uncertain emotion classification problem and apply fuzzy-rough theory for solving it. Fuzzy-rough set was first proposed by Dubois and Prade in 1990 which based on Rough set [4] and Fuzzy set theory [7] and combined their advantages together [2]. Now, it has been widely used in many uncertain classification problems. Sun and Ma [6] gave an approach to decision making problem by combining the soft set with fuzzy-rough theory.

Although fuzzy-rough theory has been widely used in many applications, there is no existing literature on uncertain emotion classification for social media.

3   Preliminary

Let SM be a social media text document consisted of m sentences, i.e. $SM=\{s_1, s_2, ..., s_m\}$, where s_i (i=1, 2, ..., m) be ith sentence and contains n_i sentiment words. The sentiment word set $W_i=\{w_1, w_2, ..., w_{n_i}\}$ in s_i expresses a multi-labeled emotion set with different intensity $E_i=\{e_ji_j\}$, where e_j (j=1, 2, ..., 8) represents ith emotion in 8 emotions: *joy, hate, love, sorrow, anxiety, surprise, anger,* and *expect*, respectively, and i_j represents the intensity of e_j in s_i. Our task is to detect every emotion $e_j \in E_i$ and calculate its intensity $i_j \in E_i$ for s_i according to recognized W_i. Moreover, for a paragraph containing multi sentences and a document containing multi paragraphs, our task is to do so for every paragraph and the document.

In this paper we intend to apply fuzzy-rough set method for solving the challenge problem.

Definition 1: U and W are two finite and nonempty universes. Suppose that R is an arbitrary relation from U to W, the triple (U, W, R) is called a generalized fuzzy approximation space. For any set $A \in F(U)$, the upper and lower approximations of A, $\overline{R}(A)$ and $\underline{R}(A)$, are defined respectively as:

$$\overline{R}(A) = \vee_{y \in W}[R(x, y) \wedge A(y)], x \in U \qquad (1)$$

$$\underline{R}(A) = \vee_{y \in W}[1 - R(x, y) \vee A(y)], x \in U \qquad (2)$$

The pair $(\overline{R}(A), \underline{R}(A))$ is referred to as a generalized fuzzy rough set, and R is referred to as upper and lower generalized fuzzy rough approximation operators.

We regard the emotion between [0, 1] as fuzzy degree, and use fuzzy-rough set to deal with emotion intensity analysis problem. Our plan is to calculate approximations of each sentiment value to the strongest A, and propose an improved algorithm of fuzzy-rough set.

4   Improving Fuzzy-Rough Set for Emotion Intensity Prediction

Suppose the following $E=\{e_1, e_2 ..., e_8\}$ is the set of emotions, and $W=\{w_1, w_2, w_3, w_4, w_5\}$ is the set of sentiment words in a sentence. We first mark the intensity as 0 or 1 in Table 1. The analysis methods are as follows.

Table 1. The Examples of Relationship between Multiple Emotions and Different Words

w\i\e	e_1	e_2	e_3	e_4	e_5	e_6	e_7	e_8
w_1	0	1	0	1	0	1	0	0
w_2	1	0	1	0	0	1	0	0
w_3	0	0	1	1	1	0	0	0
w_4	1	0	1	0	1	0	0	0
w_5	1	0	0	0	0	1	0	0

We have this following result: $F(e_1)=\{w_2, w_4, w_5\}$, $F(e_2)=\{w_1\}$, $F(e_3)=\{w_2, w_3, w_4\}$, $F(e_4)=\{w_1, w_3\}$, $F(e_5)=\{w_3, w_4\}$, and $F(e_6)=\{w_1, w_2, w_5\}$. This means that three words in this sentence has the emotion *hate*. Similarly, we can describe the features for other emotions in universe U.

Here we introduce eight basic emotion categories from [5]: *joy, hate, love, sorrow, anxiety, surprise, anger* and *expect*. Each word is associated with eight emotions and ten intensities, and most sentences have more than one emotion word in this situation. We need a new algorithm to estimate the multi-labeled emotional attributes of the whole sentence, paragraph, or document.

Definition 2: Let (F^1, W) be a fuzzy set over E, the triple relation (E, W, F^1) is called as the fuzzy approximation space. For any $A \in F(W)$, the upper and lower approximations of A, $\overline{F}(A)$ and $\underline{F}(A)$ with respect to the fuzzy approximation space (E, W, F^1) are fuzzy sets of U, whose membership functions are defined as:

$$\overline{F}(A)(x) = \vee y \in W[(F^{-1}(x)(y)) \wedge A(y)], x \in E \qquad (3)$$

$$\underline{F}(A)(x) = \wedge y \in W[(1 - F^{-1}(x)(y)) \vee A(y)], x \in E \tag{4}$$

In order to show the importance of key words, we divide the [0,1] range at 0.5, For any $A \in F(W)$, the upper approximation and lower approximation of A, $\overline{F}(A)$ and $\underline{F}(A)$, if $F(e)(w)=0$ then $\overline{F}(A)=0$ and $\underline{F}(A)=0$, if $F(e)(w) \neq 0$ then

$$\overline{F}(A)(e) = \vee_{w \in W}[F(e)(w) \wedge A(w)], e \in E \tag{5}$$

$$\underline{F}(A)(e) = \begin{cases} \wedge_{w \in W}[(1 - F(e)(w)) \vee A(w)], & e \in E, F(e)(w) \in [0.5,1] \\ \wedge_{w \in W}[F(e)(w) \wedge A(w)], & e \in E, F(e)(w) \in [0,0.5) \end{cases} \tag{6}$$

In sentiment analysis problem defined in this paper, due to the fuzzy mathematics, all the emotional intensities of words are marked between 0 and 1. Here A is very important, and we choose the strongest emotion intensity of each sentiment word as the value of A. Because it is the highest value of each emotion, if other words is more approximate to A, the stronger of the emotion intensity of the sentiment words is. So we only need to calculate the approximation of each word. Then we construct the decision object A on the evaluation of the words universe W.

Take one sentence for example: "我在6点之前就睡不着了，因为激动，马上就能领略九寨沟的秀丽风光了". The emotional value mark of this sentence is: joy=0.8, love=0.5 and expect=0.5. We calculate $\overline{F}(A)$ and $\underline{F}(A)$ with Formula (5) and (6), the calculated value of our algorithm is the same order as the result annotated by human which is showed in Table 2.

Table 2. The Examples of Weight Computation Based on Our Improving Fuzzy-Rough Set

	激动	领略	秀丽	风光	睡	$\overline{F}(A)$	$\underline{F}(A)$	value
Joy	0.6	0.6	0	0	0.5	0.6	0.5	1.1
Hate	0	0	0	0	0	0	0	0
Love	0	0.6	0.6	0.4	0	0.6	0.4	1
Sorrow	0	0	0	0	0	0	0	0
Anxiety	0	0	0	0	0	0	0	0
Surprise	0	0	0	0	0	0	0	0
Anger	0	0	0	0	0	0	0	0
Expect	0.5	0	0	0	0.5	0.5	0.5	1
A	0.6	0.6	0.6	0.4	0.5			

In sentiment analysis, we always focus on the extraction of sentiment words, such as adjectives/adverbs, verbs used to qualify nouns. In Table 2, there are eight emotions, and each sentiment word may have more than one basic emotions. The value of word is between 0 and 1, and 0 means the words do not have the emotional attribute. As the value growing approach to 1, the intensity of the emotion of the word is getting bigger. This feature of the value applies to fuzzy set theory, so we can use our improving fuzzy-rough set to do our work.

Our improving method has several advantages. Firstly, it is suitable to calculate the intensity of the text which between [0, 1]. Secondly, it considers the weight of the emotion value not only from the value of the words but also the influence of the whole words. Thirdly, it focuses on a human logic that emotion strength is not simply the sum. For example, three weak sentiment words usually neither form a powerful sentiment value nor get a weaker one.

As we explained above, our method is suitable to solve this problem. In Table 2, we get the result in [0, 2] (e.g. 1.1), but the emotion value of the sentence annotated by people is between [0, 1]. So we still need to build a regression algorithm to make our predictive method completed. In this paper, we choose Linear Regression Algorithm to approach our target. In summary, we give the algorithm for Intensity Prediction based on our improving fuzzy-rough set model as follows.

Algorithm 1. Intensity Prediction Algorithm

Input: fuzzy set (F, W) // W is the sentiment words set, and F is the fuzzy relation
Output: emotion intensity
Description:
1. Read every sentiment word $w \in W$;
2. Compute strongest emotional intensity object A as:
$$A = \sum_{i=1}^{W} \frac{maxF(w_i)}{w_i}, w_i \in W, i.e., A(w_i) = max\{F(e_j)(w_i) | e_j \in E\}$$
3. Calculate the improving fuzzy-rough upper approximation $\overline{F}(A)$ and fuzzy-rough lower approximation $\underline{F}(A)$ // see formula (5) and (6);
4. Calculate the choice value $m = \overline{F}(A)(e_i) + \underline{F}(A)(e_i)$, $e_i \in E$;
   // E is the universe of the emotion
5. Normalize m into [0, 1] with Linear Regression Algorithm;
6. Return m as emotion intensity;

---

5 Experiments

5.1 Dataset and Evaluation Metric

We use Changqin Quan's [5] dataset to evaluate our proposed method. The corpus contains 1,487 documents, with 11,953 paragraphs, 38,051 sentences, and 971,628 Chinese words. And each of them is annotated with 8 kinds of emotion [5]. We use cross-validation to validate our algorithm. The whole dataset is divided into 5 parts on average, and each time there are four training sets and one testing set.

In this experiment, we use 4 evaluation metrics based on [8] to test our method. Let $S = \{(x_i, Y_i) | 1 \leq i \leq p)\}$ be the test set and h be the learned multi-label classifier.

$$\textbf{Subset Accuracy:} \quad subsetacc_s(h) = \frac{1}{p} \sum_{i=1}^{p} [| h(x_i = Y_i) |] \tag{7}$$

The subset accuracy evaluates the fraction of correctly classified examples, i.e. the predicted label set is identical to the ground-truth label set. It tends to be overly strict especially when the size of label space is large.

$$\textbf{Hamming Loss:} \quad hloss_s(h) = \frac{1}{p}\sum_{i=1}^{P}\frac{1}{q}|h(x_i)\Delta Y_i| \tag{8}$$

where Δ stands for the symmetric difference between two sets. The hamming loss evaluates the fraction of misclassified instance-label pairs, i.e. a relevant label is missed or an irrelevant is predicted. In this paper, we need to improve the evaluation metric. In the formula below, we try to measure the difference between the value we predict and the real one:

$$hloss_s(h) = \frac{1}{p}\sum_{i=1}^{P}\frac{1}{q}|h(x_i)-\Delta Y_i| \tag{9}$$

$$\textbf{One-error:} \quad One-error_s(h) = \frac{1}{p}\sum_{i=1}^{P}\frac{1}{q}[[\arg\max_{y\in Y}f(x_i,y)]\notin Y_i] \tag{10}$$

The one-error evaluates the fraction of examples whose top-ranked label is not in the relevant label set. In our paper, it means the strongest emotion of the eight we predict is wrong. As we can see, the value of the One-error is the fewer, the better.

Average Precision

$$avgprec_s(h) = \frac{1}{p}\sum_{i=1}^{P}\frac{1}{|Y_i|}\sum_{y\in Y_i}\frac{|\{y'|rank_f(x_i,y')\leq rank_f(x_i,y),y'\in Y_i\}|}{rank_f(x_i,y)} \tag{11}$$

The average precision evaluates the average fraction of relevant labels ranked higher than a particular label $y\in Y_i$. In this paper, it means whether the descending order of the value of each emotion is right. This metric is the larger, the better.

5.2   Experiment Results

As few methods were proposed for multi-labeled emotion intensity, we compare our method with the classic Fuzzy union one. It is defined as: $(A\cup B)(x)=\max(A(x),B(x))$ for all $x\in X$. Taking the value of *joy* from Table 2 for example: *Joy*(Fuzzy union)=max(激动(*joy*), 领略(*joy*), 睡(*joy*))=max(0.6, 0.6, 0.5)=0.6.

In our experiment, it means each kind of the 8 emotions takes its strongest value of all words. And we have also introduced a Naïve Bayes method to compare with our algorithm. In NB method, we assumed that every emotion is independent from each other. Taking the value of *joy* from Table 2 for example: *Joy*(Naïve Bayes)=P(激动|*joy*)*P(领略|*joy*)*P(睡|*joy*)*P(*joy*)=6/17*6/17*5/17*17/43=0.01448.

Firstly, we evaluate the label prediction accuracy with Subset Accuracy. Using our method, the percent of emotions figured out in the paper is showed in Table 3.

Table 3. Label Prediction Accuracy

	Document	Paragraph	Sentence
Subset Accuracy	99.76%	98.61%	96.38%

Next we compare our multi-labeled method with two algorithms, the first one is multi-label learning method (with no label intensity), and the other one is Fuzzy union. As we all know, in multi-label learning setting, the emotion value of any sentiment word is 0 or 1. In the fuzzy mathematics, fuzzy union is a very commonly used method. In this experiment it means that we use the max value of each emotion in every sentiment words in the sentence. We take an average result values of 5 cross validation, and the results at three different textual levels are shown in Table 4.

Table 4. Emotion Label Intensity Analysis of Hamming Loss

	Hloss (single-labeled)	*Hloss* (Fuzzy union)	*Hloss* (NB)	*Hloss* (ours)
Document	0.09845	0.06853	0.21445	**0.03929**
Paragraph	0.10659	0.05117	0.15133	**0.02598**
Sentence	0.11179	0.03927	0.10320	**0.01767**

Hamming Loss is a measure of the value of general accuracy. *Hloss* is the fewer, the better. So as we can see, our method is better than the others.

One-error is a method to measure whether the maximum value we predict is one of the final results, and the results we get are shown as Table 5. Average precision is widely used in many areas. According to the measure of Average precision [8], the results of the experiments at three levels of text are showed in Table 6.

Table 5. Label Intensity Analysis of One-error

	One-error (Fuzzy union)	*One-error* (NB)	*One-error* (ours)
Document	0.78649	0.76689	**0.04764**
Paragraph	0.58395	0.58807	**0.06903**
Sentence	0.84421	0.37633	**0.07377**

Table 6. Label Intensity Analysis of Average precision

	Avgprec (Fuzzy union)	*Avgprec* (NB)	*Avgprec* (ours)
Document	0.67249	0.36975	**0.74374**
Paragraph	0.71452	0.44310	**0.77940**
Sentence	0.70985	0.58617	**0.95494**

According to Formula (10) and (11), one-error is the fewer, the better. In contrast, average precision is the larger, the better. So we can see our algorithm is significantly better than others.

5.3   Discussion

Our method has several advantages: Firstly, the traditional multi-label learning methods only consider about whether the sentiment words in this sentence have this emotion. It cannot show the emotion intensity. The fuzzy union only consider the strongest one emotion, which means it has to ignore some of sentiment words. This strategy is less objective. In Naïve Bayes method, we presume that 8 emotions are independent, but the human emotional logic is not completely independent or simply multiplied. Secondly, our method not only considers every emotion value in each sentiment word, but also improves fuzzy-rough logic to fix them together. Thirdly, using the method, we both considered the value of the sentiment and the importance of the words. Specifically, although the emotion value in one kind of emotion may be the same, but with the influence of other kinds of emotion which belonging to different words, the sentiment value of the test result may be different.

On the other hand, the fuzzy logic is more suitable for this situation not only because the value of emotion area is between 0~1, but also considered the human language logic. As we know, a sentence may have some key emotion words, and the common feature of these key words is that their emotion intensities are always the largest. Maybe there are many other sentiment words that have different intensities, but finally, the key words dominate the emotions. Taking this sentence for example: "我游览了沈阳，中街的热闹，世园会的恬静，怪坡的奇妙使我爱上了这座城市". In this sentence, these three Chinese words "热闹", "恬静", "奇妙" all get sentiment of *joy*, the value of them are *joy*=0.5, *joy*=0.3, *joy*=0.4. But the key word is "爱", whose value is *love*=0.9. Although *joy* has larger sum of words and emotion value, we can still see that the main emotion of this sentence is *love*. This is the human logic. It is neither the sum of words nor the sentiment value. Our improved method inherits the advantage of fuzzy logic, which perfectly matches the emotion logic in this situation.

6   Conclusion and Future Work

In this paper, we propose a new way to solve multi-label emotions intensity analysis problem. For this new field of emotion analysis, we leverage the advanced fuzzy-rough set theory. The proposed algorithm could simultaneously detect multi-labeled emotions and their intensity at sentence, paragraph, and document level. The experiment results demonstrate the effectiveness of our proposed model and algorithm.

In the future, we would like to build the model of multi-labeled emotion and intensity analysis on microblog, website reviews, or other social media. The adverbs and negative words can be further taken into consideration that can improve the result.

References

1. Boutell, R., Luo, J., Shen, X., Brown, C.: Learning multi-label scene classification. Pattern Recognition (PR) **37**(9), 1757–1771 (2004)
2. Dubois, D., Prade, H.: Rough fuzzy sets and fuzzy rough sets. International Journal of General Systems **17**(2–3), 191–209 (1990)
3. Elisseeff, A., Weston, J.: A kernel method for multi-labelled classification. In: NIPS, pp. 681–687 (2001)
4. Pawlak, Z.: Rough sets. International Journal of Parallel Programming (IJPP) **11**(5), 341–356 (1982)
5. Quan, C., Ren, F.: A blog emotion corpus for emotional expression analysis in Chinese. Computer Speech & Language (CSL) **24**(4), 726–749 (2010)
6. Sun, B., Ma, W.: Soft fuzzy rough sets and its application in decision making. Artif. Intell. Rev. (AIR) **41**(1), 67–80 (2014)
7. Zadeh, L.: Fuzzy sets information and control. IANDC **8**(3), 338–353 (1965)
8. Zhang, M., Zhou, Z.: A review on multi-label learning algorithms. IEEE Trans. Knowl. Data Eng. (TKDE) **26**(8), 1819–1837 (2014)

What Causes Different Emotion Distributions of a Hot Event? A Deep Event-Emotion Analysis System on Microblogs

Yanyan Zhao[1]([✉]), Bing Qin[2], Zhenjiang Dong[3], Hong Chen[3], and Ting Liu[2]

[1] Department of Media Technology and Art, Harbin Institute of Technology,
Harbin, China
yyzhao@ir.hit.edu.cn
[2] Department of Computer Science and Technology, Harbin Institute of Technology,
Harbin, China
{bqin,tliu}@ir.hit.edu.cn
[3] ZTE Corporation, Shenzhen, China
{dong.zhenjiang,chen.hong}@zte.com.cn

Abstract. Current online public opinion analysis systems can explore lots of hot events and present the public emotion distribution for each event, which are useful for the governments and companies. However, the public emotion distributions are just the shallow analysis of the hot events, more and more people want to know the hidden causation behind the emotion distributions. Thus, this paper presents a deep Event-Emotion analysis system on Microblogs to reveal what causes different emotions of a hot event. We here use several related sub-events to describe a hot event in different perspectives, accordingly these sub-events combined with their different emotion distributions can be used to explain the total emotion distribution of a hot event. Experiments on 15 hot events show that the above idea is reasonable to exploit the emotion causation and can help people better understand the evolution of the hot event. Furthermore, this deep Event-Emotion analysis system also tracks the amount treads and emotion treads of the hot event, and presents the deep analysis based on the user profile.

Keywords: Emotion causation · Emotion distribution · Hot event · Event-emotion analysis · Microblog

1 Introduction

Nowadays, the social media (e.g., Twitter, Facebook, Sina Microblog, Dianping) significantly influence our life. A large number of users are involved in the activities on the social media, such as posting a tweet or image, commenting on a hot event and so on. Thus the Web contains a considerable amount of user-generated content that describes the opinions of customers regarding the events in the form of reviews, blogs, tweets, and so on. These reviews are valuable for

© Springer International Publishing Switzerland 2015
J. Li et al. (Eds.): NLPCC 2015, LNAI 9362, pp. 453–464, 2015.
DOI: 10.1007/978-3-319-25207-0_42

assisting governments in social management, helping customers in making purchasing decisions and for guiding companies in pursuing their business activities. However, browsing the extensive collection of available reviews to search for useful information is a time-consuming and tedious task. Therefore, automatically analyzing and processing the reviews, such as detecting and tracking the hot event, analyzing the public emotion distributions of a hot event, is necessary.

To date, many online public opinion analysis systems are built to exploit the hot events that people focus on and to present the public emotion distributions for each hot event. For example, a search-based demonstration system, called Meme-Tube [7], is presented to summarize the sentiments of microblog messages in an audiovisual manner. C-Feel-It [6] is another system which can categorizes tweets pertaining to a search string as positive, negative or objective and gives an aggregate sentiment score that represents a sentiment snapshot for a search string. Social Event Radar [5] is a new social networking-based service platform, that aim to alert as well as monitor any merchandise flaws, food-safety related issues, unexpected eruption of diseases or campaign issues towards to the Government.

Although so many applications can provide the hot events and their related emotion distributions, there is no system showing the reasons what causes the emotion distributions. Apparently, exploiting the emotion causations hidden behind the hot event is more valuable and important for the governments and companies, comparing to just exploring the emotion distributions of a hot event. For example, Figure 1 is an example representation of the emotion distributions for a hot event "埃博拉病毒" (Ebola virus). Many people want to know why some people are "喜悦" (happy) about "Ebola virus" which people actually feel fear and sad about. In this paper, we propose a new and important task, **emotion causation analysis**, which aims to reveal the causations that are hidden behind the total emotion distribution (such as the distribution in Figure 1) of a hot event.

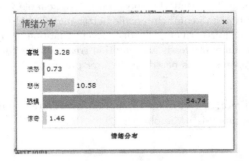

Fig. 1. Example of the emotion distribution for a hot event "埃博拉病毒" (Ebola virus).

What can be used to represent the emotion causation is the key question for emotion causation analysis. We use the sub-events extracted from the Hashtags of the hot event to represent the emotion causation. For example, we can use the sub-event "埃博拉病毒出现" ("Appearance of Ebola virus") to explain why so many people feel fear, and use the sub-event "疫苗研制成功" ("success of vaccine study") to explain why there is "happy" emotion appearing in the emotion distribution. To extract the sub-events of the hot event, a hierarchical clustering method is applied. Then, we can classify all the hot event related microblogs into these sub-events. The extracted sub-events combined with their new emotion distributions can be used to reveal the reasons that causes the total emotion distribution of the hot event. For example, the two different emotion distributions in Figure 2 corresponds to two sub-events of the hot event "Ebola virus". The left one is the "Appearance of Ebola virus" sub-event, in which the most prominent emotion is "fear". Thus this sub-event can explain the "fear" emotion in the total emotion distributions in Figure 1. The right one in Figure 2(b) is the "success of vaccine study" sub-event, in which the most prominent emotion is "happy". So this sub-event can explain why there is "happy" emotion for the event "Ebola virus".

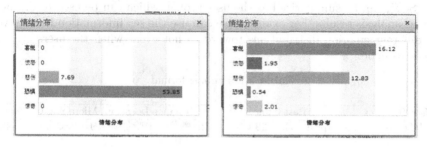

Fig. 2. Different emotion distributions for two sub-event of the hot event "埃博拉病毒"(Ebola virus).

We analyze 15 hot events (e.g., "MERS virus", "QingAn shooting"), including more than one million microblogs. We observe that using the sub-events and their emotion distributions is a good way to exploit the hidden causation of the public emotion distributions of a hot event. Experiments on the 15 hot events show that the hierarchical clustering based sub-event extraction method can achieve an accuracy of 66.7%, which can reveal most of the reasons. Besides, the deep Event-Emotion analysis system also tracks the hot events over time and displays various results.

The remainder of this paper is organized as follows. Section 2 introduces the framework of the Event-Emotion analysis system. Section 3 presents the experiments and results. Section 4 introduces the related work. Finally we conclude this paper in Section 5.

2   Framework of Event-Emotion Analysis System

Event-Emotion analysis system can be divided into several steps, namely, microblog fetcher, hot event detection, emotion predictor, emotion causation analysis, other analysis result display. Since microblog fetcher, hot event detection and emotion predictor are the techniques that are usually used in the related systems shown in Section 4, we will briefly introduce them. Specially, emotion causation analysis related steps, including emotion causation detection and causation related microblog collection, are new proposed tasks. We will introduce them in detail.

2.1 Microblog Fetcher

We crawl and store about 120 messages each second because the Sina Weibo API allows only the retrieval of a subset of messages. Thus we can store about 4,000,000 messages each day. The format for each microblog we record is shown in Table 2. We mainly use the microblog contents to exploit the hot events and their emotion distributions. The other information, such as the "User ID" or the "Sex", can be summarized as the user profile, which can be used to display various analysis results. For example, we can use the sex information to observe the different emotion distributions towards a hot event when the users are all female or male.

Table 1. Microblog format.

User ID	Sex	Post Time	Province	Location	Microblog Content
2054436031	男	Tue Apr 30 00:01:29 CST 2013	黑龙江	哈尔滨	为什么受伤的总是我

2.2 Hot Event Detection

Because of the characteristics of microblogging, detecting hot event in social media is different from detecting them from news, which has been studied in the previous work [1,3]. Messages that report such event are usually teemed with meaningless "babbles." Moreover, event detection algorithm should be scalable given the sheer amount of messages [13]. Considering the above works, the task of hot event detection in this study contains the following steps:

- **Event Detection:** Considering that hashtags can cover almost all events, hashtags that appear in the messages can be extracted after a filtering model as the event set. The biggest advantage of using hashtags as events is that hashtags themselves are perfectly organized, topic-related phrases, which are short, simple, and easy-to-understand. They are much more easy and accurate than the phrases extracted from the main body of the messages.
- **Event Clustering:** The events (hashtags) are sponsored by different users. Thus, some events describe the same issues. Event clustering aims to solve this problem by clustering the events into different clusters. And each cluster is factually a meaningful event.
- **Event Popularity Ranking:** Each event cluster has different popularity. To exploit the hot events, the event clusters should be ranked according to their popularity. Then we can extract the top events as the hot events.

2.3   Emotion Predictor

Five kinds of emotions are used to tag the sentiments of the messages, namely, "happy," "sad," "surprise," "angry," and "fear". This task aims to classify each message into one of the five emotions. Many researchers focused on the emotion classification task [9], which is a typical task in the sentiment analysis.

Among all the algorithms on this task, machine learning based method is the most effective one. In our Event-Emotion analysis system, we use the features of the best system in the "Sentiment analysis in twitter" task in SemEval [9,12] and the word embedding features. We trained a SVM model based on these features to predict the emotion for each microblog.

Figure 3 shows the total emotion distribution for the hot event "QingAn Shooting", in which different color refers to different emotion. We can find that many people feel "angry" and "sad" about this event. This is reasonable. However, there are still some people who feel "happy" about this event. This is very intriguing and worth exploring.

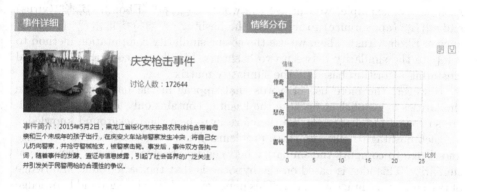

Fig. 3. The total emotion distribution for the hot event "QingAn Shooting".

2.4   Emotion Causation Analysis

What is Used as the Emotion Causation

The first necessary question that needs to be solved is what should be used as the emotion causation. One obvious way is to use some key words to represent the emotion causation. For example, MoodLens system [15] extracted the top 5 bi-gram terms of high frequency to detect the abnormal event, which is a little similar to the emotion causation task. However, the key words are hard to be understood by users, if the users do not know the background of the events. Thus, using the key words as the emotion causation is not a good way.

Instead, we observe that the different emotions for a same hot event are caused by some sub-events. In detail, most of the hot events can last for several days and can derive several related sub-events. The public emotion for the hot event can change with the evolution of the sub-events, such as the examples in Figure 1 and Figure 2. Based on the above, using the sub-events as the emotion causation looks like a good way. Therefore, the emotion causation analysis can be converted to exploit the sub-events and obtain all the related microblogs for each sub-event. Finally, the sub-events combined with their emotion distributions can be used to reveal the hidden emotion causation of the hot event.

In summary, the emotion causation analysis can be divided into two subtasks: Sub-event Detection and Sub-event related Microblogs Collection.

Sub-event Detection

Similar to the event detection task in Section 2.2, we also extract some typical hashtags to represent the sub-events, since the hashtag is user-generated content and easy to understand. Thus the sub-event detection can also be considered as a clustering task. We can use various kinds of clustering algorithms. This paper employs hierarchical clustering algorithm as a case of study. Similarity computation between two hashtags is the key technique during the clustering process. We can initially segment the hashtag and use a word vector to represent each hashtag. For example, we can use the words "埃博拉" (Ebola), "病毒" (virus) and "出现" (appearance) to represent the hashtag "埃博拉病毒出现" (Appearance of Ebola virus). Then, we use the cosine similarity computation method to compute the similarity between two hashtags. Finally, we conduct hierarchical clustering algorithm based on the similarity matrix.

However, the mere use of words that appear in the hashtag to compute similarities is insufficient because the hashtag contains only a few words, which are so sparse for clustering. We therefore introduce the background knowledge for each hashtag to alleviate these problems. This means that we can use more knowledge to compute the similarity between two hashtags besides their literal similarity. This idea is based on the hypothesis that the background knowledge of two similar hashtags is similar. This paper expands the background knowledge for each hashtag by importing all the messages that contain the hashtag.

Fig. 4. Sub-events extracted for the hot event "QingAn Shooting".

The hierarchical clustering algorithm is conducted based on this similarity. First, we suppose each hashtag as a cluster, noted as c_1, ..., c_i, ..., c_n. Next, we compute the similarity between each pair of clusters. If the similarity between c_i and c_j is the maximum and greater than a threshold θ, c_i and c_j are merged into a new cluster. The process is repeated until the amount of the clusters remains constant. Finally, each cluster can be considered as a sub-event. We simply rank the clusters according to their messages and capture the top n clusters as the typical sub-events. We use the longest hashtag in the cluster to represent the sub-event.

Figure 4 shows the five extracted sub-events for the hot event "QingAn Shooting", in which different color refers to different sub-event.

Sub-event Related Microblogs Collection

When we get the top clusters as the sub-events of the hot event, there are also many clusters left. Since lots of them are related to the sub-events, we need to classify these clusters into the sub-events. In this step, we adopt a simple method of training a classifier based on bag-of-words to classify the left clusters. Then we can obtain all the related microblogs for each sub-event.

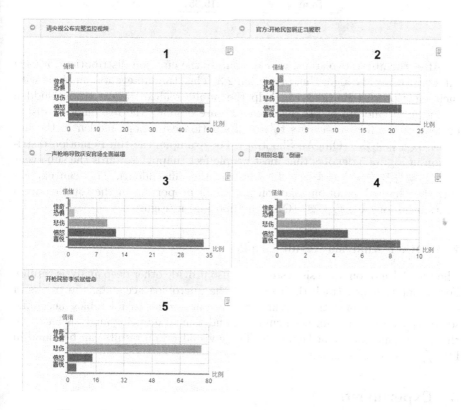

Fig. 5. Emotion causations for the hot event "QingAn Shooting".

Table 2. Statistics of 15 hot events.

ID	Hot Event	# of microblogs	# of hashtags
1	毕福剑不雅视频事件	121,204	1,988
2	云南女导游辱骂游客	18,338	605
3	四川女司机被当街暴打	324,775	4,091
4	男子暴打扫地小男孩	111,080	1,611
5	庆安枪击事件	172,644	1,930
6	王思聪范冰冰骂战	42,252	1,408
7	吴镇宇发飙	76,366	1,049
8	携程被黑	6,137	508
9	故宫女模裸照	7,198	255
10	MERS入侵广东	237,520	1,341
11	东方之星游轮倾覆	556,008	7,396
12	边策吸毒坠楼身亡	14,896	500
13	肃宁县特大枪击案	22,128	578
14	微博炫腹大赛	18,533	720
15	贵州自尽男童	30,848	1,930
	Sum	1,219,351	18,715

After the above two steps, we can analyze the emotion distribution for each sub-event using the algorithm in Section 2.3. Thus the sub-events combined with their emotion distributions can help the public realize the causations hidden behind each emotion of the hot event. Figure 5 shows the five emotion distributions of the five sub-events to reveal emotion causations hidden in the hot event "庆安枪击案" (QingAn Shooting). For example, the third emotion distribution in Figure 5 denotes that most people feel "happy" towards the sub-event "一声枪响导致安庆官场全面崩塌" (destroy the officialdom). This can explain why the "happy" emotion takes on not small proportion in the total emotion distribution of "庆安枪击案" (QingAn Shooting) in Figure 3.

2.5 Other Analysis Result Display

The Event-Emotion analysis system can also provide other deep analysis results. For example, it can track the trends of the microblog accounts, the trends of emotion distributions, the province ranking according to microblog accounts, some typical examples for each emotion, and some interesting analysis based on the user profile and event keywords. These visualization results can be found in http://qx.8wss.com/zkmap.com.

3 Experiments

In this part, we just have conducted the experiments on the emotion causation analysis task. The reason is that the hot event detection and emotion analysis are both popular tasks in natural language processing, lots of researchers have focused on these tasks and lots of algorithms are studied and compared. We just adopt the-state-of-the-art method to complete the two steps. In comparison, the emotion causation analysis task is firstly proposed in this paper.

3.1 Dataset

We collect 15 hot events in the recent half years, the statistics of which are listed in Table 1. We can find that these events are all discussed and commented by millions of people. And meanwhile, each hot event contains thousands of hashtags, it is possible to detect the sub-events from these hashtags.

3.2 Results of the Emotion Causation Analysis

We use two steps, sub-event detection and sub-event related microblogs collection to analyze the emotion causation hidden in a hot event.

Table 3. Performance of sub-event detection.

Method	P@5
Sub-event detection	66.7%

A hierarchical clustering based method is used to detect the sub-events. Table 3 shows the performance, which can reach an *accuracy* of 66.7%. Table 4 illustrates two hot events and their detected sub-events, where the standard sub-events are manually annotated. We can observe that most detected sub-events are correct causations.

In the sub-event related microblogs collection step, a simple classifier is used. Experiments show that we can obtain an average *F-score* of 68.74%, which is not high enough.

Table 4. Examples of detected sub-events.

Hot event	Standard sub-events	Detected sub-events
庆安枪击事件	请央视公布完整监控视频 官方:开枪民警属正当履职 一声枪响导致庆安官场全面崩塌 真相别总靠"倒逼" 开枪民警李乐斌偿命	庆安县副县长董国生被停职 官方:庆安开枪民警属正当履职 监控还原庆安枪击案全程 视频还原庆安枪击案全过程:徐纯合夺棍连砸警察 "黑龙江庆安火车站事件"调查结果公布
东方之星游轮倾覆	载客458人已救起8人 沉船内部有生命迹象 乘客家属收到诈骗短信 "东方之星"客船翻沉事件幸存游客口述 沉船存在生还者可能性已十分渺茫	载客458人已救起8人 沉船内部有生命迹象 乘客家属收到诈骗短信 "东方之星"客船翻沉事件幸存游客口述 大圣炽盛光如来拥护咒轮

It should be noted that this paper aims to try some simple algorithms for each step to validate the idea of using sub-events combined with their emotion distributions to explain the emotion causation of a hot event. Although the performance for each step is not very ideal, the treads or the ratios of the five emotions for a hot event are reasonable. We will still polish every step of our framework in future.

4   Related Work

A major contribution of Web 2.0 is the explosive rise of user-generated content. The importance of social media is constantly growing. Lots of systems or platforms are constructed to share mainstream media news and comment on the related events. In this manner, readers can follow the main events happening in the world, both from the perspective of mainstream as well as social media and the public's perception on them. Balahur et al. [4] build a system that links the main events detected from clusters of newspaper articles to tweets related to them, detects complementary information sources from the links they contain and subsequently applies sentiment analysis to classify them into positive, negative and neutral. Hsieh et al. [5] propose a Social Event Radar platform, which is used as a realtime risk control management technology to assist monitoring huge amount of new media related information and giving a warning for utility users' sake in efficiency way. Joshi et al. [6] build a C-Feel-It platform, which is a web-based system predicting sentiment in microblogs on Twitter. Li et al. [7] build a sentiment-based audiovisual system, which is called MemeTube, to recognize the sentiments of messages and display related music melody. Zhao et al. [14] present a Social Sentiment Sensor (SSS) system on Sina Weibo to detect daily hot topics and analyze the sentiment distributions toward these topics. Osborne et al. [10] introduce ReDites, a system for realtime event detection, tracking, monitoring and visualization.

From all the related systems, we can observe that topic detection/tracking and sentiment analysis are two key techniques. Topic detection and tracking (TDT) [2,16] refers to the automatic techniques for finding topically related material in streams of data (e.g., newswire or social media data), in which topic detection involves detecting the occurrence of a new event such as a plane crash, a murder, a jury trial result, or apolitical scandal in a stream of news stories from multiple sources, and topic tracking is the process of monitoring a stream of news stories to find those that track (or discuss) the same event as one specified by a user. Sentiment analysis (also known as opinion mining) [8,11] aims to determine the attitude of a speaker or a writer with respect to some topic or the overall contextual polarity of a document.

For both techniques, lots of related algorithms have been exploited. They are indeed very useful and important in many social media applications. However, to date, almost all the public opinion analysis results are the emotion or sentiment orientation ("like" or "dislike") distributions for the hot events, which can be considered as kind of shallow analysis of the hot events. More and more people want to know the hidden causation behind the emotion distributions. Unfortunately, no system is built to automatically analyze the causation of the hot event emotion distribution. Thus in our knowledge, we are the first to propose the emotion causation analysis task and focus on this task.

5   Conclusion and Future work

In this paper, we present an Event-Emotion analysis system. Different from other public opinion analysis systems, our Event-Emotion system can not only show the hot events and their public emotion distributions like other common systems, but also reveal what causes these emotion distributions. The main contributions of this paper can be concluded as follows:

- We are the first to propose the emotion causation analysis task, which can help the users deeply analyze the social media.
- We propose to use the sub-events combined with their emotion distributions to represent the emotion causation. Experiments show that this idea is reasonable.
- We built a Chinese public opinion analysis system: Event-Emotion system, which fused the deep emotion causation analysis.

Although we have designed and built the architecture of the Event-Emotion system, the performance for each technique in this system is not ideal enough. As future work, we would like to polish each technique to make the system more practical.

Acknowledgments. We thank the anonymous reviewers for their helpful comments. This work was supported by National Natural Science Foundation of China (NSFC) via grant 61300113, 61133012 and 61273321, the Fundamental Research Funds for the Central Universities via grant No.HIT.NSRIF.2013090 and ZTE Research Project.

References

1. Allan, J., Carbonell, J., Doddington, G., Yamron, J., Yang, Y.: Topic detection and tracking pilot study: final report. In: Proceedings of the DARPA Broadcast News Transcription and Understanding Workshop, Lansdowne, VA, USA, pp. 194–218, February 1998
2. Allan, J.: Introduction to topic detection and tracking. In: Topic Detection and Tracking, pp. 1–16. Kluwer Academic Publishers, Norwell (2002). http://dl.acm.org/citation.cfm?id=772260.772262
3. Allan, J. (ed.): Topic Detection and Tracking: Event-based Information Organization. Kluwer Academic Publishers, Norwell (2002)
4. Balahur, A., Tanev, H.: Detecting event-related links and sentiments from social media texts. In: Proceedings of the 51st Annual Meeting of the Association for Computational Linguistics: System Demonstrations. Association for Computational Linguistics, Sofia, Bulgaria, pp. 25–30, August 2013. http://www.aclweb.org/anthology/P13-4005

5. Hsieh, W.T., Wu, C.M., Ku, T., Chou, S.C.T.: Social event radar: a bilingual context mining and sentiment analysis summarization system. In: Proceedings of the ACL 2012 System Demonstrations. Association for Computational Linguistics, Jeju Island, Korea, pp. 163–168, July 2012

6. Joshi, A., Balamurali, A.R., Bhattacharyya, P., Mohanty, R.: C-feel-it: a sentiment analyzer for micro-blogs. In: Proceedings of the ACL-HLT 2011 System Demonstrations. Association for Computational Linguistics, Portland, Oregon, pp. 127–132, June 2011. http://www.aclweb.org/anthology/P11-4022

7. Li, C.T., Wang, C.Y., Tseng, C.L., Lin, S.D.: Memetube: a sentiment-based audio-visual system for analyzing and displaying microblog messages. In: Proceedings of the ACL-HLT 2011 System Demonstrations. Association for Computational Linguistics, Portland, Oregon, pp. 32–37, June 2011. http://www.aclweb.org/anthology/P11-4006

8. Liu, B.: Sentiment Analysis and Opinion Mining. Synthesis Lectures on Human Language Technologies. Morgan & Claypool Publishers (2012)

9. Mohammad, S., Kiritchenko, S., Zhu, X.: NRC-canada: building the state-of-the-art in sentiment analysis of tweets. In: Proceedings of the Seventh International Workshop on Semantic Evaluation Exercises (SemEval-2013), Atlanta, Georgia, USA, June 2013

10. Osborne, M., Moran, S., McCreadie, R., Von Lunen, A., Sykora, M., Cano, E., Ireson, N., Macdonald, C., Ounis, I., He, Y., Jackson, T., Ciravegna, F., O'Brien, A.: Real-time detection, tracking, and monitoring of automatically discovered events in social media. In: Proceedings of 52nd Annual Meeting of the Association for Computational Linguistics: System Demonstrations. Association for Computational Linguistics, Baltimore, Maryland, pp. 37–42, June 2014. http://www.aclweb.org/anthology/P14-5007

11. Pang, B., Lee, L.: Opinion mining and sentiment analysis. Found. Trends Inf. Retr. **2**(1–2), 1–135 (2008)

12. Rosenthal, S., Ritter, A., Nakov, P., Stoyanov, V.: Semeval-2014 task 9: sentiment analysis in twitter. In: Proceedings of the 8th International Workshop on Semantic Evaluation (SemEval 2014). Association for Computational Linguistics and Dublin City University, Dublin, Ireland, pp. 73–80, August 2014. http://www.aclweb.org/anthology/S14-2009

13. Weng, J., Lee, B.S.: Event detection in twitter (2011). http://www.aaai.org/ocs/index.php/ICWSM/ICWSM11/paper/view/2767

14. Yanyan, Z., Bing, Q., Ting, L., Duyu, T.: Multimedia tools and applications, pp. 1–18, August 2014

15. Zhao, J., Dong, L., Wu, J., Xu, K.: Moodlens: an emoticon-based sentiment analysis system for chinese tweets. In: Proceedings of the 18th ACM SIGKDD International Conference on Knowledge Discovery and Data Mining. KDD 2012, pp. 1528–1531. ACM, New York (2012). http://doi.acm.org/10.1145/2339530.2339772

16. Zhao, X., Shu, B., Jiang, J., Song, Y., Yan, H., Li, X.: Identifying event-related bursts via social media activities. In: Proceedings of the 2012 Joint Conference on Empirical Methods in Natural Language Processing and Computational Natural Language Learning. Association for Computational Linguistics, Jeju Island, Korea, pp. 1466–1477, July 2012. http://www.aclweb.org/anthology/D12-1134

Deceptive Opinion Spam Detection Using Deep Level Linguistic Features

Changge Chen[1,2], Hai Zhao[1,2]($\boxtimes$), and Yang Yang[1,2]

[1] Department of Computer Science and Engineering, Shanghai Jiao Tong University,
Shanghai 200240, China
changge.chen.cc@gmail.com, {zhaohai,yangyang}@cs.sjtu.edu.cn
[2] Key Laboratory of Shanghai Education Commission for Intelligent Interaction and
Cognitive Engineering, Shanghai Jiao Tong University, Shanghai 200240, China

Abstract. This paper focuses on improving a specific opinion spam detection task, deceptive spam. In addition to traditional word form and other shallow syntactic features, we introduce two types of deep level linguistic features. The first type of features are derived from a shallow discourse parser trained on Penn Discourse Treebank (PDTB), which can capture inter-sentence information. The second type is based on the relationship between sentiment analysis and spam detection. The experimental results over the benchmark dataset demonstrate that both of the proposed deep features achieve improved performance over the baseline.

Keywords: Spam detection · Shallow discouese parsing · Sentiment analysis

1 Introduction

In nowadays, online reviews of products and services have increasingly large impact on consumer purchasing decisions[1]. Accordingly, there comes an increasing potential to gain money through deceptive opinion spam - fraudulent reviews that are deliberately written to deceive readers.

While various studies have been carried out in recent years, most of them focused on extracting scattered syntactic features which capture local information. In this work, we consider more deep features on the basis of syntactic

This work was supported in part by the National Natural Science Foundation of China under Grants 60903119, 61170114, and 61272248, the National Basic Research Program of China under Grant 2013CB329401, the Science and Technology Commission of Shanghai Municipality under Grant 13511500200, the European Union Seventh Framework Program under Grant 247619, the Cai Yuanpei Program (CSC fund 201304490199, 201304490171, and the art and science interdiscipline funds of Shanghai Jiao Tong University (a study on mobilization mechanism and alerting threshold setting for online community, and media image and psychology evaluation: a computational intelligence approach under Grant 14X190040031 (14JCRZ04).

[1] http://www.conecomm.com/2011coneonlineinfluencetrendtracker

J. Li et al. (Eds.): NLPCC 2015, LNAI 9362, pp. 465–474, 2015.
DOI: 10.1007/978-3-319-25207-0_43

feature set. Specifically, we consider two types of features. Firstly, we explore the possibility of integrating results generated by a shallow discourse parser into opinion spam detection. Instead of treating the whole text as an unordered set of sentences or terms, from a high level of viewing, we inspect the text consisting of sentences that are glued together by discourse relations in a systematic way. Automatic discourse parsing is considered as one of the most challenging Natural Language Processing tasks [1]. In this work, we integrate the most frequent type of discourse relations - *'Explicit'* relation, and its corresponding sense into our module. Secondly, we design an extensive feature set to capture sentiment cues for further performance improvement.

This work will be strictly evaluated on a public golden standard dataset. Ott et al.[2] constructed a negative deceptive opinion spam dataset. Combining with the positive opinion spams, a dataset consists of both negative and positive reviews is available. This dataset includes 1,600 reviews with four categories: positive truthful, positive deceptive, negative truthful, and negative deceptive.

The rest of paper is organized as follows: in Section 2, we discuss and compare recent related works. In Section 3, we give a description of the construction and analyze the dataset. In Section 4, we describe features employed by our detection methodology. In Section 5, we present and discuss our experiment results. Finally, conclusions are given in Section 6.

2   Related Work

Opinion spam detection has attracted growing interests from academia and industry in recent years. While enormous researches focus on mining opinions, the source of reviews are seldom concerned. Jindal and Liu [3] were the first to study the trustworthiness of opinions in reviews according to our knowledge. They categorized the spam reviews into three types: a) untruthful opinion spam, b) reviews on brands, and c) non-reviews. While they found that it is easy to distinguish non-reviews, it is difficult to disambiguate the first and second types of spams due to absence of annotated data. By the observation that spams often appear many times, they annotated duplicated reviews as spams. Based on this novel assumption, they got positive results by comparing area under receiver operating characteristic curve (ROC)[2].

Ott et al.[4] constructed the first public DECEPTIVE opinion dataset by hiring experienced online writers through crowd-sourcing service. The paid writers were asked to simulate staff working in 20 hotels located in Chicago. Then the 'staff' were requested by their manager to write positive reviews towards their 'own' hotels in order to promote their reputation. Based on several filtering rules, they solicited 400 deceptive positive reviews. To construct a truthful reviews, they collected reviews from TripAdvisor[3]. After filtering unqualified reviews and balancing the number with deceptive reviews, they got 400 truthful reviews. One concern would be that reviews from TripAdvisor may also contain

---

[2] http://en.wikipedia.org/wiki/Receiver_operating_characteristic
[3] http://tripadvisor.com

opinion spams. Since we focus on the detection of opinion spam, the impurity of truthful dataset only demonstrates the performance of the classifier. Ott et al.[2] later constructed another dataset in the same manner. But at this time, both truthful reviews and deceptive reviews are in negative attitude.

Various studies have been carried out based on Ott's dataset. Xu et al.[5][6] exploited linguistic features generated from dependency parsing tree. Li et al.[7] proposed a generative LDA-based topic modeling approach for fake review detection. They introduced a semi-supervised manifold ranking algorithm for this task. Feng et al.[8] investigated syntactic stylometry for deception detection. Banerjee et al.[9] developed a linguistic framework to distinguish between genuine and deceptive reviews based on their readability, genre and writing style.

Howerer, all these works treat the text as an unordered set of terms. Beyond that, we exploit discourse relations that hold the text together to extract information from inter-sentence level. Furthermore, on the availability of the dataset with both positive and negative emotion polarities, we also study the relation between sentiment analysis and deceptive detection.

3 Construction of Dataset

As this work focuses on deceptive spam that is very hard to be exactly identified even by human observation, corpus or dataset should be carefully selected. We adopt a public dataset with necessary extensions for our evaluation, i.e., the corpus annotated by Ott et al.[2][4]. According to our best knowledge, it is the first public dataset on deceptive spam detection. In this section, a brief description is given to the procedure of data collection and annotation. A simple rule-based deceptive spam detection method is deployed to compare with human judge on this dataset.

3.1 Truthful Dataset

The truthful reviews are from TripAdvisor concerning 20 most popular Chicago hotels. Filtering rules are given for better quality control. Reviews thus must be:

- 5-star reviews;
- Only English reviews;
- More than 150 characters (To comply with deceptive reviews);
- Written by non-first-time authors;

To balance truthful and deceptive opinion reviews, 400 truthful reviews are selected randomly.

3.2 Deceptive Dataset

While truthful opinions are ubiquitous online, deceptive opinions are difficult to obtain without resorting to heuristic methods [3]. Through the crowd-sourcing service provided by Amazon Mechanical Tuck (AMT)[4], Ott created a pool of 400 Human Intelligence Tasks (HIT) to solicit golden standard positive, deceptive opinion spam toward the 20 chosen hotels. With a reward of 1$ for each review, they restricted their task to Tuckers located in United States with an approval rating of at least 90%. The time duration for each task should be between 1 and 30 minutes. Each Tucker was presented with the name and website of the hotel. They were asked to assume that they were the staff of the hotel's marketing department with the mission to write positive, realistic sounding reviews for their own hotel. At the end, after filtering out unqualified reviews (e.g., unreasonably short, plagiarized and so on), they obtained 400 golden deceptive opinions. Later, they constructed 400 golden negative deceptive opinions using the same way [2].

3.3   Human Judge

Ott et al.[4] adopted a skeptical meta judge which labels a review as spam if any of the judges believes so. However, native speakers' performance on detecting deceptive reviews was slightly over random guess. Xu et al.[5] demonstrated that non-native speakers did even worse than natives, which means that opinion spam may do more harm for non-native speakers.

While reading a couple of deceptive reviews, we find that the names of the hotels that the turkers were asked to review on, continuously appear in the first line of the crafted reviews. We investigate this phenomenon in all the corpus. The result is given in Table 1. About 73% of the 800 deceptive reviews have the hotels' name in the first line, which is much higher than that in true reviews (46.8%, 374/800) This phenomenon is not sensitive to sentiment polarity, as there are 486 and 472 reviews fall in positive and negative categories respectively. We can see that this phenomenon is apparent in deceptive reviews. This difference can be ascribed to the difference of motivation between spam writer and true consumers. With reward in minds, spammers need to craft a positive or negative comment on the target as soon as possible, and at the same time, complying with the requirements the HIT asked. Whereas true customers just want to express their own feelings, pleasant or unpleasant, after their personal experiences. Based on this observation, we set up a simple rule-based method which labels a review as deceptive if the hotel name appears in the first line of the review. Otherwise we adopt the rule to label it as truthful. We compare the result with human judges [4] in Table 2.

Table 1. Hotel names appearance in the golden standard English Dataset.

	Total	Positive	Negative
Deceptive	584	284	300
Truthful	374	202	172

---

[4] http://mturk.com

Table 2. Performance of native speaker, non-native speaker, and rule on deceptive detection.

	Accuracy(%)	Precision(%)	Recall(%)	F-score(%)
Rule-based	**66**	**61.0**	**73.0**	**66.4**
Native speaker	60.6	60.5	61.3	60.9
Non-native speaker	58.1	56.3	61.5	58.8

We can see that a simple rule outperforms both native and non-native judge. It recalls 73% of the all opinion spam reviews, which demonstrates our observation. The overall F-score outperforms native speaker by almost 6 percents which shows the vulnerability of human exposed of opinion spam. Therefore, we integrate this syntactic cue into our feature set.

4 Deep Features for Deceptive Spam Detection

4.1 Shallow Syntactic Features

We view the spam detection task as a text categorization problem with the following features.

a) Bag-of-words (baseline) A model only with bag of word features may outperform human judge by achieving a F-score of 88.3% [2] [4]. This serves as our baseline.

b) POS-n-gram Since the frequency distribution of part-of-speech (POS) tag in a text often depends on the genre of the text [10], and POS tag bigram will not only show frequency information of POS, but also the structure of the sentence, we therefore adopt the POS-unigram and POS-bigram.

b) Punctuation This feature indicates the appearance of exclamation and question marks.

d) Hotel name This feature indicates whether the hotel name is appear in the first sentence.

4.2 Discourse Parsing Features

A typical text consists of sentences that are glued together in a systematic way to form a coherent discourse. Shallow discourse parsing [11] is to parse a piece of text into a set of discourse relations between two adjacent or non-adjacent discourse units. Thus discourse relations convey the infomation of the structure of the article. Combining sentences together enables us to represent a text from an inter-sentence level. PDTB [11] defines five types of discourse relations: 'Explicit', 'Implicit', 'AltLex', 'EntRel', 'NoRel'. 'Explicit' relation, as its name indicates, is the relation signaled by a connective explicitly. There are 100 connectives annotated by PDTB. Senses have been annotated in the form of sense tags for 'Explicit' and 'Implicit' connectives, and 'AltLex' relations. Sense tags provide a semantic description of the relation between the arguments of connectives.

The tag set of senses is organized hierarchically as in Table 3. The top level has four tags representing four major semantic classes: *'Temporal'*, *'Contigency'*, *'Comparison'* and *'Expansion'*. For each class, a second level of types is defined to further refine the semantics levels.

Table 3. Hierarchy of sense tags.

First Level	Second Level	Third Level
TEMPORAL	Asynchronous	–
	Synchronous	precedence, succession
COMPARISON	Constrast	juxtaposition, opposition
	Pragmatic Contrast	–
	Concession	exception, contra-exception
	Pragmatic Concession	–
CONTINGENCY	Cause	reason, result
	Pragmatic Cause	justification
	Condition	hypothetical, general, unreal present, unreal past, factual present, factual past
	Pragmatic Condition	relevance, implicit assertion
EXPANSION	Conjunction	-
	Instantiation	-
	Alternative	conjunction, disjunction, chosen alternative

The following example shows a *'Condition'* relation. And the underlined spans are explicitly signaled by the connective *'if'*.

– *The Treasury said the U.S. <u>will default on Nov. 9</u> **if** <u>Congress doesn't act by them</u>* .

The shallow discourse parsing can be divided into two parts: explicit and non-explicit. Giving the fact that parsing the non-explicit relation is relatively hard at this time, we only deploy the explicit part here. Noting not all the connectives that appear in the text represent discourse relations, we use a classifier to disambiguate those connectives that convey a relation from those not. In addition, the classifier also determines the sense the connective conveys. Seven features generated from constituent parsing tree are employed to train such a classifier:

a) **Self Category** The highest dominated node which covers the connective.

b) **Parent Caterogy** The category of the parent of the self category.

c) **Left Sibling Category** The syntactic category of immediate left sibling of the self-category. It would be *'NONE'* if the connective is the leftmost node.

d) **Right Sibling Category** The immediate right sibling of the self category. It also would be assigned *'NONE'* if the self-category has been the rightmost node.

e) **VP Existence** The binary feature is to indicate whether the right sibling contains a VP.

f) **Connective** In addition to those features proposed by Pilter et al. [12], we introduce connective feature. The potential connective itself would be a strong sign of its function. Thus we use the POS tag of the candidate connective, its preceding and following word.

The explicit discourse parser is trained and evaluated on the dataset provided by CoNLL 2015 shared task[6]. With an explicit discourse parser, we may adopt related features derived from the parser outputs for the task of deceptive spam detection. In detail, all the 100 connectives defined by PDTB, their frequencies and corresponding senses are used as features. Considering the difficulty of disambiguating all the three levels of senses, in this work we only adopt the first level senses, i.e., 'Temporal', 'Contingency', 'Comparison' and 'Expansion'.

4.3 Sentiment Features

Ott et al.[2] gave a brief description between the sentiment and deceptive detection. They claimed that fake negative reviewers over-produced negative emotion terms relative to the truthful reviews in the same way that fake positive reviewers over-produced the positive emotion terms. Based on their assumption, we design a set of sentiment features to express the sentiment information like polarity, intensity, and so on. Similar to what Zhou et al.[13] did, we use term frequency to quantify these features. Note that a negation word that is adjacent to a sentiment word can change its emotional polarity. For any sentiment words within a window following a negation word, we reverse its sentiment polarity from positive to negative, or vice versa. Based on annotation provided by Inquirer Dictionaries[7], we construct a sentimental dictionary consisting of positive, negative, and negation terms. A partial list of the dictionary is given in Table 4.

Table 4. A partial list of positive, negative, and negative words.

Positive terms	Negative terms	Negation terms
accurate, agile, apt, boost,confident, ...	*abandon, blame, stubborn, torment, ...*	*not, never, none, little, few, seldom, ...*

5 Experiments

After combining the positive and negative dataset we have a total of 1,600 reviews. The dataset is divided into four categories: a) positive, deceptive reviews; b) positive, truthful reviews; c) negative, deceptive reviews; d) negative, truthful reviews. For every category, there are 20 reviews toward each of the 20 hotels. All the 400 reviews in each category are divided evenly into 5 folds. Each fold contains 80 reviews toward the corresponding 4 hotels.

We use a 5-fold nested cross validation [14] for evaluation. 4 folds are to do model selection. The selected model is then tested on the left out fold. The division of folds guarantees learned models are always evaluated on reviews from unseen hotels.

---

[6] http://www.cs.brandeis.edu/~clp/conll15st/index.html
[7] http://www.wjh.harvard.edu/~inquirer/homecat.htm

For basic language processing tools, BasePOS [15][16][17] is used to do POS tagging. We train our explicit discourse parser based on features extracted from constituent tree. The phrase structure parsing tree is predicted by the Berkeley Parser[8]. The parser outputs are converted by Standford dependency converter[9]. All the approaches we describe in Section 3 are used on a maximum entropy classifier[18][19].

6   Results and Discussion

As shown in Table 5, our explicit discourse parser obtains a good enough result. This prevents aggravating unnecessary errors into the following steps.

Table 5. Performance of explicit discourse parser.

	Precision(%)	Recall(%)	F-score(%)
Explicit discourse Classifier	91.2	89.1	90.1

Table 6. Performance of our approaches based on 5-fold nested cross validation.

Features (%)	Accuracy	Precision	Recall	F-score
Bigram[+] + LIWC (baseline)	88.4	89.1	87.5	88.3
baseline + Sentiment	88.6	89.2	88.2	88.7
baseline + Syntactic	89.1	89.1	89.3	89.2
baseline + Syntactic + Discourse Parsing	**89.5**	**89.4**	**89.8**	**89.6**

The result is given in Table 6. We can see that integrating sentiment features improves the recall by 0.8 percents, whereas the precision stays almost the same as the baseline. Based on annotation provided by Inquirer Dictionaries[7], the distribution of emotion terms in the dataset is shown in Table 7. We have two interesting findings. First, contrary to the analysis by Ott et al.[2], we find that the emotion intensity of deceptive reviews are tend to be milder than truthful reviews: Negative deceptive reviewers use less negative terms than truthful reviewers. And positive deceptive reviewers use less positive terms than truthful reviewers. Second, to our surprise, we find that in negative reviews, both deceptive and truthful reviews over-produce positive terms relative to negative terms. One reason would be that a negation word that is adjacent to a sentiment word can change its emotional polarity. We thus collect a list of negation words. The distribution across different categories is given in Table 8. We can see that there are more negation words in negative reviews.

The syntactic feature continues improving the recall with a slight drop on precision. Finally, combining with discourse parsing features, the model improves the accuracy by 1.1 percents.

---

[8] http://nlp.cs.berkeley.edu/
[9] http://nlp.stanford.edu/software/stanford-dependencies.shtml
[7] http://www.wjh.harvard.edu/~inquirer/homecat.htm

Table 7. The number of sentiment words per review across different categories.

Positive reviews	Deceptive reviews	Truthful reviews
Positive terms	7.22	7.42
Negative terms	3.26	2.77
Negative reviews	Deceptive reviews	Truthful reviews
Positive terms	7	6.89
Negative terms	6.08	6.3

Table 8. The number of negation words per review across different categories.

Negation Terms	Deceptive reviews	Truthful reviews
Positive reviews	2.49	2.49
Negative reviews	4.44	4.64

7   Conclusion

In this work, we extend the existing standard deceptive opinion spam dataset. Then, we analyze its sentiment term distribution across different categories. Contrary to previous work, we find deceptive reviews tend to have a milder emotion than truthful reviews. We also receive a useful observation that paid-writers tend to demonstrate their truthfulness by stressing their presence as early as possible. Based on this discovery, a rule-based method can be effectively used for deceptive spam annotation, which demonstrates a better performance than human judge. To build a better model for deceptive spam detection, we consider two types of deep level linguistic features that are respectively given by an explicit discourse parser built from constituent parsing tree and a sentiment polarity classifier. After integrating the proposed features, our model gives a performance improvement by 1.1 percents. This result verifies the effectiveness of the proposed techniques.

References

1. Ghosh, S., Johansson, R., Tonelli, S.: Shallow discourse parsing with conditional random fields. In: Proceedings of the 5th International Joint Conference on Natural Language Processing, Chiang Mai, Thailand, pp. 1071–1079, November 2011
2. Ott, M., Cardie, C., Hancock, J.T.: Negative deceptive opinion spam. In: Proceedings of NAACL-HLT 2013, Atlanta, Georgia, pp. 497–501, June 2013
3. Jindal, N., Liu, B.: Opinion spam and analysis. In: Proceedings of the 2008 International Conference on Web Search and Data Mining, Palo Alto, California, USA, pp. 219–230, February 2008
4. Ott, M., Choi, Y., Cardie, C., Hancock, J.T.: Finding deceptive opinion spam by any stretch of the imagination. In: Proceedings of the ACL-2011, Stroudsburg, PA, USA, pp. 309–319, June 2011
5. Xu, Q., Zhao, H.: Using deep linguistic features for finding deceptive opinion spam. In: Proceedings of the 23th International Conference on Computational Linguistics, Mumbai, pp. 1341–1350, December 2012

6. Zhang, J., Zhao, H., Lu, B.L.: A comparative study on two large-scale hierarchical text classification tasks' solutions, ICMLC-2010, pp. 3275–3280 (2010)

7. Jiwei, L., Cardie, C., Sujian, L.: Topicspam: a topic-model-based approach for spam detection. In: Proceedings of ACL-2013, Sofia, Bulgaria, pp. 217–221, August 2013

8. Feng, S., Banerjee, R., Choi, Y.: Syntactic stylometry for deception detection. In: Proceedings of ACL-2012, Jeju, Republic of Korea, pp. 171–175, July 2012

9. Banerjee, S., Chua, A.Y.: A linguistic framework to distinguish between genuine and deceptive online reviews. In: Proceedings of the International Multi Conference of Engineers and Computer Scientists, vol. 1, Hong Kong, pp. 501–506, March 2014

10. Newman, M.L., Pennebaker, J.W., Berry, D.S., Richards, J.M.: Lying words: Predicting deception from linguistic styles. Personality and Social Psychology Bulletin 29(5), 665–675 (2003)

11. Lin, Z., Ng, H.T., Kan, M.Y.: A PDTB-styled end-to-end discourse parser. Natural Language Engineering 20, 151–184 (2014)

12. Pitler, E., Nenkova, A.: Using syntax to disambiguate explicit discourse connectives in text. In: Proceedings of ACL-IJCNLP 2009, Suntec, Singapore, pp. 13–16, August 2009

13. Xiang, B., Zhou, L.: Improving twitter sentiment analysis with topic-based mixture modeling and semi-supervised training. In: Proceedings of ACL-2014, Baltimore, Maryland, USA, pp. 434–439, June 2014

14. Quadrianto, N., Smola, A.J., Caetano, T.S., Le, Q.V.: Estimating labels from label proportions. The Journal of Machine Learning Research 10, 2349–2374 (2009)

15. Zhao, H., Kit, C.: Parsing syntactic and semantic dependencies with two single-stage maximum entropy models. In: Proceedings of the Twelfth Conference on Computational Natural Language Learning, Manchester, pp. 203–207, August 2008

16. Zhao, H., Chen, W., Kazama, J., Uchimoto, K., Torisawa, K.: Multilingual dependency learning: exploiting rich features for tagging syntactic and semantic dependencies. In: Proceedings of CoNLL-2009, Boulder, Colorado, pp. 61–66, June 2009

17. Zhao, H., Zhang, X., Kit, C.: Integrative semantic dependency parsing via efficient large-scale feature selection. Journal of Artificial Intelligence Research 46, 203–233 (2013)

18. Jia, Z., Wang, P., Zhao, H.: Grammatical error correction as multiclass classification with single model. In: Proceedings of CoNLL- 2013, Sofia, Bulgaria, pp. 74–81, August 2013

19. Wang, P., Jia, Z., Zhao, H.: Grammatical error detection and correction using a single maximum entropy model. In: Proceedings of CoNLL-2014, Baltimore, Maryland, USA, pp. 74–82, July 2014

Multi-sentence Question Segmentation and Compression for Question Answering

Yixiu Wang[1], Yunfang Wu[1(✉)], and Xueqiang Lv[2]

[1] Key Laboratory of Computational Linguistics, Peking University, MOE, Beijing, China
{labyrinth,wuyf}@pku.edu.cn
[2] Beijing Key Laboratory of Internet Culture and
Digital Dissemination Research, Beijing, China
lxq@bistu.edu.cn

Abstract. We present a multi-sentence question segmentation strategy for community question answering services to alleviate the complexity of long sentences. We develop a complete scheme and make a solution to complex-question segmentation, including a question detector to extract question sentences, a question compression process to remove duplicate information, and a graph model to segment multi-sentence questions. In the graph model, we train a SVM classifier to compute the initial weight and we calculate the authority of a vertex to guide the propagating. The experimental results show that our method gets a good balance between completeness and redundancy of information, and significantly outperforms state-of-the-art methods.

Keywords: Question answering · Question compression · Question segmentation · Complex-question analysis

1 Introduction

In a CQA service, users usually have a high tendency to ask multi-sentence questions. The complexity of multi-sentence questions reflects on following three aspects: first, a complete question usually contains several sub-sentences; secondly, inner redundancies exist in sentences; thirdly, the misuse of comma and period marks makes sentence more complicate. Figure 1 shows an example question in a Chinese tourism CQA. In Figure 1, the Description gives a complement to the Title, which consists of five sentences. The Title and Sentence 4 in the question thread are actually expressing the same meaning. In the Description, the Sentence 5 is a polite saying. Sentence 2 serves as a context of Sentence 3, which is a question, and Sentence 4 needs Sentence 1 as a complement.

Complex questions in CQAs are very difficult to analyze by traditional approaches. Multi-sentence question segmentation provide with an effective method to alleviate the complexity of long sentences. Our work develops a complete scheme on complex question segmentation. In Section 2, we provide an overview of our approach. In Section 3, and 4, we describe our approach of initial weight calculation, and the propagation of linking scores, respectively. In Section 5, we report the experimental results. Section 6 gives a comparison with previous work. Finlay Section 7 draws conclusions.

J. Li et al. (Eds.): NLPCC 2015, LNAI 9362, pp. 475–483, 2015.
DOI: 10.1007/978-3-319-25207-0_44

Title: 从马经新加坡回中国需不需要提前获得新加坡的签证? *(Whether should I obtain a visa of Singapore in advance if I return to china via Singapore from Malaysia?)*

Description:

Sentence 1: 我从马来亚经过新加坡回北京, 有马来西亚的旅游签证 *(I return back to Beijing via Singapore from Malaysia, and I own a tourism Visa of Malaysia)*

Sentence 2: 之前看到网上新加坡有过境96小时落地签这个东西, 只要有96小时 内离开新加坡的机票就可以获得. *(I have learned from the internet that Singapore used to provide a landing check of 96 hours, as long as a flight leaving Singapore in 96 hours is processed.)*

Sentence 3: 但不知道现在还有没有? *(But whether such policy is still valid now?)*

Sentence 4: 如果没有了, 是不是如果要在新加坡转机, 就需要提前获得新加坡 的签证了呢? *(If not, then if I want to transit in Singapore, whether I have to obtain a visa of Singapore in advance?)*

Sentence 5: 请各位知道的驴友们回答一下, 谢谢! *(Looking forward to your answers, thank you!)*

Fig. 1. An example question in tourism CQA

2   Overview of Approach

Our method transfers the question thread into a directed graph (V, E) in the following steps shown in Figure 2.

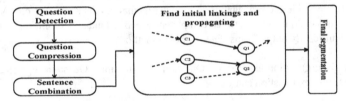

Fig. 2. A diagram view of our approach

In our method, the elementary unit (EU) to cope with is a sub-sentence segmented by a comma, instead of a full sentence denoted by a period.

A question detector is designed to find the question sentences in the question thread. The sentences are divided to two different sets: C (context) and Q (question and command). In addition, the meaningless sentences (greetings, polite sayings) that provide no useful information for answer detection are deleted. We use a rule and regex based question detector, which achieves an F score of 92% on Chinese text.

The question compression removes some duplicate information via similarity calculation. We conduct question compression separately on question sub-sentences and non-question sub-sentences. For each sub-sentence, we calculate its similarity with other sub-sentences and sentences to remove the duplication in the question thread. We consider the following factors in calculating the similarity:

- Extended longest common substring
- Number of same word
- Cosine Similarity

For each sub-sentence, we calculate its similarity with other sub-sentence and natural sentence to compress the question thread. If $Sim(S_1, S_2) > \theta$, then the shorter one will be deleted from the set of vertex. We set θ by experience, which is set to 1.3 in our method. Then we connect the remained continuous non-question sentences in a sentence together to make a longer context.

We then model the thread into a directed acyclic graph, and the initial weight is calculated using a SVM classifier in the building of the DAG. We add a propagation of the weight of edge to the graph. Finally, we split the long question thread into a few short sentences associated with the context of each question.

3   Finding Linkings Between Sentences

The question thread is modeled to a directed graph (V, E). We find the relationship between sentences by adding edges between vertexes. An edge $u \to v$ demonstrates that vertex v has dependence on u, or sentence u is the context of sentence v, meaning the information represents by v becomes more complete because of the existence of u.

In our model, both a question sentence, and a non-question sentence, can serve as a context of a question. Notably, we only allow edges of $c \to c, c \to Q$ and $Q \to Q$. The edge $Q \to C$ is considered as meaningless, because when people are talking, a question usually does not motivate a context. In addition, if a question Q_2 appears later than Q_1, only the dependence $Q_1 \to Q_2$ is allowed, as usually the earlier question provides information for the later one. Every edge in our model is considered as directed.

We calculate the weight of edges between each pair of possible combination of edges by exploiting various lexical and structure features:

- **KL-divergence**
Given two sentences u and v, we separately construct unigram language model of u as M_u and v as M_v. The KL-divergence between languages M_u and M_v is computed as follow:

$$D_{KL}(M_u \| M_v) = \sum_w p(w | M_u) \log \frac{p(w | M_u)}{p(w | M_v)} \tag{1}$$

KL-divergence shows the difference of the probability distributions of M_u and M_v under the same vector space. The KL-divergence is asymmetry.

- **Correlation**
Given two sentences u and v, the correlation between u and v is defined as the degree of their similarity. We use Word2Vec to calculate the similarity between words.

$$Co(u, v) = \sum_{w_i \in u, w_j \in v} \cos(w_i, w_j) \tag{2}$$

- **Coherence**
Usually, the existence of conjunctions and linking words indicate a relationship between sentences. Some Chinese conjunctions appear in pairs, whose order of appearance suggests the dependence between u and v. The conjunction pairs are found in a mandarin Chinese dictionary.

Other features used in out method are listed as follows:

- *if_same_sen: whether two sub-sentences are in a natural sentence*
- *if_pron: whether any of the two sentences contains a pronoun*
- *if_num: whether any of the two sentences contains a number word*
- *if_ns: whether any of the two sentences contains a location name*
- *if_time: whether any of the two sentences contains a time word*
- *if_v: whether any of the two sentences contains a verb*
- *short_length: the shorter length of the two sentences*
- *ratio: the length ratio of a shorter sentence to a longer sentence*
- *word_pair: whether two sub-sentence contains frequent word pair*

Using the above features, a SVM classifier is trained to calculate the initial weight of edges.

4   Propagating the Linking Scores

Our SVM classifier provides an initial weight of edges in the graph (V, E). The initial weight of edges presents a direct relationship between sub-sentences; however, indirect relationship exists between sentences. For example, if C_1 is the context of Q_1, and Q_1 provides information for Q_2, then C_1 is possibly a context of Q_2. In such situations, hidden dependences between $C_1 \rightarrow Q_1$, and $Q_1 \rightarrow Q_2$ should also be considered, while the weight of SVM classifier cannot directly detect the relationship. Correspondingly, given a $C_1 \rightarrow Q_1$, although the weight of the edge E ($C_1 \rightarrow Q_1$) is not significant enough for a relationship, but if a vertex Q_2 makes chain $C_1 \rightarrow Q_2 \rightarrow Q_1$ exist, then vertex Q_2 should strengthen the potential hidden relationship of $C_1 \rightarrow Q_1$.

In our propagating method (shown in Figure 3), the weight of $C_1 \rightarrow Q_1$ is possibly refreshed by the existence of another vertex V, where $C_1 \rightarrow V, V \rightarrow Q_1$ exist. Whether the refreshing of edge $C_1 \rightarrow Q_1$ is efficient is determined by the significant degree of V. Thus, the authority of vertex V is crucial in propagating. We adopt an algorithm similar with page-rank to calculate the authority for each vertex.

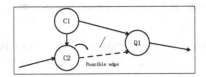

Fig. 3. Example of weight propagating

4.1   Calculating the Authority

The authority of vertex in the entire question graph is associated with the sematic information provided by the vertex, and the position information of the vertex in the graph. Our algorithm calculates the in-Degree, out-Degree as the position information, and a High-Frequency salient word count as a measure of the sematic information.

The in-Degree measure of a vertex presents the degree of independence of a vertex. A vertex that owns a high in-Degree refers to a processing of more context, and the owning of more contexts indicate that the vertex itself provides less information for others. Therefore, the in-Degree measure is a negative factor for the authority of the vertex. In-Degree is computed by:

$$in - Degree(v) = e^{-\lambda \times \frac{in(v)}{|V|}} \tag{3}$$

The out-Degree measure of a vertex shows the importance of the vertex. A vertex with high out-Degree indicates the vertex is considered as the context of many other vertexes. Out-Degree is computed by:

$$out - Degree(v) = \frac{out(v)}{|V|} \tag{4}$$

In a question, some salient words carry very important information for answer detection. We count the ratio of salient words in each sentence as $salient(v)$.
The authority of a vertex is calculated as follow:

$$Au(v) = \alpha In - Degree(v) + \beta Out - Degree(v) + \gamma Salient(v) \tag{5}$$

α, β, γ are set to 0.3, 0.3 and 0.4. We add a normalization method like PageRank algorithm. When computing the authority of v, we take all the valid generators of v into consideration:

$$Au(v) = \sum_{u \in G(v)} Au(u) \times e^{\mu \times weight(u,v)} \tag{6}$$

(6) shows that the generators of v's contribution to v differed by weight (u, v). As the weight is only the initial weight, we give a very small μ to reduce the influence of weight. We suppose the propagating chain to be brief and short, thus the PageRank–like algorithm stops after the first layer. We add final normalization as following:

$$Au(v) = |V| \times \frac{Au(v)}{\sum_{u \in V} Au(u)} \tag{7}$$

4.2   Propagating the Scores

Our propagating method refreshes the weight in the following algorithm:

$$w(u,v) \to \frac{w(u,v)}{2} + \lambda Au(a)\sqrt{w(u,a) * w(a,v)} \tag{8}$$

We only refresh $w(u,v)$ when it becomes larger. λ is a damping factor to reduce the propagating chain from growing longer. We repeat the calculation until no changes of weight occurs in the graph.

4.3   Getting Final Segmentation

We use a dynamic algorithm to determine whether the relationship exists between two sentences. The edges are sorted in descending order. We successively deal with each edge E_i, and the algorithm stops if the $weight_i$ is below 0.5, or

$$weight_{i-1} - weight_i > \frac{1.5(weight_1 - weight_i)}{i} \tag{9}$$

5   Experiments

Our data comes from XieCheng tourism forum, which is a famous traveling CQA forum in China. We randomly choose 1,200 sentences with more than five sub-sentences as our training data. An under-graduate student annually annotated the direct relation between contexts and questions. We train our Word2Vec on a 100M tourism corpus. We train a SVM classifier to compute an initial weight of the relations (Section 3), and then we add a further propagation to refresh the closeness scores (Section 4).

We evaluated the effectiveness of our methods by user tests. We proposed an evaluation metric via four different aspects:

- *Total redundancy (TR):, which shows whether the segmentation results have redundancy information between different segments.(1 point)*
- *Total completeness (TC), which shows whether the segmentation results present all the question information in the initial question thread.(1 point)*
- *Segment redundancy (SR), which shows the degree of the appearance of un-related information in each segment.(2 points)*
- *Segment completeness (SC), which shows whether each question segment is complete and explicit to find answers.(2 points)*

In our evaluation metric, redundancy and completeness are complementary. When a segmentation result has a high degree of completeness, usually it has a high tendency of containing more information that is redundant.

We set up two kinds of baselines. First, we use the natural sentence segmentation (NSS) as a baseline, which use the question detector to find the question sub-sentence, and then cut the question sentence by the full stop as the segmentation result. Second, we employ the algorithm of MQS [6] as our baseline. Our method in this paper is denoted as Compression and Propagation Segmentation (CPS).

We randomly choose 200 sentences outside the training data in the same CQA forum as the test data. Three systems (NSS, MQS, and CPS) are employed to run on the test data to get three different segmentation results. Finally, according to the above-mentioned metrics, two human annotators are required to give points to each segmentation result of three methods, without knowing which system generates the result. The evaluation results are shown in Table 1.

From Table 1, our system CPS obtains the best subjective performance by both annotators. Our method achieves an average total point of 4.896 (0.000~6.000), which considerably outperforms the other two baseline systems. The simple NNS method rivals the previous MQS approach in total score.

Comparing with NNS method, our algorithm notably raises the point of SC, with only a small decrease in the other three metrics. On the other hand, our algorithm CPS significantly outperforms the previous MQS approach both in TR and in SR, with only a little drop in CT and SC. The previous approach MQS has a high tendency to joint short sentences to a longer one, so the system is likely to cover more information and shows better completeness but resulting in much redundancy. Our method finds a better way to balance the completeness and redundancy, and the experimental results demonstrate its effectiveness.

Table 1. Evaluation results of Segmentation

Systems		TR	TC	SR	SC	Total
NNS	Tester 1	0.839	0.964	1.668	1.197	4.668
	Tester 2	0.865	0.99	1.616	1.161	4.601
MQS	Tester 1	0.275	0.984	0.980	1.766	4.005
	Tester 2	0.269	0.979	0.102	1.792	4.062
CPS	Tester 1	0.777	0.881	1.554	1.684	**4.896**
	Tester 2	0.746	0.891	1.518	1.632	**4.777**

6   Related Work

Sentence segmentation is a basic method to alleviate the complexity of multi-sentence questions. Many previous researches have studied on the direction of long sentence segmentation, and the methods can be divided into two genres: chunking-based sentence segmentation and segmentation based on finding relations. The chunking may use rule and regex method [1], decision tree [2], a maximum entropy model, and conditional random field (CRF) [3] have been proposed to deal with segmentation. Takechi [3] proposed a method of combining unigram and bigram word features to reach better segmentation result. The question segmentation has a strong dependence on question detection, and the question detection can be vector space model [3], language model [4], translation model [5], syntactic tree matching model [6] and the recently proposed convolutional neural network model [7]. However, in CQAs, most sentences are long and complex, and the context of a question is often not adjacent to the question itself, which make chunking-based methods less effective.

Among segmentation based on finding relations, Wang et al. (2010) [8] address the problem of multi-sentence question segmentation towards further analysis of question sentences. They build a question detector to extract question sentences and context sentences, and propose a simple graph based approach to segment multi-sentence questions with simple propagating method.

Wang's technique and ours differ in the following points: our method provides a robust algorithm of finding semantic relations between sentence by expanding word2vector, and use synonyms to deal with lexical gap; we exploit an effective propagation method with an authority-calculating algorithm, which reduce the redundancy of propagating significantly. Many studies worked on calculating the authority of vertex in an acyclic graph to find crucial vertexes [9, 10, 11, 12, 13]. In addition, we add a sentence compression to decrease the duplication in complex CQA sentences.

7   Conclusion

In this paper, we propose a robust method to split multi-sentence questions. The sentence compression method effectively reduces the duplication in the question sentence. Our method exploits various lexical and structure features, applies Word2Vec on finding related words, and use synonyms to deal with lexical gap. We also propose an effective propagating method to refresh the relation by a graph algorithm. Our method balances the information completeness and redundancy of multi-sentence segmentation and our result outperform the state-of-art. Our further study will focus on the application of question segmentation on multi-sentences question retrieval, and employ the sentence segmentation on complex non-question sentences.

Acknowledgement. This work is supported by Humanity and Social Science foundation of Ministry of Education (13YJA740060), National High Technology Research and Development Program of China (2015AA015403), Key Program of Social Science foundation of China (12&ZD227), and the Opening Project of Beijing Key Laboratory of Internet Culture and Digital Dissemination Research (ICDD201302).

References

1. Ang, J., Liu, Y., Shriberg, E.: Automatic dialog act segmentation and classification in multiparty meetings. In: Proc. Int. Conf. Acoust Speech, Signal Process (2005)
2. Takechi, M., Tokunaga, T., Matsumoto, Y.: Chunking-based question type identification for multi-sentence queries. In: SIGIR (2007)
3. Duan, H., Cao, Y., Lin, C.-Y., Yu, Y.: Searching questions by identifying question topic and question focus. In: HLT-ACL (2008)
4. Jeon, J., Croft, W.B., Lee, J.H.: Finding similar questions in large question and answer archives. In: CIKM (2005)
5. Riezler, S., Vasserman, A., Tsochantaridis, I., Mittal, V., Liu, Y.: Statistical machine translation for query expansion in answer retrieval. In: ACL (2007)
6. Wang, K., Ming, Z., Chua, T.-S.: A syntactic tree matching approach to finding similar questions in community-based qa services. In: SIGIR (2009)
7. Yinh, W., He, X., Meek, C.: Semantic parsing for single-relation question answering. In: ACL (2014)
8. Wang, K., Ming, Zh., Hu, X., Chua, T.: Segmentation of multi-sentence questions: towards effective question retrieval in cQA services. In: SIGIR (2010)
9. Cong, G., Wang, L., Lin, C., Song, Y., Sun, Y.: Finding question-answer pairs from online forums. In: SIGIR (2008)
10. Mihalcea, R.: Unsupervised large-vocabulary word sense disambiguation with graph-based algorithms for sequence data labeling. In: EMNLP (2005)

11. Navigli, R., Lapata, M.: Graph connectivity measures for unsupervised word sense disambiguation. In: IJCAI (2007)
12. Sinha, R., Mihalcea, R.: Unsupervised graph-based word sense disambiguation using measures of word semantic similarity. In: ICSC (2007)
13. Hessami, E., Mahmoud, F., Jadidinejad, A.: Unsupervised graph-based word sense disambiguation using lexical relation of WordNet. In: IJCSI (2011)

A User-Oriented Special Topic Generation System for Digital Newspaper

Xi Xu[1,2,3(✉)], Mao Ye[2], Zhi Tang[1,2], Jian-Bo Xu[2], and Liang-Cai Gao[1]

[1] Institute of Computer Science and Technology, Peking University, Beijing, China
{xuxi}@founder.com.cn, 10548887@pku.edu.cn
[2] State Key Laboratory of Digital Publishing Technology,
Peking University Founder Group Co. LTD., Beijing, China
{yemao.apb,tommie,xu.jb}@founder.com.cn
[3] Postdoctoral Workstation of the Zhongguancun Haidian Science Park, Beijing, China

Abstract. With the coming of digital newspaper, user-oriented special topic generation becomes extremely urgent to satisfy the users' requirements both functionally and emotionally. We propose an applicable automatic special topic generation system for digital newspapers based on users' interests. Firstly, extract subject heading vector of the topic of interest by filtering out function words, localizing Latent Dirichlet Allocation (LDA) and training the LDA model. Secondly, remove semantically repetitive vector component by constructing a synonymy word map. Lastly, organize and refine the special topic according to the similarity between the candidate news and the topic, and the density of topic-related terms. The experimental results show that the system has both simple operation and high accuracy, and it is stable enough to be applied for user-oriented special topic generation in practical applications.

Keywords: Digital newspaper · User-oriented special topic · Latent Dirichlet allocation · Synonymy word map

1 Introduction

In the day of information and network, digital reading is increasingly important and people have unprecedentedly strong demand for news. The demand mainly embodies the quick picking up of the latest, the deepest and the most comprehensive news. The future of the digital newspaper has arrived. News can be custom-made to suit your own requirements. All these will make news expression more interactive than ever. Thus, it is necessary to innovate and develop the reorganization of digital news.

Many researchers have made attempts to organize and generate special topic, while there are few achievements of user-oriented automatic special topic generation. In Ref. 1, a set of topic-related news is extracted by textual retrieval using query expansion algorithm. Ref. 2 gets the candidate set of subject headings by integrating the result of meaningful string recognition algorithm into the subject heading vocabulary compiled in advance, and then filter the candidate set and calculate the weight of each subject heading in accordance with certain heuristic rules. Incremental clustering

© Springer International Publishing Switzerland 2015
J. Li et al. (Eds.): NLPCC 2015, LNAI 9362, pp. 484–491, 2015.
DOI: 10.1007/978-3-319-25207-0_45

algorithm and named entity are introduced to detect and represent the topics in Ref. 3, and Support Vector Machine (SVM) [4] is also adopted to organize and track the topics. Ref. 5 represents each piece of news as a sequence of words by Vector Space Model (VSM) [6] and trains self-defined topic model on documents around the same topic to estimate whether a piece of news belongs to the topic.

All above-mentioned methods not only can not generate fine-grained special topic automatically according to users' interests but also have two major defects. One is that a subject heading vocabulary needs to be compiled manually in advance. The other is that the impacts of polysemant and synonyms are not taken into account and reduced. In application, as newspapers contain a large amount of information and update very fast, there is hardly any subject heading vocabulary and it is a fussy and high subjective task to add subject headings manually. The number of polysements is small, but the probability of their occurrence is very high. Synonyms are sometimes the relationship between meanings because of polysemants. Polysemants and synonyms increase the difficulty of semantic similarity computation of texts.

To overcome above weaknesses and improve interactivity, we designed and implemented a user-oriented special topic generation system for digital newspaper. We first filter out function words based on part of speech tagging, localize Latent Dirichlet Allocation (LDA) [7] and train the LDA model to extract subject headings of the topic which a user is interested in. The function word filter and localized LDA model prevents us from compiling subject heading vocabulary manually or recognizing named entity. Then, we construct a synonymy word map according to Chinese thesaurus [8] to filter synonym words out of the subject headings. The removing of semantically repetitive subject headings effectively reduces the impacts of polysemant and synonyms. At last, we develop an algorithm to compute the similarity between a text and a topic and organize the special topic according to priority transformed from similarity. The similarity calculation algorithm can globally expresses similarity under localized LDA and subject heading vector.

2   Subject Heading Extraction

To avoid compiling a subject heading vocabulary beforehand, key words of the samples are extracted based on LDA as subject headings. In practice, the samples are selected by the user and are on the same subject that the user is interested in.

2.1   Text Preprocessing

Although there are many elements of news, such as the "press", "title" and "content", only the "content" is used to extract subject headings in our research. First, the "content" has to be segment by word segmenter according to Chinese grammatical and lexical analysis. Considering that not all parts of speech are qualified for subject headings, content word is much better than function word, the word segmenter must tag part of speech. Thus, the processing of "content" includes Chinese word segmentation and part of speech tagging.

There are many Chinese word segmenters, such as IKAnalyzer, Ansj and so on. Ansj is selected because of its high accuracy and function of part of speech tagging.

It is a Java implementation of the ictclas [9]. As the candidate terms of subject headings should keep in full meaning as far as possible and new terms always appear in news, coarse-grained method in Ansj, NlpAnalysis, is used in our research. User-defined dictionary is also added to improve the performance of new term discovery and the reliability of word segmentation.

Not all terms obtained from word segmenting are qualified for subject headings. Firstly, filter the function words from the terms within parts of speech. Specifically, stop preposition, conjunction, auxiliary, punctuation and so on. Secondly, filter the insignificant words from the rest of the terms within a stop-word list. Stop-words are words which are filtered out before or after processing of natural language text [10]. In our research, the stop-word is compiled based on the feature of the style of writing in news, such as "本报记者" (our reporter) and importing Chinese stop-word list. It covers a wide number of stop-words without getting too aggressive and including too many words which might point to the subject of news.

Eventually, the "content" of each piece of news is condensed into a sequence of content words and the candidate word set of subject headings are obtained.

2.2   Key Word Extracting Based on LDA

Topic model is introduced to our research to semantically extract the subject information according to the demand of special topic generation for digital newspaper. In natural language processing, a topic model is a type of statistical model for discovering the abstract "topics" that occur in a collection of documents [11].

One of the most classic topic models is LDA. It was first presented as a probabilistic graphical model for topic discovery by David Blei et al in 2003. The basic idea and more details of LDA are described in Ref.12.

In digital newspapers, neologisms and new topics are constantly emerging. Thus, it is quite difficult to provide adequate corpus for LDA estimator. Furthermore, the topics attracting users are always fine-grained and the performance of LDA inferencer is not in sufficient reliability to satisfy demand. All these led us to apply LDA in an extremely novel way that is reducing the dimensionality of topic and using just LDA estimator. The basic idea of LDA can be simply represent by the following formula:

$$p(word \mid doc) = \sum_{topic} p(word \mid topic) \times p(topic \mid doc) \ . \tag{1}$$

If set the number of topic the minimum, one, all documents belong to the same topic. Here one piece of news, which is composed of the sequence of content words obtained in Section 2.1, is assumed as a document and all documents around one topic are chosen by the user in real time. This method can be extended to two or more topics, but the size of corpus and the number of topics should keep small. The Dirichlet and conditional multinomial parameters for the LDA model that is α and β are iteratively computed with initial value using Expectation Maximization (EM) until convergence. Then the probability of word w_n under topic z_n can be calculated by multinomial distributions $p(w_n \mid z_n)$ with known α and β.

2.3   Remove Semantically Repetitive Subject Headings

As mentioned in Section 1, polysemant and synonym are common phenomenon, and therefore two different subject headings obtained above may have the same meaning especially when they are in the same text. Furthermore, the significance of the meaning, which more than one subject heading have, should be enhanced in the topic.

We construct a synonymy word map according to Chinese thesaurus to filter synonymy words out of the subject headings. The thesaurus was firstly compiled by Mei J.-J. et al. in 1983 [13] and was expanded by Information Retrieval Laboratory of Harbin Institute of Technology. Pertinent experiment has shown that although the expanded Chinese thesaurus is already covering a large number of Chinese words, there are still many words, especially proper nouns such as organization name and so on, appeared in the digital newspaper but not in the thesaurus. Thus, we once again supplemented the Chinese thesaurus based on neologism and abbreviations (e.g. "马航" and "马来西亚航空公司" (Malaysia Airlines)) in popular usage in newspaper.

In the synonymy word map, each meaning is identified by a character string composed of numbers and letters, named the meaning code. Thus, every monosemy is mapped to one character string and every polysemant is mapped to two or more meaning codes. We iteratively remove semantically repetitive subject headings using the synonymy word map. When two different subject headings have the same meaning code, the one with lower probability is removed and its probability is added to the other one with higher probability. The experiments show that the removing can reduce the subject headings by 10 to 15 percent. 20 percent of the highest probability words (terms in Section 2.1) of a topic are retained as subject headings considering the features of natural language and writing style in news. After that the probability of each subject heading is normalized within the topic. Until now, the condensed subject headings altogether construct a subject heading vector denoted by *topicwords* = (*tterm*$_1$, *tterm*$_2$, $\cdots$, *tterm*$_M$), where *tterm*$_m$ is the mth dimension of the vector. Moreover, the probability of *tterm*$_m$, denoted by p_m, is also obtained and

$$\sum_{i=1}^{M} p_i = 1 \quad . \tag{2}$$

As the probability of the subject heading, expressed by different words with the same meaning, is increased, the subject heading vector represents the topic more accurately.

3   Special Topic Generation

To meet the demand of user-oriented special topic generation, we use LDA estimator in a small corpus to extract subject headings and their corresponding probability. Obviously, the LDA model here is just local model because the global one is not suitable for our application, so LDA inferencer is not applicable any more. A new method is proposed to compute similarity between a new "content" and the topic to infer that whether a piece of news belongs to a topic. The similarity is actually between the key words of the new "content" and the subject headings of the topic.

3.1   Compute Similarity Between News and a Topic

In the practical application, there is a huge amount of news and the news is updated at very high speed. Moreover, many neologisms always appear in news. Therefore, it is quite difficult to train word vector under the corpus of news and we have to calculate the similarity in another way. For a piece of news to be observed, firstly, preprocess its "content" using the method described in Section 2.1 to obtain the sequence of content words, denoted by $seqterms = (term_1, term_2, \cdots, term_L)$, where $term_l$ is the lth word in the sequence. And compute the probabilities of the words in $seqterms$ using the method "computeArticleTfidf", which is based on TF-IDF (term frequency-inverse document frequency) and supplied by Ansj. Secondly, remove semantically repetitive words by the method proposed in Section 2.3 and update the corresponding probabilities. Thirdly, sort the words in descending order by probability and the result is denoted by $keywords = (kterm_1, kterm_2, \cdots, kterm_N)$, where $kterm_n$ is the nth word in $keywords$. Finally, set the weight of $kterm_n$, denoted by w_n,

$$w_n = 1 - \frac{n-1}{2N} ,$$

(3)

where N is the total number of terms in $keywords$ and n is the sequence number of $kterm_n$. Obviously, $1 \le n \le N$ and $0.5 < w_n \le 1$.

Now, we can compute the similarity between a piece of news to be observed and a certain topic. As semantically repetitive terms have already been removed in the $topicwords$ using the synonymy word map (detailed in Section 2.3), the algorithm starts iteration by $topicwords$ to calculate the similarity:

- Step 1: Choose a term in $topicwords$, denoted by $tterm_i$. Search $keywords$ in descending order by probability using the synonymy word map until a term with the same meaning as $tterm_i$ is found or all terms are compared.
- Step 2: Mark $tterm_i$ processed. If no term is found, repeat step 1 until all terms in $topicwords$ are processed. Else, denote the term found in Step 1 by $kterm_j$ and compute the $tterm_i$ component of similarity S_i, using the following equation:

$$S_i = p_i w_j ,$$

(4)

where p_i is the probability of $tterm_i$ and w_j is the weight of $kterm_j$.
- Step 3: Remove $kterm_j$ from $keywords$. If $keywords$ is not empty, repeat step 1 until all terms in $topicwords$ are processed.
- Step 4: Define the similarity between a piece of news and a certain topic as

$$S = \sum_{i=1}^{M} S_i ,$$

(5)

where M is the total number of terms in $topicwords$., We can deduce $0 \le S < 1$ according to Eqs. (2), (3) and (4).

The similarity calculation algorithm proposed in our research is aiming at a global similarity expression under localized LDA and subject heading vector.

3.2   Organize and Refine the Special Topic

To generate a special topic, it is essential to choose a candidate set from the database of news with restrictions such as the "date" and so on. As this operation is not the main point, here we just assume that the candidate set has already been offered.

For each piece of news in the candidate set, the similarity S between its "content" and the certain topic is calculated in by Eq. (5). A similarity threshold, denoted by θ, is introduced as a parameter between 0 and 1. Its value is determined by user and the default setting is 0.3. As only the piece of news which satisfied the inequality $S > \theta$ is collected to construct a rough special topic, the value of θ can adjust the grain size of the special topic. Finally, we obtain a rough special topic based on S and θ.

To increase the accuracy of the special topic, we filter each piece of news in the rough special topic by the density of topic-related terms. Firstly, construct the set of topic-related terms from all $kterm_j$, iteratively found in Step 2 of similarity calculation algorithm, and their synonyms in $keywords$ (obtained in Section 3.1). Secondly, calculate the maximum distance d_{max} between any two topic-related terms in $seqterms$ (obtained in Section 3.1). Lastly, if d_{max} is smaller than half of the total number of the terms in $seqterms$, remove this piece of news from the rough special topic.

After the filtering process, all pieces of news remained in the rough special topic are organized based on S to construct the final special topic. For any piece of news in the rough special topic, set its priority level $prior$ the integer (100-100S). The smaller the $prior$ is, the higher the priority level becomes. Then, sort all news remained in ascending order by $prior$. For all news with the same $prior$, first, choose a piece of news, denoted by $refnew$, and extract its subject heading vector using the same method proposed in Section 2; second, calculate the similarity S_{local} between $refnew$ and any other piece of news with the same $prior$; last, set $refnew$ is the standard news of this $prior$ and group together $refnew$ and all news satisfied $S_{local} > 0.8$. Repeat above steps until all pieces of news with the same $prior$ are grouped.

Eventually, we generate the fine-grained special topic. All pieces of news are organized by the priority level and repetitive pieces under the same priority level are grouped together according to the similarity. As $prior$ becomes bigger that is the priority level becomes lower, the grain size of the special topic gradually increases.

4   Experiments

There is no public database created for testing the performance of special topic generation methods, so it is difficult to do quantitative comparison experiments with other special topic generation methods and we evaluated the proposed system using real digital newspapers data. To verify our system (simply as UOSTG) has the better performance than muti-keyword retrieval (simply as MKR) when it is very difficult to accurately describe the subject by limited and certain keywords, we generated a special topic according to one piece of news focused on how foreign media report "2014 Kunming attack". Two thousand pieces of news was randomly chosen from twenty-one thousand pieces reported by People's Daily, Guangming Daily, Yunnan Daily and Global Times during two months, from Mar. 1st, 2014 to Apr. 30th, 2014. The statistical result of the experiment is shown in Table 1. In muti-keyword retrieval, key terms are "昆明" (Kunming), "恐怖" (terror),"国外" (foreign) and "媒体" (press).

Table 1. Recall and accuracy of muti-keyword retrieval and user-oriented special topic generation.

Method	Recall (%)	Accuracy (%)
MKR	8.70	50.00
UOSTG	82.61	90.48

Table 1 shows that the user-oriented special topic generation has brought about huge improvement both in recall and accuracy. The recall of MKR is particularly lower because multi- keyword retrieval cannot accurately describe the topic.

To validate the advantage of removing semantically repetitive subject headings using synonymy word map, we generated another special topic focused on sea search for Malaysia Airlines incident. Three thousand pieces of news was randomly chosen from thirty thousand pieces of news reported by People's Daily, Guangming Daily, Beijing Youth Daily and Beijing Morning Post, different presses from the above experiment to make result analysis easy, but during the same period. The subject heading vector without the synonymy word map (or simply as NoSWM) was {海域 (sea area) = 0.0432, 飞机 (airplane) = 0.0305, 客机 (airliner) = 0.0029, 马来西亚 (Malaysia) = 0.0208, 救援 (rescue) = 0.0203, ⋯}, while with the synonymy word map (also simply as UOSTG), {海域 (sea area) = 0.0468, 飞机 (airplane) = 0.0336, 舰 (warship) = 0.0318, 救援 (rescue) = 0.0289, 搜寻 (search) = 0.0275, ⋯}. Table 2 demonstrates the experimental result.

Table 2. Recall and accuracy with and without the synonymy word map.

Method	Recall (%)	Accuracy (%)
NoSWM	84.87	87.16
UOSTG	90.79	89.61

As shown in table 2, both recall and accuracy have risen because the subject heading vector refined by the synonymy word map can represent the topic more accurately. Most of the news, which appeared in the special topic but were regarded as extraneous pieces, related to Malaysia Airlines incident, but not to sea search.

Table 3 shows the experiment result of another five topics using user-oriented special topic generation system to further prove the universality of the system.

Table 3. Recall and accuracy of another five different topics.

Topic ID	Recall (%)	Accuracy (%)
1	85.96	92.45
2	93.90	79.38
3	84.73	88.80
4	76.47	92.85
5	95.52	87.67

5   Conclusion

To satisfy the users' requirements both functionally and emotionally, a user-oriented special topic generation system is designed, implemented and evaluated in our research. It has the following advantages. The function word filter based on part of speech tagging and localized LDA model prevents us from compiling subject heading vocabulary manually or recognizing named entity. The removing of semantically repetitive subject headings using the synonymy word map effectively reduces the impacts of polysemant and synonyms. Furthermore, the similarity calculation algorithm proposed in our research can globally expresses similarity under localized LDA and subject heading vector.

The experimental results show that the subject heading vector extracted by our methods can more accurately describes a topic than multi-keywords and the system has not only high recall but also high accuracy. The system is simple to use and stable enough to be applied for user-oriented special topic generation for digital newspaper in practical applications.

References

1. Fan, J.-R.: Research on Topic Generation and Retrieval of News Video Based on Text. Institute of Computing Technology, Chinese Academy of Science, Beijing (2008)
2. Li, H.-X., Zhang, H.-P.: Internet hot topic detection based on topic words. In: Proceedings of the 5th China Information Retrieval Conference, Shanghai (2009)
3. Wang, Z.-M.: Research on Web News Topic Organization and Acquisition System. College of Information Science & Engineering, Central South University (2008)
4. Cui, J.-M., Liu, J.-M., Liao, Z.-Y.: A Research of Text Categorization Based on Support Vector Machine. Computer Simulation 30(2), 294–299 (2013)
5. Tan, H., Jia, Z.-Y., Shi, Z.-Z.: How to Organize and Generate News Topics with Great Efficiency. Science & Technology Review 7, 48–51 (2004)
6. Erk, K., Padó, S.: A structured vector space model for word meaning in context. In: Proceedings of the Conference on Empirical Methods in Natural Language Processing. Association for Computational Linguistics (2008)
7. Biggers, L.R., Bocovich, C., Capshaw, R., Eddy, B.P., Etzkorn, L.H., Kraft, N.A.: Configuring Latent Dirichlet Allocation Based Feature Location. Empirical Software Engineering 19(3), 465–500 (2014)
8. He, D.: Retrospect of and Prospect for Chinese Thesaurus. Information Studies Theory & Application (2010)
9. Feng, G.-H., Zhen, Z.: Review of Chinese Automatic Word Segmentation. Library and Information Service 55(2), 41–45 (2011)
10. Rajaraman, A., Ullman, J.D.: Mining of Massive Datasets. Cambridge University Pr., pp. 1–17 (2011)
11. David, M.B.: Probabilistic Topic Models. Communications of the ACM 55(4), 77–84 (2012)
12. David, M.B., Andrew, Y.N., Michael, I.J.: Latent Dirichlet Allocation. Journal of Machine Learning Research 3, 993–1022 (2003)
13. Mei, J.-J., Zhu, Y.-M., Gao, Y.-Q.: Cilin-thesaurus of Chinese words. Shanghai Lexicographic Publishing House, Shanghai (1983)

Shared Task (Long Papers)

Exploiting Heterogeneous Annotations for Weibo Word Segmentation and POS Tagging

Jiayuan Chao, Zhenghua Li$^{(\boxtimes)}$, Wenliang Chen, and Min Zhang

School of Computer Science and Technology, Soochow University,
Suzhou 215006, Jiangsu, China
chaojiayuan.china@gmail.com, {zhli13,wlchen,minzhang}@suda.edu.cn

Abstract. This paper describes our system designed for the NLPCC 2015 shared task on Chinese word segmentation (WS) and POS tagging for Weibo Text. We treat WS and POS tagging as two separate tasks and use a cascaded approach. Our major focus is how to effectively exploit multiple heterogeneous data to boost performance of statistical models. This work considers three sets of heterogeneous data, i.e., Weibo (*WB*, 10K sentences), Penn Chinese Treebank 7.0 (*CTB7*, 50K), and People's Daily (*PD*, 280K). For WS, we adopt the recently proposed coupled sequence labeling to combine *WB*, *CTB7*, and *PD*, boosting F1 score from 93.76% (baseline model trained on only *WB*) to 95.58% (+1.82%). For POS tagging, we adopt an ensemble approach combining coupled sequence labeling and the guide-feature based method, since the three datasets have three different annotation standards. First, we convert *PD* into the annotation style of *CTB7* based on coupled sequence labeling, denoted by PD^{CTB}. Then, we merge $CTB7$ and PD^{CTB} to train a POS tagger, denoted by $Tag_{CTB7+PD^{CTB}}$, which is further used to produce guide features on *WB*. Finally, the tagging F1 score is improved from 87.93% to 88.99% (+1.06%).

1 Introduction

Chinese word segmentation (WS) and part-of-speech (POS) tagging are two fundamental tasks in Chinese language processing. In past decades, supervised approaches have gained extensive progress on canonical texts, especially on texts from domains or genres similar to existing manually labeled data[1]. However, the upsurge of web data imposes great challenges on existing techniques. The performance of the state-of-the-art systems degrades dramatically on informal web texts, such as micro-blogs, product comments, and so on. Driven by this challenge, NLPCC 2015 organizes a shared task with an aim of promoting WS and POS tagging on Weibo (*WB*, Chinese pinyin of micro-blogs) text [10].

This work was supported by National Natural Science Foundation of China (Grant No.61432013, 61273319) and Jiangsu Planned Projects for Post-doctoral Research Funds (No.1401075B).

[1] Please refer to http://zhangkaixu.github.io/bibpage/cws.html for a long list of related papers.

© Springer International Publishing Switzerland 2015
J. Li et al. (Eds.): NLPCC 2015, LNAI 9362, pp. 495–506, 2015.
DOI: 10.1007/978-3-319-25207-0_46

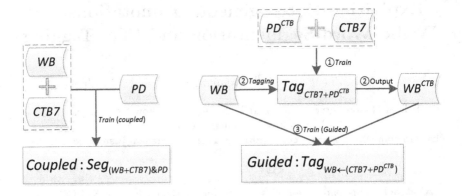

Fig. 1. Our final system for **Fig. 2.** Our final system for the POS tagging subtask the WS subtask

This paper describes our system designed for the shared task in detail. We treat WS and POS tagging as two separate tasks and use a cascaded approach. We treat both WS and POS tagging as sequence labeling problems, and build our model based on the standard conditional random field (CRF) [5] with bigram features. Our major focus is how to effectively exploit multiple heterogeneous data to boost performance of WS and POS tagging on WB ($10K$ sentences). The third-party heterogeneous resources used in the work are Penn Chinese Treebank 7.0 ($CTB7$, $50K$) and People's Daily (PD, $280K$).

Figure 1 illustrates our final system for the WS subtask. We adopt the recently proposed coupled sequence labeling to combine WB, $CTB7$, and PD [6]. Strictly speaking, the three datasets have three different annotation standards. However, we empirically find that the annotation standard of WB and $CTB7$ are very close while PD is different from both of them. Therefore, in our final system, we assume that WB and $CTB7$ have the same annotation standard while PD follows a different annotation standard. Then, we train a coupled sequence labeling model on the three datasets with two sets of WS labels. The baseline model trained on the single WB achieve an F1 score of 93.76%, whereas our best model has an F1 score of 95.58%, which is an absolute improvement of 1.82%. Our WS system ranks the third place among all 17 systems (5 participating systems in the open track, 4 semi-open, 8 closed).

Figure 2 illustrates our final system for the POS tagging subtask. The three datasets have three different annotation standards. Meanwhile, the coupled sequence labeling model has an inefficient problem due to the large bundled POS tag set. Therefore, due to time limitation for the shared task, we adopt an ensemble approach combining coupled sequence labeling and the guide-feature based method [1]. First, we apply the coupled model described in Li et al. (2015) [6] to convert PD into the annotation style of $CTB7$, denoted by PD^{CTB}. Then, we merge $CTB7$ and PD^{CTB} to train a POS tagger, denoted by $Tag_{CTB7+PD^{CTB}}$, which is further used to produce guide features on WB. Finally, we train a POS

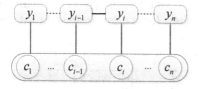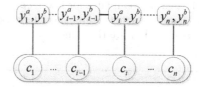

Fig. 3. Graphical structure of **Fig. 4.** Graphical structure of the coupled CRF the traditional CRF

tagger on the WB with guide features. The tagging F1 score is improved from 87.93% (baseline model trained on only *WB*) to 88.99% (+1.06%). Our POS tagging system ranks the second place among all 12 participating systems (3 systems in the open track, 3 semi-open, 6 closed).

This paper is organized as follows. Since Li et al. (2015) [6] describe coupled sequence labeling for POS tagging, we introduce the traditional CRF model and the coupled CRF model for WS in Section 2 and Section 3 respectively. Section 4 presents experiments. We discuss closely related works in Section 5 and conclude this paper in Section 6.

2   Traditional CRF for WS

We adopt CRF for the WS subtask, which treats WS as a sequence labeling problem. We adopt the $\{B, M, E, S\}$ tag set, indicating the beginning of a word, the middle of a word, the end of a word and a single-character word [16].

Figure 3 shows the graphical structure of the coupled model. Given an input sentence, which is a sequence of n characters, denoted by $\mathbf{x} = c_1...c_n$, WS aims to determine the best tag sequence $\mathbf{y} = y_1...y_n$, where $y_i \in \{B, M, E, S\}$ is a segmentation label for c_i. As a log-linear model, CRF defines the probability of a tag sequence as:

$$P(\mathbf{y}|\mathbf{x}; \theta) = \frac{exp(Score(\mathbf{x}, \mathbf{y}; \theta))}{\sum_{\mathbf{y}'} exp(Score(\mathbf{x}, \mathbf{y}'; \theta))}$$

$$Score(\mathbf{x}, \mathbf{y}; \theta)) = \sum_{1 \le i \le n+1} \theta \cdot \mathbf{f}(\mathbf{x}, i, y_{i-1}, y_i)$$

$$(1)$$

where $Score(\mathbf{x}, \mathbf{y}; \theta)$ is a scoring function; $\mathbf{f}(\mathbf{x}, i, y_{i-1}, y_i)$ is the feature vector at the ith character and θ is the feature weight vector. Please note that c_0 and c_{n+1} are two pseudo characters marking the beginning and end of the sentence. We use the features described in zhang et al. (2014) [21], as shown in Table 1.

Suppose the training data is $\mathcal{D} = \{(\mathbf{x}_i, \mathbf{y}_i)\}_{i=1}^{N}$, where $\mathbf{y}_i$ is the provided gold-standard tag sequence for $\mathbf{x}_i$. Then the log likelihood of $\mathcal{D}$ is:

$$\mathcal{L}(\mathcal{D}; \theta) = \sum_{i=1}^{N} \log P(\mathbf{y}_i|\mathbf{x}_i; \theta)$$

$$(2)$$

Table 1. Feature templates for $\mathbf{f}(\mathbf{x}, i, y_{i-1}, y_i)$. $T(c_i)$ returns the type of the character c_i (time, number, punctuation, special symbols, else). $I(c_i, c_j)$ judges whether the two characters c_i and c_j are the same.

Unigram		Bigram
01: $y_i \circ c_k$	$i-2 \leq k \leq i+2$	09: $y_{i-1} \circ y_i$
02: $y_i \circ c_{k-1} \circ c_k$	$i-1 \leq k \leq i+2$	10: $y_{i-1} \circ y_i \circ c_i$
03: $y_i \circ c_{k-1} \circ c_k \circ c_{k+1}$	$i-1 \leq k \leq i+1$	11: $y_{i-1} \circ y_i \circ c_{i-1} \circ c_i$
04: $y_i \circ T(c_k)$	$i-1 \leq k \leq i+1$	
05: $y_i \circ T(c_{k-1}) \circ T(c_k)$	$i \leq k \leq i+1$	
06: $y_i \circ T(c_{i-1}) \circ T(c_i) \circ T(c_{i+1})$		
07: $y_i \circ I(c_i, c_k)$	$i-2 \leq k \leq i+2, k \neq i$	
08: $y_i \circ I(c_{i-1}, c_{i+1})$		

The training objective is to find an optimal θ to maximize the log likelihood, which can be resolved using the standard gradient descend algorithms. We omit the details for space limitation.

3 Coupled CRF for WS

In this section, we introduce the coupled model proposed in Li et al. (2015) [6], which can learn and predict two heterogeneous annotations simultaneously. Figure 4 shows the graphical structure of the coupled CRF. We can see that the only difference from the traditional CRF is that in the coupled model, two tags $[y_i^a, y_i^b]$ are simultaneously assigned for one character, which we name as a pair of *bundled tags*. Then, the score of a bundled tag sequence $[\mathbf{y}^a, \mathbf{y}^b]$ is defined as:

$$Score(\mathbf{x}, [\mathbf{y}^a, \mathbf{y}^b]; \theta) = \sum_{1 \leq i \leq n+1} \theta \cdot \begin{bmatrix} \mathbf{f}(\mathbf{x}, i, [y_{i-1}^a, y_{i-1}^b], [y_i^a, y_i^b]) \\ \mathbf{f}(\mathbf{x}, i, y_{i-1}^a, y_i^a) \\ \mathbf{f}(\mathbf{x}, i, y_{i-1}^b, y_i^b) \end{bmatrix} \qquad (3)$$

where the first item of the extended feature vector is called *joint features*, which can be obtained by replacing y_i with bundled tags $[y_i^a, y_i^b]$ in Table 1; the second and third items are called *separate features*, which are based on single-side tags. The coupled model provides us the flexibility of using both kinds of features.

3.1 Creating Bundled Tags

The key challenge for the idea of coupled sequence labeling is that the heterogeneous datasets are non-overlapping and each contains only one-side tags. Therefore, the problem is how to construct training data for the coupled model. Li et al. (2015) [6] study how to construct bundled POS tags and show that best performance can be achieved using a complete mapping between two tag sets. In this work, we use two tag sets $\{B, M, E, S\}$ and $\{b, m, e, s\}$ to distinguish the labels in two heterogeneous data. Then, we map a one-side tag into a

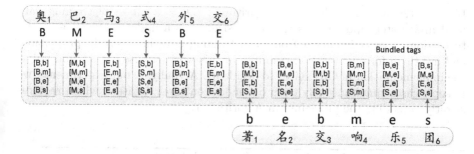

Fig. 5. An example to illustrate how to obtain bundled tags from one-side tags of *CTB7* (above) and *PD* (below).

set of bundled tags by considering all tags at the missing side, as illustrated in Figure 5. In the resulting training data, each sentence has multiple possible tag sequences as its gold-standard reference, which is known as *ambiguous labelings*. In the coupled model, the tag space contains $4 \times 4 = 16$ bundled tags.

3.2 Training Objective with Ambiguous Labelings

After our transformation, each character in a sentence has four bundled tags as gold-standard references, known as ambiguous labels, as shown in Figure 5. Given a sentence $\mathbf{x}$, we denote the set of ambiguous tag sequences as $\mathcal{V}$. The probability of $\mathcal{V}$ is the sum of probabilities of all tag sequences contained in $\mathcal{V}$.

$$P(\mathcal{V}|\mathbf{x}; \theta) = \sum_{\mathbf{y} \in \mathcal{V}} P(\mathbf{y}|\mathbf{x}; \theta) \qquad (4)$$

Suppose the training data is $\mathcal{D} = \{(\mathbf{x_i}, \mathcal{V}_i)\}_{i=1}^{N}$. Then the log likelihood of is:

$$\mathcal{L}(\mathcal{D}; \theta) = \sum_{i=1}^{N} \log P(\mathcal{V}_i|\mathbf{x}_i; \theta) \qquad (5)$$

Following Li et al. (2015) [6], we can derive the gradient of the likelihood function and apply standard gradient descent algorithms to learn θ. We omit the details for space limitation.

3.3 Stochastic Gradient Descent (SGD) Training with Two Datasets

In this work, we adopt SGD to train both the traditional and coupled CRF models. The basic idea is to derive the gradient w.r.t. θ using a small batch of sampled training data, and use it update θ at each step. However, we need to train a model on two separate data, and one dataset may be overwhelmed by another due to imbalance in scale if directly merging two datasets into one (see table 2). Therefore, we adopt the corpus-weighting strategy proposed in

Li et al. (2015) [6], as shown in Algorithm 1. D_k^b is a subset of training data used in k^{th} step update; b is the batch size; η_k is a update step. The idea is to randomly select instances from each training data in a certain proportion, and use them for one-iteration training. In this work, we use $b = 30$, and follow the implementation in CRFsuite[2] to decide η_k. Actually, in our experiments, we in some cases simultaneously train our model on three datasets, which requires slightly extending the procedure in Algorithm 1.

Algorithm 1. SGD training with two labeled datasets

Input: Two labeled datasets: $\mathcal{D}^1 = \{(x_i^1, v_i^1)\}_{i=1}^N$, $\mathcal{D}^2 = \{(x_i^2, v_i^2)\}_{i=1}^M$;
        Parameters: I(Iterations) , N', M', b
Output: θ
1 **Initialization:** $\theta_0 = 0, k = 0$;
2 **for** $i = 1;\ i <= I;\ i + +$ **do**
3      $\mathcal{D}_i = N'$ instances from $\mathcal{D}^1 + M'$ instrances from $\mathcal{D}^2$ (randomly select) ;
4   Shuffle $\mathcal{D}_i$;
5   **while** *Traverse $\mathcal{D}_i$ and $\mathcal{D}_k^b \subseteq \mathcal{D}_i$* **do**
6          $\theta_{k+1} = \theta_k + \eta_k \frac{1}{b} \bigtriangledown \mathcal{L}(\mathcal{D}_k^b; \theta_k)$;
7          $k = k + 1$;
8   **end**
9 **end**

---

4   Experiments

4.1   Datasets

Tabel 2 shows the datasets explored in this work. For labeled data, the NLPCC 2015 shared task organizer provides us only a training dataset consisting $10,000$ WB sentences. To develop our model and tune the parameter settings, we randomly split the training data into two sets: a train set to learn model parameters and a dev set for model evaluation [10]. However, for the final submission, we use the whole training data to train our model. Actually, the organizer also provides a large set of unlabeled WB text, which is not considered in this work.

For the third party resources, we follow the suggestion in the data description guideline for data split for CTB7, and use the data of Li et al. (2015) [6] for PD.

4.2   Experiments on WS

In this section, we conduct a lot of experiments to find the best strategy to utilize multiple heterogeneous data that performs best on WB.

Table 2. Data statistics

Dataset	Partition	Sentences	Words	Characters
	train	8,000	172,800	279,676
WB	dev	2,000	42,372	68,535
	test	5,000	106,741	172,339
	train	46,572	1,039,774	168,2485
CTB7	dev	2,079	59,955	100,316
	test	2,796	81,578	134,149
	train	283,862	6,797,623	11,308,278
PD	dev	4,965	116,769	194,641
	test	9,913	236,234	393,185

Coupled Models on *CTB7* and *PD*: For the first time, we apply the recently proposed coupled sequence labeling to the task of Chinese WS. We would like to know how the approach performs on canonical data. Table 3 shows the performance of the coupled CRF on *CTB7* using both *CTB7* and *PD* training data, denoted as $Seg_{CTB7\&PD}$. We reimplement the guide-feature based method of Jiang et al. [1], using *PD* to produce guide feature on *CTB7*, denoted as $Seg_{CTB7\leftarrow PD}$. The baseline model Seg_{CTB7} is a traditional CRF trained on *CTB7*. The results show that *the coupled model outperforms both the baseline and the guide-feature based method on the task of WS*.

Model Selection on *WB*-dev: We then try to find the best strategy of utilizing the three datasets that performs best on *WB*. We experiment with different methods of combining the three training datasets (*WB*, *CTB7*, and *PD*) and report F1 scores on the corresponding dev dataset. The results are shown in Table 4.

The first major row refers to baseline CRF models, which are trained on a single training dataset, and evaluated against all three dev datasets. We can see that all models performs best on the dev dataset that corresponds to its training dataset. Moreover, we notice that Seg_{CTB7} has an F1 score on *WB*-dev which is very close to that of Seg_{WB}, indicating that the WS guidelines in *WB* and *CTB7* may be close.

In the second major row, we assume that two training datasets follow the same guideline, and directly combine the two datasets to train traditional CRF models. Algorithm 1 is adopted for training on two datasets with $N' = 5K$

Table 3. WS F1 scores on *CTB7*

Approaches	dev	test
Baseline: Seg_{CTB7}	95.39	94.95
Guide-Feature: $Seg_{CTB7\leftarrow PD}$	95.80 (+0.41)	95.29 (+0.34)
Coupled: $Seg_{CTB7\&PD}$	95.81 (+0.42)	95.45 (+0.50)

Table 4. WS F1 scores on *WB/CTB7/PD* dev datasets

Approaches		*WB*	*CTB7*	*PD*
Baseline	Seg_{WB}	**93.13**	87.90	85.80
	Seg_{CTB7}	92.31	**95.39**	89.33
	Seg_{PD}	89.72	90.23	**97.06**
Merge two datasets	$Seg_{WB+CTB7}$	**94.81**	95.46	89.79
	Seg_{WB+PD}	93.70	90.84	**96.38**
	$Seg_{WB+PD^{CTB}}$	94.78	93.43	92.75
Coupled	$Seg_{WB\&CTB7}$	95.06	95.14	–
	$Seg_{WB\&PD}$	95.13	–	**96.96**
	$Seg_{(WB+CTB7)\&PD}$	**95.54**	**95.52**	96.54
	$Seg_{WB\&(CTB7+PD^{CTB})}$	95.14	95.04	–

and $M' = 5K$. For the results, we can see that the model trained on combined *WB* and *CTB7*, denoted as $Seg_{WB+CTB7}$, achieves much higher performance on *WB*-dev than the baseline Seg_{WB} (+1.68%). This clearly shows that *CTB7 can be directly used as extra training data for WS on WB*. The model also slightly outperforms the baseline Seg_{CTB7} on *CTB7*-dev (+0.07%). Directly combining *WB* and *PD* also lead to improved F1 score on *WB*-dev over the baseline model, but is much inferior the case of *CTB7*. PD^{CTB} refer to a modified *PD* data which is converted into the *CTB7* WS annotation using $Seg_{CTB7\&PD}$ (described in Table 3). As described in Li et al. (2015) [6], the coupled model can be naturally used to conduct annotation conversion, based on constrained decoding, given one-side tags. $Seg_{WB+PD^{CTB}}$ achieve very close F1 score to $Seg_{WB+CTB7}$, indicating the *PD can also effectively help WS on WB*. Moreover, all these results also suggest that the WS guidelines in *WB* and *CTB7* may be close.

In the third major row, we conduct experiments with the coupled models trained on multiple datasets. The coupled model trained on *WB* and *CTB7*, denoted as $Seg_{WB\&CTB7}$, achieves slightly higher F1 score on *WB*-dev than $Seg_{WB+CTB7}$ (0.25%), indicating *there still exists some differences in the guidelines of two dataset*. $Seg_{WB\&(CTB7+PD^{CTB})}$ is trained on three dataset, i.e., *WB*, *CTB7*, and PD^{CTB}, assuming the latter two having the homogeneous annotations. However, the model only achieves slightly higher accuracy when using only *CTB7* or PD^{CTB} as the second extra training dataset. The best-performing model is $Seg_{(WB+CTB7)\&PD}$, which assumes that *WB* and *CTB7* are under the same annotation guideline and uses *PD* as the second training dataset. Therefore, we choose $Seg_{(WB+CTB7)\&PD}$ as our final system for result submission. Our next plan is to directly train a coupled model on the three datasets that uses three-side bundled tags.

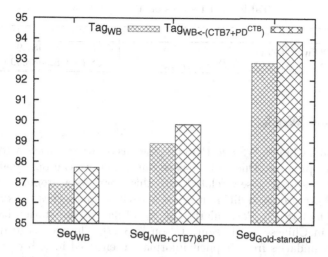

Fig. 6. POS F1 scores on *WB*-dev, with three different settings of WS results

4.3   Experiments on POS Tagging

Our initial plan is to combine *WB*, *CTB7*, and *PD* with the coupled sequence labeling approach, which has an inefficient problem due to the large bundled POS tag set. Due to time limitation, we adopt an ensemble approach, as illustrated in Figure 2. PD^{CTB} refers to a modified *PD* converted into the annotation guideline of *CTB* using the coupled model described in Li et al (2015) [6]. Please note that the coupled model of Li et al (2015) [6] is built on *CTB5* and *PD*.

To better comprehend the effect of different WS approaches on POS F1 scores, we experiment with three settings of WS results on *WB*-dev. In the first setting, *WB*-dev is automatically segmented by the baseline Seg_{WB}; the second setting uses the best-performing model $Seg_{(WB+CTB7)\&PD}$ to produce WS results on *WB*-dev; the third setting uses gold-standard WS results of *WB*-dev. Figure 6 shows the results. We can see that our guide-feature based model $Tag_{WB\leftarrow(CTB7+PD^{CTB})}$ consistently outperforms the baseline model Tag_{WB}.

4.4   Final Results on Test Data

For the final submission, we re-train our models on the whole *WB* training data, using the iteration number that leads to best performance on *WB*-dev during model development. Table 5 shows the results on the test data used for final evaluation. Our effort leads to an improvement on WS F1 score from 93.76% to 95.58% (+1.82%), and on POS F1 score from 87.93% to 88.99% (+1.06%).

Table 5. F1 scores on WB-test data

	WS F1	POS F1
Baseline: $Seg_{WB} + Tag_{WB}$	93.76	87.93
Our best: $Seg_{(WB+CTB7)\&PD} + Tag_{WB \leftarrow (CTB7+PD^{CTB})}$	95.58 (**+1.82**)	88.99 (**+1.06**)

5   Related Work

In recent years, Chinese WS and POS tagging has drawn extensive attention and made a lot of progress. Due to the space limitation, we can only make limited references to research closely related with this work. As a promising direction, researchers have proposed different semi-supervised approaches to explore unlabeled data [9,14,18,19]. Particularly, the use of natural annotation has become a hot topic in Chinese word segmentation [3,8,17]. The idea is to derive segmentation boundaries from implicit information encoded in web texts, such as anchor texts and punctuation marks, and use them as partially labeled training data in sequence labeling models.

In the line of exploiting multiple labeled data, Jiang et al. (2009) [1] propose the simple yet effective guild-feature based method, which is further extended in [7,12,13]. Qiu et al. (2013) [11] propose a model that performs heterogeneous Chinese word segmentation and POS tagging and produces two sets of results following CTB and PD styles respectively. Their model is based on linear perceptron, and uses approximate inference. Instead, this work is based on the coupled sequence labeling model of Li et al. (2015) [6], which is a natural extension of CRFs.

Although we adopt a cascaded framework for WS and POS tagging, it is more popular to jointly solve the two tasks [2,4,23]. For recent works on Chinese lexical analysis on web data, please refer to [15,20,22].

6   Conclusion

We have participated in the NLPCC 2015 shared task on Chinese WS and POS tagging for Weibo Text. For the first time, we apply the recently proposed coupled sequence labeling to the task of WS, and find that coupled sequence labeling also slightly outperforms the guide-feature based method on $CTB7$ with extra PD. Furthermore, we experiment with different strategies to combine the three datasets (plus WB), and achieve promising results. Due to the high time complexity of coupled POS tagging, we instead adopt the guide-feature based method for the POS tagging subtask.

For future work, we plan to advance our progress on exploiting multiple heterogeneous resources in the following directions. First, we will try to experiment with coupled sequence labeling for POS tagging on WB, and make comparison with the guide-feature based methods. Second, we would like to seek effective ways to further incorporate unlabeled WB texts. Finally, we find that our baseline WS system lags behind by large margin other systems in the same closed track, which motivates us to investigate the issue and improve our final performance through enhancing our baseline system.

References

1. Jiang, W., Huang, L., Liu, Q.: Automatic adaptation of annotation standards: Chinese word segmentation and POS tagging - a case study. In: Proceedings of ACL, pp. 522–530 (2009)
2. Jiang, W., Huang, L., Liu, Q., Lü, Y.: A cascaded linear model for joint Chinese word segmentation and part-of-speech tagging. In: Proceedings of ACL 2008: HLT, pp. 897–904 (2008)
3. Jiang, W., Sun, M., Lü, Y., Yang, Y., Liu, Q.: Discriminative learning with natural annotations: word segmentation as a case study. In: Proceedings of ACL, pp. 761–769 (2013)
4. Kruengkrai, C., Uchimoto, K., Kazama, J., Wang, Y., Torisawa, K., Isahara, H.: An error-driven word-character hybrid model for joint Chinese word segmentation and POS tagging. In: Proceedings of ACL-AFNLP 2009, pp. 513–521 (2009)
5. Lafferty, J., McCallum, A., Pereira, F.C.: Conditional random fields: probabilistic models for segmenting and labeling sequence data. In: Proceedings of the 18th International Conference on Machine Learning (ICML 2001), pp. 282–289 (2001)
6. Li, Z., Chao, J., Zhang, M., Chen, W.: Coupled sequence labeling on heterogeneous annotations: pos tagging as a case study. In: Proceedings of the 53rd Annual Meeting of the Association for Computational Linguistics and the 7th International Joint Conference on Natural Language Processing, vol. 1: Long Papers, pp. 1783–1792. Association for Computational Linguistics, Beijing july, 2015
7. Li, Z., Che, W., Liu, T.: Exploiting multiple treebanks for parsing with quasisynchronous grammar. In: ACL, pp. 675–684 (2012)
8. Liu, Y., Zhang, Y., Che, W., Liu, T., Wu, F.: Domain adaptation for CRF-based Chinese word segmentation using free annotations. In: Proceedings of EMNLP, pp. 864–874 (2014)
9. Qiu, X., Huang, C., Huang, X.: Automatic corpus expansion for Chinese word segmentation by exploiting the redundancy of web information. In: Proceedings of COLING, pp. 1154–1164 (2014)
10. Qiu, X., Qian, P., Huang, X.: Overview of the nlpcc 2015 shared task: chinese word segmentation and pos tagging for micro-blog texts (2015). arXiv preprint arXiv:1505.07599
11. Qiu, X., Zhao, J., Huang, X.: Joint Chinese word segmentation and POS tagging on heterogeneous annotated corpora with multiple task learning. In: Proceedings of EMNLP, pp. 658–668 (2013)
12. Sun, W.: A stacked sub-word model for joint chinese word segmentation and part-of-speech tagging. In: Proceedings of ACL, pp. 1385–1394 (2011)
13. Sun, W., Wan, X.: Reducing approximation and estimation errors for Chinese lexical processing with heterogeneous annotations. In: Proceedings of ACL, pp. 232–241 (2012)
14. Sun, W., Xu, J.: Enhancing chinese word segmentation using unlabeled data. In: Proceedings of EMNLP, pp. 970–979 (2011)
15. Wang, A., Kan, M.Y.: Mining informal language from chinese microtext: joint word recognition and segmentation. In: Proceedings of the 51st Annual Meeting of the Association for Computational Linguistics, vol. 1: Long Papers, pp. 731–741. Association for Computational Linguistics, Sofia, August 2013

16. Xue, N., et al.: Chinese word segmentation as character tagging. Computational Linguistics and Chinese Language Processing 8(1), 29–48 (2003)
17. Yang, F., Vozila, P.: Semi-supervised Chinese word segmentation using partial-label learning with conditional random fields. In: Proceedings of EMNLP, pp. 90–98 (2014)
18. Zeng, X., Wong, D.F., Chao, L.S., Trancoso, I.: Graph-based semi-supervised model for joint chinese word segmentation and part-of-speech tagging. In: Proceedings of the 51st Annual Meeting of the Association for Computational Linguistics, vol. 1: Long Papers, pp. 770–779. Association for Computational Linguistics, Sofia, August 2013
19. Zhang, L., Wang, H., Sun, X., Mansur, M.: Exploring representations from unlabeled data with co-training for Chinese word segmentation. In: Proceedings of EMNLP, pp. 311–321 (2013)
20. Zhang, L., Wang, H., Sun, X., Mansur, M.: Improving Chinese word segmentation on micro-blog using rich punctuations. In: Proceedings of ACL: Short Papers (2013)
21. Zhang, M., Zhang, Y., Che, W., Liu, T.: Character-level Chinese dependency parsing. In: Proceedings of ACL, pp. 1326–1336 (2014)
22. Zhang, M., Zhang, Y., Che, W., Liu, T.: Type-supervised domain adaptation for joint segmentation and POS-tagging. In: Proceedings of COLING, pp. 588–597 (2014)
23. Zhang, Y., Clark, S.: Joint word segmentation and POS tagging using a single perceptron. In: Proceedings of ACL 2008: HLT, pp. 888–896 (2008)

Entity Recognition and Linking
in Chinese Search Queries

Jinwei Yuan[1], Yan Yang[1(✉)], Zhen Jia[1], Hongfeng Yin[2], Junfu Huang[1], and Jie Zhu[3]

[1] School of Information Science and Technology,
Southwest Jiaotong University, Chengdu, China
{502966377,990504422}@qq.com, {yyang,zjia}@swjtu.edu.cn
[2] DOCOMO Innovations Incorporation, Palo Alto, USA
hongfeng_yin@yahoo.com
[3] Department of Computer Science, Tibet University, Tibet, China
trocky.jie@gmail.com

Abstract. Aiming at the task of Entity Recognition and Linking in Chinese Search Queries in NLP&CC 2015, this paper proposes the solutions to entity recognition, entity linking and entity disambiguation. Dictionary, online knowledge base and SWJTU Chinese word segmentation are used in entity recognition. Synonyms thesaurus, redirect of Wikipedia and the combination of improved PED (Pinyin Edit Distance) algorithm and LCS (Longest Common Subsequence) are applied in entity linking. The methods of suffix supplement and link value computation based on online encyclopedia are adopted in entity disambiguation. The experiment results indicate that the proposed solutions in this paper are effective for the case of short queries and insufficient contexts.

Keywords: Entity recognition · Entity linking · Entity disambiguation · Suffix supplement · Online encyclopedia

1 Introduction

With the wide application of knowledge graph technology in information retrieval, user modeling, human-computer interaction, question answering and knowledge reasoning, knowledge linking based on structured knowledge base has become a significant task. Entity recognition and linking in Chinese search queries is the task that recognizes all possible entities from queries and links these entities to the target entities in the given knowledge base. It has many difficulties in this task because queries are too short (e.g: in Baidu search engine, queries are limited within 38 words) and often contain a lot of noises (such as spelling mistakes, abbreviations, Internet slangs, nicknames). Several methods are proposed in this paper to solve entity recognition and linking in Chinese search queries. Word dictionary, word popularity, and synonymous thesaurus are constructed based on several online encyclopedias such as Hudongbaike, Baidubaike, and Wikipedia. The solution for entity recognition is mainly based on dictionary, online encyclopedias and SWJTU Chinese word segmentation system [26]. Entity linking is based on synonyms thesaurus, redirection search, and algorithm of combining

© Springer International Publishing Switzerland 2015
J. Li et al. (Eds.): NLPCC 2015, LNAI 9362, pp. 507–519, 2015.
DOI: 10.1007/978-3-319-25207-0_47

improved PED (Pinyin Edit Distance) with LCS (Longest Common Subsequence). Entity disambiguation uses suffix supplement and link value computation methods. The experiment data are acquired from the sample data provided in the task and the queries are extracted from Sogou search logs. Experiment results show that most of the entities in queries can be recognized and can be linked to the entities in the given knowledge base. The average-F1 is 73.3% which ranks first in NLP&CC 2015 evaluation task of entity recognition and linking in Chinese search queries.

2   Related Works

Named entity recognition is the foundation of natural language processing [1] and an important task in many areas [14], including information retrieval, human-machine interaction, machine translation, and question answering. Al-Rfou et al. [2] used online content analysis algorithm to recognize entities. However, this method is only suitable for long texts but not for short texts. Zhao et al. [3] proposed the method of corpus annotation and hierarchical Hidden Markov Model (HMM) to process the recognition of product named entity. Michal et al. [4] introduced a method which uses latent semantic of entities to recognize named entities. Joel et al. [5] recognized named entity by using the content and structure of Wikipedia.

Entity linking refers to linking a given entity to an existing knowledge base [6], [7], [15,16]. Zhu et al. [8] used improved pinyin edit distance and suffix vocabulary matching method to link entities, but this method only suit for Microblog texts. Chen et al. [17] proposed linking with collaborative ranking by using integrated information of entities. But this method is not suitable for short queries because the short texts lack comprehensive information or regularization. Nadeau et al. [25] and Zhang et al. [18] used a supervised machine learning method for entity linking. However, this method is too dependent on semantic information that once extended to other types of corpus, the performance will be affected seriously by noises [19]. Gattani et al. [9] adopted online encyclopedia such as Wikipedia to implement entity linking of social data.

Entity disambiguation can solve the problem that an entity mention refers to multiple real-world things [8], [20], [21]. Yang et al. [10] proposed the method of person name disambiguation based on dependency features of webpage text. Nguyen et al. [11] extracted the contents and characteristics of the corresponding pages from the Wikipedia to carry out the entity disambiguation. Zhu et al. [8] combined entity clustering disambiguation and similar entity disambiguation based on Baidu encyclopedia. Meng et al. [22] adopted improved VSM to get textual similarity for entity disambiguation. Kataria et al. [23] and Sen [24] proposed the methods based on topic models to achieve entity disambiguation, but these methods are not suitable for short texts.

3   Methods

3.1   Problem Analysis and Method Procedure

By analyzing the given sample data and knowledge base in the task, problems are mainly as follow:

1) In some queries, there exist English words and a variety of punctuations.
2) Queries are too short and have no contexts.
3) Entity mentions in queries may be aliases or abbreviations of the target entities in the knowledge base.
4) Queries contain wrongly written characters.
5) Many different entities have the same mentions in the knowledge base. For example, entity mention "步步惊心"(Startling by Each Step) may refer to a book or a TV series.

To solve these problems, this paper proposes the following solutions:

1) Preprocess queries and remove noises in them.
2) Build the word dictionary and use SWJTU Chinese word segmentation system to recognize entities of short phrases in open domains.
3) Construct a synonyms thesaurus to identify aliases or abbreviations of entities.
4) Optimize the knowledge base and remove noises in it.
5) Use LCS algorithm and online encyclopedias to recognize the abbreviations of entities.
6) Combine improved edit distance algorithm [8], [13] with LCS to recognize wrongly written characters.
7) Apply word frequency in Baidu encyclopedias and suffix supplement method to disambiguate entities.

The pipeline of the methods is shown in Figure 1.

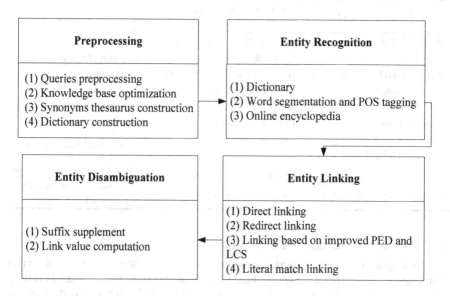

Fig. 1. Method pipeline

As shown in Figure 1, the methods pipeline includes preprocessing, entity recognition, entity linking and entity disambiguation. Preprocessing consists of query preprocessing, knowledge base optimization, synonyms thesaurus construction and dictionary construction. Entity recognition is mainly based on dictionary, word segmentation and POS tagging, and online encyclopedias. Entity linking has four steps carried out in order to find candidate entities in the knowledge base. Entity disambiguation uses suffix supplement and link value computation method on basis of online encyclopedias.

3.2   Preprocessing

3.2.1   Knowledge Base Preprocessing

It contains many noises in the knowledge base which is provided in the evaluation task, such as disunification in uppercase and lowercase letters in English, mixed use of Chinese and English punctuation, lack of values in knowledge triples. In order to improve accuracy of entity linking, several solutions to remove noises in knowledge base are taken. Firstly, English words are rewritten in capitalization to ensure it's unified. Then, all the punctuations in the knowledge base are turned into unified Chinese punctuation. Finally, the knowledge triples that some values are absent are removed. Examples are as shown below in Table 1.

Table 1. Knowledge base preprocessing

The original knowledge	The knowledge after preprocessing	remark
林纳斯·托瓦兹 employer Linux 基金会 (Linus Torvalds employer Linux foundation)	林纳斯·托瓦兹 EMPLOYER LINUX基金会 (Linus Torvalds EMPLOYER LINUX foundation)	Rewritten English words in capitalization.
南京!南京! (Nanjing! Nanjing!)	南京！南京！ (Nanjing ! Nanjing ！)	Turn English punctuation into Chinese punctuation.
中华人民共和国abstract null (the People's Republic of China)		Remove this knowledge triple.
《步步惊心》 (《Startling by Each Step》)	步步惊心 (Startling by Each Step)	Delete book title mark punctuations which are in the start or end of knowledge entries

3.2.2   Synonyms Thesaurus Construction

For the length limitation in queries, many entity mentions appear in form of aliases, abbreviations and Internet slangs. In fact, these mentions with variable names refer to one entity. For example, "习大大" (Xi Dada), and "习主席" (Chairman Xi) are different mentions but represent one person. A large number of synonyms are extracted from Hudong encyclopedia, Baidu Encyclopedia, Douban and Wikipedia for constructing a synonyms thesaurus which in total reaches 184,430 synonyms.

3.2.3   Dictionary Construction

The dictionary contains the words are extracted from the knowledge base given in the task, Wikipedia, Baidu encyclopedia, Hudong encyclopedia and Douban. The words collected in the dictionary include nouns, academic terms, idioms, times, aliases, abbreviations, Internet slangs which in total reach about more than 800,000 words.

3.2.4   Query Preprocessing

Before recognizing entities for the given queries, preprocessing need to be done to get the large amount of noise removed. Firstly, filtering punctuations and turning lowercase letters to capital letters in English. Then, using SWJTU Chinese word segmentation system to process queries and remaining the words which are tagged as noun, academic term, abbreviation, idiom and time[1]. Query preprocessing is described in detail by taking one query as example, as shown below in Table 2.

Example Query:

《Linkin Park》歌曲的演唱者麦克·信田 (《Linkin Park》song's singer Mike Shinoda)

3.3   Entity Recognition

3.3.1   Entity Recognition Difficulties

In general, entity recognition is carried out in such cases as entities have full names, but many entities in queries appear in form of aliases or abbreviations because of length limitation. At present, named entity recognition algorithms with better effect are domain related, while the sample data provided in this evaluation task are from open domain, thus it cannot reach good result using traditional machine learning methods(such as CRF) based on labeled training data. To solve this problem, we apply a simple method of combining the existing Chinese word segmentation, dictionary, and online encyclopedia.

3.3.2   Named Entity Recognition

For the given query, word segmentation and POS tagging are performed in the step of query preprocessing. The NLP processing tool is SWJTU Chinese word segmentation system. This system supports the user-defined dictionary, so we add the dictionary

---

[1] The POS tags are Entity, n, t, j, nnt, nrf etc. Entity is a user-defined POS that refers to the words in the dictionary and possible entities. Other POS descriptions are in the website http://ics.swjtu.edu.cn.

Table 2. Main steps of query preprocessing

Step No.	Preprocessing	Example
1	punctuations filtering and letters unification	LINKIN_PARK歌曲的演唱者麦克·信田 (LINKIN_PARK song's singer Mike Shinoda)
2	Word segmentation and POS tagging	LINKIN_PARK/Entity 歌曲/n 的/ude1 演唱者/nnt 麦克·信田/nrf(LINKIN_ PARK/Entity song/n s/ude1 singer/nnt Mike Shinoda/nrf)
3	POS filtering	LINKIN_PARK/Entity 歌曲/n 演唱者/nnt 麦克·信田/nrf (LINKIN_PARK/Entity song/n singer/nnt Mike Shinoda/nrf)

into the system and the words in the dictionary are tagged as "Entity". This system supports two segmentation ways that are coarse-grained and fine-grained segmentations. Coarse-grained segmentation uses longest matching algorithm, but fine-grained segmentation can split the words into smaller units. For example, "西南交通大学" (Southwest Jiaotong University) is cut into one entity in coarse-grained segmentation "西南交通大学/ntu", but three words in fine-grained segmentation "[西南/ns 交通/n 大学/nis]/ntu"([Southwest/ns Jiaotong/n University/nis]/ntu).

Firstly, the query is segmented in coarse-grained segmentation. If the POS tags of the words are "Entity" (user-defined POS tag), nr (person name), ns (place) or j (abbreviation), they are viewed as entities. While if the POS tags are nt (organization), nz (proper noun), l (idiom), and n (noun), the following steps are applied to determine whether the words are entities.

(1) If the word has redirect in Wikipedia, it can be recognized as an entity directly.
(2) If the word has no redirect and the POS tag is nt or nz, it should be segmented in fine-grained segmentation to smaller unit. If some of the units are in the dictionary, these units are determined named entities.

Here are some detail examples about entities recognition, as shown below in Table 3.

3.4   Entity Linking

The pipeline of entity linking is shown below in Figure 2.

The pipeline includes four main steps which are carried out in order: direct linking, redirect linking, linking based on improved PED and LCS, and literal match linking. The detail process is as follows.

Table 3. Entity recognition examples

Result of word segmentation	Detailed description
赵本山/Entity 否认/v 离婚/n[2] (Benshan Zhao/Entity denied/ divorce/n)	The POS of 赵本山(Benshan Zhao) is **Entity**, so it can be reserved as an entity directly. 离婚(divorce) is tagged as **n**, but redirect in Wikipedia cannot be found, so it be removed.
湖北/ns 石首/ns群体性事件/nz (Hubei/ns Shishou/ns group event/nz)	湖北(Hubei) and 石首(Shishou) are marked as **ns**, so they are reserved as entities directly. 群体性事件(group event) is marked as **nz**, and the redirect is骚乱(riot), so it is reserved as an entity.
巴比伦通天塔/nz (Babylon tower/nz)	巴比伦通天塔(Babylon tower)is tagged as **nz**, but it's redirect cannot be found, then it is segmented using fine-grained segmentation into [巴比伦/**nsf** 通天塔/**nz**]/nz ([Babylon/nsf tower/nz]/nz). The two words "Babylon" and "tower" are in the dictionary, so both of them are recognized as entities.

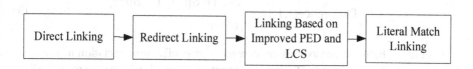

Fig. 2. Entity linking pipeline

1) Direct Linking

Direct linking means searching an entity or its synonyms in the knowledge base directly which is the first step. For improving linking accuracy, when the entity is recognized as person name, especially foreigner name, we first search this name in Hudong encyclopedia to find the name with highest popularity. Namely, the most frequently visited and concerned entry. In Hudong encyclopedia, when users search some entries, the most popular entries will be returned. For example, when "科比" (Kobe) is searched, "科比·布莱恩特"(Kobe Bryant) will be returned. In the general case, "科比" refers to "科比·布莱恩特". Then we link the full name to the knowledge base.

2) Redirect Linking

If the entity cannot be found in the knowledge base after the step above, we search this entity or its synonyms in Wikipedia to get the redirect. If the redirect exists, we use it to search in the knowledge base.

---

[2] The notation after "/" is POS tag.

3) Linking Based on Improved PED and LCS

Some entities contain wrongly written characters or inconsistence in English to Chinese transliteration so they cannot be linked to the knowledge base. We adopt the method of combining the improved PED(Pinyin Edit Distance) algorithm [8] [13] with LCS to solve this problem.

Firstly, we use PED algorithm to find the candidates set of the entity in the knowledge base, as shown in (1).

$$e = \{e_1, e_2, e_3, \ldots, e_i\} \tag{1}$$

Where e_i is the i-th candidate entity of e.

In PED algorithm, Spell is a set of consonants and vowels with similar pronunciation as follows.

$$Spell = \{(l,n),(l,r),(z,zh),(c,ch),(an,ang),(en,eng),(in,ing),(ang,ong),(si,ci)\} \tag{2}$$

Difference between character I and I' is computed according to the consonants and vowels of them, as shown in (3) [8], [13].

$$Difference(I, I') = \begin{cases} 0.5 , & I \in Spell_i, I' \in Spell_i \\ 1, & other \end{cases} \tag{3}$$

Difference between consonants or vowels with similar pronunciation in the same $Spell_i$ is set to 0.5. Difference of other consonants and vowels which are not in Spell set is set to 1.

If the difference between each character of two entity mentions is less than 1, then the entity in knowledge base will be a candidate entity. For example, for the entity "诺维茨基"(Nowitzki) in queries pronouncing nuo wei ci ji in Chinese and the entity in the knowledge base "诺维斯基" pronouncing nuo wei si ji, the differences between each character is less than 1, then "诺维斯基" is put into the candidate set.

If the difference between characters of two entities is no less than 1, the entity cannot be added to the candidate set. For example, the entity in queries "周杰伦"(Jay Chou) pronounces zhou jie lun in Chinese and "周杰峰" in the knowledge base pronounces zhou jie feng, and the difference between "轮"(lun) and "峰"(feng) reaches 1, so "周杰峰" is not the candidate of "周杰伦".

After acquiring the candidate set of the entities, we choose the one with the highest literal similarity as the linking result in the knowledge base. Literal similarity is computed by LCS. The principle idea of LCS is that with the same length, the more the same words are, the higher the similarity will be. The formula is as shown below.

$$Sim_i = \frac{len(m, e_i)}{len(m)} \tag{4}$$

where Sim_i is the similarity between the linking entity m and the candidate entity e_i; len(m) indicates the length of the linking entity m; $len(m, e_i)$ represents the number of same characters between linking entity m and candidate entity e_i.

4) Literal Match Linking

Literal match linking is to solve the problem of abbreviations. For example, "中科院" (CAS) in the query is often used to refer to "中国科学院" (Chinese Academy of Sciences). In this paper, when the literal matching degree between two mentions is no less than 0.6, the two entity mentions will be linked.

3.5   Entity Disambiguation

3.5.1   Difficulties in Entity Disambiguation

For the variable meanings of the same word in different contexts, the traditional word disambiguation determines the exact meaning according to the contexts [12]. However, search queries are too short to provide complete contexts of entities, which brings difficulties in disambiguation, mainly shown as follow points:

1) Diversity of Entity Mentions

An entity may have various mentions, such as full name, abbreviation, aliases, etc. For example, the famous American basketball player "科比·布莱恩特" (Kobe Bryant) can be called as "Kobe", "黑曼巴" (Black Mamba), "小飞侠" (Peter Pan), etc.

2) Ambiguity of Entity in Different Contexts

An entity has different meanings in different contexts. For example, the word "Apple" may refer to a fruit apple or Apple Company.

In the task, we may get several candidate entity links in the knowledge base after the step of entity linking. To disambiguate these candidate entities, we propose the methods of suffix supplement and link value computation based on online encyclopedias.

3.5.2   Suffix Supplement

Some online encyclopedias uses suffix to disambiguate the entities which have the same names. Since the knowledge base is from Wikipedia, it is effective to use the suffix of online encyclopedia to pad and disambiguate entities. First, we search the entities in the Hudong encyclopedia, which can return the most popular entry name with suffix according to visiting frequency of users. The suffix of entry name will be added into the entity. Then, by calculating the similarity between the new entity mentions with suffix and candidate entities in the knowledge base, the candidate entity with the highest similarity will be returned as the final linking result.

3.5.3   Link Value Computation

This method includes three steps:

Step 1, computing word weight of each candidate entities in knowledge base based on Wikipedia.

Step 2, getting popularity weight based on Baidu encyclopedia.

Step 3, computing link value to decide which candidate entity is the linking result.

1) Word Weight Computation

Suppose E is the entity recognized in queries, C is one of the candidate entities of E linking to the knowledge base.

Firstly, we search C in Wikipedia and get the search result webpage of C. Then, we count the word frequency of C sum(C) in the webpage. We else count the word frequency sum(E) in the webpage. If C contains E, that is to say, if E is part of C, we don't count. The word weight W_i of C is calculated using the following formula:

$$W_i = \frac{sum(E)}{sum(C)} \tag{5}$$

2) Popularity Weight Acquisition

Second, we search C in Baidu encyclopedia to get the visiting frequency and determine its popularity. A popularity weight value λ is assigned to each C according to its visiting frequency which can be estimated by experience, as shown in Table 4:

Table 4. Weight λ of popularity

Popularity ranking	1	2	3	4	5
Weight λ	1.0	0.9	0.8	0.7	0.6

3) Link Value Computation

According to the word weight and popularity weight of the candidate entity C, we can calculate the link value of C, the calculation method is as shown in formula (6):

$$S_i = \lambda \times W_i \tag{6}$$

where S_i indicates the link value of the candidate linked entities.

Select the candidate linked entity with largest value of S_i as the linking result.

4   Experiments

For the sample data provided in this evaluation task is only 159 queries, we extract 1000 queries from the Sogou search logs, which are similar to the sample data. We manually label the entities and linking results of each query according to the knowledge base. We use these corpus as experiment data to design entity recognition, linking and disambiguating algorithms. The entity recognition result is shown in Table 5. The linking result is shown in Table 6. The entity recognition and linking result for sample data is shown in Table 7.

Table 5. Entity Recognition Result

Marked entity	Precision	Recall	F1
1159	0.844	0.886	0.865

The result in Table 5 shows that using SWJTU Chinese word segmentation and dictionary to recognize entities is effective.

Table 6. Entity Linking Result

Marked entity	Precision	Recall	F1
1159	0.905	0.905	0.905

From Table 6, we noted that the effect is very good using the above entity linking method. That is to say, most of the entities link right.

Table 7. Entity recognition and linking

Precision	Recall	F1
0.742	0.779	0.761

From Table 7, the precision of entity recognition and linking is not particularly high. But it is also noted that the precision of entity recognition has a great effect on the final results.

The testing data is provided by CCF in the evaluation task, among which includes 3849 search queries. The final evaluation result is shown below as Table 8.

Table 8. Evaluation result of entity recognition and linking in Chinese search queries

NO.	Precision	Recall	F1	Average-F1
1	**0.724**	**0.736**	**0.73**	**0.733**
2	0.562	0.695	0.621	0.611
3	0.497	0.704	0.583	0.569

The result of NO.1 represents the performance of our system. Compared with other systems in this evaluation task, our system achieves higher precision, recall, F1, and average-F1 than other systems, which shows the method proposed in this paper is effective.

5   Conclusions and Outlook

In this paper, the method used in the evaluation task of entity recognition and linking in Chinese search queries in NLP&CC2015 is introduced. The result shows that it works fine in entity linking, but the entity recognition and disambiguation need to be improved. In future work, we will use and merge more information, including the results of Internet search engine, to expand the length of Chinese search queries. And then we will conduct semantic analysis of the queries to further improve the effect of named entity recognition and disambiguation.

Acknowledgement. This work is supported by National Natural Science Foundation of China under Grants No.61170111, No. 61202043, No. 61262058, and Fundamental Research Funds for the Central Universities under Grants No. 2682015CX070.

References

1. Lei, J., Tang, B., Lu, X., et al.: A comprehensive study of named entity recognition in Chinese clinical text. Journal of the American Medical Informatics Association **21**(5), 808–814 (2014)
2. Al-Rfou, R., Skiena, S.: Speedread: a fast named entity recognition pipeline (2013). arXiv preprint arXiv: 1301.2857
3. Zhao, J., Liu, F.: Product named entity recognition in Chinese text. Language Resources & Evaluation **42**(2), 197–217 (2008)
4. Konkol, M., Brychcín, T., Konopík, M.: Latent semantics in Named Entity Recognition. Expert Systems with Applications **42**(7), 3470–3479 (2015)
5. Nothman, J., Ringland, N., Radford, W., et al.: Learning multilingual named entity recognition from wikipedia. Artificial Intelligence **194**, 151–175 (2013)
6. Hachey, B., Radford, W., Nothman, J., et al.: Evaluating Entity Linking with wikipedia. Artificial Intelligence **194**(194), 130–150 (2013)
7. Shen, W., Wang, J., Han, J.: Entity Linking with a Knowledge Base: Issues, Techniques, and Solutions. IEEE Transactions on Knowledge and Data Engineering **27**(2), 443–460 (2015)
8. Zhu, M., Jia, Z., Zuo, L., et al.: Research on Entity Linking of Chinese Micro Blog. Acta Scientiarum Naturalium Universitatis Pekinensis **1**, 73–78 (2014). (in Chinese)
9. Gattani, A., Lamba, D.S., Garera, N., et al.: Entity extraction, linking, classification, and tagging for social media: a wikipedia-based approach. Proceedings of the VLDB Endowment **6**(11), 1126–1137 (2013)
10. Yang, X., Li, P., Zhu, Q.: Name Disambiguation Based on Dependency Feature in Web Page Text. Computer Engineering **38**(19), 133–136 (2012). (in Chinese)
11. Nguyen, H.T., Cao, T.H.: Exploring wikipedia and text features for named entity disambiguation. In: Nguyen, N.T., Le, M.T., Świątek, J. (eds.) ACIIDS 2010. LNCS, vol. 5991, pp. 11–20. Springer, Heidelberg (2010)
12. Zhao, J.: A Survey on Named Entity Recognition, Disambiguation and Cross-Lingual Coreference Resolution. Journal of Chinese information Processing **23**(2), 3–13 (2009). (in Chinese)

13. Cao, J., Wu, X., Xia, Y., et al.: Pinyin-indexed method for approximate matching in Chinese. Journal of Tsinghua University (Science and Technology) **49**(S1), 1328–1332 (2009). (in Chinese)

14. Cucerzan, S.: Large-scale named entity disambiguation based on wikipedia data. In: Proceedings of the 2007 Joint Conference on Empirical Methods in Natural Language Processing and Computational Natural Language Learning, pp. 708–716 (2007)

15. Zheng, Z., Li, F., Huang, M., et al.: Learning to link entities with knowledge base. In: Human Language Technologies: The 2010 Annual Conference of the North American Chapter of the Association for Computational Linguistics, pp. 483–491 (2010)

16. Han, X., Sun, L., Zhao, J.: Collective entity linking in web text: a graph-based method. In: Proceedings of the 34th International ACM SIGIR Conference on Research and Development in Information Retrieval, pp. 765–774 (2011)

17. Chen, Z., Ji, H.: Collaborative ranking: a case study on entity linking. In: Proceedings of the Conference on Empirical Methods in Natural Language Processing, pp. 771–781 (2011)

18. Zhang, W., Sim, Y.C., Su, J., Tan, C.L.: Entity linking with effective acronym expansion, instance selection and topic modeling. In: Proceedings of the Twenty-Second International Joint Conference on Artificial Intelligence, vol. 3, pp. 1909–1914 (2011)

19. Zou, X., Sun, C., Sun, Y., Liu, B., Lin, L.: Linking entities in tweets to wikipedia knowledge base. In: Zong, C., Nie, J.-Y., Zhao, D., Feng, Y. (eds.) NLPCC 2014. CCIS, vol. 496, pp. 368–378. Springer, Heidelberg (2014)

20. Davis, A., Veloso, A., Silva, A.S.D., et al.: Named entity disambiguation in streaming data. In: Proceedings of the Conference on European Chapter of the Association for Computational Linguistics, vol. 1, pp. 815–824 (2012)

21. Han, X., Zhao, J.: Named entity disambiguation by leveraging wikipedia semantic knowledge. In: Proceedings of the 18th ACM Conference on Information and knowledge management, pp. 215–224 (2009)

22. Meng, Z., Yu, D., Xun, E.: Chinese microblog entity linking system combining wikipedia and search engine retrieval results. In: Zong, C., Nie, J.-Y., Zhao, D., Feng, Y. (eds.) NLPCC 2014. CCIS, vol. 496, pp. 449–456. Springer, Heidelberg (2014)

23. Kataria, S.S., Kumar, K.S., Rastogi, R.R., et al.: Entity disambiguation with hierarchical topic models. In: Proceedings of the 17th ACM SIGKDD International Conference on Knowledge Discovery and Data Mining, pp. 1037–1045 (2011)

24. Sen, P.: Collective context-aware topic models for entity disambiguation. In: Proceedings of the 21st International Conference on World Wide Web, pp. 729–738 (2012)

25. Nadeau, D., Turney, P.: A supervised learning approach to acronym identification. In: The Eighteenth Canadian Conference on Artificial Intelligence (2005)

26. SWJTU Chinese Word Segmentation System. http://ics.swjtu.edu.cn

BosonNLP: An Ensemble Approach for Word Segmentation and POS Tagging

Kerui Min[✉], Chenggang Ma, Tianmei Zhao, and Haiyan Li

BosonData, Inc., Shanghai, China
{kerui.min,chenggang.ma,tianmei.zhao,haiyan.li}@bosondata.com.cn

Abstract. Chinese word segmentation and POS tagging are arguably the most fundamental tasks in Chinese natural language processing. In this paper, we show an ensemble approach for segmentation and POS tagging, combining both discriminative and generative methods to get the advantage of both worlds. Our approach achieved the F1-score of 96.65% and 91.55% for segmentation and tagging respectively in the contest of NLPCC 2015 Shared Task 1, obtained the 1st place for both tasks.

1 Introduction

Chinese word segmentation and POS tagging are arguably the most fundamental tasks in Chinese natural language processing [4]. The problem has been studied for a long time in formal language domains such as newspapers and radio reports. Recently, informal text domain such as micro-blog started to attract researchers' attention, due to its practical and industrial value. As a result, the Shared Task 1 in NLPCC 2015 tries to provide a platform for evaluation [1].

In addition to other widely used dataset such as *the Chinese TreeBank (CTB)*, the difficulty of the task comes from the following:

1. Relatively small corpus for training, which only contains 10,000 sentences and 215,027 words.
2. High rate of OOV (out-of-vocabulary) words: it is estimated that the OOV rate for the test data is 7.25%.
3. Sentences contain informal words such as "萌萌哒", *e.g.*, 【/PU 美声/NN 版/NN 《/PU 小/JJ 苹果/NN 》/PU 传递/VV 萌萌哒/JJ 关爱/NN 】/PU..
4. The POS tag-set defines two types of adjectives: JJ and VA, distinguishing a normal adjective from predicative adjective, which requires longer dependency syntactic information for inference.

In this paper, we show an ensemble approach for segmentation and POS tagging, combining both discriminative and generative methods to get the advantage of both worlds. Evaluated on the Sina Weibo data, our approach achieved the F1-score of 96.65% and 91.55% for segmentation and tagging respectively, in the contest of NLPCC 2015 Shared Task 1, obtained the 1st place for both tasks.

© Springer International Publishing Switzerland 2015
J. Li et al. (Eds.): NLPCC 2015, LNAI 9362, pp. 520–526, 2015.
DOI: 10.1007/978-3-319-25207-0_48

Type	Template
UNIGRAM	$C[-2], C[-1], C[0], C[1], C[2]$
BIGRAM	$C[-2.. - 1], C[-1..0], C[0..1], C[1..2]$
TRIGRAM	$C[-2..0]$
SKIPGRAM	$C[-1]C[1]$
CHARTYPE	$T[-2..2]$
PUNCTUATION	TRUE if $C[0]$ is a punctuation
LASTCHAR	TRUE if $C[0]$ is the last char of **x**
REPEAT-1	TRUE if $C[0] = C[-1]$
REPEAT-2	TRUE if $C[0] = C[-2]$

Fig. 1. Feature template for character-level features.

2 Backbone Algorithm

As usual, we treat both word segmentation and POS tagging as sequence labeling tasks. We use conditional random fields (CRF) as the backbone algorithm for ensemble learning, combining results from different models.

In this section, we describe our approach for the backbone algorithm in three steps: pre-processing, statistical modeling, and post-processing. In the next section, we show how to modify this algorithm to build an ensemble model, where the latter one will be used for open track evaluation.

2.1 Pre-processing

The training data was given in the format that each character was associated with one of the four states $\{\mathbb{B}, \mathbb{M}, \mathbb{E}, \mathbb{S}\}$ for word segmentation, representing the beginning, inside, ending, isolation of a word respectively. In literature, this is usually referred as the 4-tag labeling [3]. In addition to the 4-tag labeling, we also generated the following formats:

1. 3-tag: $\{\mathbb{M}, \mathbb{E}, \mathbb{S}\}$, where the state "$\mathbb{M}$" represents both the beginning and inside states.
2. 5-tag: $\{\mathbb{B}, \mathbb{C}, \mathbb{M}, \mathbb{E}, \mathbb{S}\}$: where "$\mathbb{C}$" indicates the second character in a word.
3. 6-tag: $\{\mathbb{B}, \mathbb{C}, \mathbb{D}, \mathbb{M}, \mathbb{E}, \mathbb{S}\}$: where "$\mathbb{D}$" indicates the thrid character in a word.

Larger tag-sets allow more detailed feature representation, at the cost of potentially higher variance. This is the well-known *Bias-Variance Tradeoff* [5]. In our experiments, we found that 5-tag consistently provides the best results for the given corpus, often 0.1% to 0.2% higher than other representations. Therefore, we use the 5-tag representation for both segmentation and POS tagging. As the shared task requires to submit evaluation results using the 4-tag representation, we simply convert the 5-tag to 4-tag before submission.

2.2 Statistical Modeling

As we have mentioned earlier, we use the second-order linear-chain conditional random fields (CRF) as the backbone algorithm, as exact inference can be done in

Type	Template
PREFIX	P-k=TRUE if $C[0..k-1]$ is k-prefix $(2 \le k \le 10)$
SUFFIX	S-k=TRUE if $C[-k+1..0]$ is k-suffix $(2 \le k \le 10)$
SINGLETON	TRUE if $C[0] \in \mathfrak{D}$
PREFIXTAG	P-k-t=TRUE if $C[0..k-1]$ is k-prefix and $t \in \mathfrak{D}[C[0..k-1]]$ $(2 \le k \le 10)$
SUFFIXTAG	S-k-t=TRUE if $C[1-k..0]$ is k-suffix and $t \in \mathfrak{D}[C[1-k..0]]$ $(2 \le k \le 10)$
MORPHOLOGY1	M-t-$C[k]$=TRUE $t \in \mathfrak{D}[C[0..k-1]]$ and $C[0..k] \notin \mathfrak{D}$
MORPHOLOGY2	M-$C[0]$-t=TRUE $t \in \mathfrak{D}[C[1..k]]$ and $C[0..k] \notin \mathfrak{D}$

Fig. 2. Feature template for dictionary features.

polynomial time. Specifically, the conditional probability of the hidden sequence **y**, given the observation sequence **x**, is defined as

$$p(\mathbf{y}|\mathbf{x}) = \frac{\exp\left(\sum_i \sum_k \lambda_k f_k(\mathbf{y}_{i-1}, \mathbf{y}_i, \mathbf{x})\right)}{Z_\mathbf{x}}, \tag{1}$$

where $Z_\mathbf{x}$ is the normalization factor.

Notice that from the above equation, the observation **x** influences **y** through feature functions $\{f_k(\cdot)\}_k$. The problem boils down to good feature extraction $\{f_k(\mathbf{y}_{i-1}, \mathbf{y}_i, \mathbf{x})\}_{k=1}^m$, which has been well-studied. Similar to previous work, we use $C[i]$ to indicate the character of the i-th position, centered at the current position, and $T[i]$ to indicate the type of $C[i]$. Below, five types of character are defined.

1. $T[i] = P$: if $C[i]$ is either an English or Chinese punctuation.
2. $T[i] = N$: if $C[i]$ is either an Arabic digit or Chinese number, *e.g.*, "0123 $\cdots$ 零一二...".
3. $T[i] = A$: if $C[i]$ is in alphabet set, $[a-z][A-Z]$, and full-width characters like "A B C D $\cdots$" are included.
4. $T[i] = D$: if $C[i]$ is in the following set "年月日时分秒".
5. $T[i] = O$: for any other characters.

Given the above definition, we are able to define effective feature template as in Fig 1. To deal with OOV words with repeat-character patterns, we define the REPEAT-1, REPEAT-2 to distinguish word of "AABB" type from "ABAB" type, where the former one is more likely an adjective and the later one a verb [2].

Notice that the above features only require character-level information, which is good for its simplicity. However, in order to get relatively accurate parameter estimation for weights $\{\lambda_k\}_{k=1}^m$, a minimal occurrence number of ~ 10 is often required for *maximum-likelihood estimation*. In other words, if a word, say $C[-1..0]$, occurred only a few times in training, it is unlikely to obtain accurate and unbiased estimation for it. Therefore, if we know $C[-1..0]$ can be a word, we would like to incorporate the information that $C[-1..0]$ occurred as a word in training directly, in addition to the BIGRAM features.

Let $\mathfrak{D}$ be the dictionary extracted from training corpus. We say a substring $C[i..i+k-1]$ is a k-prefix if $C[i..i+k-1] \in \mathfrak{D}$; a substring $C[i-k+1..i]$ is a k-suffix if $C[i - k + 1..i] \in \mathfrak{D}$. Furthermore, we define $\mathfrak{D}[s]$ to be the set of all possible POS tags of the given word s in our dictionary. If $s \notin \mathfrak{D}$, we have $\mathfrak{D}[s] = \emptyset$.

Here we introduce another family of feature to alleviate the difficulty of OOV words. Although Chinese is not a *morphologically rich language* (MRL) [6], we do observe several common patterns that might be useful for segmentation and tagging. For example, there are a number of words has the structure of "NN+ 们", such as "童鞋们/NN 动物们/NN 亲们/NN 鼠狗们/NN". During evaluation, if we ever encounter an OOV word "设计师们", we can infer that the it is likely to be a NN word if NN NN $\in \mathfrak{D}[$设计师$]$. Inspired by it, we designed the morphological features, as listed in Fig 2.

In the experiment, we will see that the combination of both character-level features and dictionary features provides effective information for our model to produce accurate prediction.

2.3   Post-processing

From the formulation of Eq (1), the character-based CRF model is a probabilistic model. We impose the following rules as post-processing to further improve the accuracy of the prediction.

1. the prediction of the last character of $\mathbf{y}$ must end with $\{\mathbb{S}, \mathbb{E}\}$.
2. never mark the end of a number if the next character is also a number.
3. never mark the end of a English word if the next character is in alphabet.

The above post-processing rules make the final prediction more coherent. We also tried to use *regular expression* to recognize URLs, which gave limited improvement, since URLs are very rare in the corpus.

3   Ensemble Model for Open Track

The backbone algorithm described in Section 2 is able to produce relatively accurate prediction, as will be found in the experiment section. However, due to the small training size, the variance of the model parameters is high. In the open track of the shared task, it is allowed to use other information and tools to improve the prediction. In addition to the backbone algorithm, we use the output from the following models:

– *Model A*: a HMM-based model trained on People's daily corpus[1].
– *Model B*: a CRF-based model trained on *the Chinese TreeBank (CTB) 7.0* [7]. The features used in this model are the same as in Section 2.
– *Model C*: a third-order discriminative sequence model trained on People's daily corpus, plus 100,000 sentences of more recent news collected and annotated by BosonData, Inc. This model will be released at www.bosonnlp.com for industrial developers and researchers to use for free.

---

[1] Newspaper text from People's Daily 1998.

Dataset	Sents	Words	Chars	Word Types	Char Types	OOV Rate
Training	10,000	215,027	347,984	28,208	39,71	-
Test	5,000	106,327	171,652	18,696	3,538	7.25%
Total	15,000	322,410	520,555	35,277	4,243	-

Fig. 3. Statistical information of dataset.

For a given input $\mathbf{x}$, denote the output of the above models as $\mathbf{y}_A, \mathbf{y}_B$ and $\mathbf{y}_C$. For each feature described in Section 2, we concatenate the original feature with the output from the above models during feature extraction, *i.e.* using $\{f_k (\mathbf{y}_{i-1}, \mathbf{y}_i, \mathbf{x}, \mathbf{y}_A, \mathbf{y}_B, \mathbf{y}_C)\}_k$. For example, for the original TRIGRAM feature, $C[-2..0]$, we enhance it by the following features:

$$C[-2..0] \circ \mathbf{y}_A[i], C[-2..0] \circ \mathbf{y}_B[i], C[-2..0] \circ \mathbf{y}_C[i],$$

where i is the current position for feature extraction. Finally, we train the same model with these extra information to get better prediction. We call it the ensemble model. In the next section, we will show that although different corpus has its own segmentation and POS tagging standard, the ensemble model could benefit from them.

4 Experiments

As we have mentioned earlier, the dataset used for this shared task is relatively informal, collected from *Sina Weibo*. The training and test data consist of microblogs from various topics, such as finance, sports, entertainment, and so on. Basic statistics of the dataset can be found in Fig 3.

Next, we show the performance of our models, BosonNLP, with other top competitors for this shared task. Notice that these evaluation scores were released from the official program committee of NLPCC 2015.

4.1 Closed Track Evaluation

During the closed track, competitors are not allowed to use external datasets or tools. We therefore apply the backbone algorithm described in Section 2 on the given training corpus. Notice that the dictionary $\mathfrak{D}$ can be obtained from the training coropus. The evaluation result can be found in Fig 4.

From the result, one can see that the proposed algorithm is very effective. Our result is on a par with the best solution (with 0.09% difference) for segmentation, and obtained the best result for POS tagging with 0.74% higher in terms of the F-1 score.

4.2 Open Track Evaluation

As shown in Fig 5, the evaluation result for the open track clearly demonstrate the advantage of our ensemble approach, compared with other teams.

Task	Team	Precision	Recall	F-1
POS	BosonNLP (1st)	88.91	88.95	88.93
POS	XUPT (2nd)	88.54	87.83	88.19
POS	WHU (3rd)	88.28	87.67	87.97
SEG	NJU (1st)	95.14	95.09	95.12
SEG	BosonNLP (2nd)	95.03	95.03	95.03
SEG	BUPT (3rd)	94.78	94.42	94.60

Fig. 4. Closed track result for NLPCC 2015 Shared Task 1.

Task	Team	Precision	Recall	F-1
POS	BosonNLP (1st)	91.42	91.68	91.55
POS	SZU (2nd)	88.93	89.05	88.99
POS	BJTU (3rd)	79.85	83.51	81.64
SEG	BosonNLP (1st)	96.56	96.75	96.65
SEG	NJU (2nd)	96.03	96.15	96.09
SEG	SZU (3rd)	95.52	95.64	95.58

Fig. 5. Open track result for NLPCC 2015 Shared Task 1.

We obtained the *1st* place for both segmentation and POS tagging, with significantly higher F-1 score (2.56% higher than the second place team). For the POS tagging task, we obtained the F-1 score of 91.55%, which is the only team with the score above 90%.

As an example, we compare the output from our proposed algorithm for the closed track, our ensemble algorithm, and *Model C* output, in order to get some insights.

- **Input:** 据气象部门预告，哈尔滨有瞬时风速6-7级左右的大风，并可能伴有短时强降水，雷电或冰雹等强对流天气。
- *Backbone algorithm* output: 据/P 气象/NN 部门/NN 预告/VV ，/PU 哈尔滨/LOC **有瞬/VV 时风速/NN** 6-7/CD 级/NN 左右/LC 的/DSP 大风/NN ，/PU 并/AD 可能/MV 伴有/VV 短时/JJ 强/JJ 降水/NN ，/PU **雷电/PER** 或/CC 冰雹/NN 等/ETC 强对流/NN 天气/NN 。/PU
- *Ensemble model* output: 据/P 气象/NN 部门/NN 预告/VV ，/PU 哈尔滨/LOC 有/VV **瞬时/NN 风速/NN** 6-7/CD 级/NN 左右/LC 的/DSP 大风/NN ，/PU 并/AD 可能/MV 伴有/VV 短时/JJ 强/JJ 降水/NN ，/PU **雷电/NN** 或/CC 冰雹/NN 等/ETC 强对流/NN 天气/NN 。/PU
- *Model C* output: 据/p 气象/n 部门/n 预告/v ，/wd 哈尔滨/ns 有/vyou **瞬时/t** 风速/n 6/m -/wp 7/m 级/q 左右/m 的/ude 大风/n ，/wd 并/c 可能/v 伴有/v 短时/b 强/a 降水/n ，/wd 雷电/n 或/c 冰雹/n 等/udeng 强/a 对流/n 天气/n 。/wj

We highlighted the difference between models. Observe that the backbone algorithm made a segmentation mistake at "有瞬/VV 时风速/NN", which was a typical OOV word mistake. The result was corrected by the ensemble model to be "瞬时/NN 风速/NN". Notice that although the POS tagging standard, including the word segmentation standard for *Model C* is different ("瞬时/t 风速/n"), the output is helpful to correct the backbone algorithm output. The ensemble model proposed in Section 3 provides a probabilistic model to *fuse* the information.

5 Conclusion

In this paper, we described our approach for the word segmentation and POS tagging approach for the NLPCC 2015 Shared Task 1. Although some of the features and techniques were designed for this particular task, they are also useful for general word segmentation and POS tagging tasks with varying standards. We plan to release our solution at www.bosonnlp.com, to allow other researchers and developers to exploit.

References

1. Qiu, X., Qian, P., Yin, L., Huang, X.: Overview of the NLPCC 2015 Shared Task: Chinese Word Segmentation and POS Tagging for Micro-blog Texts (2015). http://arxiv.org/abs/1505.07599
2. Huang, C.-T.J., Li, Y.-H.A., Li, Y.: The Syntax of Chinese (Cambridge Syntax Guides). Cambridge University Press (2009)
3. Zhao, H., Huang, C.-N., Li, M., Lu, B.-L.: Effective tag set selection in chinese word segmentation via conditional random field modeling. In: The 20th Pacific Asia Conference on Language, Information and Computation (PACLIC-2006) (2006)
4. Wong, K.-F., Li, W., Xu, R., Zhang, Z.-S.: Introduction to Chinese Natural Language Processing (Synthesis Lectures on Human Language Technologies). Morgan & Claypool Publishers (2009)
5. Bias-variance decomposition. In: Sammut, C., Webb, G.I. (eds.) Encyclopedia of Machine Learning. Springer (2011)
6. Sarikaya, R., Kirchhoff, K., Schultz, T., Hakkani-Tur, D.: Introduction to the special issue on processing morphologically rich languages. IEEE Transactions on Audio Speech, and Language Processing (2009)
7. Xue, N., Xia, F., Chiou, F.-D., Palmer, M.: The Penn Chinese TreeBank: Phrase structure annotation of a large corpus. Journal of Natural Language Engineering (2005)

Research on Open Domain Question Answering System

Zhonglin Ye[1], Zheng Jia[1(✉)], Yan Yang[1], Junfu Huang[1], and Hongfeng Yin[2]

[1] School of Information Science and Technology,
Southwest Jiaotong University, Chengdu, China
zhonglin_ye@foxmail.com, {zjia,yyang}@swjtu.edu.cn,
990504422@qq.com
[2] DOCOMO Innovations Incorporation, Palo Alto, USA
hongfeng_yin@yahoo.com

Abstract. Aiming at open domain question answering system evaluation task in the fourth CCF Natural Language Processing and Chinese Computing Conference (NLPCC2015), a solution of automatic question answering which can answer natural language questions is proposed. Firstly, SPE (Subject Predicate Extraction) algorithm is presented to find answers from the knowledge base, and then WKE (Web Knowledge Extraction) algorithm is used to extract answers from search engine query result. Experimental data provided in the evaluation task includes the knowledge base and questions in natural language. The evaluation result shows that MRR is 0.5670, accuracy is 0.5700, and average F1 is 0.5240, and indicates the proposed method is feasible in open domain question answering system.

Keywords: Automatic question answering · Open domain · Natural language understanding · Information extraction

1   Introduction

Google, Baidu, Bing and other search engines return hyperlinks containing the keywords of user queries, which do not give users a simple and direct answer, and users need to browse web pages to find answers they need. Although search engines can help users find answers to a certain extent, but users may need to click many hyperlinks of pages. Driven by the mobile Internet, the automatic question answering system based on domain knowledge base can directly get an intuitive and accurate answer, so it becomes an important focus of research.

This paper presents a solution of open domain question answering (QA). The method mainly includes three parts: (1) SPE (Subject Predicate Extraction) algorithm; (2) WKE (Web Knowledge Extraction) algorithm; (3) answer format standardization. SPE algorithm extracts subject and predicate in questions, searching answers from the knowledge base according to the subject and predicate of questions. SPE solves the problem of QA in limited domain, such as People, Time and Geography. For other

J. Li et al. (Eds.): NLPCC 2015, LNAI 9362, pp. 527–540, 2015.
DOI: 10.1007/978-3-319-25207-0_49

domains, WKE algorithm is used to extract answers from the unstructured texts of search engine results, and can effectively solves the problem of QA in open domain. In order to improve the accuracy of question answering system, answers are turned into a standardized format. The knowledge base is from Baidubaike, and the quantity of it is about 4 million triples which are in the format of <Subject, Predicate, Object>. Testing data are 1000 natural language questions which are from Microsoft's Bing query log and generated from the knowledge base.

The rest chapters of this paper are arranged as follows: the second chapter is about the related work. The third chapter introduces the proposed method. The fourth chapter is the experiment. The last chapter concludes the whole paper and looks forward to the future research.

2   Related Work

QA is to provide a quick, direct and accurate answer for the natural language questions [1]. The existing QA technologies can be divided into three categories according to the source of answers: (1) QA based on search engine; (2) QA based on community, (3) QA based on knowledge base [1]. In QA based on search engine, the answers are extracted from the web pages of search result [2]. QA based on community calculates the similarity between the questions raised by users and questions stored in database asked in the past to get results [3]. The main works of QA based on knowledge base is semantic analysis [4,5,6,7,8,9] and knowledge base building [10], such as Poon [11], Yahya [12] and Berant [13,14,15] put forward a method to build the QA system based on semantic analysis. First, they extracted the subject entities and predicate words of questions. Then the questions are converted into SPARQL structured query language. Finally, answers are acquired through retrieving knowledge base by SPARQL. Bordes [16] and Yao [17] proposed a way of information retrieval to build QA system. Fader [18] and Kushman [19] presented a method of building QA system based on open information extraction. Bao [20] and Comas [21] raised an approach of constructing QA system based on translating question into answer on the basis of knowledge base.

3   Methods

The procedure of open domain QA is shown in Figure 1 as follows:

As shown in Figure 1, the system is mainly composed of four parts: question classification, SPE algorithm, and WKE algorithm and answer format standardization. Firstly, natural language questions are classified. Then, SPE algorithm is applied to extract subject and predicate of the questions, and retrieve the answers from the knowledge base. If there are no answers, WKE algorithm is applied to search the questions in search engine to get the web pages containing the questions, to analyze the texts of search results, and extract answers from the texts. Finally, the answers are turned into a standard format.

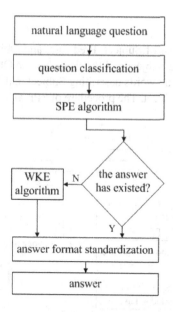

Fig. 1. Procedure of open domain question answering

3.1   Question Classification

The questions are divided into four categories: 人物|People[1], 地点|Geography, 时间|Time and Others. In this paper, the category of questions is got through feature words matching. Some of feature words for each category are described in Table 1.

Table 1. Feature word examples

Category	Feature words
人物|People	谁|who, 名字|name, 姓名|name, 中文名|Chinese name, 英文名|English name, 别名|alias, 叫什么|what's called
地点|Geography	在哪|where, 哪里|where, 地方|address, 位置|position
时间|Time	哪一年|which year, 什么时候|when, 时间|time, 诞生|born, 时期|date, 哪年|which year, 何时|when
Other	无|null

If any question contains the feature words, the category of the question can be acquired. The category has two functions: one is used to identify the predicate of the question in SPE algorithm, and the other is to extract answers in WKE algorithm.

---

[1] The content after "|" is the English translation of Chinese words.

3.2 SPE Algorithm

The knowledge triples consist of Subject, Predicate, and Object. The questions usually contain the words describing Subject and Predicate but the Object is absent. The goal of SPE algorithm is to extract the words describing Subject and Predicate and to search the knowledge base to find the Object. The procedure of SPE algorithm is shown in Figure 2:

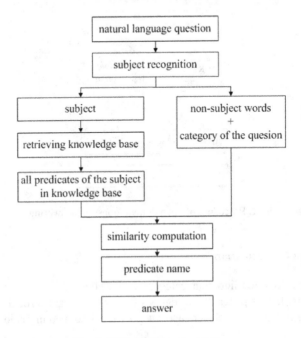

Fig. 2. SPE algorithm description

Illustrated as Figure 2, the steps of SPE algorithm are shown as below.

(1) Identify the subject entity name based on CRF (Conditional Random Field) algorithm and SWJTU Chinese Word Segmentation System.

(2) If the question contains a subject, then retrieve knowledge base to find all triples of the subject.

(3) The words except the subject which are called non-subject words are combined with category of the question.

(4) Compute the similarity [22] between each predicate of the subject and the combination of non-subject words and categories. Select predicate name which similarity value is the highest.

(5) According to the subject and predicate, the object can be got from triples in the knowledge base and the object is the answer of the question.

For example, the question is 侏罗纪世界什么时候上映|When will Jurassic World release.

The feature word is 什么时候|when, so the category of the question is 时间|Time. The subject is recognized as 侏罗纪世界|Jurassic World based on CRF. The non-subject word is上映|release. The combination of category and non-subject word is 上映时间|release time. There is a predicate named "上映时间|release time" in the triple <侏罗纪世界|Jurassic World, 上映时间|release time, 2015-06-12> of the subject 侏罗纪世界|Jurassic World, so the object "2015-06-12" can be found as the answer of the question. The detailed descriptions of main steps are as follows.

3.2.1 Subject Recognition

We firstly use SWJTU Chinese Word Segmentation System [23] to recognize the entities in questions. This system can achieve good performance on recognizing several categories of entities such as Person, Place, and Organization etc. However it cannot recognize many other categories such as Film and Music etc. To realize good performance of entity recognition, then we use the CRF algorithm [24] to recognize the entities of Film, Music, Book, Game and Application categories.

CRF named entity recognition algorithm requires the establishment of a training model which is used to predict the new entities. Because the better the quality of training data is, the higher performance the algorithm becomes, it needs to collect a lot of high-quality questions manually labeled as training data. We collected a large number of questions through crawling on the Internet or writing by hand, and labeled these questions manually. The training data are got mainly from the following resources.

(1) Web. We crawled the data from Baidu Knows, Douban, 360 search engine and so on.
(2) Question collection online system. We developed a question collection online system and about more than 300 students participated in writing and labeling questions.
(3) Mobile assistant voice data..We bought the mobile assistant voice data from Datatang Corporation. The number of the questions is 145371 and we labeled the data manually.

The total number of the questions is about more than 30000. The average F-measure value of recognizing subject named entities for the five categories based on CRF is 92.44%.

3.2.2 Retrieve Knowledge Base

Once the subject entity in questions is recognized, the triples in knowledge base containing this subject can be acquired by retrieving knowledge base. A predicate dictionary of the subject can be created according to the triples.

For example, the question is 梁启超的生日是什么时候|When is Qichao Liang's birthday. The subject is recognized as a person梁启超|Qichao Liang. By retrieving the knowledge base, we can get the knowledge triples as shown in Table 2.

The predicate dictionary of Qichao Liang is created from triples as follows:

[中文名|Chinese name, 外文名|English name, 别名|alias, 国籍|nationality, 民族|volk, 出生地|birthplace, 出生日期|birthdate, 逝世日期|death time, 职业|career, 信仰|faith, 主要成就|achievement, 代表作品| representative works].

Table 2. The triple examples of subject "梁启超|Qichao Liang"

No.	<Subject, Predicate, Object>	No.	<Subject, Predicate, Object>
1	<梁启超|Qichao Liang, 中文名|Chinese name, 梁启超|Qichao Liang >	7	<梁启超|Qichao Liang, 出生日期|birthdate, 1873年2月23日|February 23, 1873 >
2	<梁启超|Qichao Liang, 外文名|English name, Qichao Liang >	8	<梁启超|Qichao Liang, 逝世日期|death time, 1929年1月19日|January 19, 1929 >
3	<梁启超|Qichao Liang, 别名|alias, 卓如|Zhuo Ru >	9	<梁启超|Qichao Liang, 职业|career, 思想家|thinker>
4	<梁启超|Qichao Liang, 国籍|nationality, 中国|China>	10	<梁启超|Qichao Liang, 信仰|faith, 儒学|Confucianism >
5	<梁启超|Qichao Liang, 民族|volk, 汉族|Han>	11	<梁启超|Qichao Liang, 主要成就|achievement, 公车上书| Gong Che Shang Shu 、戊戌变法|Wu Xu Reformation>
6	<梁启超|Qichao Liang, 出生地|birthplace, 广东省新会市|Xinhui city, Guangdong province>	12	<梁启超|Qichao Liang, 代表作品| representative works, 《中国近三百年学术史》|《scholastic history of the past 300 years in China》、《中国历史研究法》|《historiography of Chinese history》 >

3.2.3   Similarity Computation

There are many different expressions for a predicate in natural language questions. After acquiring the subject in questions and the predicate dictionary in knowledge base, the similarity between the words in questions and predicates in dictionary is computed to determine the predicate in questions. We combine the non-subject words with the category of the questions into new words.

For example, 生日|birthday is the non-subject word of the question "梁启超的生日是什么时候|When is Qichao Liang's birthday" and the question category is 时间|Time. These two words are combined into a new word "生日时间|birthday time". Then we use the method proposed in [22] to compute the semantic similarity between "生日时间|birthday time" and the words in the predicate dictionary of "梁启超|Qichao Liang", and we choose the word with the highest similarity and more than 0.5 as the predicate in the question. When the predicate of a question has been determined, the object which is the answer can be found,

3.3   WKE Algorithm

WKE algorithm is a real-time knowledge extraction algorithm based on search engine. It mainly consists of two parts: 1) structured knowledge extraction from Baidu Zhixin; 2) unstructured text knowledge extraction from web pages. Knowledge extraction procedure is shown in Figure 3.

Firstly, we use extract structured knowledge from Baidu Zhixin. If we cannot get the answer from Baidu Zhixin or Baidu Zhixin doesnot return any result in search engine, we combine the title and abstract of web pages in search engine result into an unstructured text and extract answers from it.

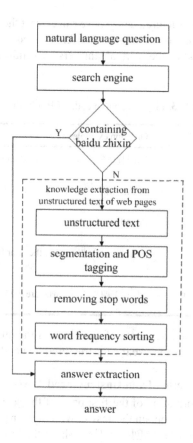

Fig. 3. WKE knowledge extraction procedure

3.3.1 Knowledge Extraction from Baidu Zhixin

Baidu Zhixin is a QA system integrated into Baidu search engine, which can answer simple semantic question. For example, if input the natural language question "成都有多少人口|What is the population of Chengdu" in Baidu search engine, the search result of Baidu Zhixin is shown in Figure 4:

成都人口：

1417万（2013年）

成都，简称"蓉"，别称"锦官城"，四川省省会，副省级市，国家区域中心城市（西南），1993年被国务院确定为西南地区科技、商贸、金融中心和交通、通... 详情>>

来自百度百科 ｜ 报错

Fig. 4. Baidu Zhixin search result

Baidu Zhixin is a semi-structured data, so we can extract the answer by analyzing the structure of Baidu Zhixin. The structure is different according to different types of queries. Some flags marking the start of answers in Baidu Zhixin source code are shown in Table 3.

Table 3. Flags in source code of Baidu Zhixin

Flags	Question examples
op_exactqa_s_answer	成都有多少人口| what is the population of Chengdu
c-text c-text-important	春秋五霸是谁|who are the Five Overloads in Spring and Autumn Period
op_exactqa_body	京城四美是谁|who are the Four Beauty in Capital City
op_definitive_answer_po_item_img	海贼王四皇是谁| who are the Four Emperor in One Piece

Baidu Zhixin can only answer some of questions. If the questions cannot be solved by Baidu Zhixin, we need to analyze the unstructured text and extract answers from it.

3.3.2   Knowledge Extraction from Unstructured Texts

The title and abstract information of the first page of the search engine are formed an unstructured text which may contain the answer with high probability. For example, the search result of the question "谷歌创始人是谁|who is the founder of Google" is shown in Figure 5.

Fig. 5. Search result example

The process of extracting knowledge from unstructured text is as follows:

(1) Get the title and abstract content of each link in the first page.
(2) Segment and tag POS for the unstructured text.
(3) Remove stop words.
(4) Count word frequency and sort the words
(5) Extract the word with the highest frequency according to the category of questions.

We use the term frequency-inverse document frequency (TF-IDF) to count word frequency at step 4. To gain the useful feature of answer, WKE algorithm defines the categories of questions as Time, Geography, People and Others, which are also the types of answers. When the category of the question is Time, the answer of the question is a time. When the category is People, the answer of the question is a person name. We extract the answer according to the POS tag of words. The POS tag of Time is "t", the tag of People is "nr" and the tag of Geography is "ns". Other category is not limited to any POS tag. Therefore, the answer extraction and features selection rules are summarized as following:

(1) In Geography category questions, if multiple words have "ns" POS tag and appear in the text continuously, they can be combined into a complete geographic name. Otherwise the word tagged as "ns" with the highest frequency is extracted as the answer.
(2) In People category questions, if the question contains "五霸|The Five Wverloads" , "四杰|Four Outstanding Poets", "四大|Four Celebrities" , "四少|Four Talented People", "六君子|Six Nobles",, and the words in the text have "nr" POS tag and appear in successive, these person names are extracted as the answer. Otherwise, the word tagged as "nr" with the highest frequency is extracted as the answer.
(3) In Time category questions, if the unstructured text contains the words with "t" POS tag, the word with the highest frequency is extracted. If the unstructured text doesn't contain the words with "t", the word with the POS tag of "m" which refers to quantifiers and the highest frequency is extracted as the answer.
(4) For the above three categories, if there is not any part of speech needed, any word with the highest frequency is as the answer.
(5) Other question category, the word with the highest frequency is the answer.

3.4   Answer Format Standardization

Answer format standardization refers to change the format of answer which is tagged as Time or Geography according to the words contained in questions.

For the Geographic category questions, if the questions contain the characters such as "省|province", "国家|country", "州|state", or "市|city", the part in answers after these above characters is removed.

For the Time category questions, if the questions contain about the characters such as "年|year", the part in answers after the above character is removed.

4   Experiment

4.1   Data Set

The data set used in this paper is published by NLPCC2015 evaluation task and includes knowledge base and testing questions. The knowledge base is huge which has about 48 million triples given in the form as shown in Table 4:

Table 4. The triple knowledge information in Knowledge base

Subject	Predicate	Object
成都|Chengdu	中文名称|Chinese name	成都|Chengdu
成都|Chengdu	外文名|English name	chengdu
成都|Chengdu	别名|alias	蓉城| Rong city、芙蓉城| lotus city、锦官城|the city of brocade、天府之国| the land of abundance
成都|Chengdu	所属地区|subordinate regions	西南地区| southwest
成都|Chengdu	面积|area	12390平方公里|12390 square kilometers
成都|Chengdu	人口|population	1417万（2013年）|1.417 millions (2013)
成都|Chengdu	常住人口| resident population	1435万（2013年）|1.435 millions (2013)
成都|Chengdu	气候条件| climate condition	亚热带季风性湿润气候|subtropical monsoon humid climate
-----	-----	-----

The number of the testing questions is 1000, which contains 421 questions of People category, 179 questions of Geography category, 168 questions of Time category, and the remains belong to Others category. Some of question examples of each category are shown in Table 5:

Table 5. Question examples of each category

Category	Question example	Category	Question example
People	初唐四杰是谁？|who is the four distinguished poet in Primary Tang?	Time	什么时候开始的南水北调工程？ | when did the South-North Water Transfer Project start?
Geography	武当山在哪里？|where is the Wudang Mountain?	Other	新加坡的国花是什么？|what is the national flower of Singapore？

4.2   Evaluation Metric

This evaluation uses the MRR (Reciprocal Rank Mean), accuracy and averaged F1 values to measure the performance of open domain QA, and the related calculation formula is defined as following:

$$MRR = \frac{1}{|Q|} \sum_{i=1}^{|Q|} \frac{1}{rank_i} \tag{1}$$

$|Q|$ denotes the total number of testing questions set, $rank_i$ denotes position of the first correct answer, if the all answer is not correct, $\frac{1}{rank_i}$ is set to 0.

$$Accuracy = \frac{1}{|Q|} \sum_{i=1}^{|Q|} \delta(C_i, A_i) \tag{2}$$

C_i denotes test answer, A_i denotes standard answer, if A_i contains one answer of C_i at least, $\delta(C_i, A_i)$ equals to 1, otherwise it equals to 0.

$$AveragedF1 = \frac{1}{|Q|} \sum_{i=1}^{|Q|} F_i \tag{3}$$

If A_i are not fully existed in C_i, F_i equals to 0, otherwise, F_i can be calculated by using the following formula:

$$F_i = \frac{2 * \frac{\#(C_i, A_i)}{|C_i|} * \frac{\#(C_i, A_i)}{|A_i|}}{\frac{\#(C_i, A_i)}{|C_i|} + \frac{\#(C_i, A_i)}{|A_i|}} \tag{4}$$

$\#(C_i, A_i)$ denotes the total number of same answer between C_i and A_i, $|C_i|$ and $|A_i|$ is the total number of the answers in $|C_i|$ and $|A_i|$.

4.3   Experimental Results

The methods presented in this paper are described below: (A) SPE algorithm based on knowledge base; (B) SKEBZ (Structured Knowledge Extraction Based on Baidu Zhixin); (C) UKEWP (Unstructured Knowledge Extraction Based on Web Page). After the QA system receives a natural language question, firstly, we firstly use method (A) to answer the question, and then uses method (B), finally, uses method (C). In order to evaluate the performance of each method, the averaged F1, accuracy and MRR values of each algorithm are given for the 1000 testing data, the results is shown in Table 6:

Table 6. Performance comparison of the three methods proposed in this paper

Question answering methods	MRR	Accuracy	Average F1

SPE	0.1601	0.1730	0.1312
SKEBZ	0.3600	0.3690	0.3314
UKEWP	0.4841	0.4970	0.4479
SPE+SKEBZ	0.3914	0.4022	0.3692
SPE+UKEWP	0.5317	0.5540	0.5066
SKEBZ+UKEWP	0.5390	0.5481	0.5211

The results of the methods proposed in the evaluation task in NLPCC2015 are shown in Table 7:

Table 7. Evaluation results in the task

Registered team	MRR	Accuracy	Average F1
Team-1	0.1430	0.1660	0.1196
Team-2	**0.5675**	**0.5700**	**0.5240**
Team-3	0.3360	0.4130	0.2990

Our methods achieve the best performance that MRR is 0.5675, accuracy is 0.57, and average F1 is 0.5240 in all teams.

5   Conclusion

Aiming at NLPCC2015 open domain QA evaluation task of complex sentence structure, the paper presents a solution of open domain QA. Firstly, question is divided into "People", "Time", "Geography" and "Other" category, and then SPE algorithm is proposed to search answers in knowledge base. If the answer is not retrieved in knowledge base, WKE algorithm is adopted to answer the questions. In addition, the paper studies how to use question context information to standardize the answer format in order to enhance the accuracy of QA. The proposed methods achieve good performance that MRR is 0.5672, accuracy is 0.57 and average F1 value is 0.5240 in NLPCC2015 open domain evaluation, and the evaluation results fully proves the feasibility of method proposed in this paper.

Acknowledgement. This work is supported by National Natural Science Foundation of China under Grants No. 61170111, No.61202043, No.61262058, and Fundamental Research Funds for the Central Universities under Grants No. 2682015CX070.

References

1. Mooney, R.J.: Learning for semantic parsing. In: Gelbukh, A. (ed.) CICLing 2007. LNCS, vol. 4394, pp. 311–324. Springer, Heidelberg (2007)
2. Filman, R.E., Sangam, P.: Searching the Internet. IEEE Internet Computing 2(4), 21–23 (1998)
3. Jeon, J., Croft, W.B., Lee, J.H.: Finding simliar question in large question and answer archives. In: Proceedings of the ACM Fourteen Conference on Inference and Knowledge Management, pp. 84–90 (2005)

4. Zettlemonyer, L.S., Collins, M.: Learning to map sentence to logical form: Structured classification with probabilistic catagorical grammars. In: Proceedings of the 21th Conference on Uncertainty in Artificial Intelligence, pp. 658–666 (2005)
5. Wong, Y.W., Mooney, R.J.: Learning for semantic parsing with statistical machine translation. In: Proceedings of the Main Conference on Human Language Technology Conference of the North American Chapter of the Association of Computational Linguistics, pp. 439–446 (2006)
6. Wong, Y.W., Mooney, R.J.: Generation by inverting a semantic parser that uses statistical machine translation. In: Proceeding of the Human Language Technology Conference of the North American Chapter of the Association for Computational Linguistics, Rochester, pp. 172–179 (2007)
7. Zettlemoyer, L., Collions, M.: Online learning of relaxed CCG grammars for parsin to logical form. In: Proceeddings of ENNLP, pp. 678–687 (2007)
8. Kwiatkowski, T., Luke, S., ZettleMoyer, S.G.: Lexical generalization in CCG grammer induction for semantic parsing. In: Proceeddings of ENNLP, pp. 1512–1523 (2011)
9. Kwiatkowski, T., Luke, S., ZettleMoyer, S.G.: Inducing probabilistic CCG grammar form with higherorder unification. In: Proceeddings of EMNLP, pp. 1223–1233(2010)
10. Freebase (2015). http://www.freebase.com
11. Poon, H.F., Domingos, P.: Unsupervised semantic parsing. In: Proceeddings of EMNLP, pp. 1–10 (2009)
12. Yahya, M., Berberich, K., Elbassuoni, S.: Natural language question for the web of data. In: Proceedings of EMNLP, pp. 379–390 (2012)
13. Berant, J., Chou, A., Roy, F., et al.: Freebase QA: Information extraction or semantic parsing. In: The 2014 Conference on Empirical Methods on Natural Language Processing, pp. 1511–1527 (2014)
14. Berant, J., Chou, A., Roy, F., et al.: Semantic parsing on freebase from question-answer pairs. In: The 2013 Conference on Empirical Methods on Natural Language Processing, pp. 153–1544 (2013)
15. Berant, J., Liang, P.: Semantic parsing via paraphrasing. In: The 52nd Annual Meeting of the Association for Computational Linguistics, pp. 479-485 (2014)
16. Bordes, A., Chopra, S., Weston, J.: Question answering with subgraph embeddings. In: The 2014 Conference on Empirical Methods on Natural Language Processing, pp. 1535–1545 (2014)
17. Yao, X., Durme, B.: Information extraction over structured data: question answering with freebase. In: The 52nd Annual Meeting of the Association for Computational Linguistics, pp. 753–770 (2014)
18. Fader, A., Luke, Z., Oren, E.: Open question answering over curated and extracted knowledge bases. In: Proceedings of the Conference on Knowledge Discovery and Data Mining (KDD), pp. 1256–1265 (2014)
19. Kushman, N., Artzi, Y., Luke, Z., et al.: Learning to automatically solve algebra word problems. In: The 52nd Annual Meeting of the Association for Computational Linguistics, pp. 1042–1061 (2014)
20. Bao, J., Duan, N., Zhou, M., et al.: Knowledge-based question answering as machine translation. In: The 52nd Annual Meeting of the Association for Computational Linguistics, pp. 1272–1294 (2014)

21. Cristina, E.B., Pere, R., Comas.: Full machine translation for factoid question answering. In: Proceedings of EACL, pp. 20–29. ACL, Stroudsburg (2012)
22. Wei, C.Y., Zhan, Q., Fan, X.Z., et al.: Event Information Enhanced Question Semantic Representation for Chinese Question Answering System. Journal of Chinese Information Processing 1(29), 147–154 (2015)
23. SWJTU Chinese Word Segmentation System. http://ics.swjtu.edu.cn
24. Cho, H.C., Okazaki, N., Miwa, M., et al.: Named Entity Recognition with Multiple Segment Representation. Information Processing Management 49(4), 954–965 (2013)

Overview of the NLPCC 2015 Shared Task: Chinese Word Segmentation and POS Tagging for Micro-blog Texts

Xipeng Qiu[✉], Peng Qian, Liusong Yin, Shiyu Wu, and Xuanjing Huang

School of Computer Science, Fudan University,
825 Zhangheng Road, Shanghai, China
{xpqiu,pqian11,lsyin14,sywu13,xjhuang}@fudan.edu.cn

Abstract. In this paper, we give an overview for the shared task at the 4th CCF Conference on Natural Language Processing & Chinese Computing (NLPCC 2015): Chinese word segmentation and part-of-speech (POS) tagging for micro-blog texts. Different with the popular used newswire datasets, the dataset of this shared task consists of the relatively informal micro-texts. The shared task has two sub-tasks: (1) individual Chinese word segmentation and (2) joint Chinese word segmentation and POS Tagging. Each subtask has three tracks to distinguish the systems with different resources. We first introduce the dataset and task, then we characterize the different approaches of the participating systems, report the test results, and provide a overview analysis of these results. An online system is available for open registration and evaluation at http://nlp.fudan.edu.cn/nlpcc2015.

1 Introduction

Word segmentation and Part-of-Speech (POS) tagging are two fundamental tasks for Chinese language processing. Benefiting from the developments of the machine learning techniques and the large scale shared corpora, such as Chinese Treebank [9], Chinese word segmentation and POS tagging have achieved a great progress. The state-of-the-art method is to regard these two tasks as sequence labeling problem [6,8], which can be handled with supervised learning algorithms such as Maximum Entropy (ME) [1], averaged perceptron [2], Conditional Random Fields (CRF)[4]. However, their performances are still not satisfying for the practical demands to analyze Chinese texts, especially for informal texts. The key reason is that most of annotated corpora are drawn from news texts. Therefore, the system trained on these corpora cannot work well with the informal or specific-domain texts.

In this shared task, we focus to evaluate the performances of word segmentation and POS tagging on relatively informal micro-texts.

© Springer International Publishing Switzerland 2015
J. Li et al. (Eds.): NLPCC 2015, LNAI 9362, pp. 541–549, 2015.
DOI: 10.1007/978-3-319-25207-0_50

2   Data

Different with the popular used newswire dataset, we use relatively informal texts
from Sina Weibo[1]. The training and test data consist of micro-blogs from various
topics, such as finance, sports, entertainment, and so on. Both the training and
test files are UTF-8 encoded.

The information of dataset is shown in Table 1. The out-of-vocabulary (OOV)
rate is slight higher than the other benchmark datasets. For example, the OOV
rate is 5.58% in the popular division [10] of the Chinese Treebank (CTB 6.0)
dataset [9], while the OOV rate of our dataset is 7.25%.

Table 1. Statistical information of dataset.

Dataset	Sents	Words	Chars	Word Types	Char Types	OOV Rate
Training	10,000	215,027	347,984	28,208	39,71	-
Test	5,000	106,327	171,652	18,696	3,538	7.25%
Total	15,000	322,410	520,555	35,277	4,243	-

There are total 35 POS tags in this dataset. A detailed list of POS tags is
shown in Table 2.

Table 2. Statistical information of POS tags.

词性(POS)		Labels	Occurrences
名词		NN	84,006
实体名	人名	PER	3,232
	机构名	ORG	2,578
	地名	LOC	9,701
	其他	NR	550
	邮件	EML	3
	型号名	MOD	34
	网址	URL	11
副词	疑问副词	ADQ	340
	副词	AD	26,155
形貌	形容词	JJ	9,477
	形谓词	VA	3,339
动词	动词	VV	51,294
	情态词	MV	3,700
	趋向动词	DV	781
	被动词	BEI	927
	把动词	BA	600
时间短语		NT	5,881

词性(POS)		Labels	Occurrences
代词	人称代词	PNP	4,903
	疑问代词	PNQ	492
	指示代词	PNI	834
连词	并列连词	CC	2,725
	从属连词	CS	866
数量	数词	CD	10,764
	量词	M	7,917
	序数词	OD	1,219
助词	方位词	LC	4,725
	省略词	ETC	673
	语气词	SP	1,076
	限定词	DT	3,579
	叹词	IJ	20
	标点	PU	52,922
	结构助词	DSP	13,756
	介词	P	9,488
	时态词	AS	3,382

---
[1] http://weibo.com/

2.1 Background Data

Besides the training data, we also provide the background data, from which the training and test data are drawn. The purpose of providing the background data is to find the more sophisticated features by the unsupervised way.

3 Description of the Task

In recent years, word segmentation and POS tagging have undergone great development. In this shared task, we wish to investigate the performances of Chinese word segmentation and POS tagging for the micro-blog texts.

3.1 Subtasks

This task focus the two fundamental problems of Chinese language processing: word segmentation and POS tagging, which can be divided into two subtasks:

1. **CWS** subtask: The first subtask is Chinese word segmentation (CWS). Word is the fundamental unit in natural language understanding. However, Chinese sentences consists of the continuous Chinese characters without natural delimiters. Therefore, Chinese word segmentation has become the first mission of Chinese natural language processing, which identifies the sequence of words in a sentence and marks the boundaries between words.
2. **S&T** subtask: The second subtask is joint Chinese word segmentation and POS tagging.

3.2 Tracks

Each participant will be allowed to submit the three runs for each subtask: **closed track** run, **semi-open track** run and **open track** run.

1. In the **closed** track, participants could only use information found in the provided training data. Information such as externally obtained word counts, part of speech information, or name lists was excluded.
2. In the **semi-open** track, participants could use the information extracted from the provided background data in addition to the provided training data. Information such as externally obtained word counts, part of speech information, or name lists was excluded.
3. In the **open** track, participants could use the information which should be public and be easily obtained. But it is not allowed to obtain the result by the manual labeling or crowdsourcing way.

4   Participants

Sixteen teams have registered for this task. Finally, there are 27 qualified submitted results from 10 teams. A summary of qualified participating teams are shown in Table 3.

Table 3. Summary of the participants.

	CWS			S&T		
	closed	open	semi-open	closed	open	semi-open
NJU	√	√	√			
BosonNLP	√	√		√	√	
CIST	√		√	√		√
XUPT	√			√		
CCNU	√	√				
ICT-NLP	√					
BJTU	√	√	√	√	√	√
SZU		√		√		
ZZU			√			
WHU				√		√

5   SubTask 1: Chinese Word Segmentation

The evaluation measures are precision, recall, and an evenly-weighted F1.

5.1   Baseline Systems

Currently, the mainstream method of word segmentation is discriminative character-based sequence labeling. Each character is labeled as one of {B, M, E, S} to indicate the segmentation. {B, M, E} represent *Begin, Middle, End* of a multi-character segmentation respectively, and S represents a *Single* character segmentation.

For the joint word segmentation and POS tagging, the state-of-the-art method is also based on sequence learning with cross-labels, which can avoid the problem of error propagation and achieve higher performance on both subtasks[5]. Each label is the cross-product of a segmentation label and a tagging label, e.g. {B-NN, I-NN, E-NN, S-NN, ...}. The features are generated by position-based templates on character-level.

Sequence labeling is the task of assigning labels $\mathbf{y} = y_1, \ldots, y_n$ to an input sequence $\mathbf{x} = x_1, \ldots, x_n$. Given a sample $\mathbf{x}$, we define the feature $\Phi(\mathbf{x}, \mathbf{y})$. Thus, we can label $\mathbf{x}$ with a score function,

$$\hat{\mathbf{y}} = \arg\max_{\mathbf{y}} F(\mathbf{w}, \Phi(\mathbf{x}, \mathbf{y})), \tag{1}$$

where $\mathbf{w}$ is the parameter of function $F(\cdot)$.

For sequence labeling, the feature can be denoted as $\phi_k(y_i, y_{i-1}, \mathbf{x}, i)$, where i stands for the position in the sequence and k stands for the number of feature templates.

Here, we use two popular open source toolkits for sequence labeling task as the baseline systems: FNLP[2] [7] and CRF++[3]. Here, we use the default setting of CRF++ toolkit with the feature templates as shown in Table 4. The same feature templates are also used for FNLP.

Table 4. Templates of CRF++ and FNLP.

unigram feature	$c_{-2}, c_{-1}, c_0, c_{+1}, c_{+2}$
bigram feature	$c_{-1} \circ c_0, c_0 \circ c_{+1}$
trigram feature	$c_{-2} \circ c_{-1} \circ c_0, c_{-1} \circ c_0 \circ$
	$c_{+1}, c_0 \circ c_{+1} \circ c_{+2}$

5.2   Participant Systems

In CWS subtask, the best F1 performances are 95.12, 95.52 and 96.65 for closed, semi-open and open tracks respectively. The best system outperforms the baseline systems on closed track. The best system on semi-open track is better than that on closed track. Unsurprisingly, the performances boost greatly on open track.

Table 5. Performances of word segmentation.

Systems	Precision	Recall	F1	Track
CRF++	93.3	93.2	93.3	baseline, closed
FNLP	94.1	93.9	94.0	
NJU	95.14	95.09	95.12	closed
BosonNLP	95.03	95.03	95.03	
CIST	94.78	94.42	94.6	
XUPT	94.61	93.85	94.22	
CCNU	93.95	93.45	93.7	
ICT-NLP	93.96	92.91	93.43	
BJTU	89.49	93.55	91.48	
CIST	95.47	95.57	95.52	semi-open
NJU	95.3	95.31	95.3	
BJTU	90.91	94.46	92.65	
ZZU	85.36	85.25	85.31	
BosonNLP	96.56	96.75	96.65	open
NJU	96.03	96.15	96.09	
SZU	95.52	95.64	95.58	
CCNU	93.68	93.09	93.38	
BJTU	91.79	94.92	93.33	

---

[2] https://github.com/xpqiu/fnlp/
[3] http://taku910.github.io/crfpp/

The participant systems are briefly described as follows.

- The **ZZU** system uses sequence labeling for CWS with CRF model. Besides the traditional discrete feature templates, the dense representation for every character (called character vector) is also used as feature of the character. They report that the features that learned from background data automatically is weaker than the features artificially designed.
- The **CIST** system opts for the 6-tag set and corresponding six n-gram character features. Accessor variety (AV) [3] features are used to measure the possibility of whether a substring is a Chinese word. They report that the ability of OOV detection can be improved by integrating unsupervised global features extracted from the provided background data.
- The **SZU** system exploits multiple heterogeneous data to boost performance of statistical models. The system considers three sets of heterogeneous data, i.e., Weibo (WB, 10K sentences), Penn Chinese Treebank 7.0 (CTB7, 50K), and People's Daily (PD, 280K). With the additional datasets, the F1 score is boosted from 93.76% (baseline model trained on only WB) to 95.58% (+1.82%).
- The **BJTU** system uses the CRF model to integrate several features, including normal features, dictionary features (named entity, hot micro-blog words and symbols) and branch entropy (BE) features. They also use the error-driven rule learning method to expand the training data set in order to improve the accuracy of the system, and improve the adaptability of the system.
- The **ICT-NLP** system adopts the character sequence labelling model and is trained with the averaged perceptron algorithm. To further improve the performance, they combine rules to deal with numbers and English strings and use internal dictionary (extracted from the training data) to do post-processing.
- The **NJU** system applies a word-based perceptron algorithm to build the base segmenter. They also use a bootstrap aggregating model of bagging which improves the segmentation results consistently on the three tracks of closed, semi-open and open test. Besides the basic features, they also use mutual information and accessor variety features.
- The **BosonNLP** system adopts an ensemble approach by combining both discriminative and generative methods. They find that 5-tag labeling consistently provides the best results. They also use several common patterns that might be useful for segmentation and tagging, such as "NN+们". For the open track, they use several models (HMM, CRF) trained on the other corpora (People's daily corpus and Chinese TreeBank 7.0).

6 SubTask 2: Joint Chinese Word Segmentation and POS Tagging

The evaluation measures are precision, recall, and an evenly-weighted F1.

In the joint word segmentation and POS tagging, the best performances are 88.93, 88.69 and 91.55 for closed, semi-open and open tracks respectively.

Table 6. Performances of joint word segmentation and POS tagging.

Systems	Precision	Recall	F1	Track
BosonNLP	88.91	88.95	88.93	closed
XUPT	88.54	87.83	88.19	
BJTU	88.28	87.67	87.97	
CIST	88.09	87.76	87.92	
BJTU	80.64	85.1	82.81	
CIST	88.64	88.73	88.69	semi-open
WHU	88.59	87.96	88.27	
BJTU	81.76	85.82	83.74	
BosonNLP	91.42	91.68	91.55	open
SZU	88.93	89.05	88.99	
BJTU	79.85	83.51	81.64	

- The **CIST** system takes segmented inputs which are produced by the word segmenter used in CWS task, and we assign POS tags on a word-by-word basis, making use of features in the surrounding context (word-based). POS tagger for closed and semi-open track differ only in the segmentation step, closed and semi-open tagger receives word sequences from closed and semi-open segmenter respectively.
- The **SZU** system also adopts a cascaded approach for POS tagging. They use an ensemble approach combining coupled sequence labeling and the guide-feature based method. Same to CWS subtask, they use three datasets to boost the performance. Finally, the tagging F1 score is improved from 87.93% to 88.99% (+1.06%).
- The **BosonNLP** system for the S&T subtask is same to CWS, which obtains the best results in closed and open track of the S&T subtask.

7   Analysis

The analyses of the participant systems are as follows.

1. All participant systems adopt both pre-processing and post-processing. The major purpose is to remove noises and regular patterns in micro text, such as username, URL, Email, expression symbols and the other special symbols. These processings improve the final performances greatly.
2. All the top systems adopt the ensemble based method. The improvement is obtained by combining several models which are trained on other heterogenous annotated datasets.

3. The background data can improve the performance of CWS by about 1%. Some statistical features are extracted from the background data, such as branch entropy and mutual information.

4. For the POS tagging subtask, most systems adopt the pipeline method: first word segmentation, then POS tagging. The reason behind this is that they does their utmost to optimize the performances of CWS. These optimization cannot be applied for joint segmentation and POS tagging.

Since this is the first time for us to organize the shared task, there are a few points needed to be improved.

1. Most of the annotated data are obtained from the micro-blogs of the official news accounts. These texts are relatively more formal than the real micro-texts. Therefore, more informal micro-texts should be added in the future evaluation.

2. The noises in micro-texts should be removed with the same standard, which can reduce the differences of the participant systems with different pre-processings and post-processings.

8   Conclusion

After years of intensive researches, Chinese word segmentation and POS tagging have achieved a quite high precision. However, the performances of the state-of-the-art systems are still relatively low for the informal texts, such as micro-blogs, forums. The NLPCC 2015 Shared Task on Chinese Word Segmentation and POS Tagging for Micro-blog Texts focuses on the fundamental research in Chinese language processing. It is the first time to use the micro-texts to evaluate the performance of the state-of-the-art methods.

In future work, we hope to run an online evaluation system to accept open registration and submission. Currently, a simple system is available at http://nlp.fudan.edu.cn/nlpcc2015. The system also gives the leaderboards for the up-to-date results under the different tasks and tracks. Besides, we also wish to extend the scale of corpus and add more informal texts.

Acknowledgement. We are very grateful to the students from our lab for their efforts to annotate and check the data. We would also like to thank the participants for their valuable feedbacks and comments. This work was partially funded by the National Natural Science Foundation of China (61472088), National High Technology Research and Development Program of China (2015AA015408), Shanghai Science and Technology Development Funds (14ZR1403200).

References

1. Berger, A.L., Della, V.J.: Pietra, and S.A. Della Pietra. A maximum entropy approach to natural language processing. Computational Linguistics **22**(1), 39–71 (1996)
2. Collins, M.: Discriminative training methods for hidden markov models: theory and experiments with perceptron algorithms. In: Proceedings of the 2002 Conference on Empirical Methods in Natural Language Processing (2002)
3. Feng, H., Chen, K., Deng, X., Zheng, W.: Accessor variety criteria for chinese word extraction. Computational Linguistics **30**(1), 75–93 (2004)
4. Lafferty, J.D., McCallum, A., Pereira, F.C.N.: Conditional random fields: probabilistic models for segmenting and labeling sequence data. In: Proceedings of the Eighteenth International Conference on Machine Learning (2001)
5. Ng, H.T., Low, J.K.: Chinese part-of-speech tagging: one-at-a-time or all-at-once? word-based or character-based. In: Proceedings of EMNLP, vol. 4 (2004)
6. Peng, F., Feng, F., McCallum, A.: Chinese segmentation and new word detection using conditional random fields. In: Proceedings of the 20th International Conference on Computational Linguistics (2004)
7. Qiu, X., Zhang, Q., Huang, X.: FudanNLP: a toolkit for Chinese natural language processing. In: Proceedings of Annual Meeting of the Association for Computational Linguistics (2013)
8. Xue, N.: Chinese word segmentation as character tagging. Computational Linguistics and Chinese Language Processing **8**(1), 29–48 (2003)
9. Xue, N., Xia, F., Chiou, F.-D., Palmer, M.: The Penn Chinese TreeBank: Phrase structure annotation of a large corpus. Natural language engineering **11**(2), 207–238 (2005)
10. Yang, Y., Xue, N.: Chinese comma disambiguation for discourse analysis. In: Proceedings of the 50th Annual Meeting of the Association for Computational Linguistics: Long Papers, vol. 1, pp. 786–794. Association for Computational Linguistics (2012)

Overview of the NLPCC 2015 Shared Task: Entity Recognition and Linking in Search Queries

Yansong Feng[1]([✉]), Zhe Han[1], and Kun Zhang[2]

[1] ICST, Peking University, Beijing, China
{fengyansong,hanzhe1992}@pku.edu.cn
[2] Sogou Inc., Beijing, China
zhangkun@sogou-inc.com

Abstract. This paper provides an overview of the Shared Task at the 4th CCF Conference on Natural Language Processing and Chinese Computing (NLPCC 2015): Entity Recognition and Linking in Search Queries, where participant systems are required to recognize entity mentions from short search queries in Chinese, and further link them into a given structured knowledge base. In this paper, we introduce how the task is defined, how we collect the datasets and last we report the evaluation results with a brief analysis.

Keywords: Entity recognition · Entity linking · Knowledge base

1 Introduction

Recent years have witnessed rapid development of structure knowledge bases, and their various applications in our lives. In 2012, search engine giant Google has launched a new project, Knowledge Graph, to prompt semantic search and other related services. Now, more and more search engines start to semantically analyse queries so as to help better understand users' query intention and further provide well-structured background knowledge. One of the key techniques involved here is to discover/recognise named entities from users' queries, and later disambiguate/link them into a knowledge base, which will facilitate further information retrieval techniques, such as related search (*you may like*), query expansion and so on.

Although there have been increasing efforts devoted into this area, it is still challenging to perform accurate entity recognition and linking for informal representations, or short text, e.g., online forum discussions, tweets, search queries, etc. Among others, search queries pose the greatest difficulties: 1) there is often limited context that can be used to fully interpret and disambiguate a named entity mention; 2) as mentioned in [1], conventional but ambiguous entity mentions and limited context may lead to different interpretations for all entity mentions in the short text.

© Springer International Publishing Switzerland 2015
J. Li et al. (Eds.): NLPCC 2015, LNAI 9362, pp. 550–556, 2015.
DOI: 10.1007/978-3-319-25207-0_51

There have been efforts focusing on short-text entity linking in English [2,3], but relatively less attention to other languages, such as Chinese, either in terms of curing benchmark datasets or models/resources specific for Chinese. Thus, the Shared Task committee of NLPCC 2015 propose the task of Entity Recognition and Linking in Search Queries to evaluate the current advances of techniques in aligning Chinese named entity mentions in short search queries to entities in a reference Chinese knowledge base.

2 Task Definition

In this Entity Recognition and Linking challenge, each query consists of an ID and a short search query. A system is expected to provide a reasonable linking result for the given query, where every named entity mention in this query should be linked to the given knowledge base, if exists. For example, given a query: 射雕英雄传周迅版, a system is expected to link 射雕英雄传 into a knowledge base entity: *pk:tv: 射雕英雄传 (2003 年电视剧)* and simultaneously, to link 周迅 into a KB entity *pk:per: 周迅*, as shown in Table 1.

Table 1. The linking result for a query: 射雕英雄传周迅版

Team-tag	Run-tag	Query-id	Group-id	Entity Mention	KB-id
TeamABC	ABC-1	q-1	1	射雕英雄传	pk:tv: 射雕英雄传 (2003 年电视剧)
TeamABC	ABC-1	q-1	1	周迅	pk:per: 周迅

For ambiguous queries that may be analyzed into different scenarios, a system could enumerate up to 5 possible groups of interpretations for the query. For example, given a query: 射雕英雄传, a system is expected to provide the following linking results: *pk:bk: 射雕英雄传*, *pk:tv: 射雕英雄传 (1983 年电视剧)*, *pk:tv: 射雕英雄传 (1994 年电视剧)*, *pk:tv: 射雕英雄传 (2003 年电视剧)*, *pk:tv: 射雕英雄传 (2006 年电视剧)*, and *pk:tv: 射雕英雄传 (2008 年电视剧)* (shown in Table 2). Note that those 5 different interpretations are all possible to be correct linking results, since all those 5 knowledge base entities can be found in a surface form of 射雕英雄传, and we do not have sufficient evidences from the given context to distinguish among those candidate interpretations.

Table 2. The linking results for an ambiguous query: 射雕英雄传

Team-tag	Run-tag	Query-id	Group-id	Entity Mention	KB-id
TeamABC	ABC-1	q-1	1	射雕英雄传	pk:bk: 射雕英雄传
TeamABC	ABC-1	q-1	2	射雕英雄传	pk:tv: 射雕英雄传 (1983 年电视剧)
TeamABC	ABC-1	q-1	3	射雕英雄传	pk:tv: 射雕英雄传 (1994 年电视剧)
TeamABC	ABC-1	q-1	4	射雕英雄传	pk:tv: 射雕英雄传 (2003 年电视剧)
......	......	......	......	......	......

3   Data

An essential part of datasets used in this evaluation is the Chinese knowledge base, which we collect from the InfoBoxes of Chinese online encyclopaedias, including Wikipedia Chinese, and Baidu Baike. We also include the first paragraph of each entity's page as a summary for the entity, which also provide auxiliary context for the entity. The details for the resulting knowledge base are summarized in Table 3. Note that this knowledge base is inevitably noisy, but is relatively easier to access and indeed provides a structured description for each entity.

Table 3. The statistics of the Chinese knowledge base.

# of entities	752,743
# of predicates	18,202
# of categories	114,502
# of entity descriptions	690,533
average # predicates per entity	10.2
average # categories per entity	3.1
average length of entity description	85.0

The search query dataset is collected from the real query log of Sogou Search Engine[1], one of the major search engines in China. We randomly select 767 queries in Chinese[2] with average length of 9.1 characters per query, covering various topics, such as sports, entertainment, politics, social events, foreign affairs and so on. We split the dataset into training and evaluation sets, where the training set contains 98 queries, and the rest 669 queries for evaluation. We recruit 10 postgraduate volunteers to manually label entity mentions from search queries, disambiguate and link them into the given Chinese knowledge base. Regarding the gold-standard annotations of the evaluation set, there are averagely 1.6 entity mentions per query recognised and linked by our annotators. We list several example queries in Table 4.

4   Scoring Metrics

Note that there may be different ways to label a named entity in Chinese, e.g., both 贵州 and 贵州省 (in Table 4) should be considered as correct recognitions. We thus put our emphasis on evaluating a system's performance on how well it can discover entity mentions and accurately link them to the given knowledge base. We use Link-Precision, Link-Recall, Link-F1, and the query level Average-F1 as the evaluation metrics.

---

[1] www.sogou.com

[2] there may be English words or abbreviations inside the queries.

Table 4. Example queries from Sogou query log during March 2015.

女足世界杯维基	灵溪有没有死飞专卖店
误入女配要翻身雪茶小说	贵州省安顺市有几个县
Linkin Park 歌曲	济南梨园大戏院
2014 重大交通事故案例	我是一个在黑暗中大雪纷飞的人
邱淑贞大女儿 13 岁美貌似母亲	日照市有岛么
红米 note4g 增强版在哪买	梁思成和林徽因的资料
温州国药医药有限公司	生化危机 7 十大变革猜想重塑辉煌
香干炒肉的做法	劫夺萨宾妇女
黄晓明陈乔恩吻戏	尼古拉 -米罗蒂奇
街舞少年 tfboys	兄弟连美国电视连续剧
他其实没那么喜欢你迅雷	le onde 钢琴谱
安徽卫视宝宝爬行比赛报名	冬阳童年骆驼队

In more detail, given a query $q \in Q$ with gold-standard annotation[3] $E_q = \{e_1, e_2, ...\}$, a system outputs a set of entities $E'_q = \{e'_1, e'_2, ...\}$. The Link-Precision can be defined as:

$$\text{Link-Precision} = \frac{\sum_{q \in Q} |E_q \cap E'_q|}{\sum_{q \in Q} |E'_q|},$$

the Link-Recall can be defined as:

$$\text{Link-Recall} = \frac{\sum_{q \in Q} |E_q \cap E'_q|}{\sum_{q \in Q} |E_q|},$$

and the Link-F1 is defined as:

$$\text{Link-F1} = \frac{2 * \text{Link-Precision} * \text{Link-Recall}}{\text{Link-Precision} + \text{Link-Recall}}.$$

The query level Average-F1 is defined as the average F1 score over all queries:

$$\text{Average-F1} = \frac{1}{|Q|} \sum_{q \in Q} F1_q,$$

where the query-wise F1 score is computed by

$$F1_q = \frac{2 * P_q * R_q}{P_q + R_q}$$

where the query-wise precision is:

$$P_q = \frac{|E_q \cap E'_q|}{|E'_q|},$$

---

[3] Note that for ease of evaluation, we simply combine different groups of interpretations into one single group.

Table 5. Teams with valid submissions at the Task of Entity Recognition and Linking in Search Queries

System ID	Team ID	Institutions
1	SWJTU-TEAM-1	Southwest Jiaotong University
2	efusion	Xi'an University of Posts & Telecommunications
3	ZZUNLP	Zhengzhou University
4	Bit-1	Beijing Institute of Technology
5	LUDI-CIST	Beijing University of Posts and Telecommunications
6	createMoMo-1	Beijing University of Posts and Telecommunications
7	KGG@BNU-1	Beijing Normal University
8	SWU-1	Southwest University
9	CCNUTextMiner-NLP1	Central China Normal University
10	LZP-CIST	Beijing University of Posts and Telecommunications
11	EL-BLCU-NLPCC2015	Beijing Language and Culture University
12	firework	Beihang University
13	zutnlp	Zhongyuan University of Technology
14	NJU-AIRS	Nanjing University

Table 6. Evaluation Results

System ID	Link-Precision	Link-Recall	Link-F1	Average-F1
1	0.724	**0.736**	**0.73**	**0.733**
2	0.386	0.213	0.274	0.223
3	0.283	0.431	0.342	0.322
4	0.497	0.704	0.583	0.569
5	0.335	0.38	0.356	0.326
6	0.366	0.485	0.417	0.368
7	0.332	0.652	0.44	0.432
8	0.399	0.468	0.43	0.388
9	0.348	0.43	0.384	0.34
10	0.48	0.656	0.555	0.535
11	0.562	0.695	0.621	0.611
12	0.464	0.49	0.477	0.423
13	0.394	0.192	0.258	0.201
14	**0.731**	0.43	0.541	0.453

and the query-wise recall is:

$$R_q = \frac{|E_q \cap E_q'|}{|E_q|}.$$

5   The Teams

There are in total 24 teams registering to this shared task, and 14 of them submitted 18 valid runs of results. We list the 14 teams with valid submissions in Table 5. Note that we only record the best results for teams with multiple runs.

Most participant systems collect nicknames, convention names or abbreviations for entities from Wikipedia, Baidu Baike, Hudong Baike, and other resources, and further compute similarities between entity mentions and knowledge base entries by taking popularity and context features into account [4–6]. Some even use existing search engines to explore more general context and obtain competitive performances, but this may not be much helpful for further improvement of the task itself.

6   Results

We summarize the performances of all teams in Table 6, where we can see that Team 1 (SWJTU-TEAM-1) achieves the best performances in almost all metrics and outperforms other systems by a large margin, except that Team 14 (NJU-AIRS) obtains 0.731 in Link-Precision, slightly better than Team 1.

When we take a close look at Table 6, we could find that most systems can achieve a Link-F1 or Average-F1 score at a range of 0.3 to 0.6, which is still far from real-world applications. And the relatively lower Link-Precision scores show that the short context indeed makes the task challenging.

7   Conclusion

This Shared Task Evaluation has provided an picture for the current advances of techniques in recognizing and analyzing Chinese named entity mentions in short search queries and also techniques in aligning those entity mentions into structured knowledge base. We have released all evaluation datasets to the community[4], which we expect to work as a benchmark dataset to prompt further development of related research from not only natural language processing community, but also more diverse fields, such as information retrieval and data mining.

Acknowledgments. This work was supported by the National High Technology R&D Program of China (Grant No. 2014AA015102, 2015AA015403), National Natural Science Foundation of China (Grant No. 61272344, 61202233, 61370055) and the joint project with IBM Research.

References

1. Carmel, D., Chang, M.-W., Gabrilovich, E., Paul Hsu, B.-J., Wang, K.: ERD 2014: entity recognition and disambiguation challenge. In: SIGIR Forum Entity Recognition and Disambiguation Challenge Workshop (2014)
2. Roth, D., Ji, H., Cassidy, T.: Entity Linking and Wikification. Morgan & Claypool Publishers (2015)

---

[4] http://tcci.ccf.org.cn/conference/2015/

3. Ji, H., Nothman, J., Hachey, B.: Overview of TAC-KBP2014 entity discovery and linking tasks. In: Proceedings of Text Analysis Conference (TAC 2014) (2014)
4. Yuan, J., Yang, Y., Jia, Z., Yin, H., Huang, J., Zhu, J.: Entity recognition and linking in Chinese search queries. In: Proceedings of NLPCC (2015)
5. Li, J., Pan, J., Ye, C., Huang, Y., Wang, Z., Wen, D.: Linking entities in Chinese queries to knowledge graph. In: Proceedings of NLPCC (2015)
6. Tang, G., Guo, Y., Yu, D., Xun, E.: A hybrid re-ranking method for entity recognition and linking in search queries. In: Proceedings of NLPCC (2015)

Overview of the NLPCC 2015 Shared Task: Weibo-Oriented Chinese News Summarization

Xiaojun Wan(✉), Jianmin Zhang, Shiyang Wen, and Jiwei Tan

Institute of Computer Science and Technology, The MOE Key Laboratory
of Computational Linguistics, Peking University, Beijing 100871, China
wanxiaojun@pku.edu.cn

Abstract. The Weibo-oriented Chinese news summarization task aims to auto-matically generate a short summary for a given Chinese news article, and the short summary is used for news release and propagation on Sina Weibo. The length of the short summary is less than 140 Chinese characters. The task can be considered a special case of single document summarization. In this paper, we will introduce the evaluation dataset, the participating teams and the evaluation results. The dataset has been released publicly.

Keywords: Chinese news summarization · Document summarization · Shared task · NLPCC 2015

1 Task

With the popularity of social media web sites (e.g. Sina Weibo), news stories are usually released and propagated on social media sites in recent years. A short summary for each hot news is posted as a Weibo message, together with a URL link to the full news article. People can quickly understand the news by reading the short summary and then decide whether to read the full news article or whether to comment on the news. Till now, there are many news accounts on Sina Weibo (e.g. Breakingnews/头条新闻[1], Caijing/财经网[2], etc.) for news release. Currently, the short news summaries given by these accounts are provided by human editors. It is very interesting to investigate news summarization techniques for news release and propagation on Sina Weibo. Therefore, the shared task of Weibo-Oriented Chinese News Summarization is defined as a task of automatically generating a short summary for a given Chinese news article, and the short summary is used for news release and propagation on Sina Weibo. The length of the short summary is limited with 140 Chinese characters.

This shared task can be considered a special case of single document summarization. Since traditional news document summarization techniques have been widely explored on the DUC[3] and TAC[4] conferences [1], we encourage participants to

---

[1] http://weibo.com/breakingnews
[2] http://weibo.com/caijing
[3] http://duc.nist.gov/

© Springer International Publishing Switzerland 2015
J. Li et al. (Eds.): NLPCC 2015, LNAI 9362, pp. 557–561, 2015.
DOI: 10.1007/978-3-319-25207-0_52

develop more competitive summarization methods by considering the specialty of the shared task.

2   Data

In this shared task, we construct the evaluation dataset in an automatic way. We first collected all Chinese Weibo messages on a few news accounts on Sina Weibo, such as Renminwang/人民网[5], Beijingdaily/北京日报[6], SouthernMetropolisWeekly/南都周刊[7], Breakingnews/头条新闻, etc. All messages with a URL link to the full Chinese news article were kept, and we stored the news URLs which correspond to two different Weibo Messages. Note that there are very few URL links which correspond to more than two Weibo Messages in the real dataset. We then downloaded the web pages via the URL links and extracted the news articles from the web pages. Lastly, we obtained a number of news articles and each news article was associated with two Weibo messages. Each Weibo message was written and posted by a human editor and we consider it a human-written model summary for the associated news article. In many Weibo messages, the news titles were used as the first sentence, and we removed the news titles from the Weibo messages. In this way, we collected a dataset of Chinese news articles with reference summaries. We split the dataset into training/sample set and test set. The training set includes 140 news articles and each news article corresponds to two model summaries. The test set includes 250 news articles. The sentences in each news article have been detected and stored. The two model summaries for a news article of "training9.txt" in the training set are given below:

Model 1:

16日清晨，5名身份不详的男童，被发现死于贵州省毕节市城区一处垃圾箱内。

官方尚未披露死者具体身份及死因，据初步分析，5个小孩可能是躲进垃圾箱避寒窒息"闷死"。

目前已排除他杀的可能，死者具体身份尚不知。

Model 2:

5名男童16日清晨被发现死于贵州毕节城区一处垃圾箱内，据初步分析，可能是躲进垃圾箱避寒窒息"闷死"。

死亡男孩年龄不等，均在10岁左右。

最早发现尸体的是一位拾垃圾的老太太。

---

[4] http://www.nist.gov/tac/

[5] http://weibo.com/renminwang

[6] http://weibo.com/beijingdaily

[7] http://weibo.com/nbweekly

毕节15日最低温6°C，当夜曾下毛毛雨。

The participants can direclty use the provided sentences for summariation, and alternatively, they can segment the news text into sentences in their own way. We do not provide Chinese word semgentation results and the participates can use any toolkit for Chinese word segmentation.

3   Participants

Each team is allowed to submit at most two runs of results. The length of each summary is limited to 140 Chinese characters and longer summaries will be truncated. The participants are allowed to use any NLP resources or toolkits, but it is not allowed to crawl and use the Weibo messages from the news accounts on Sina Weibo.

There are 9 teams participating in this shared task and they submitted a total of 16 runs of results. The participating teams are shown in Table 1. Various summarization techniques have been used by the participating teams. For example, NLP@WUST uses a weighted linear combination of four feature values to evaluate a sentence: term frequency, sentence position, sentence length and the similarity between sentence and title. CIST-SUMM uses both rule-based sentence scoring and machine learning based sentence scoring. The features include sentence coverage, sentence similarity to title, sentence location, key words, named entities, and new features derived from hLDA topic tree. CCNUTextMiner uses a graph-based ranking algorithm for sentence ranking and selection. USC1 first computes the TextRank scores for sentences, and then linearly combine the TextRank score, the position based score and keywords based score to get the overall score of each sentence.

Table 1. Participating teams.

TeamID	Organization
CCNUTextMiner	Central China Normal University
CIST-SUMM	Beijing University of Posts and Telecommunications
DluflNLP	Dalian University of Foreign Languages
FLCTest4	Chongqing University of Technology
NLP@WUST	Wuhan University of Science and Technology
USC1	University of South China
zutnlptest4	Zhongyuan University of Technology
YTSC-run	Chongqing University of Technology
zzubylong	Zhengzhou University

4   Results

We adopted automatic evaluation for this shared task. The peer summaries were automatically compared with the model summaries by using the ROUGE-1.5.5 toolkit [2][8]. ROUGE-N F-measure was used as evaluation metrics. In particular, we reported the F-measure scores of ROUGE-1, ROUGE-2, ROUGE-3, ROUGE-4 and ROUGE-SU4. We slightly modified the ROUGE-1.5.5 toolkit to evaluate Chinese summaries. The recommended options for the toolkit are -c 95 -2 4 -U -r 1000 -n 4 -w 1.2 –a –l 140. Note that we adopted character-based evaluation for evaluating Chinese summary. Character-based evaluation means that we do not need to perform Chinese word segmentation when running the ROUGE toolkit. Instead, we only need to separate Chinese characters with blank spaces.

The evaluation results are shown in Table 2. The second run (run2) of NLP@WUST achieved the best results. As mentioned in Section 3, NLP@WUST makes use of four typical features to evaluate the sentences. The method is simple but effective for this shared task.

Table 2. Evaluation results.

(The results are unsorted and the best results are in bold.)

Team	Run	R-1	R-2	R-3	R-4	R-SU4
CCNUTextMiner	run1	0.44166	0.28354	0.21853	0.18420	0.26633
CIST-SUMM	run1	0.47368	0.32354	0.26122	0.22869	0.30737
	run2	0.46887	0.31584	0.25245	0.21966	0.29955
DluflNLP	run1	0.41692	0.25842	0.19569	0.16397	0.24340
	run2	0.42166	0.26136	0.19580	0.16211	0.24464
FLCTest4	run1	0.48773	0.34913	0.28898	0.25574	0.33162
	run2	0.40683	0.24148	0.17928	0.15010	0.23040
NLP@WUST	run1	0.51496	0.38847	0.33165	0.29876	0.37107
	run2	**0.52422**	**0.40035**	**0.34483**	**0.31234**	**0.38318**
USC1	run1	0.41572	0.25132	0.1865	0.15548	0.23756
	run2	0.41722	0.24817	0.18221	0.14984	0.23350
zutnlptest4	run1	0.39945	0.24392	0.18361	0.15297	0.23018
YTSC-run	run1	0.45124	0.29799	0.23561	0.20323	0.28333
	run2	0.46747	0.32056	0.25901	0.22655	0.30448
zzubylong	run1	0.32441	0.19641	0.13431	0.10510	0.17289
	run2	0.36854	0.22465	0.15649	0.12352	0.20000

---

[8] http://berouge.com/

5   Conclusion and Future Work

The evaluation dataset has been released publicly[9]. We expect more advanced summarization methods can be proposed for this special Chinese summarization task. This shared task is our first attempt at Chinese news summarization, and we will refine the task and formulate new summarization tasks for Chinese documents in the following years.

Acknowledgments. This work was supported by National Hi-Tech Research and Development Program (863 Program) of China (2015AA015403, 2014AA015102) and National Natural Science Foundation of China (61170166, 61331011).

References

1. Nenkova, A., McKeown, K.: A survey of text summarization techniques. In: Mining Text Data, pp. 43–76. Springer US (2012)
2. Lin, C.-Y.: Rouge: A package for automatic evaluation of summaries. In: Text summarization Branches out: Proceedings of the ACL 2004 Workshop, vol. 8 (2004)

---

[9] http://tcci.ccf.org.cn/conference/2015/pages/page05_evadata.html

Overview of the NLPCC 2015 Shared Task: Open Domain QA

Nan Duan[✉]

Microsoft Research, Beijing, China
nanduan@microsoft.com

Abstract. In this paper, we give the overview of the open domain Question Answering (or open domain QA) shared task in NLPCC 2015. We first review the background of QA, and then describe open domain QA shared task in this year's NLPCC, including the construction of the benchmark datasets, the auxiliary dataset, and the evaluation metrics. The evaluation results of submissions from participating teams are presented in the experimental part, together with a brief introduction to the techniques used in each participating team's QA system.

Keywords: Question answering · Knowledge base

1 Background

Question Answering (or QA) is a fundamental task in Artificial Intelligence, whose goal is to build a system that can automatically answer natural language questions. In the last decade, the development of QA techniques have been greatly promoted by both academic field and industry field.

In the academic field, with the rise of large scale curated knowledge bases, like Yago, Satori, Freebase and etc., more and more researchers pay their attentions to the open domain QA task. The state-of-the-art methods used in open domain QA can by summarized into two categories: semantic parsing-based approaches and information retrieval-based approaches. Semantic parsing-based approaches, such as [1] [2] [3] [4] [5] [6] [7], first transform a natural language question into its corresponding meaning representation, and then use it as a structured query to lookup answers from an existing KB; information retrieval-based approaches, such as [8] [9] [10] [11] [12] [13] [14] [15] [16], first define and generate the representations of answers stored in KB, and then retrieve the most relevant answers from KB by computing the similarity between input questions and the representations of answers. Recently, with the development of the open IE techniques [20] [21] [22] [23] [24] [25], some approaches, such as [17] [18] [19], build QA systems based on extracted knowledge bases, which consist of assertions extracted from unstructured text by open IE. Comparing to curated KBs, extracted KBs can be extracted from arbitrary corpus, so it is very flexible to be applied to any specific domain. But it also suffers the extraction noise issue, which is brought by open IE. In the industry field, many influential QA-related products have been built, such as

J. Li et al. (Eds.): NLPCC 2015, LNAI 9362, pp. 562–570, 2015.
DOI: 10.1007/978-3-319-25207-0_53

IBM Watson, Apple Siri, Google Now, Facebook Graph Search, Microsoft Cortana/XiaoIce and etc. These kind of systems are immerging into every user's life who is using mobile devices. Under such circumstance, in this year's NLPCC shared tasks, we call the open domain QA task, whose motivations are two-folds:

1. We expect this activity can provide more benchmark data for QA research, especially for Chinese;
2. We encourage more QA researchers to share their experiences, new techniques, and latest progress.

The remainder of this paper is organized as follows: Section 2 simply describe this year's open domain QA shared task; in Section 3, we will describe the benchmark datasets used in this year's QA evaluation, Section 4 describe the auxiliary dataset, which are crawled from semi-structured web pages and can be used as a structured database to build a Chinese QA system; in Section 5, we describe several evaluation metrics that are used to measure the QA quality of submissions generated by participating teams, and present the evaluation results of different submissions in Section 6, with a brief introduction to the techniques used by each team; Finally, we will conclude the paper in Section 7.

2   Task Description

This year's QA shared task provides two benchmark datasets, one for English and one for Chinese. For each question in each dataset, the participating teams should provide a list of answers as the prediction. We don't restrict participating teams to use any specified data for answer generation, so any data resources can be used. We evaluate the quality of the generated answers submitted from each team based on golden answers and several evaluation metrics (described in Section 5). Each team is allowed to provide multiple submissions for each dataset, but should specify one of them as their primary result.

3   Benchmark Data

Recently, two benchmark datasets for English have been released and frequently used by the academic field for the open domain QA task, including:

- *SimpleQuestions* dataset [8] consists of a total of 108,442 question-answer pairs, each of which is labeled by human English-speaking annotators based on a *single* fact from Freebase.
- *WebQuestions* dataset [6] consists of a total of 5,810 question-answer pairs. Different from SimpleQuestions, the questions are selected from the Goggle Suggest API, and then labeled by the Amazon Mechanical Turk (AMT) based on Freebase.

In order to encourage more researchers and institutions in China to devote to the QA research, in this year's NLPCC QA shared task, we specially provide a QA benchmark dataset for Chinese, together with another QA benchmark dataset for English as well. There two datasets are described as follows:

- *NLPCC15QuestionsCH* dataset consists of 1,000 question-answer pairs for Chinese. These questions are randomly sampled from a subset of queries coming from Bing China's query log based on the following rules: (1) each query's character length should be between 5 and 25; (2) each query should contain at least one n-gram contained in an answer type name list, such as '谁', '哪里', '哪一年', etc. These names are heuristically extracted and collected based on a simple statistical analysis on the entire query log; (3) each query should be answered based on an auxiliary data only. We will describe the auxiliary data in the later part. Each question is labeled by a list of answers, which should be agreed by three human annotators.
- *NLPCC15QuestionsEN* dataset consists of 68,481 question-answer pairs for English. These questions are collected in a hybrid way: some of the questions come from online QA sites like WikiAnswers and EVI; while the other questions are randomly sampled from a subset of queries coming from Bing US's query log and labeled by human annotators. The answers of the questions in the former part are crawled from the sites directly; while the answers of the questions in the latter part are labeled by based on Freebase. Unfortunately, no team submitted results for this dataset, and we encourage more institutes can leverage this dataset in the future for the QA research.

The answer annotations of all questions have already been provided to the participating teams, as the evaluation procedure has been finished. Two examples of these two datasets mentioned above are shown in Table 1, and their corresponding statistics are shown in Table 2.

Table 1. Two examples of NLPCC15QuestionsCH and NLPCC15QuestionsEN.

NLPCC15QuestionsCH	
<question id="1">	谁能百里挑一是哪个电视台的节目?
<answer id="1">	东方卫视

NLPCC15QuestionsEN	
<question id="1">	Who founded CBS?
<answer id="1">	William S. Paley

Table 2. Statistics of NLPCC15QuestionsCH and NLPCC15QuestionsEN.

	NLPCC15QuestionsCH	NLPCC15QuestionsEN
# of Questions	1,000	68,481
Averaged Question Length	12.8 (characters)	4.8466 (words)
Averaged Answer Numbers per Question	1.4	1.4

4   Auxiliary Data

Freebase is available for all Web users, so all teams can build QA systems for the English dataset based on APIs provided by Freebase. But such convenient APIs are not available for Chinese. In order to facilitate the system construction procedure for the Chinese QA task, we provide an auxiliary data resource, which plays a role as a Chinese knowledge base.

Formally, each entry stored in the auxiliary has the triple form: <Subject, Predicate, Argument>, where 'Subject' denotes an entity, 'Predicate' denotes a relation, and 'Argument' denotes either an entity or a string that gives a description of the subject entity. This data set is extracted from Baidu Baike pages.   An example is given below in Figure 1, and some statistics of this data set are shown in Table 3.

```
新还珠格格 ||| entity.primaryName ||| 新还珠格格
新还珠格格 ||| 中文名 ||| 新还珠格格
新还珠格格 ||| 外文名 ||| New my fair Princess
新还珠格格 ||| 出品时间 ||| 2011年和2014年
新还珠格格 ||| 出品公司 ||| 上海创颖文化传播有限公司
新还珠格格 ||| 制片地区 ||| 中国大陆，中国台湾
新还珠格格 ||| 拍摄地点 ||| 横店影视城
新还珠格格 ||| 发行公司 ||| 上海创颖文化传播有限公司
新还珠格格 ||| 首播时间 ||| 2011年7月16日
新还珠格格 ||| 导演 ||| 李平、丁仰国
新还珠格格 ||| 编剧 ||| 琼瑶，黄素媛
新还珠格格 ||| 主演 ||| 李晟，海陆，张睿，李佳航，潘杰明，赵丽颖，邱心志，邓萃雯，刘雪华
新还珠格格 ||| 集数 ||| 总共98集→第一部1至37集→第二部37至74集→第三部74至98集
新还珠格格 ||| 每集长度 ||| 前三部：45分钟 第四部：48分钟
新还珠格格 ||| 类型 ||| 古装，爱情，励志，喜剧
新还珠格格 ||| 上映时间 ||| 前三部：2011年07月16日至2011年9月8日第四部：2016年暑期档
新还珠格格 ||| 在线播放平台 ||| 芒果TV，PPTV，暴风影音，优酷，搜狐。
新还珠格格 ||| 总策划 ||| 杨文红，苏晓
新还珠格格 ||| 出品人 ||| 欧阳常林
新还珠格格 ||| 总监制 ||| 魏文彬
新还珠格格 ||| entity.description ||| 《新还珠格格》翻拍自琼瑶经典之作《还珠格格》，由李晟、海
```

Fig. 1. An example of auxiliary data for NLPCC15QuestionsCH.

Table 3. Statistics of auxiliary data for NLPCC15QuestionsCH.

	Statistics
# of Subject Entities	8,721,640
# of Triples	47,943,429
# of Averaged Triples per Subject Entity	5.5

Note, the answers of the questions in NLPCC15QuestionsCH are labeled based on this auxiliary data only, in order to ensure the participating teams can achieve reasonable QA quality by just using the auxiliary data we provided. Of course, other data resources are allowed to be used as well, such as other structured knowledge bases, web pages or offline documents.

5   Evaluation Metric

In this year's QA shared task, the quality of different QA systems are measured by the three evaluation metrics described below:

- Mean Reciprocal Rank (MRR)

$$MRR = \frac{1}{|Q|} \sum_{i=1}^{|Q|} \frac{1}{rank_i}$$

Where $|Q|$ denotes the total number of questions in the dataset, $rank_i$ denotes the position of the first correct answer in the generated answers C_i for the ith question Q_i. If C_i doesn't overlap with the golden answers A_i for Q_i, $\frac{1}{rank_i}$ is set to 0.

- Accuracy@N

$$Accuracy@N = \frac{1}{|Q|} \sum_{i=1}^{|Q|} \delta(C_i, A_i)$$

Where $\delta(C_i, A_i)$ equals to 1 when there is at least one answer contained by C_i occurs in A_i, and 0 otherwise.

- Averaged F1

$$AveragedF1 = \frac{1}{|Q|} \sum_{i=1}^{|Q|} F_i$$

Where F_i denotes the F1 score for question Q_i computed based on C_i and A_i. F_i is set to 0 if C_i is empty or doesn't overlap with A_i. Otherwise, F_i is computed as follows:

$$F_i = \frac{2 \cdot \frac{\#(C_i, A_i)}{|C_i|} \cdot \frac{\#(C_i, A_i)}{|A_i|}}{\frac{\#(C_i, A_i)}{|C_i|} + \frac{\#(C_i, A_i)}{|A_i|}}$$

Where $\#(C_i, A_i)$ denotes the number of answers occur in both C_i and A_i. $|C_i|$ and $|A_i|$ denote the number of answers in C_i and A_i respectively.

6   Evaluation Result

There are totally 12 teams registered for the Chinese QA task, and 7 teams registered for the English QA task. However, only 3 teams submitted final results for the Chinese QA task, and no submission is received for the English QA task. Table 4 lists some statistics of these three submissions:

Table 4. Statistics of submissions.

	# of Questions Answered	Average # of Answers per Question
Team 1	432	1.82
Team 2	1,000	1.06
Team 3	972	1.98

From Table 4 we can see that Team 2's QA system generates answers for all the 1,000 questions, and Team 3's QA system only ignored 28 questions. Team 1's QA system only generate answers for 432 questions, but this angle only cannot tell 'good' or 'bad' of a QA system, as saying no also represents an intelligence for an AI system. We can also see that the average number of answers of Team 2 is nearly one, which means they only provide their top-1 results as the predicted answer for most questions, while this number for Team 1 and Team 3 is nearly two.

We then list the evaluation results in Table 5, 6, and 7, based on MRR, Accuracy@N, and F1 score respectively. From these 3 tables we can see that Team 2 performs better than the other two teams on all three metrics.

Table 5. Evaluation results based on MRR.

	MRR
Team 1	0.1430
Team 2	0.5675
Team 3	0.3360

Table 6. Evaluation results based on Accuracy@N.

	ACC@1	ACC@2	ACC@3	ACC@4	ACC@5
Team 1	0.1270	0.1490	0.1590	0.1650	0.1660
Team 2	0.5650	0.5700	0.5700	0.5700	0.5700
Team 3	0.2640	0.3980	0.4130	0.4130	0.4130

Table 7. Evaluation results on all questions based on F1 Score.

	Precision	Recall	F1 Score
Team 1	0.1169	0.1439	0.1196
Team 2	0.5660	0.5103	0.5240
Team 3	0.2890	0.3547	0.2990

Recall that Team 1 only answer 50% of questions, and in order to compare the quality of different QA systems from more perspectives, we also compare Precision, Recall, and F1 Score on answered questions only. Evaluation results are shown in Table 8, from which we can see that Team 2's results don't change, as they generate answers for all questions; Team 3's results change a little, this is due to the reason that they only ignored 28 questions; Team 1's results become much better, almost comparable to Team 3's numbers. This is to say that Team 1 can further improve their system by enlarging the recall first.

Table 8. Evaluation results on answered questions based on F1 Score.

	Precision	Recall	F1 Score
Team 1	0.2706	0.3331	0.2769
Team 2	0.5660	0.5103	0.5240
Team 3	0.2973	0.3650	0.3076

We also investigate the oracle results of different QA systems, and compare them with the corresponding Accuracy@1 in Table 9. From Table 9 we can see that Team 3 has the largest potential to improve their system; while the potential of Team 2's system is very limited. One key reason of this finding is that, for most of questions, Team 2 just submit a single answer as output.

Table 9. Oracle results.

	ACC@1	Oracle
Team 1	0.1270	0.1660
Team 2	0.5650	0.5700
Team 3	0.2640	0.4130

In order to understand the differences between these three QA systems, we tried to find system description papers from this year's submissions during the paper review procedure, by checking whether there is any evaluation result reported based on NLPCC15QuestionsCH. Below gives a brief introduction to the techniques used in Team 2 and Team 3's QA systems. It is a pity that we failed to find the corresponding system description for Team 1.

- Team 2 leverages both triple knowledge and search engine to answer input questions. For the triple knowledge part, the SPE algorithm is used to transform a natural language question into a triple query; For search engine part, the WKE algorithm is used to extract answer candidates from both Baidu Zhixin and unstructured web texts. This method achieves the best result on the Chinese QA dataset. We also expect Team 2 can show more detailed evaluation results, to compare the impacts of triple knowledge and search engine in their QA system.

- Team 3 leverages 3 steps to predict answer candidates for a given question, including (1) question analysis, which detects the answer type of a given question based on heuristic rules; (2) multi-source retrieval, which extract answer candidates from two main resources, including knowledge triples and social QA collections (i.e. <question, answer> pairs); and (3) candidate ranking, which ranks different answer candidates based on similarity/redundancy features.

We also did some analysis on the answers predicted by three teams, and found that the evaluation results are better actually. This is because that in some cases, an answer is decided to be wrong just because it cannot match the labeled answer in an exact way. Below is an example:

- Question: 哪些城市有迪士尼乐园？
- Golden Answers: ["洛杉矶","奥兰多","东京","巴黎","香港"]
- Predicted Answers: ["美国加州迪士尼乐园","美国奥兰多迪斯尼世界","日本东京迪斯尼乐园","法国巴黎迪斯尼乐园","中国香港迪斯尼乐园"]

This is due to the fact that currently different resources are allowed to be used for answer generation, so answers extracted from different corpus may have different surface forms but identical semantic meaning. Such issue can be alleviated by using a specified KB only for answer extraction.

7   Conclusion

This paper briefly introduce the overview of this year's Open Domain QA shared task. Although there are only 3 teams that submitted results finally, we still see promising results and different techniques used. We are looking forward more organizations can take part in this yearly activity, and more benchmark data sets and techniques will be delivered to the community.

Reference

1. Wang, Y., Berant, J., Liang, P.: Building a semantic parser overnight. In: ACL (2015)
2. Pasupat, P., Liang, P.: Compositional semantic parsing on semi-structured tables. In: ACL (2015)
3. Pasupat, P., Liang, P.: Zero-shot entity extraction from web pages. In: ACL (2014)
4. Bao, J., Duan, N., Zhou, M., Zhao, T.: Knowledge-based question answering as machine translation. In: ACL (2014)
5. Yang, M.-C., Duan, N., Zhou, M., Rim, H.-C.: Joint relational embeddings for knowledge-based question answering. In: EMNLP (2014)

6. Berant, J., Chou, A., Frostig, R., Liang, P.: Semantic parsing on freebase from question-answer pairs. In: EMNLP (2013)
7. Kwiatkowski, T., Choi, E., Artzi, Y., Zettlemoyer, L.: Scaling semantic parsers with on-the-fly ontology matching. In: EMNLP (2013)
8. Bordes, A., Usunier, N., Chopra, S., Weston, J.: Large-scale simple question answering with memory network. In: ICLR (2015)
9. Weston, J., Bordes, A., Chopra, S., Mikolov, T.: Towards AI-complete question answering: a set of prerequisite toy tasks (2015). arXiv
10. Dong, L., Wei, F., Zhou, M., Xu, K.: Question answering over freebase with multi-column convolutional neural networks. In: ACL (2015)
11. Yih, W., Chang, M.-W., He, X., Gao, J.: Semantic parsing via staged query graph generation: question answering with knowledge base. ACL (2015)
12. Yao, X.: Lean question answering over freebase from scratch. In: NAACL (2015)
13. Berant, J., Liang, P.: Semantic parsing via paraphrasing. In: ACL (2014)
14. Yao, X., Durme, V.: Information extraction over structured data: question answering with freebase. In: ACL (2014)
15. Bordes, A., Weston, J., Chopra, S.: Question answering with subgraph embeddings. In: EMNLP (2014)
16. Bordes, A., Weston, J., Usunier, N.: Open question answering with weakly supervised embedding models. In: Calders, T., Esposito, F., Hüllermeier, E., Meo, R. (eds.) ECML PKDD 2014, Part I. LNCS, vol. 8724, pp. 165–180. Springer, Heidelberg (2014)
17. Yin, P., Duan, N., Kao, B., Bao, J., Zhou, M.: Answering questions with complex semantic constraints on open KBs. In: CIKM (2015)
18. Fader, A., Zettlemoyer, L., Etzioni, O.: Open question answering over curated and extracted knowledge bases. In: KDD (2014)
19. Fader, A., Zettlemoyer, L., Etzioni, O.: Paraphrase-driven learning for open question answering. In: ACL (2013)
20. Del Corro, L., Gemulla, R.: ClausIE: Clause-based open information extraction. In: WWW (2013)
21. Mausam, Schmitz, M., Bart, R., Soderland, S., Etzioni, O.: Open language learning for information extraction. In: EMNLP (2012)
22. Yahya, M., Berberich, K., Elbassuoni, S.: Natural language questions for the web of data. In: EMNLP-CoNLL (2012)
23. Fader, A., Soderland, S., Etzioni, O.: Identifying relations for open information extraction. In: EMNLP (2011)
24. Wu, F., Weld, D.S.: Open information extraction using wikipedia. In: ACL (2010)
25. Banko, M., Cafarella, M.J., Soderland, S., Broadhead, M., Etzioni, O.: Open information extraction from the web. In: IJCAI (2007)

Short Task (Short Papers)

Word Segmentation of Micro Blogs with Bagging

Zhenting Yu, Xin-Yu Dai$^{(\boxtimes)}$, Si Shen, Shujian Huang, and Jiajun Chen

State Key Laboratory for Novel Software Technology, Department of Computer
Science and Technology, Nanjing University, Nanjing 210023, China
{yuzt,shens}@nlp.nju.edu.cn, {daixinyu,huangsj,chenjj}@nju.edu.cn

Abstract. This paper describes the model we designed for the Chinese
word segmentation Task of NLPCC 2015. We firstly apply a word-based
perceptron algorithm to build the base segmenter. Then, we use a Bootstrap Aggregating model of bagging which improves the segmentation
results consistently on the three tracks of closed, semi-open and open
test. Considering the characteristics of Weibo text, we also perform rule-
based adaptation before decoding. Finally, our model achieves F-score
95.12% on closed track, 95.3% on semi-open track and 96.09% on open
track.

Keywords: Weibo · Word segmentation · Word-based perceptron
model · Bagging

1 Introduction

Since Chinese sentences are written in continuous characters without explicit
word boundaries, Chinese word segmentation (CWS) is a critical and a necessary
initial step for most NLP tasks such as syntax parsing, information extraction
and machine translation. At present, there are two main models for Chinese
word segmentation "word-based" approach and "character-based" approach. In
this paper, we prefer to use a word-baed model rather than a character based
model, because word-based model may use more contextual information (Zhang
and Clark 2011)[1].

Microblog is a new kind of broadcast medium in the form of blogging. A
microblog differs from traditional blog in its smaller size. Furthermore, microblog
text contains a large number of new words, name entities, punctuation patterns
(such as "..."), structured symbols representing conversation ("@") and topics
("#...#") etc. These characteristics take more challenges to microblog text segmentation.

In this paper, we use an averaged perceptron model as the base segmenter.
We train several segmenters on the several sampling of the training data. Then
we apply a Bootstrap Aggregating model of bagging for voting the segmentation results. We use this bagging strategy for the closed track. For the semi-
open track, we incorporate more statistic-based features in the bagging model.
For the open track, we combine a bagging model with extra lexicon features, unsupervised statistical features. Experimentally, our method gets F-score
95.12% on the closed track, 95.3% on the semi-open track and 96.09% on the
open track, respectively.

© Springer International Publishing Switzerland 2015
J. Li et al. (Eds.): NLPCC 2015, LNAI 9362, pp. 573–580, 2015.
DOI: 10.1007/978-3-319-25207-0_54

2 System Description

In this section, we describle the details of our system. We use a bagging model to combine multiple segmenters. In each segmenter, preprocessing are conducted to recognize URLs, Emails, numbers, latin letters. Optional features like PMI, MI, Lexicon features can be included to enhance the performance of our system. And the system architecture is illustrated in fig.1.

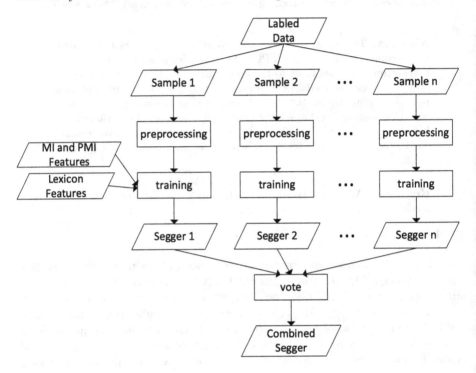

Fig. 1. System Architecture

2.1 Basic Model

In this paper, we use a word-based perceptron model. Word-based models read an input sentence from left to right and predict whether the current piece of continuous characters is a word. After a word is identified, this method moves on and searches for the next possible word. Zhang and Clark(2007)[2] firstly proposed a word-based approach for segmentation using a discriminative perceptron algorithm proposed by Michael Collins(2002)[3]. Compared with traditional character-based model, word-based perceptron model maps the CWS problem into an action sequence generation task (zhang 2012)[4] rather than a sequence labeling task (Xue 2003)[5]. There are two kinds of actions, **a** and **s**. The action a_i=**a** indicates the i-1-th character and the i-th character are in the same word. The action a_i=**s** indicates the i-1-th character and the i-th character are in two separate words. The problem of finding a best segmented sentence is transformed into generating a best action sequence.

Given a sentence X, the output $\hat{y}$ is defined as the highest scored segment among the possible segments for X, shown as Equ.1:

$$\hat{y} = \underset{y \in GEN(X)}{\operatorname{argmax}} \ Score(y) \tag{1}$$

Where $GEN(X)$ denotes the possible segments for an input sentence X and Score(y) is used to evaluate a segment.

Let us denote a possible action sequence as $A = \{a_0, a_1, ..., a_{|x|}\}$. Then Score(y) can be computed as:

$$Score(y) = \sum_{i=0}^{|X|} \phi(X, s_i) w \tag{2}$$

Where $\phi(X, s_i)$ is feature vector generated by the input X, w is a parameter vector and s_i is the current state which is composed of the current index i, the current action a_i and the previous state with an action s.

For the decoding phase, we follow the beam search method presented by Zhang and Clark(2007)[2].

2.2 Feature Templates

We now describe the features used in our word based segmentation model.

Basic Features. We define three kinds of features as basic features, including character-based features, character-type-based features and word- based features. Particularly, we use full word templates only when the current action is s. The word-based features are mainly based on Zhang and Clark(2011)[1]. First we show atom features in Table 1 and specific basic feature templates are shown in Table 2. $|w|$ is the length of w. $w[0]$ if the first character of w and $w[-1]$ is the last character of w.

Totally, there are 6 character-based features, 6 character-type-based features and 12 word-based features.

Table 1. Atom Features

Atom Feautres	Description
x_i	the i-th character in X
a_i	the current action
c_i	the character type of x_i
	(alphabet, digit, punctuation or others)
w_{-1}	the last word before current character
w_{-2}	the word before last word

Table 2. Basic features for the word segmenter

character-based	$\langle \mathbf{c\text{-}1}, x_i, a_i \rangle$, $\langle \mathbf{c\text{-}2}, x_{i-1}, a_i \rangle$, $\langle \mathbf{c\text{-}3}, x_{i+1}, a_i \rangle$
	$\langle \mathbf{c\text{-}4}, x_{i-1}x_i, a_i \rangle$, $\langle \mathbf{c\text{-}5}, x_i x_{i+1}, a_i \rangle$, $\langle \mathbf{c\text{-}6}, x_{i-1}x_i x_{i+1}, a_i \rangle$
chartype-based	$\langle \mathbf{ct\text{-}1}, c_i, a_i \rangle$, $\langle \mathbf{ct\text{-}2}, c_{i-1}, a_i \rangle$, $\langle \mathbf{ct\text{-}3}, c_{i+1}, a_i \rangle$
	$\langle \mathbf{ct\text{-}4}, c_{i-1}c_i, a_i \rangle$, $\langle \mathbf{ct\text{-}5}, c_i c_{i+1}, a_i \rangle$, $\langle \mathbf{ct\text{-}6}, c_{i-1}c_i c_{i+1}, a_i \rangle$
word-based	$\langle \mathbf{w\text{-}1}, w_{-1} \rangle$, $\langle \mathbf{w\text{-}2}, |w_{-1}| \rangle$, $\langle \mathbf{w\text{-}3}, w_{-1}, w_{-2} \rangle$
	$\langle \mathbf{w\text{-}4}, w_{-1}[0], w_{-1}[-1] \rangle$, $\langle \mathbf{w\text{-}5}, w_{-1}[0], |w_{-1}| \rangle$, $\langle \mathbf{w\text{-}6}, w_{-1}[-1], |w_{-1}| \rangle$
	$\langle \mathbf{w\text{-}7}, w_{-1}, w_{-1}[-1] \rangle$, $\langle \mathbf{w\text{-}8}, w_{-2}, w_{-2}[-1] \rangle$, $\langle \mathbf{w\text{-}9}, w_{-1}, |w_{-2}| \rangle$
	$\langle \mathbf{w\text{-}10}, w_{-2}, |w_{-1}| \rangle$, $\langle \mathbf{w\text{-}11}, w_{-1}, x_i \rangle$, $\langle \mathbf{w\text{-}12}, w_{-1}[0], x_i \rangle$

Mutual Information. In probability theory and information theory, the mutual information of two random variables is a measure of the variables' mutual dependence. The large value of mutual information indicates two consecutive strings are more probable to be combined together, while small value of mutual information often means they are unlikely to be in a word.

In this paper, we follow Sun and Xu(2011)[7]'s definition of mutual information. For two continuous charaters $x_i x_{i+1}$, the mutual information between x_i and x_{i+1} is computed as below:

$$MI(x_i, x_{i+1}) = \log \frac{p(x_i x_{i+1})}{p(x_i)p(x_{i+1})} \tag{3}$$

For each character x_i, $MI(x_i, x_{i+1})$ and $MI(x_{i-1}, x_i)$ are computed and rounded to an integer. We include these two features as additonal feature template.

Accessor Variety. A string with various linguistic environments may be a meaningful word. This idea is first proposed as Accessor Variety by Feng(2004)[6] to extract meaningful words from unlabeled corpus. This criterion of Accessor Variety is used to evaluate how independently a string is. Sun and Xu(2011)[7] define Accessor Variety as a type of statistic-based feature. In this paper, we follow this study and the features are defined as follows.

For each character x_i, we have features $L_{AV}^l(x[i : i + l - 1])$, $L_{AV}^l(x[i + 1 : i + l])$, $R_{AV}^l(x[i - l + 1 : i])$, $R_{AV}^l(x[i - l : i])$ $(l = 2, 3, 4)$. Here left accessor variety $L_A^l V(s)$ means the number of distinct characters that precede s in a corpus and right accessor variety $R_A^l V(s)$ means the number of distinct characters that succeed s.

Lexicon Features. Empirical study shows lexicon features can enhance the performance of a segmenter. We use a feature template $Lex(s)$ where s is a string and Lex is a function to indicate whether s is in a lexicon or not.

We add several word lists to our lexicon, including SogouW[1] and a few finance, sports, entertainment related word lists from sogou's lexicon sharing website[2].

2.3 Bagging Model

Bootstrap aggregating (Bagging) is a machine learning ensemble meta-algorithm to improve classification and regression accuracy. It also reduces variance and helps to avoid overfitting. Given a training set D of size n, Bagging generates m new training set D_i of size $n' \leq n$, by samppling examples from D uniformly. The m models are fitted using the above m bootstrap samples and combined by voting (for voting) or average the output (for regression).

We use a Bagging model to combine multiple segmenters. In the training phase, given a training set D of size n, our model generates m new training sets D_i of size $80\% \times n$ by sampling sentences from D without replacement. We use each D_i to train a weak segmenter. Thus we can get m weak segmenters. In the segmentation phase, the m segmenters have m segmentation results, which are further transformed into action sequences. In other words, for each position we have m **a** or **s** actions. The final segmentation is the voting result of these m actions. We set m an odd number, because when m is even there may be equal number of **a** or **s** actions.

Our bagging model is mainly based on Sun(2010)[8] and the difference between us is we only use word-based weak segmenters.

2.4 Rule-Based Adaptation

It is worthy to note that, considering the characteristics of microblog text, We adapt a rule-based preprocessing before the statistical model.

URLs (like http://www.baidu.com), Emails (like nlp@nlp.nju.edu.cn) are first recognized. The boundaries of these components are assigned to **s**, while the inner character intervals of the URLs and Emails are assigned to **a**.

Likewise, the punctuations (such as Chinese full stop and comma) are recognized and the boundaries of these are assigned to **s**. The intervals between two Arabic numbers or two Latin letters are assigned to **a**.

By using preprocessing, we can assign some fixed action a_i to the certain positions of the recognized words before the decoding phase. Thus the search space of the statistical model can be reduced.

3 Experiments

In experiments, we firstly use 5-fold cross validation for the development and use the whole dataset to train the final model for the test data. The F-score

---

[1] https://www.sogou.com/labs/dl/w.html
[2] http://pinyin.sogou.com/dict/

is used to evaluate the performance of the word segmentation system. We follow Liu(2012)[9] and evaluate the effects of preprocessing, bagging and different feature templates.

3.1   Effect of Preprocessing and Bagging

In Table 3, we compare the results from our baseline model and baseline model with preprocessing. We can see that the performance is improved with the help of preprocessing. We also compare the results from baseline model and the bagging model. The results suggest that using the bagging method can obviously improve the performance.

Table 3. Effect of preprocessing and bagging

Model	Precision	Recall	F-score
Base	0.9405	0.9401	0.9403
Base + Pre	0.9418	0.9421	0.9419
Base + Pre + Bagging	0.9465	0.9466	0.9465

3.2   Effect of Statistic-Based Features

Table 4 shows the difference of combining different statistical features which are extracted from the given background data. It can be seen that all statistical features lead to the improvement on performance. But accessor variety features contribute to more improvement than mutual information features.

Table 4. Effect of Statistic-based features

Model	Precision	Recall	F-Score
Base + Pre	0.9418	0.9421	0.9419
Base + Pre + MI	0.9431	0.9427	0.9428
Base + Pre + AV	0.9463	0.9458	0.9460
Base + Pre + MI + AV	0.9471	0.9469	0.9470

3.3   Effect of Lexicon Features

We compare the results with lexicon features and without lexicon features. The results is shown in Table 5. With lexicon features, the model get an F-score 95.43% and an improvement of 1.24% comparing to the baseline model. As is expected, lexicon features greatly improve the performance and outperform all other statistical features.

Table 5. Effect of Lexicon features

Model	Precision	Recall	F-score
Base + Pre	0.9418	0.9421	0.9419
Base + Pre + MI + AV	0.9471	0.9469	0.9470
Base + Pre + Lexicon	0.9534	0.9552	0.9543

Table 6. Final Results

Track	Configuration	Precision	Recall	F-score
Closed	Pre+Bagging	0.9514	0.9509	0.9512
Semi-Open	Pre+Bagging+MI+AV	0.9530	0.9531	0.9530
Open	Pre+Bagging+MI+AV+Lexicon	0.9603	0.9615	0.9609

3.4 Final System

For the closed track, the configuration is set as "Pre+Bagging". For the semi-open track, the configuration is set as "Pre+Bagging+MI+AV". For the open track, the configuration is set as "Pre+Bagging+MI+AV+Lexicon". And the final results for the test data is shown in Table 6.

4 Conclusion

In this paper, we describe our system of Chinese Word Segmentation on microblog data. We exploit a bagging model which ensembles multiple weak segmenters. Rule-based preprocessing, statistic-based features and lexicon features are also used to enhance the performance of our model. Finally, our model achieves F-score 95.12% on closed track, 95.3% on semi-open track and 96.09% on open track, respectively.

In the future, we will try to apply novel words detection and name entities recognition to further enhance the performance on microblog text. Also some more complex enesemble methods can be appiled to our system.

Acknowledgments. We thank the anonymous reviewers for their insightful comments. This work was supported by the NSFC (61472183, 61333014) and the 863 program(2015AA015406).

References

1. Zhang, Y., Clark, S.: Syntactic processing using the generalized perceptron and beam search. Computational Linguistics **37**(1), 105–151 (2011)
2. Zhang, Y., Clark, S.: Chinese segmentation with a word-based perceptron algorithm. In: Proceedings of ACL, Prague, pp. 840–847 (2007)
3. Collins, M.: Discriminative training methods for hidden Markov models: theory and experiments with perceptron algorithms. In: Proceedings of EMNLP, Philadelphia, PA, pp. 1–8 (2002)
4. Zhang, K., Sun, M., Zhou, C.: Word segmentation on Chinese mirco-blog data with a linear-time incremental model. In: Second CIPS-SIGHAN Joint Conference on Chinese Language Processing (2012)

5. Xue, N.: Chinese word segmentation as character tagging. International Journal of Computational Linguistics and Chinese Language Processing **8**(1) (2003)
6. Feng, H., Chen, K., Deng, X., Zheng, W.: Accessor variety criteria for Chinese word extraction. Computational Linguistics **30**(1), 75–93 (2004)
7. Sun, W., Xu, J.: Enhancing Chinese word segmentation using unlabeled data. In: Proceedings of the Conference on Empirical Methods in Natural Language Processing. Association for Computational Linguistics (2011)
8. Sun, W.: Word-based and character-based word segmentation models: comparison and combination. In: Coling 2010: Posters, Beijing, China, August, pp. 1211–1219. Coling 2010 Organizing Committee (2010)
9. Liu, Y., Che, W.: Micro blogs oriented word segmentation system. In: Second CIPS-SIGHAN Joint Conference on Chinese Language Processing (2012)

Weibo-Oriented Chinese News Summarization via Multi-feature Combination

Maofu Liu[1,2(✉)], Limin Wang[1,2], and Liqiang Nie[3]

[1] College of Computer Science and Technology,
Wuhan University of Science and Technology, Wuhan 430065, China
liumaofu@wust.edu.cn
[2] Hubei Province Key Laboratory of Intelligent Information Processing and Real-time
Industrial System, Wuhan University of Science and Technology, Wuhan 430065, China
[3] School of Computing, National University of Singapore, Singapore 117417, Singapore

Abstract. The past several years have witnessed the rapid development of social media services, and the UGCs (User Generated Contents) have been increased dramatically, such as tweets in Twitter and posts in Sina Weibo. In this paper, we describe our system at NLPCC2015 on the Weibo-oriented Chinese news summarization task. Our model is established based on multi-feature combination to automatically generate summary for the given news article. In our system, we mainly utilize four kinds of features to compute the significance score of a sentence, including term frequency, sentence position, sentence length and the similarity between sentence and news article title, and then the summary sentences are chosen according to the significance score of each sentence from the news article. The evaluation results on Weibo news document sets show that our system is efficient in Weibo-oriented Chinese news summarization and outperforms all the other systems.

Keywords: Weibo-oriented Chinese news summarization · Multi-feature combination · Sentence position · Term frequency · Sentence length · Sentence similarity

1 Introduction

With the rapid development and popularity of social media web sites, e.g. Twitter[1] and Sina Weibo[2], the news stories have been usually released and propagated via social media services in recent years. When one social media user reads his or her favorite news, he or she would like to share the news story with his or her friends via social media services by tweet or post. The social media user hence should generate a short summary for the favorite news due to the length limit of the tweet and post, and tweets or posts it together with a URL (Uniform Resource Locator) link to the full news article.

Let us consider the news "被忽视的身体权和隐私权" (the neglect of the right to body and privacy[3]) as an example, from the sample data set of the Weibo-oriented

---

[1] https://twitter.com/
[2] https://weibo.com/
[3] All examples in English version are translated from the Chinese ones.

© Springer International Publishing Switzerland 2015
J. Li et al. (Eds.): NLPCC 2015, LNAI 9362, pp. 581–589, 2015.
DOI: 10.1007/978-3-319-25207-0_55

Chinese news summarization task, the length of the body part of this news is close to four thousand Chinese characters, and it is too long to contain in one tweet or post as a whole, according to the 140 length limit of tweet or post. Therefore, the social media user has to manually edit one tweet or post from the long news story, and it is a challenge to the social media user and certainly time-consuming. On the other hand, the short summary can assist the other social media users in quickly understanding the tweeted or posted news and deciding whether to read the full news article onward or whether to comment on the news.

As mentioned above, the tweet or post is manually provided by social media users. The Weibo-oriented Chinese news summarization task is defined as a task of automatically generating a short summary for a given Chinese news article, and the short summary is used for news release and propagation on Weibo. The length of the short summary should not exceed 140 Chinese characters due to the length limit of the tweet or post.

In our work, we apply single-document summarization techniques and extractive based summarization methods to the Weibo-oriented Chinese news summarization task and implement our system using multi-feature combination. Our system automatically extract the most significant sentences from the original Chinese news article to generate the short summary. Our system allocates a significance score for each sentence of Chinese news article, taking the certain kinds of features into consideration.

The remainder of this paper is organized as follows. Section 2 reviews related work. Section 3 introduces the proposed multi-feature combination based Weibo Chinese news summarization system. Section 4 then presents evaluation results and discussions. Finally, Section 5 concludes the paper and suggests the future work.

2   Related Work

The extractive based single-document summarization has been researched and made great improvement for the past decades [1]. Wan and Xiao [2] proposed a single-document keyphrase extraction method using both the local information in the specified document and the global information in the neighbor documents. Litvak and Last [3] put forward the graph-based keyword extraction method for single-document summarization. Hirao et al [4] formulated the extractive text summarization as the Knapsack combinatorial optimization and proposed a single-document summarization method based on the trimming of a discourse tree and solved it with integer linear programming. Mendoza et al [5] proposed a method of extractive single-document summarization based on genetic operators and guided local search.

The single-document summarization has been turned to the news articles in recent years. In the first summarization conference, SUMMAC [6], single-document summaries from newswire documents were evaluated. Kastner and Monz [7] proposed single-document key fact extraction from news articles. The model proposed by Lloret and Palomar [8] outperformed the result of the state-of-the-art in single-document summarization for newswire domain. Wang and Yang [9] focused on single-document summarization for Chinese news articles. The tweets summarization has also attracted more attentions in the past several years. Wan and Zhang [10] proposed the CTSUM system to incorporate the new factor of information certainty into the news articles summarization task. Yulianti et al [11] examined the effect of tweets on the accuracy of the generated summaries from web single document.

In this paper, we regard the Weibo-oriented Chinese news summarization task as the extractive single-document summarization and use the multi-feature combination to estimate the significance score of the sentence. The most significant sentences will be selected to generate the summary for the news article in the end.

3   System Description

3.1   System Architecture

Our system consists of three main modules, i.e. data preprocessing, feature extraction and combination, and sentence selection. Figure 1 can illustrate our system architecture in detail.

In data preprocessing, the main work of the system is to segment the Chinese words and removes the stop words. We choose Stanford Chinese word segmenter[4] and select Chinese stop word list from CCF (China Computer Federation) web site[5] provided by Harbin institute of technology. Our system also deletes too short sentence, e.g. the length of the sentence less than four Chinese characters and the sentence absolutely irrelevant to the original news, in the data preprocessing phase.

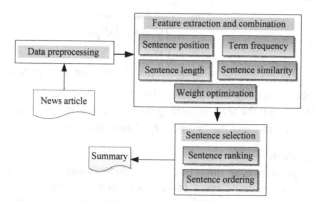

Fig. 1. System architecture

3.2   Feature Extraction and Combination

Four kinds of features at sentence level, including the term frequency in the sentence, the similarity between the sentence and news article title, the sentence position in the news article and the length of the sentence, are introduced into our system.

The term frequency is the most simple and direct kind of statistical feature to reflect the most important information content of the news article. The frequency of the term t_i in one news article can be expressed by the following formula (1).

---

[4] http://www-nlp.stanford.edu/software/segmenter.shtml
[5] http://www.ccf.org.cn/sites/ccf/ccfdata.jsp

$$tf(t_i) = \frac{m_i}{\sum\limits_{j=1}^{n} m_j} \tag{1}$$

Where n and m_i represent the total number of terms and the occurrence number of the term t_i in the news article respectively. The denominator is the total occurrence number of all the terms in the news article.

In our system, the sentence, a collection of words, is the basic information unit and the high frequency words in the sentence are more important. The significance score of the ith sentence sen_i, illustrated by formula (2).

$$TF_i = \sum_{w \in sen_i} tf(w) \tag{2}$$

Where sen_i and w represent the ith sentence in the news article and one meaningful word in the sen_i respectively, and the value of $tf(w)$ can be calculated by formula (1).

The title of the news article can represent its topic to some extent and contain the most important information. Therefore, the sentence, holding a high similarity with the news article title, will be more likely to be extracted as the summary part. Our system uses cosine similarity, showed in formula (3), to calculate the similarity between the sentence and the news article title.

$$Sim_i = \frac{\vec{s} \cdot \vec{t}}{\sqrt{\sum\limits_{j=1}^{n} s_j^2} * \sqrt{\sum\limits_{j=1}^{n} t_j^2}} \tag{3}$$

Where Sim_i denotes the similarity between the ith sentence and the news article title, $\vec{s}$ and $\vec{t}$ are the vector representation of the sentence and the news article title and the vector is built using the term frequency weighting schema, the parameter n is the dimension of the vector, and the s_j and t_j denote the ith dimension of the vectors $\vec{s}$ and $\vec{t}$ respectively.

The sentence position is an important feature for the news article. The news article is regarded as a linear arrangement of sentences and the first sentence may be the most important one. We use formula (4) to calculate sentence position feature.

$$Pos_i = \frac{n - p_i + 1}{n} \tag{4}$$

Where the parameters n and p_i denote the total number of sentences and the position of the ith sentence in the news article separately.

We hold the assumption that the extracted summary sentence should have a moderate length. If the length of a sentence is close to the average sentence length of the news article, the sentence would be assigned high significance score. Our system uses the normal distribution model to calculate the feature derived from the ith sentence length, showed in formula (5).

$$Len_i = \frac{1}{\sqrt{2\pi}\sigma} e^{-\frac{(x_i - \mu)^2}{2\sigma^2}} \tag{5}$$

Where the parameters μ and x_i denote the average sentence length and the ith sentence length of the news article respectively and σ is the standard deviation, calculated by the following formula (6).

$$\sigma^2 = \frac{\sum_{i=1}^{n}(x_i - \mu)^2}{n} \tag{6}$$

Where n is the total number of sentence in the news article.

According to the four kinds of features, we compute the final significance score $Senscore_i$ of the ith sentence by specifying a certain weight for each kind of feature, as illustrated by formula (7).

$$Senscore_i = \lambda_1 TF_i + \lambda_2 Pos_i + \lambda_3 Sim_i + \lambda_4 Len_i \tag{7}$$

Where the $\lambda_1, \lambda_2, \lambda_3$ and λ_4 represent the weights for the four kinds of features.

3.3   Sentence Selection

The following **Algorithm 1** describes the sentence selection strategy in our system in detail.

Algorithm 1

Input: Sentence set *SenSet*
Output: a short summary *Sum*
1: begin
2: for each sentence *Sen* in *SenSet*
3:      ranking(*Sen*, *Senscore*);   //Descending order
4: end for

5: for each sentence *Sen* in *SenSet*
6:      totallength+=senlength;   // the total length of the summary
7:      if(totallength<=140)
9:            choose the sentence *Sen* and add it to *Summ*
10:    else
11:          totallength-=senlength;
12:    end if
13: end for
14: for each sentence *Sen* in *Summ*
              //Ascendingly ordering according to the sentence position
15:          ordering(*Sen*, *Pos*);
16: end for
17: Output *Summ*
18: end

Our system ranks the sentences of the news article with function ranking() according to the significance score of each sentence calculated by the formula (7). We control the final length of the summary to the limit of 140 Chinese characters using the parameter *totallength* and order the selected sentence in the generated summary with function ordering() according to the sentence position feature in the end.

4   Evaluation Results and Discussions

We evaluate our multi-feature combination based extractive summarization model on the NLPCC2015 data collection for the Weibo-oriented Chinese news summarization task. The NLPCC2015 data collection contains 140 and 250 Chinese news articles in training set and testing set respectively, and all the news articles have their corresponding sentence segmentation results. All final summaries for the Chinese news articles are generated not exceeding 140 words length according to the length limit of tweet or post.

To evaluate the quality of generated summaries for the Chinese news articles, an automatic evaluation tool, called ROUGE is used. The tool presents five ROUGE values including unigram based ROUGE-1, bigram based ROUGE-2, trigram based ROUGE-3, 4-gram based ROUGE-4 and ROUGE-SU4. The ROUGE tool generates three scores for each evaluation, i.e. recall, precision and F-measure.

We submitted two formal runs for Weibo-oriented Chinese news summarization to the NLPCC2015 task organization office and the official evaluation results are listed in the Table 1. For the weight parameters $\lambda_1, \lambda_2, \lambda_3$ and λ_4 in formula (7), the weight ratios are 1:2:1:1 for "run1" and 1:2:0:0 for "run2" seperately.

Table 1. The official evaluation results of our formal runs

	ROUGE-1	ROUGE-2	ROUGE-3	ROUGE-4	ROUGE-SU4
run1	0.51496	0.38847	0.33165	0.29876	0.37107
run2	0.52422	0.40035	0.34483	0.31234	0.38318

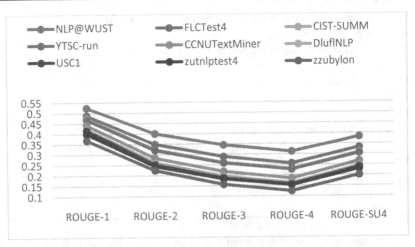

Fig. 2. ROUGE evaluation results for the formal runs of all task groups

We compare our system against all the other eight groups that participated in Weibo-oriented Chinese news summarization task and only select the best ROUGE evaluation result for each group shown in Figure 2, and we can find that our system has ranked the **first** among all nine groups according to Figure 2.

Our system tries to search the best feature combination at sentence level, i.e. the term frequency of the sentence (TF), the similarity between the sentence and the news article title (Sim), the sentence position (Pos) and the length of the sentence (Len) in the training runs. The high frequency words in the sentence can reflect that the information content can be conveyed by this sentence, and therefore our system begins the feature combination optimization from the TF feature. The evaluation results of the experiments on the testing set are illustrated in Table 2.

Table 2. The training evaluation results of our system for feature combinations

	ROUGE-1	ROUGE-2	ROUGE-3	ROUGE-4	ROUGE-SU4
TF	0.40482	0.25244	0.19493	0.16553	0.24101
TF+Sim	0.43381	0.28557	0.22591	0.19521	0.27174
TF+Pos	0.50799	0.37972	0.32348	0.29072	0.36259
TF+Pos+Sim	0.49635	0.36300	0.30425	0.27102	0.34517
TF+Pos+Sim+Len	0.49854	0.36518	0.30624	0.27293	0.34727

According to Table 2, we can find that the individual statistical feature TF is not enough to gain the good performance for our system, and our system can be effectively improved with the feature combination TF+Pos+Sim. Our system continues to combine the Len feature to avoid the long sentence favor caused by the statistical feature TF and only make a little improvement compared with the feature combination TF+Pos+Sim. To our surprise, the training evaluation results show that the feature combination TF+Pos has made the best performance, which may be affected by the genre of the news article and the feature of sentence position holding the most comprehensive influence on summary generation.

We optimize the best weights for the feature combination at sentence level in the training runs. The evaluation results of the experiments with the different weights for the feature combination on the testing set are illustrated in Table 3. In table 3, the value in front of each feature represents the weight of the feature.

Table 3. The training evaluation results of weight optimization for feature combinations

	ROUGE-1	ROUGE-2	ROUGE-3	ROUGE-4	ROUGE-SU4
TF+2Pos	0.50799	0.37972	0.32348	0.29072	0.36259
TF+2Pos+Sim+Len	0.50071	0.36859	0.30981	0.27633	0.35000

In Table 3, we optimize the weights of TF, Pos, Sim and Len and try to find out the best weight for each feature. The different weights for four kinds of features have effectively improved the evaluation results compared with the ones in Table 2.

5   Conclusions and Future Work

In this paper, we focus on the Weibo-oriented Chinese news summarization task and adopt extractive single-document summarization techniques to generate a short text for the Chinese news article. Our system has also involved the multi-feature combination at the sentence level, including the position of the sentence in the Chinese news article, the similarity to the title of the Chinese news article, the sentence length and the term frequency in the sentence. The evaluation results show that our system is efficient in generating the short summary for the Weibo-oriented Chinese news article.

In the future, we will introduce more features into the Weibo Chinese news summarization, such as syntactic, semantic and social context features. We will also utilize the event structures to represent the news article and adapt the event-based summarization method to fit in Weibo Chinese news summarization [12].

Acknowledgements. The work presented in this paper is partially supported by the National Natural Science Foundation of China under Grant No. 61100133 and the Major Projects of National Social Science Foundation of China under Grant No. 11&ZD189.

References

1. Lloret, E., Palomar, M.: Text summarisation in progress: a literature review. Artificial Intelligence Review **37**, 1–41 (2012)
2. Wan, X., Xiao, J.: Single document keyphrase extraction using neighborhood knowledge. In: Proceedings of the 23rd AAAI Conference on Artificial Intelligence, pp. 855–860 (2008)
3. Litvak, M., Last, M.: Graph-based keyword extraction for single-document summarization. In: Proceedings of the COLING Workshop on Multi-source Multilingual Information Extraction and Summarization, pp. 17–24 (2008)
4. Hirao, T., Yoshida, Y., Nishino, M., et al.: Single-document summarization as a tree knapsack problem. In: Proceedings of the 2013 Conference on Empirical Methods in Natural Language Processing, pp. 1515–1520 (2013)
5. Mendoza, M., Bonilla, S., Noguera, C., et al.: Extractive single-document summarization based on genetic operators and guided local search. Expert Systems with Applications **41**(9), 4158–4169 (2014)
6. Mani, I., House, D., Klein, G., et al.: The TIPSTER SUMMAC text summarization evaluation. In: Proceedings of the 9th Conference on European Chapter of the Association for Computational Linguistics, pp. 77–85 (1999)
7. Kastner, I., Monz, C.: Automatic single-document key fact extraction from newswire articles. In: Proceedings of the 12th Conference of the European Chapter of the Association for Computational Linguistics, pp. 415–423 (2009)
8. Lloret, E., Palomar, M.: A gradual combination of features for building automatic summarisation systems. In: Matoušek, V., Mautner, P. (eds.) TSD 2009. LNCS, vol. 5729, pp. 16–23. Springer, Heidelberg (2009)

9. Wang, J., Yang, J.: Statistical single-document summarization for Chinese news articles. In: Proceedings of 26th International Conference on Advanced Information Networking and Applications Workshops, pp. 183–188 (2012)
10. Wan, X., Zhang, J.: CTSUM: extracting more certain summaries for news articles. In: Proceedings of the 37th International ACM SIGIR Conference on Research and Development in Information Retrieval, pp. 787–796 (2014)
11. Yulianti, E., Huspi, S., Sanderson, M.: Tweet-biased summarization. Journal of the Association for Information Science and Technology, 1–17 (2015)
12. Liu, M., Li, W., Wu, M., et al: Extractive summarization based on event term clustering. In: Proceedings of the 45th Annual Meeting of the Association for Computational Linguistics, pp. 183–188 (2007)

Linking Entities in Chinese Queries
to Knowledge Graph

Jun Li[1], Jinxian Pan[2], Chen Ye[1], Yong Huang[1], Danlu Wen[1],
and Zhichun Wang[1](✉)

[1] Beijing Normal University, Beijing, China
zcwang@bnu.edu.cn
[2] Capital Normal University, Beijing, China

Abstract. This paper presents our approach for NLPCC 2015 shared task, Entity Recognition and Linking in Chinese Search Queries. The proposed approach takes a query as input, and generates a ranked mention-entity links as results. It combines several different metrics to evaluate the probability of each entity link, including entity relatedness in the given knowledge graph, document similarity between query and the virtual document of entity in the knowledge graph. In the evaluation, our approach gets 33.2 % precision and 65.2 % recall, and ranks the 6th among all the 14 teams according to the average F1-measure.

Keywords: Entity linking · Chinese query · Knowledge graph

1 Introduction

Recently, several large scale Knowledge Graphs have been developed [2][1][14][4]. One of the most important applications of knowledge graphs is to enhance web search engines' search result with semantic search information. For example, Google use its knowledge graph to provide structured and detailed information about the search topic in addition to a list of links to websites. Chinese search engines such as Baidu and Sogou also developed their own knowledge graphs and use them for semantic search.

In order to incorporate search engines with knowledge graphs, one important task is to link entities in search queries to knowledge graphs. Recently, much work has been done on the problem of entity linking in documents or tweets. The existing approaches usually use Wikipedia as a knowledge base, identify entities in text and link them to pages in Wikipedia. Only a few work has been done on the problem of entity linking in queries. Radhakrishnan et al. [10] proposed an approach for entity linking for English queries by utilizing Wikipedia inlinks. Blance et al. proposed an approach for fast and space-efficient entity linking for English queries [3]. Entity linking in queries is more difficult than traditional entity linking tasks. First, queries are usually very short texts, it is difficult to find proper context information for disambiguation of entities. Second, there is

J. Li et al. (Eds.): NLPCC 2015, LNAI 9362, pp. 590–597, 2015.
DOI: 10.1007/978-3-319-25207-0_56

a very strict time limit for process queries in search engines, so entity linking approaches for queries are supposed to run very efficiently.

In this paper, we report our approach for NLPCC 2015 shared task **Entity Recognition and Linking in Chinese Search Queries**. This task provides a reference Chinese Knowledge Graph and a small size of sample data, which contains several short Chinese queries and the sample results of entity linking. For example, entity linking in a query: "射雕英雄传刘亦菲版" is expected to get the results of linking "射雕英雄传" to a knowledge base entity "pk:tv:射雕英雄传(2006 年电视剧)" and "刘亦菲" into a KB entity "pk:per:刘亦菲".

We propose an approach that takes a query as input, and generates a ranked entity links as results. It combines several different metrics to evaluate the probability of each entity link, including entity relatedness in the given knowledge graph, document similarity between query and the virtual document of entity in the knowledge graph. In the evaluation, our approach gets 33.2% precision and 65.2% recall.

The rest of this paper is organized as follows, Section 2 describes the proposed approach in detail; Section 3 presents the evaluation results; Section 4 introduces some related work; Section 5 concludes this work.

2 The Proposed Approach

Our approach first identifies mentions in a given query, and then compute features of each possible mention-entity pairs, based on which the final results are generated.

2.1 Mention Identification

To extract entity mentions in queries, we build a mention dictionary that includes all the entity mentions in Chinese Wikipedia. In Wikipedia, an entity link is annotated by square brackets [[**entity**]] in the source data of articles. Here **entity** denotes the unique name of the referred entity. When the mentioned name of an entity is different from its unique name, the link is annotated by [[**entity | mention**]]; **mention** denotes the string tokens that actually appear in the text. In order to get all the mentions that have appeared in Wikipedia, we process all the annotated entity links in the form of [[**entity | mention**]] in Wikipedia. In addition, all the titles of articles in Wikipedia are also taken as mentions, which will be included in the mention dictionary. The mention dictionary also records the possible entities that each mention might refer to. Therefore, the dictionary can be represented as 2-tuple $D = (M, E)$, where $M = \{m_1, m_2, ..., m_k\}$ is the set of all mentions in Wikipedia, and $E = \{E_{m_1}, E_{m_2}, ..., E_{m_k}\}$ is the sets of entities corresponding to the mentions in M.

Since we are dealing with Chinese queries, word segmentation tool is used to split queries into lists of terms. Then we match the terms with mentions in the dictionary, if a term precisely matches a mention in the dictionary, we take it as a mention candidate. Each identified mention and its associated entities form a set of mention-entity pairs, which will be scored by several features.

2.2   Features of Mention-Entity Pairs

In order to decide the best entity for each identified mention in the knowledge graph, we propose to use the following features to assess the possibility of the link from a mention to an entity. Given a query q, and a set of mention-entity pairs $P = \{m_i, e_i\}_{i=1}^{n}$, the following features are computed for each mention-entity pair. For a mention-entity pair that has the same name for mention and entity, name length, entity relatedness and document similarity are computed. For other mention-entity pairs, the priori probability, entity relatedness and document similarity are computed. These features are defined as follows.

Name Length

$$f_1(m, e) = \frac{len(e) - minLen}{maxLen - minLen} \tag{1}$$

where $len(e)$ is number of characters in the entity's name. And $maxLen$ and $minLen$ represent the maximum and minimum length of entities in the given knowledge graph. This feature is defined based on the tuition that if a mention match an entity's name, then the longer of the entity's name the more possible the mention refers to the entity.

Priori Probability. This feature estimates the probability that a mention m links to an entity e:

$$f_2(m, e) = \frac{count(m, e)}{count(m)} \tag{2}$$

where $count(m, e)$ denotes the number of times that m links to e in the whole Wikipedia, and the $count(m)$ denotes the number of times that m appears in Wikipedia.

Entity Relatedness

$$f_3(m, e, P) = \begin{cases} 1 \text{ if exist an entity in } P \text{ that are linked to } e \\ \quad \text{in the knowlege graph} \\ 0 \text{ otherwise} \end{cases} \tag{3}$$

Document Similarity. We first build virtual document for each entity in the knowledge graph. The virtual document of an entity contains text information of all the other linked entities. The document similarity is calculated between the feature vectors of virtual documents. Before the similarity computation, the virtual document of each entity is represented as a vector, where the elements in the vector are weights assigned to the words in the virtual document using TF-IDF method. For a word i in virtual document j, the weight of the word is computed as

$$\omega_{ij} = tf_{ij} \cdot \lg \frac{N}{df_i} \tag{4}$$

where tf_{ij} is the number of occurrences of i in j, df_i is the number of virtual documents that contain i, and N is the total number of virtual documents. For a entity-mention pair, we compute the document similarity between the query and the virtual document of the entity. The document similarity is computed as the cosine value between their vectors:

$$f_4(m, e, q) = \frac{\sum_{i=1}^{M} w_{ie} \cdot w_{iq}}{\sqrt{\sum_{i=1}^{M} w_{ie}^2} \cdot \sqrt{\sum_{k=1}^{M} w_{iq}^2}} \tag{5}$$

where w_{ie} and w_{iq} are the ith weight in the vectors of entity document and query document. M is the total number of distinct words in all of the virtual documents.

2.3   Link Prediction

To predict links from mentions to entities in the knowledge graph, our approach computes the weighted sum of features between mentions and entities by the following score functions:

$$S_1(m, e, q, P) = w_1 \times f_1(m, e) + w_3 \times f_3(m, e, P) + w_4 \times f_4(m, e, q) \tag{6}$$

$$S_2(m, e, q, P) = w_2 \times f_2(m, e) + w_3 \times f_3(m, e, P) + w_4 \times f_4(m, e, q) \tag{7}$$

S_1 is for the mention-entity pairs that have the same names; S_2 is for the mention-entity pairs that have different names. All the candidate mention-entity pairs generated from one query are ranked by their score in descending order. And the top-k mention-entity pairs whose scores are larger than a threshold δ are the results of entity linking in the query.

In our system, all the parameters are set empirically. Their values are listed in Table 1.

Table 1. Setting of Parameters

Parameter	Value
w_1	0.2
w_2	0.2
w_3	0.4
w_4	0.4
k	3
δ	0.3

3   Evaluation Result

In this section, we will first introduce the evaluation dataset and the evaluation metrics, and then present the evaluation results of our approach.

3.1   Dataset

NLPCC 2015 shared task **Entity Recognition and Linking in Chinese Search Queries**, a file of the knowledge graph is provided by the organizer. This file contains large number of entities and relations between them. Each line represents a record in the knowledge base, and each record has 6 columns: subject ID, predicate ID, object ID, subject, predicate, object. A subject means a entity and the subject ID means the entity ID. As for the task requirement, we should link every named entity into the knowledge for each short query. And each query consists of an ID and a short search query. The answer we give should be in from of a query ID followed by an entity ID.

3.2   Evaluation Metrics

For a given query, we evaluate system performances using average F1-scores. Given a query q, the output of a system S^*, containing $|S^*|$ different groups of linking result for the named entities appearing in the given query. We compute the precision, recall and F1-score by comparing S^* with the answer the competition organizer providing S.

$$\text{F1-score} = \frac{2 \times \text{precision} \times \text{recall}}{\text{precision} + \text{recall}} \tag{8}$$

The final average F1-score is obtained by averaging over each query.

3.3   Results

Fig. 1 shows the precision and recall of each participant system. Team 7 is our team ID. Our system gets 33.2% precision and 65.2% recall. The precision of our system is not satisfied while the recall is not very low. We think the low precision is due to the manually set parameters, they might not be the optimal values. If we use the Machine Learning method to set the weight of each feature, we may get a better result. We think low precision also results from our mention identification method. Because our goal is to identify as much as mentions in the mention identification procedure, it also bring much noise which influence our precision at last. All these problems will be investigated and hopefully settled in the future. Fig. 2 shows the average F1 score of each team, our team gets a medium rank (the 6th) among all the 14 teams.

4   Related Work

In this section, we review some related work. Lots of work has been done in the problem of *Entity Linking*, which aims to identify entities in documents and link them to a knowledge base, such as Wikipedia and DBpedia.

Wikify! [8] is a system which is able to automatically perform the annotation task following the Wikipedia guidelines. Wikify! first uses a unsupervised

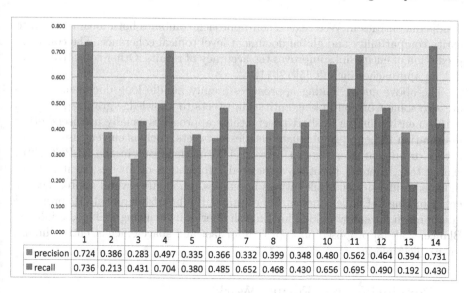

Fig. 1. Evaluation results: precision and recall

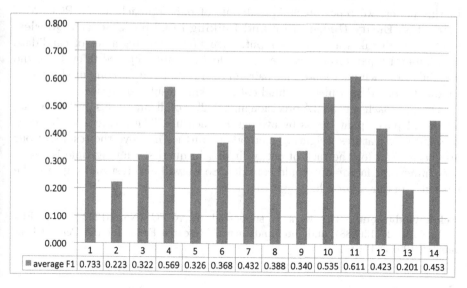

Fig. 2. Evaluation results: F1-measure

keyword extraction algorithm to identify and rank mentions; and then it combines both knowledge-based approach and data-driven method to predict the links from mentions to entities in Wikipedia. Milne et al. [9] proposed a learning based approach for linking entities in text to Wikipedia. Their approach trains a C4.5 classifier based on three features of entity-mention pairs for link disambiguation. Kaulkarni et al. [6] proposed a collective approach for annotating

Wikipedia entities in Web text. Their approach combines both local mention to entity compatibility and global document level topical coherence. The collective prediction of entity links improves the accuracy of results. Other collective entity linking approaches include [5][12][11].

The above entity linking approaches mainly handle long documents, there are also some work on linking entities in tweets to knowledge graphs [13][7]. To perform entity linking in short tweets, these approaches usually use users' other information to help disambiguate entities in tweets, such as current user's other tweets or current user's social network information. But in the NLPCC 2015 shared task, there is no other associated information of queries, so it is more difficult to identify and link entities in queries. There are several approaches for entity linking in English queries. For example, Radhakrishnan et al. [10] proposed an approach for entity linking for English queries by utilizing Wikipedia inlinks; Blance et al. [3] proposed an approach for fast and space-efficient entity linking for English queries.

5   Conclusion and Future Work

In this paper, we report technique details of our approach for NLPCC 2015 shared task **Entity Recognition and Linking in Chinese Search Queries**. Our approach takes a query as input, then first generates a set of candidate mention-entity pairs from the query; four features are proposed to evaluate the possibility of each mention-entity pair; an aggregated score is computed for each candidate, based on which the final entity linking results are drawn.

Our approach gets 33.2% precision and 65.2% recall, and ranks the 6th among all the 14 participant teams by average F1-measure. The future work includes defining new features weighting methods to further improve the results of our approach. And for the task of entity linking in queries, running time is also a very important factor we should consider. So we will also test and improve the efficiency of our approach.

Acknowledgments. The work is supported by NSFC (No. 61202246), NSFC-ANR(No. 61261130588), and the Fundamental Research Funds for the Central Universities (2013NT56).

References

1. Auer, S., Bizer, C., Kobilarov, G., Lehmann, J., Cyganiak, R., Ives, Z.G.: DBpedia: a nucleus for a web of open data. In: Aberer, K., Choi, K.-S., Noy, N., Allemang, D., Lee, K.-I., Nixon, L.J.B., Golbeck, J., Mika, P., Maynard, D., Mizoguchi, R., Schreiber, G., Cudré-Mauroux, P. (eds.) ASWC 2007 and ISWC 2007. LNCS, vol. 4825, pp. 722–735. Springer, Heidelberg (2007)
2. Bizer, C., Lehmann, J., Kobilarov, G., Auer, S., Becker, C., Cyganiak, R., Hellmann, S.: DBpedia - a crystallization point for the web of data. Web Semantics: Science, Services and Agents on the World Wide Web 7(3), 154–165 (2009)

3. Blanco, R., Ottaviano, G., Meij, E.: Fast and space-efficient entity linking for queries. In: Proceedings of the Eighth ACM International Conference on Web Search and Data Mining, WSDM 2015, pp. 179–188. ACM, New York, NY, USA (2015)

4. Bollacker, K.D., Cook, R.P., Tufts, P.: Freebase: a shared database of structured general human knowledge. In: Proceedings of the 22nd National Conference on Artificial Intelligence, vol. 2, pp. 1962–1963 (2007)

5. Han, X., Sun, L., Zhao, J.: Collective entity linking in web text: a graph-based method. In: Proceedings of the 34th International ACM SIGIR Conference on Research and Development in Information Retrieval, pp. 765–774 (2011)

6. Kulkarni, S., Singh, A., Ramakrishnan, G., Chakrabarti, S.: Collective annotation of wikipedia entities in web text. In: Proceedings of the 15th ACM SIGKDD International Conference on Knowledge Discovery and Data Mining, pp. 457–466 (2009)

7. Liu, X., Li, Y., Wu, H., Zhou, M., Wei, F., Lu, Y.: Entity linking for tweets. In: Proceedings of the 51st Annual Meeting of the Association for Computational Linguistics (ACL 2013) (2013)

8. Mihalcea, R., Csomai, A.: Wikify!: linking documents to encyclopedic knowledge. In: Proceedings of the Sixteenth ACM Conference on Conference on Information and Knowledge Management, pp. 233–242 (2007)

9. Milne, D., Witten, I.H.: Learning to link with wikipedia. In: Proceedings of the 17th ACM Conference on Information and Knowledge Management, pp. 509–518 (2008)

10. Radhakrishnan, P., Bansal, R., Gupta, M., Varma, V.: Exploiting wikipedia inlinks for linking entities in queries. In: Proceedings of the First International Workshop on Entity Recognition & #38; Disambiguation, ERD 2014, pp. 101–104. ACM, New York, NY, USA (2014)

11. Shen, W., Wang, J., Luo, P., Wang, M.: LIEGE: link entities in web lists with knowledge base. In: Proceedings of the 18th ACM SIGKDD International Conference on Knowledge Discovery and Data Mining, pp. 1424–1432 (2012)

12. Shen, W., Wang, J., Luo, P., Wang, M.: LINDEN: linking named entities with knowledge base via semantic knowledge. In: Proceedings of the 21st International Conference on World Wide Web, pp. 449–458 (2012)

13. Shen, W., Wang, J., Luo, P., Wang, M.: Linking named entities in tweets with knowledge base via user interest modeling. In: Proceedings of the 19th ACM SIGKDD International Conference on Knowledge Discovery and Data Mining, KDD 2013, pp. 68–76. ACM, New York, NY, USA (2013)

14. Suchanek, F.M., Kasneci, G., Weikum, G.: YAGO: a core of semantic knowledge. In: Proceedings of the 16th International Conference on World Wide Web, pp. 697–706 (2007)

A Hybrid Re-ranking Method for Entity Recognition and Linking in Search Queries

Gongbo Tang[1,2], Yuting Guo[2], Dong Yu[1,2(✉)], and Endong Xun[1,2]

[1] Institute of Big Data and Language Education,
Beijing Language and Culture University, Beijing 100083, China
{tanggongbo,yudong_blcu,edxun}@126.com
[2] College of Information Science,
Beijing Language and Culture University, Beijing 100083, China
guoyuting_gyt@126.com

Abstract. In this paper, we construct an entity recognition and linking system using Chinese Wikipedia and knowledge base. We utilize refined filter rules in entity recognition module, and then generate candidate entities by search engine and attributes in Wikipedia article pages. In entity linking module, we propose a hybrid entity re-ranking method combined with three features: textual and semantic match-degree, the similarity between candidate entity and entity mention, entity frequency. Finally, we get the linking results by the entity's final score. In the task of entity recognition and linking in search queries at NLPCC 2015, the *Average-F1* value of this method achieved 61.1% in 3849 test dataset, which ranks second place in fourteen teams.

1 Introduction

Search engine is the most common way to access information, sometimes, people have to search satisfied answer in retrieval result because of information explosion. To return better retrieval result, we need to deal with the entity recognition and linking task to understand users' intents better.

Search queries are nonstandard text, containing wrong spellings and abbreviation, alias names, nick names of entity. For instance, "习大大爱着彭妈妈", obviously, "习大大" refers to "习近平"and "彭妈妈" is nick name of "彭丽媛". Meanwhile, search queries are really short, the longest query may be dozens of words. Compared with the traditional entity recognition and linking works, the context can't provide enough features to disambiguate entities.

In this paper, we show our system for entity recognition and linking in search queries. There are three main stages, entity recognition, candidate entities generation and entities disambiguation. We use the rules and entity base to filter the entity mention, then we generate candidate entities by search engine and attributes of Wikipedia pages. Finally, we utilize an entity re-ranking method to score the candidate entities, which is combined with three features: textual and semantic match-degree, the similarity between candidate entity and entity mention, entity frequency. Figure 1 shows the framework of our system.

© Springer International Publishing Switzerland 2015
J. Li et al. (Eds.): NLPCC 2015, LNAI 9362, pp. 598–605, 2015.
DOI: 10.1007/978-3-319-25207-0_57

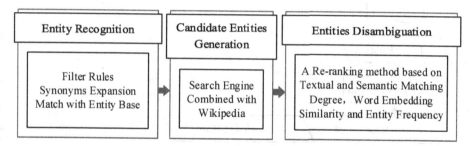

Fig. 1. The framework of our system

2   Related Work

There are two methods to recognize entities, rule based and machine learning based. Methods based on rules always recognize entity by the spelling rules, the parts of speech of entities and dictionary. Mikheev et al. [1] can recognize 79.3% entities by using a general dictionary, as some entities are ambiguous with common nouns, the remaining 20.7% entities are not recognized. In addition, methods based on machine learning always need a lot of tagged dataset to learn a model. For instance, Asahara and Matsumoto [2] utilize a support vector machine model to recognize entities.

To generate candidate entities, Han and Zhao [3] use Google API to get the retrieval result of short text, and select the entities in the title of Wikipedia pages as candidate entities. What is more, Meng et al. [4] utilize Baidu to search entity mention+ "维基百科"and entity mention+ "维基百科", and also select the entities in the title of Wikipedia pages as candidate entities. The candidate entity disambiguation need features, Dalton and Dietz [5] utilize urban dictionary to expand query. Hoffart et al. [6] use the word frequency in Wikipedia to define popularity feature. Blanco et al. [7] consider the distributional semantics of query words and entities, and train word embedding by word2vec. Shen et al. [8] make use of a SVM model, and give a rank to candidate entity for each entity mention with a linear combination of four features: entity popularity, semantic associativity, semantic similarity and global topical coherence between mapping entities.

3   Method

To reduce the system's complexity, we used some refined rules to filter and classify entity mentions. It's simple, but effective! To deal with various names, we use the synonym dictionary to expand entity mentions. And we use the search engine combined with Wikipedia to generate candidate entities. Finally, we utilize a re-ranking method to get the results. Figure 2 shows the flow of our system.

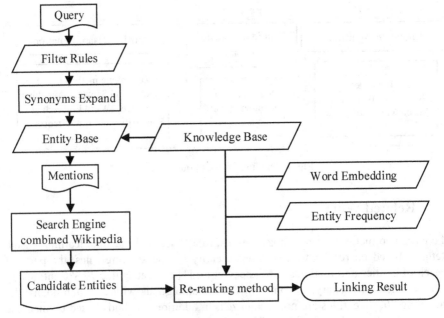

Fig. 2. The workflow of the system

3.1 Pre-processing

We use the Yebol Chinese segmentation system as word segmentation tool. Since we just need to link entities in the knowledge base, so we extract and process the entities, and then build an entity base. Table 1 shows some examples. Synonyms expansion is a simple but effective way to deal with mention variation issue. We use the same method in Meng et al. [4] to build and expand the synonym dictionary. We get "韩国" as alias of "大韩民国", and "海贼王" as a translation of "ONE_PIECE"etc. We select the entity description as feature, and expand the description with Wikipedia and Baidupedia.

Table 1. Examples of entity processing

Original	Result
"爱情公寓1", "爱情公寓2", "爱情公寓3", "爱情公寓4", "爱情公寓_(电视剧)"等	"爱情公寓"
"爱是永恒", "爱是永恒（当所爱是你）"	"爱是永恒"
"茱莉娅·罗伯茨"	"茱莉娅罗伯茨"

3.2 Named Entity Recognition

Traditional entity recognition methods are based on machine learning, which needs plenty of tagged corpus to train a model, and the types of entity in queries are various, so it's unsuitable to learn a model for entity recognition. In addition, the task only need to link entities in knowledge base, so, to simplify the process and improve the efficiency, we use the following methods to recognize entity.

1. Design refined rules by sample dataset, and then filter the entity mentions
2. Expand entity mentions by synonymy dictionary
3. Match the mentions with entities in entity base: if the mention is complete-matching, we name it identified entity mention M_i, such as "爱情公寓", otherwise, we name it unidentified entity mention M_u, such as "湖人".

3.3   Candidate Entity Generation

Search queries and entity mentions are informal text with noise, for instance, "沙糖桔" is one of misspelling format of "沙糖橘", and there is no matched word in synonym dictionary, so we can't link the entity. Fortunately, search engine provide error correction function and can convert "沙糖桔"to"沙糖橘" before searching. Therefore, we adopt the method based on search engine we used last year [4], we use search engine Baidu, and treat "entity mention" + "中文维基百科"as input query. Figure 3 shows the workflow of entity generation.

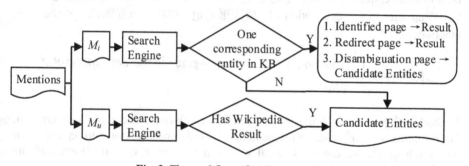

Fig. 3. The workflow of entity generation

If an identified mention has one corresponding entity in the knowledge base, and the first retrieval result is a Wikipedia page, there are three conditions:

1. An identified page, the entity in the title is our linked entity, like "诺基亚".
2. A redirect page, the entity in the title is also our linked entity.
3. A disambiguation page, all the entities in the page are candidate entities, for instance, "Angelababy"and "杨颖（作家）" are candidate entities of mention "杨颖".

If an identified mention has more than one corresponding entities in the knowledge base, all the corresponding entities are candidate entities. For instance, "倚天屠龙记", "倚天屠龙记_(1978年电影)","倚天屠龙记_(1978年电视剧)" etc. are candidate entities of mention "倚天屠龙记". For the unidentified entity mention, if the top 10 pages in the retrieval result contains Chinese Wikipedia pages, the entities in the titles are candidate entities, otherwise, we remove this mention directly.

3.4   Candidate Entity Re-ranking

The most important part of entity disambiguation is to compute the distance between candidate entities' feature and mention's feature. We compute the textual or semantic match-degree between notional words in queries context and candidate entities' description. Mikolov et al. [9] indicated that the word embedding can represent the word in syntax and semantics. So, we can compute the similarity between entity mention and candidate entities by word embedding, and we use cosine similarity. The frequency of entity can tell us the prior probability of the appearance of a candidate entity given the entity mention. Therefore, we propose a re-ranking method combining with match-degree, word embedding similarity and entity frequency to re-rank the candidate entities.

The Match-degree Between Mention Context and Candidate Entity Description
For a set of candidate entities, if there is a candidate entity's score is 1 in method 1, then every candidate entity's score is S_{match1}, otherwise, the score is S_{match2}. We assume c_i is the ith candidate entity, m is the current mention.

Method 1: If the notional word in query appears in the description of a candidate entity, for instance, query: "倚天屠龙记梁朝伟", for entity mention "倚天屠龙记", "梁朝伟" appears in the description of "倚天屠龙记_(1986年电视剧)", it scores 1, otherwise 0.

$$S_{match1}(c_i \mid m) = \begin{cases} 1 & \exists \, notional \; word \; in \; query \; appears \; in \; the \; description \; of \; c_i \\ 0 & else \end{cases} \quad (1)$$

Method 2: Compute the similarity between the search query and entity's description. For the entity's description, we use its notional words' mean vector v_d to represent; for the search query, we use its notional words' mean vector v_q to represent, and the match-degree is the cosine distance between v_d and v_q.

$$S_{match2}(c_i \mid m) = \cos(v_q, v_d) \quad (2)$$

Therefore, the final score is formula 3.

$$S_{match}(c_i \mid m) = \begin{cases} S_{match1}(c_i \mid m) & \exists i, S_{match1}(c_i \mid m) = 1 \\ S_{match2}(c_i \mid m) & else \end{cases} \quad (3)$$

The Similarity Between Candidate Entity Vector and Entity Mention Vector
We set the candidate entity vector as V_c, and the entity mention vector as V_m, so, the similarity score S_{sim} will be represented as follows,

$$S_{sim}(c_i \mid m) = \begin{cases} 0 & \exists\, i, v_{c_i} = null \ or \ v_m = null \\ \cos(V_c, V_m) & else \end{cases} \tag{4}$$

Entity Frequency

We use a Chinese Wikipedia corpus to count the frequency of entity, if a candidate entity c_i appears n_{ci} times in the corpus, and entity mention m appears n_m times in the corpus, the frequency score S_{freq} will be:

$$S_{freq}(c_i \mid m) = \frac{2 * n_{ci}}{\sum\limits_{i=1}^{n} n_{ci} + n_m} \tag{5}$$

Hence, the final score of a candidate entity is just like formula 6.

$$S_{final}(c_i \mid m) = \alpha S_{match}(c_i \mid m) + \beta S_{sim}(c_i \mid m) + \gamma S_{freq}(c_i \mid m) \ \ (i=1,2,3..n) \tag{6}$$

And $\alpha=1$, $\beta=0.7$, $\gamma=0.2$, these parameters are decided by the sample dataset.

We set threshold δ to 0.005, if the difference between highest score and the second highest score is greater than δ, we choose the candidate entity with highest scores as the final linking entity. Otherwise, the second highest candidate entity is also chosen as linking entity. Process in sequence until the adjacent entities' difference is greater than δ, or the linking entity number reaches 5.

4   Experiments

4.1   Dataset

Wikipedia is a high quality encyclopedia containing a wide coverage of named entities, massive knowledge about notable named entities, so it is fit for entity linking work. We download the newest Chinese Wikipedia from wiki dump, and get 707 MB Chinese Wikipedia corpus after processing, which is used to train word embedding and get entity statistical dataset. We use CBOW model [10], [11] in word2vec to train word embedding, and set the dimension to 100.

4.2   Experiment Result and Analysis

Our system score is 61.1% in *Average-F1*, which ranks the second place in fourteen systems. Table 2 shows experiment results. We can see that our system is a little lower than other systems in *Link-Recall*, but our *Link-Precision* is 6.5 percent higher than the third system, and 16.2 percent gap with the first, it shows that our system is promising in entity linking.

Table 2. Part of evaluation results

system	*Link-Precision*	*Link-Recall*	*Link-F1*	*Average-F1*
NO.1	0.724	0.736	0.73	0.733
Ours	0.562	0.695	0.621	0.611
NO.3	0.497	0.704	0.583	0.569

Short context, nonstandard text and various entity representations in search queries make this task difficult. In addition, entity linking task is based on entity recognition, and there will be error in entity recognition inevitably, which may cause error accumulation and pull precision in entity linking down. We concluded that there are mainly three types of error:

1. We adopt a coarse-grained method to recognize entity, for example, there are "北京", "北京交通大学" and "威海" three entities in query "北京交通大学威海校区", and we missed "北京".
2. Some words has no word embedding because of the data sparsity problem. For instance, the candidate entities of mention "天涯明月刀" are "天涯·明月·刀"and "天涯明月刀_(电视剧)" in query "天涯明月刀不删档", while these two candidate entities do not appear in training dataset.
3. The re-ranking method is not precise enough. The linking result of mention "爱情公寓" in query"爱情公寓里的小黑是谁"is "爱情公寓_(电视剧)". However, for "爱情公寓1","爱情公寓2" and "爱情公寓_(电视剧)" etc. Their score of matchdegree, similarity and frequency are extraordinary close, so all of them are selected as linking entity.

5   Conclusion

This paper introduces our entity recognition and linking system in search queries. We use a rules based method to recognize entity, then generate candidate entities by search engine and Wikipedia page attributes. Finally, we utilize an entity re-ranking method to score the candidate entities, and get the linking result by the entity score. The results of the experiment shows that our method is effective. In future work, we will optimize the word segmentation result, recognize entity in fine-grained and improve the entity re-ranking method in entity linking.

Acknowledgements. The research work is partially funded by the Natural Science Foundation of China (No.61300081, 61170162), and the Fundamental Research Funds for the Central Universities in BLCU (No. 15YJ03006).

References

1. Mikheev, A., Moens, M., Grover, C.: Named entity recognition without gazetteers. In: Proceedings of the Eacl (1999)
2. Asahara, M., Matsumoto, Y.: Japanese named entity extraction with redundant morphological analysis. In: Naacl Proceedings of the Conference of the North American Chapter of the Association for Co. (2003)
3. Han, X., Zhao, J.: Nlpr-kbp in tac 2009 kbp track: A two-stage method to entity linking. In: Proceedings of Test Analysis Conference (2009)
4. Meng, Z., Yu, D., Xun, E.: Chinese microblog entity linking system combining wikipedia and search engine retrieval results. In: Zong, C., Nie, J.-Y., Zhao, D., Feng, Y. (eds.) NLPCC 2014. CCIS, vol. 496, pp. 449–458. Springer, Heidelberg (2014)
5. Dalton, J., Dietz, L.: UMass CIIR at TAC KBP 2013 entity linking: query expansion using urban dictionary. In: Text Analysis Conference (2013)
6. Hoffart, J., Yosef, M.A., Bordino, I., et al.: Robust disambiguation of named entities in text. In: Proceedings of the Conference on Empirical Methods in Natural Language Processing. Association for Computational Linguistics, pp. 782–792 (2011)
7. Blanco, R., Ottaviano, G., Meij, E.: Fast and space-efficient entity linking for queries. In: Proceedings of the Eighth ACM International Conference on Web Search and Data Mining, pp. 179–188. ACM (2015)
8. Shen, W., Wang, J., Luo, P., et al.: LINDEN: linking named entities with knowledge base via semantic knowledge. In: Proceedings of the 21st International conference on World Wide Web. ACM (2012)
9. Mikolov, T., Sutskever, I., Chen, K., et al.: Distributed representations of words and phrases and their compositionality. In: Advances in Neural Information Processing Systems, pp. 3111–3119 (2013)
10. Mikolov, T., Yih, W., Zweig, G.: Linguistic regularities in continuous space word representations. In: HLT-NAACL (2013)
11. Mikolov, T., Chen, K., Corrado, G., Dean, J.: Efficient estimation of word representations in vector space. In: ICLR Workshop (2013)

Author Index

Bai, Hao   171
Bao, Feilong   347

Cai, Guoyong   159
Cai, Rui   270
Cao, Yixin   325
Cao, Yujie   79
Chao, Jiayuan   495
Chen, Changge   465
Chen, Hong   453
Chen, Jiajun   353, 573
Chen, Junwen   436
Chen, Lingfeng   257
Chen, Miaohong   270
Chen, Wei   388
Chen, Wenliang   495
Chen, Yijiang   182, 404
Cheng, Xueqi   113
Cui, Yihui   229

Dai, Hao   285, 299
Dai, Xin-Yu   353, 573
Dong, Jian   113
Dong, Zhenjiang   453
Duan, Nan   562

Fan, Rui   134
Feng, Shi   444
Feng, Yansong   550

Gao, Guanglai   347
Gao, Liang-Cai   484
Gao, Shengxiang   372
Gong, Zhengxian   396
Guo, Yuting   598

Han, Zhe   550
Hou, Lei   325
Hou, Yuexian   242
Huang, Junfu   507, 527
Huang, Minlie   79
Huang, Shujian   353, 573

Huang, Shuman   182
Huang, Tao   325
Huang, Xuanjing   12, 541
Huang, Yong   590

Jia, Zheng   507, 527
Jiang, Bo   146
Jiang, Yupeng   347
Jin, Xiaoqiang   102
Jing, Li   25

Kartbayev, Amandyk   421

Leng, Bing   313
Li, Ang   428
Li, Haiyan   520
Li, Jing   436
Li, Jingfei   242
Li, Jun   590
Li, Mingyang   413
Li, Mu   49
Li, Ning   339
Li, Qinglin   49
Li, Sheng   3
Li, Xiaoyun   209
Li, Zhenghua   495
Liang, Qi   339
Liang, Zhongping   313
Lin, Hongfei   90
Lin, Lei   102
Lin, Rui   3, 49
Liu, Bingquan   123
Liu, Feng   123
Liu, Huali   90
Liu, Maofu   581
Liu, Ming   123
Liu, Shujie   3, 49
Liu, Ting   453
Liu, Yiqun   257
Liu, Yongbin   413
Lou, Renjie   285, 299
Lu, Weiming   285, 299
Lv, Xueqiang   36, 475

Ma, Chenggang 520
Ma, Shaoping 257
Ma, Weizhi 257
Min, Kerui 520

Nie, Liqiang 581

Ouyang, Chunping 209

Pan, Jinxian 590
Peng, Baolin 25
Peng, Xin 339
Peng, Zhiyong 229

Qian, Peng 541
Qin, Bing 453
Qiu, Xipeng 12, 541

Sha, Ying 146
Shen, Junhui 134
Shen, Si 573
Shi, Xiaodong 404
Shi, Yao 413
Song, Dawei 242
Song, Wei 229
Su, Chang 182, 404
Su, Xiangdong 347
Sun, Chengjie 102, 123

Tan, Jiwei 557
Tan, Wei 134
Tang, Gongbo 598
Tang, Zhi 484

Wan, Xiaojun 557
Wang, Chu 444
Wang, Daling 444
Wang, Houfeng 36, 270
Wang, Lihong 146
Wang, Limin 581
Wang, Xiaojie 64, 313
Wang, Xiaolong 102, 123
Wang, Yixiu 475
Wang, Zhichun 590
Wang, Zhigang 413
Wei, Baogang 285, 299
Wei, Hao 436
Wei, Jinmao 194
Wei, Yang 194

Wei, Zhongyu 436
Wen, Danlu 590
Wen, Shiyang 557
Wong, Kam-Fai 25, 436
Wu, Jing 347
Wu, Shiyu 541
Wu, Yunfang 475

Xia, Binbin 159
Xia, Qing 372
Xia, Yunqing 219
Xie, Zhipeng 428
Xie, Zhongda 219
Xiong, Jinhua 113
Xu, Bo 388
Xu, Hengpeng 194
Xu, Jian-Bo 484
Xu, Xi 484
Xun, Endong 598

Yan, Xin 372
Yang, Liang 90
Yang, Muyun 3
Yang, Shansong 285, 299
Yang, Yan 507, 527
Yang, Yang 465
Yao, Kaisheng 25
Ye, Chen 590
Ye, Mao 484
Ye, Tengju 428
Ye, Zhonglin 527
Yin, Hongfeng 507, 527
Yin, Liusong 541
Yu, Dong 598
Yu, Ying 209
Yu, Zhengtao 372
Yu, Zhenting 573
Yuan, Caixia 64, 313
Yuan, Jinwei 507

Zhan, Xueyan 134
Zhang, Chunyue 380
Zhang, Jianmin 557
Zhang, Kun 550
Zhang, Mengdi 325
Zhang, Min 257, 495
Zhang, Peng 242
Zhang, Qiao 113
Zhang, Shaowu 90

Zhang, Shuiyuan 113
Zhang, Yifei 444
Zhang, Zhenyu 285, 299
Zhao, Hai 465
Zhao, Kangfei 436
Zhao, Lin 339
Zhao, Tianmei 520
Zhao, Tiejun 3, 380
Zhao, Yanyan 453
Zhao, Yuming 102
Zheng, Lvexing 36

Zhong, Ziming 64
Zhou, Guodong 396
Zhou, Ming 49
Zhou, Qiang 219
Zhou, Xin 362
Zhou, Yichu 353
Zhu, Chenxi 12
Zhu, Jie 507
Zhu, Peiyan 134
Zhu, Tingting 404
Zhu, Xiaoyan 79